# STUDY GUIDE AND WORKBOOK: AN INTERACTIVE APPROACH

for Starr's

# BIOLOGY

*Concepts and Applications*

THIRD EDITION

## JOHN D. JACKSON

*North Hennepin Community College*

## JANE B. TAYLOR

*Northern Virginia Community College*

**Wadsworth Publishing Company**

I(T)P® An International Thomson Publishing Company

Belmont, CA • Albany, NY • Bonn • Boston • Cincinnati • Detroit • Johannesburg
London • Madrid • Melbourne • Mexico City • New York • Paris
San Francisco • Singapore • Tokyo • Toronto • Washington

**Biology Publisher:** Jack Carey
**Assistant Editor:** Kristin Milotich
**Editorial Assistant:** Kerri Abdinoor
**Project Editor:** Jennie Redwitz
**Print Buyer:** Stacey Weinberger
**Permissions Editor:** Peggy Meehan
**Art Editor:** Roberta Broyer
**Copy Editor:** Denise Cook-Clampert
**Cover:** Gary Head
**Compositor:** Joan Olson
**Printer:** Courier Companies, Inc.
**Cover Photo:** Art Wolfe/Tony Stone Images.

Printed in the United States of America
4  5  6  7  8  9  10

For more information, contact Wadsworth Publishing Company, 10 Davis Drive, Belmont, CA 94002, or
electronically at http://www.thomson.com/wadsworth.html

International Thomson Publishing Europe
Berkshire House 168–173
High Holborn
London, WC1V 7AA, England

International Thomson Editores
Campos Eliseos 385, Piso 7
Col. Polanco
11560 México D.F. México

Thomas Nelson Australia
102 Dodds Street
South Melbourne 3205
Victoria, Australia

International Thomson Publishing Asia
221 Henderson Road
#05-10 Henderson Building
Singapore 0315

Nelson Canada
1120 Birchmount Road
Scarborough, Ontario
Canada M1K 5G4

International Thomson Publishing Japan
Hirakawacho Kyowa Building, 3F
2-2-1 Hirakawacho
Chiyoda-ku, Tokyo 102, Japan

International Thomson Publishing GmbH
Königswinterer Strasse 418
53227 Bonn, Germany

International Thomson Publishing Southern Africa
Building 18, Constantia Park
240 Old Pretoria Road
Halfway House, 1685 South Africa

ISBN 0-534-50448-5

 *This book is printed on acid-free recycled paper.*

# CONTENTS

# PREFACE

*Tell me and I will forget, show me and I might remember, involve me and I will understand.*
—Chinese Proverb

The proverb outlines three levels of learning, each successively more effective than the method preceding it. The writer of the proverb understood that humans learn most efficiently when they *involve* themselves in the material to be learned. This study guide is like a tutor; when properly used it increases the efficiency of your study periods. The interactive exercises actively involve you in the most important terms and central ideas of your text. Specific tasks ask you to recall key concepts and terms and apply them to life; they test your understanding of the facts and indicate items to reexamine or clarify. Your performance on these tasks provides an estimate of your next test score based on specific material. Most important, though, this biology study guide and text together help you make informed decisions about matters that affect your own well-being and that of your environment. In the years to come, human survival on planet Earth will require administrative and managerial decisions based on an informed biological background.

## HOW TO USE THIS STUDY GUIDE

Following this preface, you will find an outline that shows you how the study guide is organized and that will help you use it efficiently. Each chapter begins with a title and an outline list of the 1- and 2-level headings in that chapter. The Interactive Exercises follow, wherein each chapter is divided into sections of one or more of the main (1-level) headings. These sections are labeled 1 - I, 1 - II, and so on. The Interactive Exercises begin with a list of Selected Words (other than boldfaced terms) selected by the author as those that are most likely to enhance understanding. In the text chapters, these words appear in italics, quotation marks, or roman type. This is followed by a list of Boldfaced, Page-Referenced Terms, which appear boldfaced in the text. These terms are essential to understanding each study guide section of a particular chapter. Space is provided by each term for you to formulate a definition in your own words. Next is a series of different types of exercises that includes completion, short answer, true/false, fill-in-the-blanks, matching, choice, dichotomous choice, label and match, crossword puzzles, problems, labeling, sequencing, multiple-choice, and completion of tables.

A Self-Quiz immediately follows the Interactive Exercises. This quiz is composed primarily of multiple-choice questions, although sometimes we present another examination device or some combination of devices. Any wrong answers in the quiz indicate portions of the text you need to reexamine. A series of Chapter Objectives/Review Questions follows each Self-Quiz. These are tasks that you should be able to accomplish if you have understood the assigned reading in the text. Some objectives require you to compose a short answer or long essay, while others require drawing a sketch or supplying correct words.

The final part of each chapter is named Integrating and Applying Key Concepts. It invites you to try your hand at applying major concepts to situations in which there is not necessarily a single pat answer, and so none is provided in the chapter answer section (except for a problem in Chapter 9). Your text generally will provide enough clues to get you started on an answer, but this part is intended to stimulate your thought and provoke group discussions. A separate publication titled *Critical Thinking Exercises for Starr's Biology: Concepts and Applications,* Third Edition, is available. Corresponding with the text chapters, these exercises present problem situations that concentrate on the critical and higher-level thinking skills used by scientists. Solving these problems requires you to apply chapter information to form new perspectives, analyze data, draw conclusions, make predictions, and identify basic assumptions.

*A person's mind, once stretched by a new idea, can never return to its original dimension.*
—Oliver Wendell Holmes

STRUCTURE OF THIS STUDY GUIDE

The outline below shows how each chapter in this study guide is organized.

Chapter Number ─────────────────────────➤ **8**

Chapter Title ───────────────────────────➤ **MEIOSIS**

Chapter Outline ─────────────────────────➤ **COMPARISON OF ASEXUAL AND SEXUAL REPRODUCTION**

**MEIOSIS AND THE CHROMOSOME NUMBER**
Think "Homologues"
Two Divisions, Not One

**A VISUAL TOUR OF THE STAGES OF MEIOSIS**

**KEY EVENTS OF MEIOSIS I**
Prophase I Activities
Metaphase I Alignments

**FROM GAMETES TO OFFSPRING**
Gamete Formation in Plants
Gamete Formation in Animals
More Gene Shufflings at Fertilization

**MEIOSIS AND MITOSIS COMPARED**

Interactive Exercises ────────────────────➤ The interactive exercises are divided into numbered sections by titles of main headings and page references. Each section begins with a list of author-selected words that appear in the text chapter in italics, quotation marks, or roman type. This is followed by a list of important boldfaced, page-referenced terms from each section of the chapter. Each section ends with interactive exercises that vary in type and require constant interaction with the important chapter information.

Self-Quiz ───────────────────────────────➤ Usually a set of multiple-choice questions that sample important blocks of text information.

Chapter Objectives/Review Questions ──────➤ Combinations of relative objectives to be met and questions to be answered.

Integrating and Applying Key Concepts ─────➤ Applications of text material to questions for which there may be more than one correct answer.

Answers to Interactive Exercises ──────────➤ Answers for all interactive exercises can be
and Self-Quiz found at the end of this study guide by chapter and title, and the main headings with their page references, followed by answers for the Self-Quiz.

# 1

# METHODS AND CONCEPTS
# IN BIOLOGY

ORGANIZATION IN NATURE
    Levels of Biological Organization
    Metabolism: Life's Energy Transfers
    Interdependencies Among Organisms

SENSING AND RESPONDING TO THE
    ENVIRONMENT

CONTINUITY AND CHANGE
    Perpetuating Heritable Traits
    Mutations—Source of Variations in Heritable
        Traits

SO MUCH UNITY, YET SO MANY SPECIES

AN EVOLUTIONARY VIEW OF LIFE'S DIVERSITY
    Evolution Defined
    Natural Selection Defined

THE NATURE OF BIOLOGICAL INQUIRY
    On Scientific Methods
    About the Word "Theory"

*Focus on Science:* DARWIN'S THEORY AND DOING
    SCIENCE

THE LIMITS OF SCIENCE

---

## Interactive Exercises

Note: In the answer sections of this book, a specific molecule is most often indicated by its abbreviation. For example, adenosine triphosphate is ATP.

## 1 - I. ORGANIZATION IN NATURE (pp. 2 - 5)

*Selected Words: energy transfers*

In addition to the boldfaced terms, the text features other important terms essential to understanding the assigned material. "Selected Words" is a list of these terms, which appear in the text in italics, in quotation marks, and occasionally in roman type. Latin binomials found in this section are underlined and in roman type to distinguish them from other italicized words.

*Boldfaced, Page-Referenced Terms*

The page-referenced terms are important; they were in boldface type in the chapter. Write a definition for each term in your own words without looking at the text. Next, compare your definition with that given in the chapter or in the text glossary. If your definition seems inaccurate, allow some time to pass and repeat this procedure until you can define each term rather quickly (how fast you can answer is a gauge of your learning effectiveness).

(4) energy _____

_____

(4) DNA _____

_____

(4) cell _____

_____

(4) multicelled organism _____

_____

(4) population _____

_____

(4) community _____

_____

(4) ecosystem _____

_____

(4) biosphere _____

_____

(5) metabolism _____

_____

(5) photosynthesis _____

_____

(5) ATP _____

_____

(5) aerobic respiration _____

_____

(5) producers _____

_____

(5) consumers _____

_____

(5) decomposers _____

_____

## Matching

Choose the most appropriate answer for each term.

1. ___ organ system
2. ___ cell
3. ___ community
4. ___ ecosystem
5. ___ molecule
6. ___ DNA
7. ___ organelle
8. ___ population
9. ___ subatomic particle
10. ___ tissue
11. ___ biosphere
12. ___ energy
13. ___ multicelled organism
14. ___ organ
15. ___ atom

A. One or more tissues interacting as a unit
B. A proton, neutron, or electron
C. A well-defined structure within a cell, performing a particular function
D. All of the regions of Earth where organisms can live
E. A capacity to make things happen, to do work
F. The smallest unit of life
G. Two or more organs whose separate functions are integrated to perform a specific task
H. Two or more atoms bonded together
I. All of the populations interacting in a given area
J. The smallest unit of a pure substance that has the properties of that substance
K. A special molecule; sets living things apart from the nonliving world
L. A community interacting with its nonliving environment
M. An individual composed of cells arranged in tissues, organs, and often organ systems
N. A group of individuals of the same species in a particular place at a particular time
O. A group of cells that work together to carry out a particular function

## Fill-in-the-Blanks

(16) _____ refers to the cell's capacity to extract and transform energy from its surroundings and use energy to maintain itself, grow, and make more cells. Leaves contain cells that carry on the process of (17) _____ by trapping energy from the sun and using it to produce molecules of the energy carrier called (18) _____. These molecules transfer energy to sites inside the cell where enzymes put together sugars, starch, and other substances. Cells store excess energy. In most organisms, stored energy is released and transferred to ATP by way of a metabolic process known as aerobic (19) _____. Plants and other photosynthetic organisms are food (20) _____ and serve as an energy entry point for the world of life. Animals feed directly or indirectly on energy stored in tissues of the photosynthesizers; they are known as (21) _____. (22) _____ are bacteria and fungi that feed on tissues or remains of other organisms. Thus, there are (23) _____ among organisms based on a one-way flow of energy through them and a cycling of materials among them.

## 1 - II. SENSING AND RESPONDING TO THE ENVIRONMENT (p. 6)
## CONTINUITY AND CHANGE (pp. 6 - 7)

*Selected Words:* *variations, hemophilia*

## Boldfaced, Page-Referenced Terms

(6) receptors _____

_____

(6) homeostasis _____

_____

(6) reproduction _____

_____

(7) inheritance _____

_____

(7) mutation _____

_____

(7) adaptive trait _____

_____

## Fill-in-the-Blanks

Molecules and structures called (1) _____ permit organisms to detect specific information about the environment and make controlled responses to them. (2) _____ is the capacity to maintain rather constant physical and chemical conditions inside an organism within some tolerable range, even when external conditions vary. The production of offspring is known as (3) _____. (4) _____ means that parent organisms transmit specific DNA instructions for duplicating their traits to offspring. DNA assures that offspring will resemble their parents but also permit (5) _____ in the details of most traits. A person born with six fingers on each hand represents the outcome of a (6) _____, a molecular change in DNA. An (7) _____ trait is one that improves the survival and reproduction of an organism in a certain environment.

## 1 - III. SO MUCH UNITY, YET SO MANY SPECIES (pp. 8 - 9)

*Selected Words:* family, order, class, phylum, kingdoms, prokaryotic, eukaryotic

## Boldfaced, Page-Referenced Terms

(8) species _____

_____

(8) genus _____

_____

(8) monerans _____

_____

(8) protistans _____

_____

(9) fungi _____

_____

(9) plants _____

_____

(9) animals _____

_____

## *Fill-in-the-Blanks*

Different "kinds" of organisms are referred to as (1) _____. A (2) _____ is the first of a two-part name of each organism, and it encompasses all the species having perceived similarities to one another. The pronghorn antelope is known by the two-part name *Antilocapra americana; Antilocapra* is the (3) _____ name and *americana* is the (4) _____ name.

## *Complete the Table*

5. Fill in the table below by entering the correct name of each kingdom of life described.

| Kingdom | Description |
|---------|-------------|
| a. | Eukaryotic, multicelled, photosynthetic producers |
| b. | Prokaryotic, single-celled producers, consumers, or decomposers |
| c. | Eukaryotic, mostly multicelled, decomposers and consumers |
| d. | Eukaryotic, diverse multicelled consumers, typically move about |
| e. | Eukaryotic, single-celled species and multicelled forms |

## *Sequence*

Arrange in correct hierarchical order with the largest, most inclusive category first and the smallest, most exclusive category last. This exercise classifies a plant with the common name of "false Solomon's seal." Refer to page 8 in the text and Appendix II.

6. ___      A. Class: Monocotyledonae

7. ___      B. Family: Liliaceae

8. ___      C. Genus: *Smilacina*

9. ___      D. Kingdom: Plantae

10. ___      E. Order: Liliales

11. ___      F. Division: Anthophyta

12. ___      G. Species: *racemosa*

## 1 - IV. AN EVOLUTIONARY VIEW OF LIFE'S DIVERSITY (p. 10)

### *Boldfaced, Page-Referenced Terms*

(10) evolution _____

_____

(10) artificial selection _____

_____

(10) natural selection _____

_____

### *True/False*

If the statement is true, write a T in the blank. If the statement is false, correct it by changing the underlined word and writing the correct word(s) in the blank.

_____ 1. Evolution means that one or more of the traits that characterize a population are <u>stabilizing</u> through successive generations.

_____ 2. Pigeon breeding is a case of <u>artificial</u> selection.

_____ 3. If some organisms in a population inherit traits that lend them a survival advantage, they will be <u>more</u> likely to produce offspring.

_____ 4. If a trait is <u>adaptive</u>, it means that it improves the chances of surviving and reproducing.

_____ 5. Natural selection is the outcome of <u>similarities</u> in survival and reproduction that have occurred among individuals that differ in one or more traits.

## 1 - V. THE NATURE OF BIOLOGICAL INQUIRY (p. 11)
### *Focus on Science:* DARWIN'S THEORY AND DOING SCIENCE (p. 12)
### THE LIMITS OF SCIENCE (p.13)

*Selected Words:* *breadth of application, alternative hypotheses, subjective, internal conviction*

### *Boldfaced, Page-Referenced Terms*

(11) hypotheses _____

_____

(11) prediction _____

_____

(11) test _____

_____

(11) theory _____

_____

(12) experiment _____

_____

(12) control group _____

(12) key variable _____

## Sequence

Arrange the following steps of the scientific method in correct chronological sequence. Write the letter of the first step next to 1, the letter of the second step next to 2, and so on.

1. ___    A. Develop one or more hypotheses about what the solution or answer to a problem might be.

2. ___    B. Devise ways to test the accuracy of predictions drawn from the hypothesis (use of observations, models, and experiments).

3. ___    C. Repeat or devise new tests (different tests might support the same hypothesis).

4. ___    D. Make a prediction using the hypothesis as a guide.

5. ___    E. If the tests do not provide the expected results, check to see what might have gone wrong.

6. ___    F. Objectively report the results from tests and the conclusions drawn.

7. ___    G. Identify a problem or ask a question about nature

## Labeling

Assume that you have to identify what object is hidden inside a sealed, opaque box. Your only tools to test the contents are a bar magnet and a triple-beam balance. Label each of the following with an O (for observation) or a C (for conclusion).

8. ____ The object has two flat surfaces.

9. ____ The object is composed of nonmagnetic metal.

10. ____ The object is not a quarter, a half dollar, or a silver dollar.

11. ____ The object weighs x grams.

12. ____ The object is a penny.

## Complete the Table

13. Complete the following table of concepts important to understanding the scientific method of problem solving. Choose from scientific experiment, variable, prediction, control group, hypothesis, and theory.

| Concept | Definition |
|---|---|
| a. | An educated guess about what the answer (or solution) to a scientific problem might be |
| b. | A statement of what one should be able to observe in nature if one looks; the "if-then" process |
| c. | A related set of hypotheses that, taken together, form a broad explanation of a fundamental aspect of the natural world |
| d. | A carefully designed test that manipulates nature into revealing one of its secrets |
| e. | Used in scientific experiments to evaluate possible side effects of a test being performed on an experimental group |
| f. | The control group is identical to the experimental group except for the key factor under study |

## Completion

14. Questions that are _____ in nature do not readily lend themselves to scientific analysis.
15. Scientists often stir up controversy when they explain a part of the world that was considered beyond natural explanation—that is, belonging to the "_____."
16. The external world, not internal _____, must be the testing ground for scientific beliefs.

---

# Self-Quiz

___ 1. About 12 to 24 hours after a meal, a person's blood-sugar level normally varies from about 60 to 90 mg per 100 ml of blood, though it may attain 130 mg/100 ml after meals high in carbohydrates. That the blood-sugar level is maintained within a fairly narrow range despite uneven intake of sugar is due to the body's ability to carry out _____.
   a. predictions
   b. inheritance
   c. metabolism
   d. homeostasis

___ 2. Different species of Galapagos Island finches have different beak types to obtain different kinds of food. One species removes tree bark with a sharp beak to forage for insect larvae and pupae while another species has a large, powerful beak capable of crushing and eating large, heavy coated seeds. These statements illustrate _____.
   a. adaptation
   b. metabolism
   c. puberty
   d. homeostasis

___ 3. A boy is color-blind just as his grandfather was, even though his mother had normal vision. This situation is the result of_____.
   a. adaptation
   b. inheritance
   c. metabolism
   d. homeostasis

___ 4. The digestion of food, the production of ATP by respiration, the construction of the body's proteins, cellular reproduction by

cell division, and the contraction of a muscle are all part of _____.
a. adaptation
b. inheritance
c. metabolism
d. homeostasis

___ 5. Which of the following does *not* involve using energy to do work?
a. atoms bonding together to form molecules
b. the division of one cell into two cells
c. the digestion of food
d. none of these

___ 6. The experimental group and control group are identical except for _____.
a. the number of variables studied
b. the variable under study
c. the two variables under study
d. the number of experiments performed on each group

___ 7. A hypothesis should *not* be accepted as valid if _____.
a. the sample studied is determined to be representative of the entire group
b. a variety of different tools and experimental designs yield similar observations and results
c. other investigators can obtain similar results when they conduct the experiment under similar conditions
d. several different experiments, each without a control group, systematically eliminate each of the variables except one

___ 8. The principal point of evolution by natural selection is that_____ .
a. it measures the difference in survival and reproduction that has occurred among individuals who differ from one another in one or more traits
b. even bad mutations can improve survival and reproduction of organisms in a population
c. evolution does not occur when some forms of traits increase in frequency and others decrease or disappear with time
d. individuals lacking adaptive traits make up more of the reproductive base for each new generation

___ 9. Which match is incorrect?
a. Kingdom Animalia—multicelled consumers, most move about
b. Kingdom Plantae—mostly multicelled producers
c. Kingdom Monera—relatively simple, multicelled organisms
d. Kingdom Fungi—mostly multicelled decomposers
e. Kingdom Protista—many complex single cells, some multicellular

___ 10. The least inclusive of the taxonomic categories listed is_____ .
a. family
b. phylum
c. class
d. order
e. genus

# Chapter Objectives/Review Questions

This section lists general and detailed chapter objectives that can be used as review questions. You can make maximum use of these items by writing answers on a separate sheet of paper. Fill in answers where blanks are provided. To check for accuracy, compare your answers with information given in the chapter or glossary.

| Page | Objectives/Questions |
|---|---|
| (4) | 1. _____ interactions among molecules bind the parts of all structures together—they hold a rock together and they hold a frog together. |
| (4) | 2. A special molecule called _____ acid, or DNA, sets living things apart from the nonliving world. |
| (4) | 3. The _____ is an organized unit that can survive and reproduce on its own, given DNA, raw materials, and inputs of energy. |
| (4) | 4. Distinguish between single-celled organisms and multicelled organisms. |

(4)     5.  Arrange in order, from smallest to largest, the levels of organization that occur in nature. Define each as you list it.
(5)     6.  _____ means energy transfers within the cell.
(5)     7.  Organisms use a molecule known as _____ to transfer chemical energy from one molecule to another.
(5)     8.  By the process of aerobic _____, cells can release stored energy in food molecules and produce ATP molecules.
(5)     9.  Explain how the actions of producers, consumers, and decomposers create an interdependency among organisms.
(5)    10.  Describe the general pattern of energy flow through Earth's life forms and explain how Earth's resources are used again and again (cycled).
(6)    11.  _____ are certain molecules and structures that can detect specific kinds of information about the environment.
(6)    12.  A fairly constant level of physical and chemical conditions inside an organism represents a state of _____.
(6)    13.  _____ means the production of offspring.
(7)    14.  Explain the origin of trait variations that function in inheritance.
(7)    15.  An _____ trait is any trait that helps an organism survive and reproduce under a given set of environmental conditions.
(8)    16.  Explain the use of genus and species names by considering your Latin name, *Homo sapiens*.
(8)    17.  Arrange in order, from greater to fewer organisms included, the following categories of classification: class, family, genus, kingdom, order, phylum, and species.
(8)    18.  Distinguish these terms: *prokaryotic, eukaryotic*.
(8)    19.  List the five kingdoms of life; briefly describe organisms placed in each.
(10)   20.  As organisms move through time in successive generations, the character of populations change; this is called _____.
(10)   21.  Darwin used _____ selection as a model for natural selection.
(10)   22.  Define *natural selection* and briefly describe what is occurring when a population is said to evolve.
(10)   23.  Explain what is meant by the term *diversity* and speculate about what caused the great diversity of life forms on Earth.
(11)   24.  Define what is meant by *theory*; cite an actual example.
(11)   25.  Distinguish a theory from a prediction and the hypothesis.
(12)   26.  Tests performed to reveal nature's secrets are called _____.
(12)   27.  Generally, members of a control group should be identical to those of the experimental group except for the key factor under study, the _____.
(13)   28.  Explain the advantages of the "uncertainty" related to scientific endeavors.
(13)   29.  Explain how the methods of science differ from answering questions by using subjective thinking and systems of belief.

---

## Interpreting and Applying Key Concepts

1.  Humans have the ability to maintain body temperature very close to 37°C.
    a. What conditions would tend to make the body temperature drop?
    b. What measures do you think your body takes to raise body temperature when it drops?
    c. What conditions would cause body temperature to rise?
    d. What measures do you think your body takes to lower body temperature when it rises?
2.  Do you think that all humans on Earth today should be grouped in the same species?
3.  What sorts of topics are usually regarded by scientists as untestable by the kinds of methods that scientists generally use?

# 2

# CHEMICAL FOUNDATIONS FOR CELLS

## Interactive Exercises

### 2 - I. REGARDING THE ATOMS (pp. 16 - 18)
*Focus on Science:* **USING RADIOISOTOPES TO DATE FOSSILS, TRACK CHEMICALS, AND SAVE LIVES** (p. 19)

*Selected Words:* net, *half-life*, charge, *radiation therapy*, PET (Position-Emission Tomography)

In addition to the boldfaced terms, the text features other important terms essential to understanding the assigned material. "Selected Words" is a list of these terms, which appear in the text in italics, in quotation marks, and occasionally in roman type. Latin binomials founds in this section are underlined and in roman type to distinguish them from other italicized words.

## Boldfaced, Page-Referenced Terms

The page-referenced terms are important; they were in boldface type in the chapter. Write a definition for each term in your own words without looking at the text. Next, compare your definition with that given in the chapter or in the text glossary. If your definition seems inaccurate, allow some time to pass and repeat this procedure until you can define each term rather quickly (how fast you can answer is a gauge of your learning effectiveness).

(18) atoms _____

_____

(18) protons _____

_____

(18) neutrons _____

_____

(18) electrons _____

_____

(18) atomic number _____

_____

(18) mass number _____

_____

(18) isotopes _____

_____

(18) radioisotopes _____

_____

(19) tracers _____

_____

## Matching

Choose the most appropriate answer for each term. Some letters may not be used.

1. ___ atoms
2. ___ protons
3. ___ neutrons
4. ___ electrons
5. ___ atomic number
6. ___ mass number
7. ___ isotopes
8. ___ radioisotopes
9. ___ tracers
10. ___ half-life
11. ___ radiation therapy

A. Radioisotopes used with scintillation counters to reveal the pathway or destination of a substance
B. Subatomic particles with a negative charge
C. Positively charged subatomic particles within the nucleus
D. The time it takes for half of a given amount of a radioactive element to decay into a different element
E. Atoms of a given element that differ in the number of neutrons
F. Refers to the number of protons in an atom
G. Radioactive isotopes
H. Destroys or impairs living cancer cells
I. The number of protons and neutrons in the nucleus of one atom
J. Smallest units that retain the properties of a given element
K. Subatomic particles within the nucleus carrying no charge

## 2 - II. WHAT IS A CHEMICAL BOND? (pp. 20 - 21)

**Selected Words:** *formulas, chemical equations, law of conservation of mass, lowest available energy level, higher energy levels, "compounds," "mixture"*

### Boldfaced, Page-Referenced Terms

(20) chemical bonds _____

_____

(20) orbitals _____

_____

(20) shell model _____

_____

(21) molecule _____

_____

### Fill-in-the-Blanks

The expression $12H_2O$ plus $6CO_2$ ———> $6O_2$ plus $C_6H_{12}O_6$ plus $6H_2O$ is known as the chemical (1) _____ for photosynthesis. $H_2O$ is the (2) _____ for water. The arrow in the expression above means (3) _____ . The (4) _____ are to the left of the arrow and the (5) _____ are to the right of the arrow. In the expression one can count 12 hydrogen atoms on the left side of the arrow, then one should be able to count (6) _____ hydrogen atoms on the right side of the arrow. According to the law of conservation of (7) _____ , the atoms are not lost in a chemical reaction but are rearranged. Expressions used to represent cellular reactions must be (8) _____ in this way because no atoms are lost.

### Matching

Choose the most appropriate answer for each term.

9. ___ shells
10. ___ lowest energy level
11. ___ orbitals
12. ___ chemical bond
13. ___ higher energy levels

A. Regions of space around an atom's nucleus where electrons are likely to be at any one instant
B. An energy relationship
C. A series of orbitals arranged around the nucleus
D. Energy of electrons farther from the nucleus than the first orbital
E. Energy of electrons in the orbital closest to the nucleus

## Complete the Table

14. Complete the following table (refer to Figure 2.6, p. 21, in the text) by entering the name of the element and its symbol in the appropriate spaces.

| | | | | Electron Distribution | | |
| Element | Symbol | Atomic Number | First Shell | Second Shell | Third Shell | Fourth Shell |
| --- | --- | --- | --- | --- | --- | --- |
| a. | | 20 | 2 | 8 | 8 | 2 |
| b. | | 6 | 2 | 4 | | |
| c. | | 17 | 2 | 8 | 7 | |
| d. | | 1 | 1 | | | |
| e. | | 11 | 2 | 8 | 1 | |
| f. | | 7 | 2 | 5 | | |
| g. | | 8 | 2 | 6 | | |

## Identification

15. Following the model below (number of protons and neutrons shown in the nucleus), identify the indicated atoms of the elements illustrated below by entering appropriate electrons in this form: (2e-).

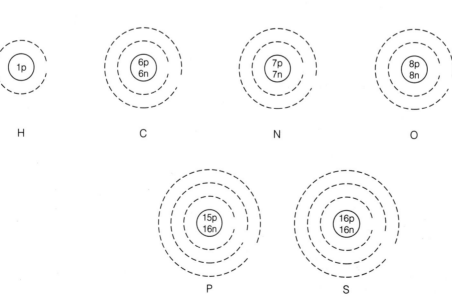

# 2 - III. IMPORTANT BONDS IN BIOLOGICAL MOLECULES (pp. 22 - 23)

*Selected Words:* *net* charge, *nonpolar* covalent bond, *polar* covalent bond

## Boldfaced, Page-Referenced Terms

(22) ion _____

_____

(22) ionic bond _____

_____

(22) covalent bond _____

_____

(23) hydrogen bond _____

_____

## Identification

1.  Following the model below, complete the diagram by adding arrows to identify the transfer of electron(s), showing how positive magnesium and negative chlorine ions form ionic bonds to create a molecule of $MgCl_2$ (magnesium chloride).

MODEL:

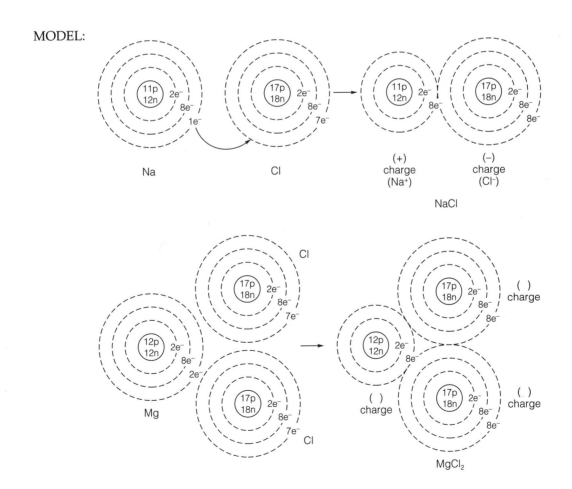

## Identification

2. Following the model of hydrogen gas below, complete the sketch by placing electrons (as dots) in the outer shells to identify the nonpolar covalent bonding that forms oxygen gas; similarly identify polar covalent bonds by completing electron structures to form a water molecule.

MODEL:

$H_2$

$O_2$

$H_2O$

## Short Answer

3. Distinguish between a nonpolar covalent bond and a polar covalent bond. Cite an example (p. 23).

_____

_____

4. Describe one example of a large biological molecule within which hydrogen bonds exist (p. 23).

_____

_____

## 2 - IV. PROPERTIES OF WATER (pp. 24 - 25)

*Selected Words:* net charge, *liquid* water, "spheres of hydration"

### Boldfaced, Page-Referenced Terms

(24) hydrophilic _____

_____

(24) hydrophobic _____

_____

(24) temperature _____

_____

(24) evaporation _____

_____

(25) cohesion _____

_____

(25) solutes _____

_____

## Fill-in-the-Blanks

The (1) _____ of water molecules allows them to hydrogen-bond with each other. Water molecules hydrogen-bond with polar substances, which are (2) _____ (water-loving). Polarity causes water to repel oil and other nonpolar substances, which are (3) _____ (water-dreading). (4) _____ is a measure of the constant motion of molecules. Water changes its temperature more slowly than air because of the great amount of heat required to break the (5) _____ bonds between water molecules; this property helps stabilize temperature in cells. The escape of water molecules from a fluid surface to the surrounding air is known as (6) _____. Below 0°C, water molecules become locked in the less dense bonding pattern of (7) _____, which is less dense than water. (8) _____ is the property of water that explains how insects walk on water and how long, narrow water columns rise to the tops of tall trees. Water is an excellent (9) _____ in which ions and polar molecules readily dissolve. Substances dissolved in water are known as (10) _____. A substance is (11) _____ in water when spheres of (12) _____ form around its individual ions or molecules.

## 2 - V. WATER, DISSOLVED IONS, AND THE WORLD OF CELLS (pp. 26 - 27)

*Selected Words:* *neutrality,* *acidic* solutions, *basic* solutions, *acidosis*

## Boldfaced, Page-Referenced Terms

(26) hydrogen ions _____

_____

(26) hydroxide ions _____

_____

(26) pH scale _____

_____

(26) acids _____

_____

(26) bases _____

_____

(27) salts _____

_____

(27) buffer _____

_____

### Fill-in-the-Blanks

Hydrogen ions are the same thing as free (1) _____. At any given instant in liquid water, a few

molecules are breaking apart into (2) _____ ($H^+$) ions and (3) _____ ($OH^-$) ions. This

ionization of water is the basis for the (4) _____ scale. At 25°C, pure water has just as many $H^+$ as

$OH^-$ ions, a condition representing (5) _____ on the pH scale. A pH of 8 has an $H^+$ concentration

100 times higher than a pH of (6) _____. The (7) _____ scale is used to express the

(8) _____ ion concentration of aqueous solutions. When acids dissolve in water, they release

(9) _____ ions; when bases dissolve in water, they release (10) _____ ions. Acidic solutions

such as vinegar have more (11) _____ ions than (12) _____ ions. Basic solutions such as

baking soda have fewer (13) _____ ions than (14) _____ ions. Most living cells maintain an

$H^+$ concentration close to pH (15) _____, or neutral. A (16) _____ is formed when an acid

reacts with a base. (17) _____ are molecules that combine with or release hydrogen ions to prevent

rapid shifts in pH. (18) _____ ($HCO_3-$) helps restore pH when blood becomes too acidic by

combining with $H^+$ to form (19) _____ acid ($H_2CO_3$). In one form of (20) _____, the body

can't eliminate carbon dioxide, so carbonic acid and $H^+$ increases in blood. Breathing becomes difficult

and weakens the body. Proteins remain dispersed in cellular fluid (rather than settling) through

electrical interactions with (21) _____ and ions. This gives (22) _____ access to molecular

surfaces as stages for life-sustaining tasks.

### Complete the Table

23. Complete the following table by consulting Figure 2.12, page 26, in the text.

| Fluid | pH Value | Acid/Base |
|---|---|---|
| a. Blood | | |
| b. Saliva | | |
| c. Urine | | |
| d. Stomach Acid | | |

## 2 - VI. PROPERTIES OF ORGANIC COMPOUNDS (pp. 28 - 29)

### Selected Words: "polymer," "monomer"

### Boldfaced, Page-Referenced Terms

(28) organic compounds _____

_____

(28) hydrocarbon _____

_____

(28) functional groups _____

_____

(28) alcohols _____

_____

(29) enzymes _____

_____

(29) functional-group transfer _____

_____

(29) electron transfer _____

_____

(29) rearrangement _____

_____

(29) condensation _____

_____

(29) cleavage _____

_____

(29) hydrolysis _____

_____

## Labeling

Study the following structural formulas of organic compounds. By reference to Figures 2.14 and 2.15, pages 28 and 29 in the text, identify the circled functional groups (sometimes repeated) by entering the correct name in the blanks with matching numbers below the diagrams.

1. _____

2. _____

3. _____

4. _____

5. _____

6. _____

7. _____

## Short Answer

8. State the general role of enzymes as they relate to the chemistry of life (p. 29). _____

_____

_____

## Complete the Table

9. Complete the following table summarizing five categories of reactions that are mediated by enzymes.

| Reaction Category | Reaction Description |
|---|---|
| a. | Through covalent bonding, two molecules combine to form a larger molecule |
| b. | One or more electrons stripped from one molecule are donated to another molecule |
| c. | A molecule splits into two smaller ones |
| d. | One molecule gives up a functional group, which another molecule accepts |
| e. | A juggling of internal bonds converts one type of organic compound into another |

## Identification

10. The structural formulas of two adjacent amino acids are shown below. Identify how enzyme action causes formation of a covalent bond and a water molecule (through a condensation reaction) by circling an H atom from one amino acid and an -OH group from the other amino acid. Also circle the covalent bond that formed the dipeptide.

amino acid          amino acid                    dipeptide

## Short Answer

11. Describe hydrolysis through enzyme action for the molecules in exercise 10 (refer to Figure 2.26, p. 35, in the text). _____

_____

_____

_____

_____

# 2 - VII. CARBOHYDRATES (pp. 30 - 31)

*Selected Words:* disaccharides

## Boldfaced, Page-Referenced Terms

(30) carbohydrate _____

_____

(30) monosaccharides _____

_____

(30) oligosaccharides _____

_____

(30) polysaccharides _____

_____

## Identification

1.  In the diagram below, identify condensation reaction sites between the two glucose molecules by circling the components of the water removed that allow a covalent bond to form between the glucose molecules. Note that the reverse reaction is hydrolysis and that both condensation and hydrolysis reactions require enzymes in order to proceed.

## Complete the Table

2. In the table below, enter the name of the carbohydrate described by its carbohydrate class and functions.

| Carbohydrate | Carbohydrate Class | Function |
|---|---|---|
| a. | Oligosaccharide (disaccharide) | Most plentiful sugar in nature; transport form of carbohydrates |
| b. | Monosaccharide | Five-carbon sugar occurring in DNA |
| c. | Monosaccharide | Main energy source for most organisms; precursor of many organic organisms |
| d. | Polysaccharide | Structural material of plant cell walls |
| e. | Monosaccharide | Five-carbon sugar occurring in DNA |
| f. | Oligosaccharide (disaccharide) | Sugar present in milk |
| g. | Polysaccharide | Main structural material in some external skeletons and other hard body parts of some animals and fungi |
| h. | Branched polysaccharide | Animal starch |
| i. | Polysaccharide | Sugar storage form in plants |

# 2 - VIII. LIPIDS (pp. 32 - 33)

**Selected Words:** *unsaturated, saturated*

## Boldfaced, Page-Referenced Terms

(32) lipids _____

_____

(32) fatty acids _____

_____

(32) triglycerides _____

_____

(33) phospholipid _____

_____

(33) sterols _____

_____

(33) waxes _____

_____

## Labeling

1. In the appropriate blanks, label the molecules shown below as saturated or unsaturated.

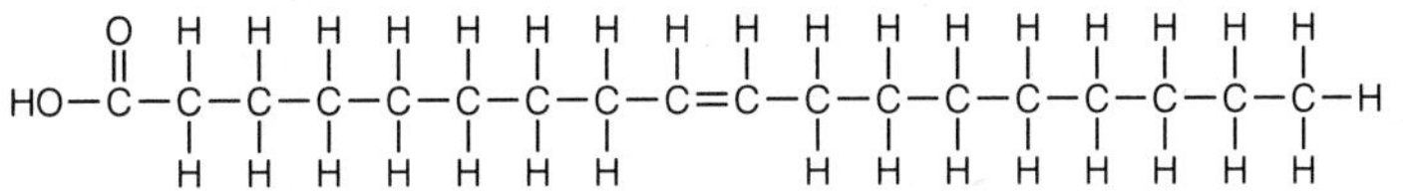

oleic acid

a. _____

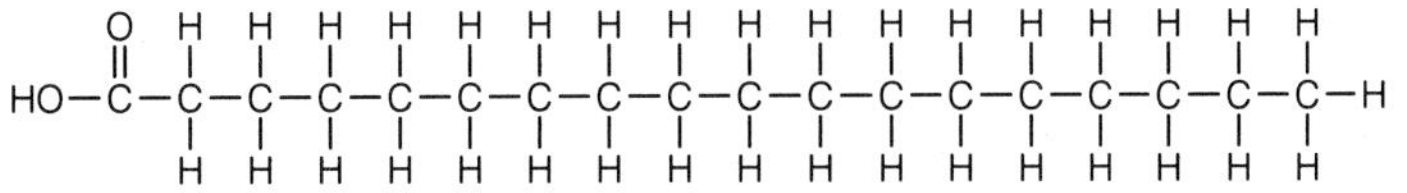

stearic acid

b. _____

## Identification

2. Combine glycerol with three fatty acids below to form a triglyceride by circling the participating atoms that will identify three covalent bonds. Also circle the covalent bonds in the triglyceride.

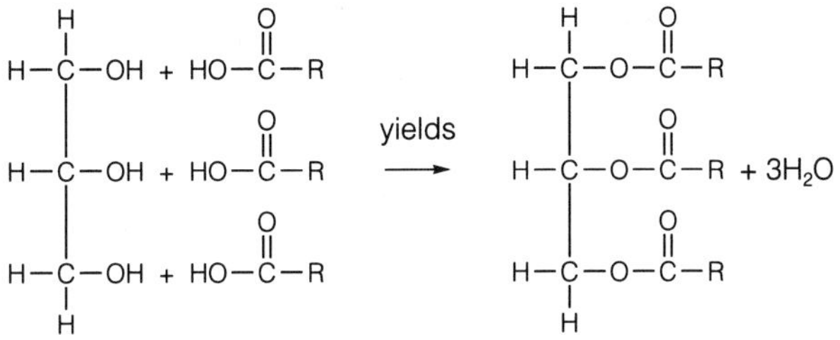

glycerol      three fatty
              acids

yields

triglyceride
(a complete fat
molecule)

## Short Answer

3. Describe the structure and biological functions of phospholipid molecules (p. 33). _____

_____

_____

_____

## Matching

Choose the most appropriate answer for each item. Some letters may be used more than once.

4. ___ richest source of body energy
5. ___ honeycomb material
6. ___ cholesterol
7. ___ saturated tails
8. ___ butter and lard
9. ___ main cell membrane component
10. ___ plant cuticles
11. ___ triglycerides
12. ___ precursors of testosterone, estrogen, and bile salts
13. ___ unsaturated tails
14. ___ vegetable oil
15. ___ vertebrate insulation

A. Fatty acids
B. Neutral fats
C. Phospholipids
D. Waxes and/or cuticle
E. Sterols

## 2 - IX. AMINO ACIDS AND PROTEINS (pp. 34 - 35)
## SOME EXAMPLES OF FINAL PROTEIN STRUCTURE (p. 36)

*Selected Words:* *peptide* bonds, *primary* structure, *secondary* structure, *tertiary* structure, *quaternary* structure

### Boldfaced, Page-Referenced Terms

(34) proteins _____

_____

(34) amino acid _____

_____

(34) polypeptide chain _____

_____

(36) hemoglobin _____

_____

(36) lipoproteins _____

_____

(36) glycoproteins _____

_____

(36) denaturation _____

_____

## Matching

For exercises 1, 2, and 3, match the major parts of *every* amino acid by entering the letter of the part in the blank corresponding to the number on the molecule.

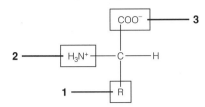

1. ___
2. ___
3. ___

A.  R group (a symbol for a characteristic group of atoms that differ in number and arrangement from one amino acid to another)
B.  Carboxyl group (ionized)
C.  Amino group (ionized)

## Identification

4.  In the illustration of four amino acids in cellular solution (ionized state) below, circle the atoms and ions that form water to allow identification of covalent (peptide) bonds between adjacent amino acids to form a polypeptide. On the completed polypeptide below, circle the newly formed peptide bonds.

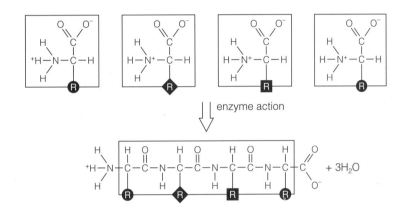

## Matching

Choose the most appropriate answer for each term.

5. ___ amino acid

6. ___ peptide bond

7. ___ polypeptide chain

8. ___ primary structure

9. ___ secondary structure

10. ___ tertiary structure

11. ___ quaternary structure

12. ___ lipoproteins

13. ___ glycoproteins

14. ___ denaturation

A.  A coiled or extended pattern of protein structure caused by regular intervals of H bonds
B.  Three or more amino acids joined in a linear chain
C.  Proteins with linear or branched oligosaccharides covalently bonded to them; found on animal cell surfaces, in cell secretions, or on blood proteins
D.  Folding of a protein through interactions among R groups of a polypeptide chain
E.  Form when freely circulating blood proteins encounter and combine with cholesterol, or phospholipids
F.  The type of covalent bond linking one amino acid to another
G.  Hemoglobin, a globular protein of four chains, is an example
H.  Breaking weak bonds in large molecules (such as protein) to change its shape so it no longer functions
I.  Lowest level of protein structure; has a linear, unique sequence of amino acids: an acid group, a hydrogen atom, and an R group
J.  A small organic compound having an amino group, an acid group, a hydrogen atom, and an R group

# 2 - X. NUCLEOTIDES AND NUCLEIC ACIDS (p. 37)

## Boldfaced, Page-Referenced Terms

(37) nucleotides _____

_____

(37) ATP _____

_____

(37) coenzymes _____

_____

(37) nucleic acids _____

_____

(37) RNA _____

_____

(37) DNA _____

_____

## Matching

For exercises 1, 2, and 3, match the following answers to the parts of a nucleotide shown in the diagram below.

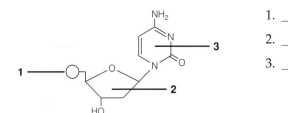

1. ____

2. ____

3. ____

A. A five-carbon sugar (ribose or deoxyribose)
B. Phosphate group
C. A nitrogen-containing base that has either a single-ring or double-ring structure

## Identification

4. In the diagram of a single-stranded nucleic acid molecule shown here, encircle as many complete nucleotides as possible. How many complete nucleotides are present?

## Matching

Choose the most appropriate answer for each term.

5. ___ adenosine triphosphate

6. ___ RNA

7. ___ DNA

8. ___ NAD⁺ and FAD

9. ___ cAMP

A. Single nucleotide strand; function in processes by which genetic instructions are used to build proteins
B. ATP, a cellular energy carrier
C. Nucleotide chemical messenger
D. Single nucleotide units; coenzymes; transport hydrogen ions and their associated electrons from one cell reaction site to another
E. Double nucleotide strand; encodes genetic instructions with base sequences

## Complete the Table

10. Complete the table below by entering the correct name of the major cellular organic compounds suggested in the "types" column (choose from carbohydrates, proteins, nucleic acids, and lipids).

*Cellular Organic Compounds* | *Types*
--- | ---
a. | Phospholipids
b. | Antibodies
c. | Enzymes
d. | Genes
e. | Glycogen, starch, cellulose, and chitin
f. | Glycerides
g. | Saturated and unsaturated fats
h. | Coenzymes
i. | Sterols, oils, and waxes
j. | Glucose and sucrose

# Self-Quiz

___ 1. A molecule is _____.
   a. a combination of two or more atoms
   b. less stable than its constituent atoms separated
   c electrically charged
   d. a carrier of one or more extra neutrons

___ 2. If lithium has an atomic number of 3 and an atomic mass of 7, it has _____ neutron(s) in its nucleus.
   a. one
   b. two
   c. three
   d. four
   e. seven

___ 3. A hydrogen bond is _____.
   a. a sharing of a pair of electrons between a hydrogen nucleus and an oxygen nucleus
   b. a sharing of a pair of electrons between a hydrogen nucleus and either an oxygen or a nitrogen nucleus
   c. formed when the weak charge of an atom of a molecule interacts weakly with a neighboring hydrogen atom that is already taking part in a polar covalent bond
   d. none of the above

___ 4.  A solution with a pH of 10 is _____ times as basic as one with a pH of 7.
a. 2
b. 3
c. 10
d. 100
e. 1,000

___ 5.  Any molecule that combines with or releases hydrogen ions, or both, and helps to stabilize pH is known as a(n) _____.
a. neutral molecule
b. salt
c. base
d. acid
e. buffer

___ 6.  Carbon is part of so many different substances because _____.
a. carbon generally forms two covalent bonds with a variety of other atoms
b. carbon generally forms four covalent bonds with a variety of atoms
c. carbon ionizes easily
d. carbon is a polar compound

___ 7.  Amino, carboxyl, phosphate, and hydroxyl are examples of _____.
a. enzymes
b. sugar units
c. functional groups
d. coenzymes

___ 8.  _____ are compounds used by cells as transportable packets of quick energy, storage forms of energy, and structural materials.
a. Lipids
b. Nucleic acids
c. Carbohydrates
d. Proteins

___ 9.  Hydrolysis could be correctly described as the _____.
a. heating of a compound in order to drive off its excess water and concentrate its volume
b. breaking of a long-chain compound into its subunits by adding water molecules to its structure between the subunits
c. linking of two or more molecules by the removal of one or more water molecules
d. constant removal of hydrogen atoms from the surface of a carbohydrate

___ 10. Genetic instructions are encoded in the bases of _____; molecules of _____ function in processes using genetic instructions to construct proteins.
a. DNA; DNA
b. DNA; RNA
c. RNA; DNA
d. RNA; RNA

---

## Chapter Objectives/Review Questions

This section lists general and detailed chapter objectives that can be used as review questions. You can make maximum use of these items by writing answers on a separate sheet of paper. Fill in answers where blanks are provided. To check for accuracy, compare your answers with information given in the chapter or glossary.

| Page | Objectives/Questions |
|---|---|
| (18) | 1. _____ are the smallest units that retain the properties of a given element. |
| (18) | 2. List and describe the three types of subatomic particles, and describe the reason that hydrogen is an exception. |
| (18) | 3. The number of protons in an atom is referred to as the _____ number of that element; the combined number of protons and neutrons in the atomic nucleus is referred to as the _____ number of that element. |
| (18) | 4. Distinguish between isotopes and radioisotopes. |
| (19) | 5. List some uses of radioisotopes. |
| (20) | 6. A chemical _____ is a union between the electron structures of atoms. |
| (20) | 7. Be able to read a chemical equation. |
| (20) | 8. What is the relationship of orbitals to shells? |

(20)  9. Be able to sketch shell models of the atoms described in the text, Figure 2.5, page 20.

(21) 10. A _____ is formed when two or more atoms bond together.

(21) 11. A _____ contains atoms of two or more elements whose proportions never vary.

(21) 12. How does a mixture differ from a compound?

(22) 13. An _____ is an atom that becomes positively or negatively charged.

(22) 14. An association of two oppositely charged ions is a(n) _____ bond.

(22) 15. In a(n) _____ bond, two atoms share electrons.

(23) 16. Explain why $H_2$ is an example of a nonpolar covalent bond and $H_2O$ has two polar covalent bonds.

(23) 17. In a(n) _____ bond, an atom of a molecule interacts weakly with a hydrogen atom that is already participating in a polar covalent bond.

(24) 18. Polar molecules attracted to water are _____; all nonpolar molecules are _____ and are repelled by water.

(24) 19. _____ is a measure of molecular motion; during _____, an input of heat energy converts liquid water to the gaseous state.

(24) 20. Describe the formation of ice in terms of hydrogen bonding.

(25) 21. Describe the property of water that allows insects to move on the water surface and that moves water from the roots to the tops of the tallest trees.

(25) 22. Distinguish a solvent from a solute.

(25) 23. Describe what happens when a substance is dissolved in water.

(26) 24. _____ are substances that release $H^+$ ions when they dissolve in water, and _____ are substances that combine with those ions.

(26) 25. Black coffee with a pH of 5 is a(n) _____ solution while baking soda with a pH of 9 is a(n) _____ solution.

(27) 26. A _____ is produced by a chemical reaction between an acid and a base.

(27) 27. Define *buffer*; cite an example and describe how it operates.

(28) 28. List the three principle elements found in organisms.

(28) 29. Each carbon atom can form as many as _____ covalent bonds with other carbon atoms as well as with atoms of other elements.

(28) 30. A _____ has only hydrogen atoms attached to a carbon backbone.

(29) 31. Hydroxyl, amino, and carboxyl are examples of _____ groups.

(29) 32. Describe the role of enzymes in the metabolism of life.

(29) 33. List the small organic compounds from which the carbohydrates, lipids, proteins, and nucleic acids are constructed.

(30) 34. Define *carbohydrates*; list their general functions.

(30) 35. The simplest carbohydrates are _____; be able to give examples and their functions.

(30) 36. An _____ is a short chain of two or more sugar monomers; be able to give examples of well-known disaccharides and their functions.

(30) 37. A _____ is a straight or branched chain of hundreds or thousands of sugar monomers, of the same or different kinds; be able to give common examples and their functions.

(32) 38. Define *lipids*; list their general functions.

(32) 39. Describe a "fatty acid"; a _____ molecule has three fatty acid tails attached to a backbone of glycerol; distinguish a saturated fat from an unsaturated fat.

(33) 40. A _____ has two fatty acid tails and a hydrophilic head attached to a glycerol backbone; _____ have long-chain fatty acids linked to long-chain alcohols or to carbon rings; list functions of these molecules.

(33) 41. Define *sterols* and describe their chemical structure.

(33) 42. List the names of some sterols and their derivatives; describe their functions.

(34) 43. Describe the primary structure of protein and cite its general functions; be able to sketch the three general parts of any amino acid.

(34–35) 44. Describe how the primary, secondary, tertiary, and quaternary structure of proteins results in three-dimensional structures.

(36) 45. Distinguish lipoproteins from glycoproteins.

(36)    46.    _____ refers to the loss of a molecule's three-dimensional shape through disruption of the weak bonds responsible for it.

(37)    47.    Describe the three parts of every nucleotide; give the general functions of DNA and RNA molecules.

## Integrating and Applying Key Concepts

1.  Explain what would happen if water were a nonpolar molecule instead of a polar molecule. Would water be a good solvent for the same kinds of substances? Would the nonpolar molecule's specific heat likely be higher or lower than that of water? Would surface tension be affected? Cohesive nature? Ability to form hydrogen bonds? Is it likely that the nonpolar molecules could form unbroken columns of liquid? What implications would that hold for trees?

2.  Humans can obtain energy from many different food sources. Do you think this ability is an advantage or a disadvantage in terms of long-term survival? Why?

3.  If the ways that atoms bond affect molecular shapes, do the ways that molecules behave toward one another influence the shapes of organelles? Do the ways that organelles behave toward one another influence the structure and function of the cells?

# 3

# CELL STRUCTURE
# AND FUNCTION

## Interactive Exercises

## 3 - I. BASIC ASPECTS OF CELL STRUCTURE AND FUNCTION (pp. 40 - 43)

*Selected Words:* "mosaic," "fluid"

*Boldfaced, Page-Referenced Terms*

(41) cell theory _____

_____

(42) cell _____

_____

(42) DNA-containing region _____

_____

(42) plasma membrane _____

_____

(42) cytoplasm _____

_____

(42) eukaryotic cells _____

_____

(42) prokaryotic cells _____

_____

(42) phospholipid _____

_____

(42) lipid bilayer _____

_____

(42) fluid mosaic model _____

_____

(43) transport proteins _____

_____

(43) receptor proteins _____

_____

(43) recognition proteins _____

_____

(43) adhesion proteins _____

_____

## Label and Match

Although cells vary in many specific ways, they are all alike in a few basic respects. Identify each part of the illustration below. Choose from cytoplasm, DNA-containing region (nucleus in eukaryotes), and cytoplasm. Complete the exercise by matching and entering the letter of the proper description in the parentheses following each label.

1. _____ _____ ( )

2. _____ ( )

3. _____ _____ ( )

A. Includes everything enclosed by the plasma membrane; a semifluid substance in which particles, filaments, and often compartments are organized
B. Thin, outermost membrane that maintains the cell as a distinct entity with metabolism occurring within
C. Molecules of heredity are here along with molecules that can read or copy hereditary instructions

## Labeling

Identify each numbered membrane protein in the illustration below.

4. _____ protein

5.–8. are types of _____ proteins

9. _____ protein

10. _____ protein

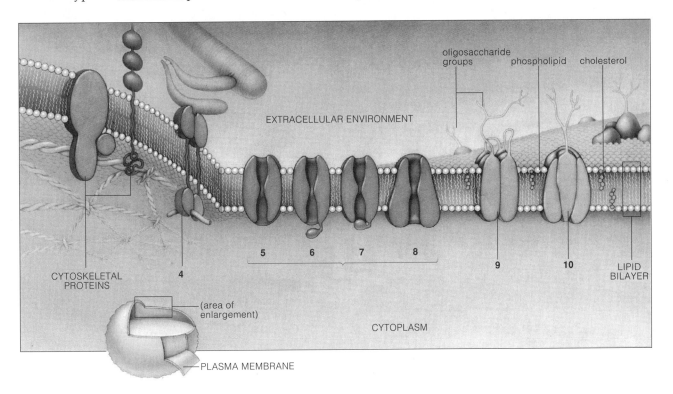

## Matching

Choose the most appropriate answer for each term.

11. ___ fluid

12. ___ phospholipid molecule

13. ___ adhesion proteins

14. ___ prokaryotic

15. ___ transport proteins

16. ___ mosaic

17. ___ recognition proteins

18. ___ eukaryotic

19. ___ fluid mosaic model

20. ___ receptor proteins

A. Allow or encourage water-soluble substances to move through their interior, thus crossing the bilayer
B. Refers to the mixed composition of lipids and proteins in cell membranes
C. Bacteria cells lacking a nucleus
D. Sticks cells of multicelled organisms together
E. Lipid and protein model of cell membrane structure
F. Bind hormones and other extracellular substances that trigger changes in cell activities
G. Cells containing distinct arrays of organelles, including the nucleus with DNA
H. Like molecular fingerprints, identifies a cell as a specific type
I. Consists of a hydrophilic head and two hydrophobic tails
J. Refers to the motions of membrane lipids and their interactions

## 3 - II. CELL SIZE AND CELL SHAPE (p. 44)
### *Focus on Science:* MICROSCOPES—GATEWAYS TO THE CELL (pp. 44 - 45)

*Selected Words:* compound light microscope, scanning tunneling microscope, transmission electron microscope, scanning electron microscope

### *Boldfaced, Page-Referenced Terms*

(44) surface-to-volume ratio _____

_____

(44) micrograph _____

_____

(44) wavelength _____

_____

### *Short Answer*

1. Explain why most cells are very small. _____

_____

_____

_____

### *Matching*

Choose the most appropriate answer for each term.

2. ___ electron wavelength

3. ___ micrograph

4. ___ transmission electron microscope

5. ___ scanning tunneling microscope

6. ___ wavelength

7. ___ compound light microscope

8. ___ scanning electron microscope

A. Glass lenses bend incoming light rays to form an enlarged image of a cell or some other specimen
B. The distance from one wave's peak to the peak of the wave behind it
C. A computer analyzes a tunnel formed in electron orbitals—the tunnel produced when voltage is applied between the tip of a needlelike probe and an atom
D. A narrow beam of electrons moves back and forth across the surface of a specimen coated with a thin metal layer
E. A photograph of an image formed with a microscope
F. About 100,000 times shorter than those of visible light
G. Electrons pass through a thin section of cells to form an image

## 3 - III. EUKARYOTIC CELLS (pp. 46 - 49)

*Selected Words: eukaryotic*

### *Boldfaced, Page-Referenced Terms*

(46) organelle _____

_____

## Short Answer

1. What are the advantages of having the organelles possessed by all eukaryotic cells? _____

_____

_____

_____

## Complete the Table

2. Complete the following table to identify eukaryotic organelles and their functions.

| Organelle or Structure | Main Function |
|---|---|
| a. | Contains most of the cell's DNA |
| b. | Synthesis of polypeptide chains |
| c. | First to modify new polypeptide chains; lipid synthesis |
| d. | Further modify polypeptide chains into mature proteins; sort and ship proteins and lipids |
| e. | Different types function in transport or storage of substances; digestion inside cells and other functions |
| f. | Efficient ATP production |
| g. | Cell shape, internal organization, and movements |

## 3 - IV. THE NUCLEUS (pp. 50 - 51)
### THE CYTOMEMBRANE SYSTEM (pp. 52 - 53)
### VESICLES THAT MOVE OUT OF AND INTO CELLS (p. 54)

*Selected Words:* *rough* ER, *smooth* ER, *exocytic* vesicles, *endocytic* vesicles

## Boldfaced, Page-Referenced Terms

(50) nucleus _____

_____

(50) nucleolus _____

_____

(51) nuclear envelope _____

_____

(51) chromatin _____

_____

(51) chromosome _____

_____

(51) cytomembrane system _____

_____

(52) endoplasmic reticulum, or ER _____

_____

(52) Golgi bodies _____

_____

(53) lysosome _____

_____

(53) peroxisomes _____

_____

(54) exocytosis _____

_____

(54) endocytosis _____

_____

## Complete the Table

1. Complete this table about the eukaryotic nucleus. Enter the name of each nuclear component described.

| Nuclear Component | Description |
| --- | --- |
| a. | One or more masses in the nucleus; sites where the protein and RNA subunits of ribosomes are assembled |
| b. | Two lipid bilayer membranes thick; proteins and protein pores span the bilayers |
| c. | A cell's total collection of DNA and its associated proteins |
| d. | An individual DNA molecule and its associated proteins in the nucleus |

## Matching

Study the illustration below and match each component of the cytomembrane system with the most correct description of function. Some components may be used more than once.

A. Spaces within smooth ER (SER)

B. Nucleus

C. Golgi body

D. Vesicles from Golgi

E. Phagocytosis

F. Vesicles from rough ER (RER)

G. Endocytosis with endocytic vesicles

H. Receptor-mediated endocytosis

I. Exocytosis with exocytic vesicles

J. Spaces within rough ER (RER)

K. Ribosomes

L. Vesicles budding from smooth ER (SER)

M. Lysosomes

N. Peroxisomes

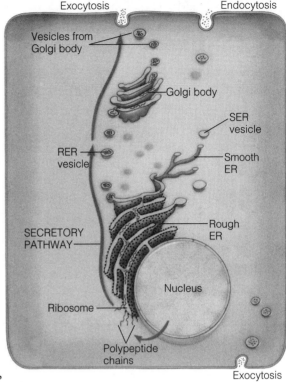

2. ___ Assembly of polypeptide chains
3. ___ Lipid assembly
4. ___ DNA instructions for building polypeptide chains
5. ___ Initiate protein modification following assembly
6. ___ Proteins and lipids take on final form
7. ___ Sort and package lipids and proteins for transport following modification
8. ___ Vesicles formed at plasma membrane transport substances into cytoplasm
9. ___ A form of endocytosis involving clusters of receptors in pits
10. ___ Sacs of enzymes that break down fatty acids and amino acids, forming hydrogen peroxide
11. ___ A startling form of endocytosis involving trapping food by temporary lobes of cytoplasm
12. ___ Special vesicle budding from Golgi body; major digestion organelle
13. ___ Transport unfinished proteins to a Golgi body
14. ___ Transport finished Golgi products to the plasma membrane
15. ___ Release Golgi products at the plasma membrane
16. ___ Transport unfinished lipids to a Golgi body

# 3 - V. MITOCHONDRIA (p. 55)
## SPECIALIZED PLANT ORGANELLES (p. 56)

### Boldfaced, Page-Referenced Terms

(55) mitochondria _____

_____

(56) chloroplasts _____

_____

(56) central vacuole _____

_____

## True/False

If the statement is true, write a T in the blank. If the statement is false, correct it by changing the underlined word and writing the correct word(s) in the blank.

_____ 1. Mitochondria <u>lack</u> DNA and can divide.

_____ 2. Mitochondria have <u>two</u> membranes, which create two distinct compartments for chemical reactions inside each mitochondrion.

_____ 3. Within a mitochondrion, enzymes mediate reactions that break down glucose; the energy released is used to form a <u>small</u> yield of ATP.

_____ 4. Infoldings of the <u>outer</u> membrane of each mitochondrion are known as cristae.

_____ 5. All <u>prokaryotic</u> cells have one or more mitochondria.

## Choice

For questions 6–15, choose from the following:

             a. chloroplasts     b. amyloplasts     c. central vacuole     d. chromoplasts

____ 6. A plastid found in living plant cells; fluid-filled

____ 7. Plastids that lack pigments

____ 8. Plastids that have an abundance of carotenoids but no chlorophylls

____ 9. Organelles that absorb sunlight energy and produce ATP

____ 10. The source of the yellow-to-red colors of many flowers, leaves, and fruits

____ 11. Plastid with internal areas known as grana and stroma

____ 12. Plastids that resemble certain photosynthetic bacteria

____ 13. Store starch grains and are abundant in cells of stems, tubers, and seeds

____ 14. The site of photosynthesis in plant cells

____ 15. Fluid-filled, stores amino acids, sugars, ions, and toxic wastes

# 3 - VI. CELL SURFACE SPECIALIZATIONS (p. 57)
## THE CYTOSKELETON (pp. 58 - 59)

**Selected Words:** *primary* cell wall, *secondary* cell wall, *tight* junctions, *adhering* junctions, *gap* junctions, "cytoplasmic streaming," <u>Paramecium</u>

## Boldfaced, Page-Referenced Terms

(57) cell wall _____

_____

(58) cytoskeleton _____

_____

(58) microtubules _____

_____

(58) microfilaments _____

_____

(58) intermediate filaments _____

_____

(59) flagellum _____

_____

(59) cilium _____

_____

(59) centrioles _____

_____

## Matching

Choose the most appropriate answer for each term.

1. ___ gap junctions

2. ___ ground substance

3. ___ primary cell wall

4. ___ plasmodesmata

5. ___ adhering junctions

6. ___ pectin

7. ___ secondary cell wall

8. ___ tight junctions

A. A pliable plant cell wall composed of bundles of cellulose strands
B. Deposits that cement adjacent plant cell walls
C. Animal cells scattered in cellular secretions of polysaccharides and protein fibers (e.g., cartilage) rich in collagen or elastin
D. Like spot welds cementing animal cells together in a tissue (e.g., skin and heart)
E. Channels linking adjacent animal cells; for passage of signals and substances
F. Link cells of animal epithelial tissues; form seals to prevent molecules freely crossing tissue linings
G. Cytoplasmic strands forming transport channels across adjacent living plant cells
H. A very rigid plant cell wall; formed after the first wall

## Dichotomous Choice

Circle one of two possible answers given between parentheses in each statement.

9. The cytoskeleton gives (prokaryotic/eukaryotic) cells their shape, internal organization, and movement.
10. (Protein/Carbohydrate) subunits form the basic components of microtubules, microfilaments, and the intermediate filaments of animal cells.
11. Microtubules consist of (tubulin/actin) subunits arranged in parallel rows.
12. Some cells move by putting out (temporary/permanent) lobes called pseudopods.
13. (Amoeboid motion/Contraction) is when myosin microfilaments repeatedly bind and release actins so that actin slides on past.
14. Cellular structures are pushed or dragged through the cytoplasm by (binding and releasing of myosin and actin/controlled assembly and disassembly of microtubule or microfilament subunits).
15. In response to the sun's position, chloroplasts flowing with attached myosin filaments in plant cells due to myosin "walking" over bundles of actin filaments is known as (amoeboid motion/cytoplasmic streaming).
16. (Flagella/Cilia) are long, whiplike motile structures on cells.
17. (Flagella/Cilia) are short but very numerous motile structures on cells.
18. The cross-sectional pattern of 9 + 2 microtubules is found in (cilia/centrioles).
19. The human respiratory tract is lined with beating (flagella/cilia).
20. Microtubules found in cilia and flagella have their origin from (basal bodies/centrosomes).
21. MTOCs are sites of dense material that generate large numbers of (microtubules/microfilaments).
22. A special type of MTOC near the cell nucleus gives rise to (microtubules/microfilaments) that move chromosomes about before a cell divides.

## 3 - VII. PROKARYOTIC CELLS—THE BACTERIA (pp. 60 - 61)

*Selected Words:* prokaryotic, <u>Escherichia</u> <u>coli</u> (<u>E. coli</u>)

### Boldfaced, Page-Referenced Terms

(60) bacterial flagella _____

_____

### Fill-in-the-Blanks

(1) _____ are the smallest and most structurally simple cells. The word "prokaryotic" means "before the (2) _____," which implies that bacteria evolved before cells possessing nuclei existed. Some bacteria have one or more long, threadlike motile structures, the (3) _____, that extend from the cell surface; they permit rapid movements through fluid environments. Most bacteria have a cell (4) _____ that surrounds the plasma membrane; this membrane controls movement of substances to and from the (5) _____. Bacterial cells have many (6) _____ where polypeptide chains are synthesized. Most of the DNA in bacterial cells is found in an irregularly shaped cytoplasmic region, the nucleoid, that lacks an enclosing membrane. Within the nucleoid area the bacterial DNA is a single, (7) _____ molecule.

## Self-Quiz

### Label and Match

Identify each indicated part of the accompanying illustrations. Complete the exercise by matching and entering the letter of the proper function description in the parentheses following each label. Some letter choices must be used more than once.

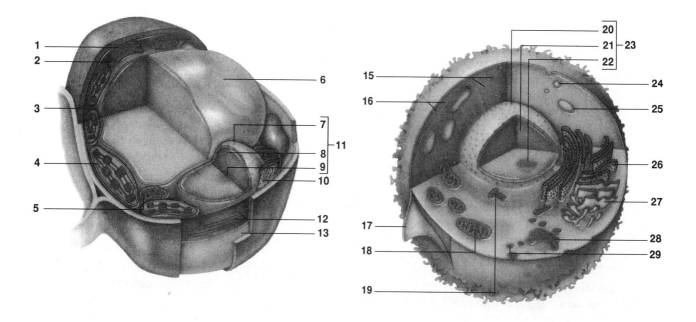

1. _____ _____ ( )
2. _____ ( )
3. _____ ( )
4. _____ ( )
5. _____ ( )
6. _____ _____ ( )
7. _____ _____ ( )
8. _____ and _____ ( )
9. _____ ( )
10. _____ _____ _____ ( )
11. _____ ( )
12. _____ ( )
13. _____ _____ ( )
14. _____ _____ ( )
15. _____ ( )
16. _____ ( )
17. _____ _____ ( )
18. _____ ( )
19. _____ ( )
20. _____ _____ ( )
21. _____ and _____ ( )
22. _____ ( )
23. _____ ( )
24. _____ ( )
25. _____ ( )
26. _____ _____ _____ ( )
27. _____ _____ _____ ( )
28. _____ _____ ( )
29. _____ ( )

A. Two-membrane structure; outermost part of the nucleus
B. Provides protection and structural support for plant cells
C. Increases cell surface area and storage of living plant cells
D. Genetic material and the other substances within the nucleus
E. Free of ribosomes; the main site of lipid synthesis in many cells
F. Formed as buds from Golgi membranes; contain enzymes for intracellular digestion
G. Small cylinders composed of triplet microtubules; produce microtubles of cilia and flagella; act as basal bodies
H. Different types function in transport or storage of substances; digestion in cell; other functions
I. Have parallel rows of tubulin subunits
J. Consist of myosin and actin subunits in muscle cells; consist of keratin subunits in hair and skin cells
K. Further modification, sorting, and shipping of proteins and lipids for secretion or for use in the cell
L. Sites where the protein and RNA subunits of ribosomes are assembled
M. A membrane-bound compartment that houses DNA in eukaryotic cells
N. Photosynthesis and some starch storage
O. Control of material exchanges; mediates cell-environment interactions
P. Site of aerobic respiration
Q. Initial modification of protein structure after formation on ribosomes

*Multiple Choice*

___ 30. _____ proteins bind hormones and other extracellular substances that trigger changes in cell activities; _____ proteins help particular kinds of cells stick together in tissues.
a. Receptor; recognition
b. Transport; adhesion
c. Receptor; adhesion
d. Adhesion; receptor

___ 31. Which of the following is *not* found as a part of prokaryotic cells?
a. Ribosomes
b. DNA
c. Nucleus
d. Cytoplasm
e. Cell wall

___ 32. Which of the following statements most correctly describes the relationship between cell surface area and cell volume?
   a. As a cell expands in volume, its diameter increases at a rate faster than its surface area does.
   b. Volume increases with the square of the diameter, but surface area increases only with the cube.
   c. If a cell grows four times in diameter, its volume of cytoplasm increases sixteen times and its surface area increases sixty-four times.
   d. Volume increases with the cube of the diameter, but surface area increases only with the square.

___ 33. Animal cells dismantle and dispose of waste materials by _____.
   a. using centrally located vacuoles
   b. several lysosomes fusing with a vesicle that encloses the wastes
   c. microvilli packaging and exporting the wastes
   d. mitochondrial breakdown of the wastes

___ 34. The nucleolus is the site where _____.
   a. the protein and RNA subunits of ribosomes are assembled
   b. the chromatin is formed
   c. chromosomes are bound to the inside of the nuclear envelope
   d. chromosomes duplicate themselves

___ 35. The _____ is free of ribosomes, curves through the cytoplasm, and is the main site of lipid synthesis.
   a. lysosome
   b. Golgi body

   c. smooth ER
   d. rough ER

___ 36. Which of the following is *not* present in all cells?
   a. Cell wall
   b. Plasma membrane
   c. Ribosomes
   d. DNA molecules

___ 37. As a part of the cytomembrane system, the _____ modify lipids and proteins to permit sorting and packaging for specific locations.
   a. endoplasmic reticulum
   b. Golgi bodies
   c. peroxisomes
   d. lysosomes

___ 38. Mitochondria convert energy stored in _____ to forms that the cell can use, principally ATP.
   a. water
   b. carbon compounds
   c. $NADPH_2$
   d. carbon dioxide

___ 39. _____ are sacs of enzymes that bud from ER; they produce potentially harmful hydrogen peroxide while breaking down fatty acids and amino acids.
   a. Lysosomes
   b. Glyoxysomes
   c. Golgi bodies
   d. Peroxisomes

## Choice

Cells of the organisms in the five kingdoms of life share the following characteristics: plasma membrane, DNA, RNA, and ribosomes. For questions 34–43, choose the kingdom(s) possessing each additional characteristic listed below. Some questions will require more than one kingdom as the answer.

a. Monera    b. Protista    c. Fungi    d. Plantae    e. Animalia

_____ 39. cytoskeletons

_____ 40. cell wall

_____ 41. nucleus

_____ 42. nucleoli

_____ 43. central vacuoles

_____ 44. photosynthetic pigments

_____ 45. endoplasmic reticulum

_____ 46. Golgi bodies

_____ 47. lysosomes

_____ 48. complex flagella and cilia

---

# Chapter Objectives/Review Questions

| Page | | Objectives/Questions |
|---|---|---|
| (42) | 1. | List and describe the three major regions that all cells have in common. |
| (42) | 2. | _____ cells contain distinct arrays of organelles; _____ cells, by contrast, have no nucleus or organelles. |
| (43) | 3. | Explain the basis for the fluid and the mosaic qualities of cell membranes. |
| (42 - 43) | 4. | Two thin sheets of _____ molecules serve as the structural framework for cell membranes; proteins situated in the two sheets or at its surface carry out most membrane functions. |
| (43) | 5. | State the major functions of the following membrane proteins: transport, receptor, recognition, and adhesion. |
| (44) | 6. | Cell size is necessarily limited because its volume increases with the _____, but surface area increases only with the _____. |
| (44 - 45) | 7. | Briefly describe the operating principles of light microscopes, scanning tunneling microscopes, transmission electron microscopes, and scanning electron microscopes. |
| (46) | 8. | In eukaryotic cells, _____ separate different incompatible chemical reactions in space and time. |
| (46 - 49) | 9. | Briefly describe the function and cellular location of the organelles typical of most eukaryotic cells: nucleus, ribosomes, endoplasmic reticulum, Golgi body, vesicles, mitochondria, and the cytoskeleton. |
| (50) | 10. | _____ are sites where the protein and RNA subunits of ribosomes are assembled. |
| (51) | 11. | Describe the nature of the nuclear envelope, and relate its function to its structure. |
| (51) | 12. | _____ is the total collection of DNA molecules and associated proteins; a _____ is an individual DNA molecule and associated proteins. |
| (51 - 53) | 13. | Explain how the endoplasmic reticulum (rough and smooth types), peroxisomes, Golgi bodies, lysosomes, and a variety of vesicles function together as the cytomembrane system. |
| (54) | 14. | Describe the function of exocytosis and exocytic vesicles; also describe the function of endocytosis and endocytic vesicles. |

(54) 15. Many times, receptors mediate _____; clusters of receptors are positioned in shallow depressions (coated pits) in the plasma membrane where they bind to lipoprotein particles.

(55) 16. In _____, energy stored in organic molecules is released by enzymes and used to form many ATP molecules in the presence of oxygen.

(56) 17. Give the function of the following plant organelles: chloroplasts, chromoplasts, amyloplasts, and central vacuole.

(56) 18. Describe the details of the structure of the chloroplast, the site of photosynthesis (include grana and stroma).

(57) 19. Distinguish a primary cell wall from a secondary cell wall in leafy plants.

(57) 20. Describe the location and function of plasmodesmata.

(57) 21. What is meant by *ground substance*? _____ junctions link the cells of epithelial tissues; _____ junctions keep cells together in tissues; _____ junctions link the cytoplasm of adjacent cells.

(58) 22. Elements of the _____ give eukaryotic cells their internal organization, overall shape, and capacity to move.

(58) 23. List the three major structural elements of the cytoskeleton.

(58) 24. Briefly define *pseudopods, contraction, amoeboid movement,* and *cytoplasmic movement*.

(59) 25. Both cilia and flagella have an internal microtubule arrangement called the _____ array.

(59) 26. Microtubules of flagella and cilia arise from _____, which remain at the base of those completed structures as basal bodies.

(60) 27. Describe the structure of a generalized bacterial cell. Include the bacterial flagellum, nucleoid, pili, capsule, cell wall, plasma membrane, cytoplasm, and ribosomes.

## Integrating and Applying Key Concepts

1. Which parts of a cell constitute the minimum necessary for keeping the simplest of living cells alive?
2. How did the existence of a nucleus, compartments, and extensive internal membranes confer selective advantages on cells that developed these features?

# 4

# GROUND RULES
# OF METABOLISM

**ENERGY AND LIFE**
    How Much Energy Is Available?
    The One-Way Flow of Energy

**DIFFUSION IN THE CELLULAR WORLD**
    Following the Gradients
    Osmosis

**MOVEMENT THROUGH TRANSPORT PROTEINS**
    Passive Transport
    Active Transport

**CHARACTERISTICS OF METABOLIC REACTIONS**
    The Direction of Metabolic Reactions
    Energy Flow and Coupled Reactions
    Metabolic Pathways

**ENZYMES**
    Characteristics of Enzymes
    Enzyme-Substrate Interactions
    Temperature, pH, and Enzyme Activity

**MEDIATORS OF ENZYME FUNCTION**
    Enzyme Helpers
    Control of Enzyme Function

**ATP—THE MAIN ENERGY CARRIER**
    Structure and Function of ATP
    The ATP/ADP Cycle

**ENERGY AND THE FLOW OF ELECTRONS**
    The Nature of Electron Transfers
    Electron Transport Systems

*Focus on Science:* YOU LIGHT UP MY LIFE—VISIBLE
    EFFECTS OF METABOLIC ACTIVITY

---

## Interactive Exercises

### 4 - I. ENERGY AND LIFE (pp. 64 - 67)

*Selected Words:* kilocalorie, "system," surroundings

*Boldfaced, Page-Referenced Terms*

(64) free radical _____

_____

(66) metabolism _____

_____

(66) energy _____

_____

(66) first law of thermodynamics _____

_____

(67) second law of thermodynamics _____

_____

(67) entropy _____

_____

## Short Answer

1. The world of life maintains a high degree of organization only because _____

_____.

## True/False

If the statement is true, write a T in the blank. If the statement is false, correct it by changing the underlined word and writing the correct word(s) in the blank.

_____ 2. The <u>first</u> law of thermodynamics states that entropy is constantly increasing in the
universe.

_____ 3. Your body steadily gives off heat equal to that from a <u>100-watt</u> light bulb.

_____ 4. When you eat a potato, some of the stored chemical energy of the food is converted
into <u>mechanical</u> energy that moves your muscles.

_____ 5. The amount of low-quality energy in the universe is <u>decreasing</u>.

## Labeling

In the blank preceding each item, indicate whether the first law of thermodynamics (I) or the second law
of thermodynamics (II) is best described.

6. ___ The cooling of a cup of coffee

7. ___ The evaporation of gasoline into the atmosphere

8. ___ A hydroelectric plant at a waterfall producing electricity

9. ___ The creation of a snowman by children

10. ___ The death and decay of an organism

## 4 - II. DIFFUSION IN THE CELLULAR WORLD (pp. 68 - 69)
## MOVEMENT THROUGH TRANSPORT PROTEINS (pp. 70 - 71)

**Selected Words:** *electric* gradient, *pressure* gradient, *isotonic* fluid, *hypotonic* fluid, *hypertonic* fluid, "bulk
flow," *turgor* pressure, *wilting*, "facilitated" diffusion, *net* movement, *cotransport* system

## Boldfaced, Page-Referenced Terms

(68) selective permeability _____

_____

(68) concentration gradient _____

_____

(68) diffusion _____

_____

(68) tonicity _____

_____

(69) osmosis _____

_____

(70) passive transport _____

_____

(70) active transport _____

_____

## Fill-in-the-Blanks

(1) _____ refers to the number of molecules (or ions) of a substance in a specified volume of fluid.

A (2) _____ means that one region of the fluid contains more molecules than a neighboring region.

The net movement of like molecules down their concentration gradient is called (3) _____.

(4) _____ is the movement of water across membranes in response to a concentration gradient, a pressure gradient, or both.

Study the illustration below and complete the table by entering the name of the membrane structure(s) involved in the transport mechanism described. Choose from lipid bilayer, gated channel protein, and channel protein.

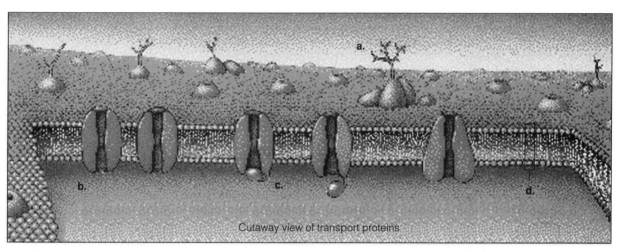

Cutaway view of transport proteins

| Membrane Structure(s) Involved | Function |
| --- | --- |
| 5. | Pumps solutes across the membrane against a gradient (energized by ATP) |
| 6. | Solute molecules pass through by facilitated diffusion |
| 7. | Simple diffusion |

## True/False

If the statement is true, write a T in the blank. If the statement is false, correct it by changing the underlined word and writing the correct word(s) in the blank.

_____ 8. Osmosis occurs in response to a concentration gradient that involves unequal concentrations of <u>water</u> molecules.

_____ 9. An animal cell placed in a <u>hypertonic</u> solution would swell and perhaps burst.

_____ 10. Physiological saline is 0.9% NaCl; red blood cells placed in such a solution will not gain or lose water; therefore, one could state that the fluid in red blood cells is <u>hypertonic</u>.

_____ 11. A solution of 80% solvent, 20% solute is <u>more</u> concentrated than a solution of 70% solvent, 30% solute.

_____ 12. Red blood cells shrivel and shrink when placed in a <u>hypotonic</u> solution.

_____ 13. Plant cells placed in a <u>hypotonic</u> solution will swell.

## Labeling

In the blank following each ion, molecule, or structure, enter the name(s) of the correct membrane transport mechanism. Choose from diffusion, osmosis, facilitated diffusion (passive transport), and active transport.

14. $H_2O$ _____

15. $CO_2$ _____

16. $Na^+$ _____

17. glucose _____

18. $O_2$ _____

19. $K^+$ _____

20. amino acids _____

21. substances pumped through interior of a transport protein; energy input required _____

22. substances move through interior of transport protein; no energy input required_____

23. lipid soluble substances _____

## 4 - III. CHARACTERISTICS OF METABOLIC REACTIONS (pp. 72 - 73)

**Selected Words:** *phenylketonuria (PKU), exergonic, endergonic, energy-releasing* reactions, *energy-requiring* reactions, *biosynthetic* pathways, *degradative* pathways

### Boldfaced, Page-Referenced Terms

(72) chemical equilibrium _____

_____

(73) metabolic pathways _____

_____

(73) substrates _____

_____

(73) intermediates _____

_____

(73) end products _____

_____

(73) enzymes _____

_____

(73) cofactors _____

_____

(73) energy carriers _____

_____

(73) transport proteins _____

_____

## Short Answer

1. Define exergonic reactions; give an example. _____

_____

_____

2. Define endergonic reactions; give an example. _____

_____

_____

3. A (3) _____ _____ is an orderly series of chemical reactions, with each reaction cat-

alyzed by a specific (4) _____.

## Labeling

Classify each of the following reactions as endergonic or exergonic.

5. _____    The product of a chemical reaction has more energy than the reactants.

6. _____    Glucose + oxygen → carbon dioxide + water + energy

7. _____    The reactants of a chemical reaction have more energy than the product.

Choose the most appropriate answer for each: A or B.

8. ___ a degradative reaction

9. ___ an endergonic reaction

10. ___ a biosynthetic reaction

11. ___ an exergonic reaction

## Matching

Study the sequence of reactions below. Identify the components of the reactions by selecting items from the following list and entering the correct letter in the appropriate blank.

| | | |
|---|---|---|
| 12. ___ | 15. ___ | A. Cofactor |
| 13. ___ | 16. ___ | B. Intermediates |
| 14. ___ | 17. ___ | C. Reactants |

D. End product

E. Reversible reaction

F. Enzymes

# 4 - IV. ENZYMES (pp. 74 - 75)
## MEDIATORS OF ENZYME FUNCTION (p. 76)

*Selected Words:* *catalytic* molecules, "energy hill," *transition* state, *fever*, "cofactors," *allosteric* control

## Boldfaced, Page-Referenced Terms

(74) activation energy _____

_____

(74) active sites _____

_____

(75) induced-fit model _____

_____

(76) coenzymes _____

_____

(76) feedback inhibition _____

_____

## Short Answer

1. List four characteristics that enzymes have in common. _____

_____

_____

## Matching

Match the items on the sketch below with the list of descriptions. Some answers may require more than one letter.

2. _____
3. _____
4. _____
5. _____
6. _____
7. _____

A. Transition state, the time of the most precise fit between enzyme and substrate
B. Complementary active site of the enzyme
C. Enzyme, a protein with catalytic power
D. Product or reactant molecules that an enzyme can specifically recognize
E. Product or reactant molecule
F. Bound enzyme-substrate complex

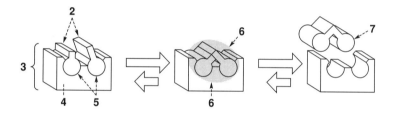

## Fill-in-the-Blanks

(8) _____ are highly selective proteins that act as catalysts, which means that they greatly enhance the rate at which specific reactions approach (9) _____. The specific substance on which a particular enzyme acts is called its (10) _____; this substance fits into the enzyme's crevice, which is called its (11) _____ _____.

The (12) _____ - _____ model describes how substrates can settle into an active site of an enzyme and change its shape. Enzymes increase reaction rates by lowering the required (13) _____ _____. (14) _____ and (15) _____ are two important factors that influence the rates of enzyme activity. Extremely high fevers can destroy the three-dimensional shape of an enzyme, which may adversely affect (16) _____ and cause death. (17) _____ enzymes have control sites where specific substances can bind and alter enzyme activity. The situation in which the end product binds to the first enzyme in a metabolic pathway and prevents product formation is known as (18) _____ _____. Nonprotein substances that aid enzymes in their catalytic task are called (19) _____; they include some large organic molecules that function as (20) _____. (21) _____ and (22) _____ are coenzymes that have roles in the breakdown of glucose and other carbohydrates. (23) _____ is a coenzyme with a central role in photosynthesis; its abbreviation is (24) _____ when it is loaded with protons and electrons. Some metal (25) _____ such as Fe$^{++}$ also serve as cofactors when they are components of cytochrome molecules that serve as transport proteins in cell membranes.

## 4 - V. ATP—THE MAIN ENERGY CARRIER (p. 77)
### ENERGY AND THE FLOW OF ELECTRONS (p. 78)
### *Focus on Science:* YOU LIGHT UP MY LIFE—VISIBLE EFFECTS OF METABOLIC ACTIVITY (p. 79)

*Selected Words:* electrons, *oxidation-reduction* reaction, <u>Pyrophorus noctilucis</u>, <u>Mycobacterium tuberculosis</u>, *luciferase*

### *Boldfaced, Page-Referenced Terms*

(77) ATP _____

_____

(77) phosphorylation _____

_____

(77) ATP/ADP cycle _____

_____

(78) electron transport system _____

_____

(79) bioluminescence _____

_____

### *Fill-in-the-Blanks*

ATP is constructed of the nitrogenous base (1) _____, the sugar (2) _____, and three (3) _____ groups. When ATP is hydrolyzed, a molecule of (4) _____ in the presence of an appropriate (5) _____ is used to split ATP into (6) _____, and a (7) _____ group; usable (8) _____ is released, which is easily transferred to other molecules in the cell. The hydrolysis of ATP provides (9) _____ for biosynthesis, active transport across cell membranes, and molecular displacements such as those required for muscle contraction. ATP directly or indirectly delivers energy to almost all (10) _____ pathways. In the (11) _____/_____ cycle, a phosphate group is linked to adenosine diphosphate, and adenosine triphosphate donates a phosphate group elsewhere and reverts back to adenosine diphosphate. Adding a phosphate to a molecule is called (12) _____. When this occurs, the molecule increases its store of (13) _____ and becomes primed to enter a specific (14) _____. The (15) _____/_____ cycle provides a renewable means of conserving and transferring energy to specific reactions.

## Labeling

Identify the molecule at the right and label its parts.

16. _____  _____

    _____

17. _____

18. _____

19. The name of this molecule is _____

    _____.

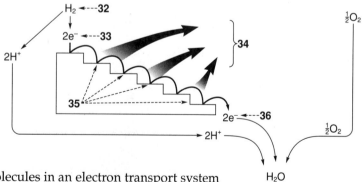

## Fill-in-the-Blanks

The release of energy from glucose in cells proceeds in controlled steps, so that (20) _____ molecules form along the route from glucose to carbon dioxide and water. At each step in a metabolic pathway, a specific (21) _____ lowers the activation energy for the formation of an intermediate compound. At each step in the pathway, only some energy is released. In chloroplasts and mitochondria, the liberated electrons released from the breaking of chemical bonds are sent through (22) _____ _____ systems; these systems consist of enzymes and (23) _____, bound in a cell membrane, that transfer electrons in a highly organized sequence. A molecule that donates electrons in the sequence is being (24) _____, while molecules accepting electrons are being (25) _____. Oxidation-reduction means an (26) _____ transfer. Electron transport systems "intercept" excited electrons and make use of the (27) _____ they release. If we think of the electron transport system as a staircase, electrons at the top of the staircase have the (28) [choose one] ❏ most ❏ least energy. As the electrons are transferred from one electron carrier to another, some (29) _____ can be harnessed to do biological (30) _____. One type of biological work occurs when energy released during electron transfers is used to bond a (31) _____ group to ADP.

## Matching

Match the lettered statements to the numbered items on the sketch below.

32. ____
33. ____
34. ____
35. ____
36. ____

A. Represent the cytochrome molecules in an electron transport system
B. Electrons at their highest energy level
C. Released energy harnessed and used to produce ATP
D. Electrons at their lowest level
E. The separation of hydrogen atoms into protons and electrons

# Self-Quiz

___ 1. An important principle of the second law of thermodynamics states that _____.
   a. energy can be transformed into matter, and because of this we can get something for nothing
   b. energy can be destroyed only during nuclear reactions, such as those that occur inside the sun
   c. if energy is gained by one region of the universe, another place in the universe also must gain energy in order to maintain the balance of nature
   d. matter tends to become increasingly more disorganized

___ 2. Essentially, the first law of thermodynamics states that _____.
   a. one form of energy cannot be converted into another
   b. entropy is increasing in the universe
   c. energy cannot be created or destroyed
   d. energy cannot be converted into matter or matter into energy

___ 3. An enzyme is best described as _____.
   a. an acid
   b. a protein
   c. a catalyst
   d. a fat
   e. both (b) and (c)

___ 4. Which is not true of enzyme behavior?
   a. Enzyme shape may change during catalysis.
   b. The active site of an enzyme orients its substrate molecules, thereby facilitating interaction of their reactive parts.
   c. All enzymes have an active site where substrates are temporarily bound.
   d. An individual enzyme can catalyze a wide variety of different reactions.

___ 5. When $NAD^+$ combines with hydrogen, the $NAD^+$ is _____.
   a. reduced
   b. oxidized
   c. phosphorylated
   d. denatured

___ 6. A substance that gains electrons is _____.
   a. oxidized
   b. a catalyst
   c. reduced
   d. a substrate

___ 7. In _____ pathways, carbohydrates, lipids, and proteins are broken down in stepwise reactions that lead to products of lower energy.
   a. intermediate
   b. biosynthetic
   c. induced
   d. degradative

___ 8. With regard to major function, $NAD^+$, FAD, and $NADP^+$ are classified as _____.
   a. enzymes
   b. phosphate carriers
   c. cofactors that function as coenzymes
   d. end products of metabolic pathways

___ 9. If a phosphate bond is linked to ADP, the bond formed _____.
   a. absorbs a large amount of free energy when the phosphate group is attached during hydrolysis
   b. is oxidized when ATP is hydrolyzed to ADP and one phosphate group
   c. is usually found in each glucose molecule; this is why glucose is chosen as the starting point for glycolysis
   d. later will release a large amount of usable energy when the phosphate group is split off during hydrolysis.

___ 10. An allosteric enzyme _____.
   a. has an active site where substrate molecules bind and another site that binds with intermediate or end-product molecules
   b. is an important energy-carrying nucleotide
   c. carries out either oxidation reactions or reduction reactions but not both
   d. raises the activation energy of the chemical reaction it catalyzes

# Chapter Objectives/Review Questions

| Page | | Objectives/Questions |
|------|---|----------------------|
| (66) | 1. | _____ is the controlled capacity to acquire and use energy for stockpiling, breaking apart, building, and eliminating substances in ways that contribute to survival and reproduction. |
| (66) | 2. | Define *energy*; be able to state the first and second laws of thermodynamics. |
| (67) | 3. | _____ is a measure of the degree of randomness or disorder of systems. |
| (66 - 67) | 4. | Explain how the world of life maintains a high degree of organization. |
| (68 - 71) | 5. | Distinguish diffusion from osmosis, and active transport from passive transport. |
| (73) | 6. | Reactions that show a net loss in energy are said to be _____; reactions that show a net gain in energy are said to be _____. |
| (72 - 73) | 7. | What is the function of metabolic pathways in cellular chemistry? |
| (73) | 8. | Give the function of each of the following participants in metabolic pathways: reactants, intermediates, enzymes, cofactors, energy carriers, transport proteins, and end products. |
| (74) | 9. | Explain the effects of enzymes on activation energy. |
| (74 - 75) | 10. | Describe the induced-fit hypothesis. |
| (75) | 11. | Explain what happens to enzymes if temperature and pH continually increase. |
| (76) | 12. | What are allosteric enzymes, and what is their function? |
| (76) | 13. | Describe the control mechanism known as feedback inhibition. |
| (76) | 14. | Cite examples of coenzymes. |
| (76) | 15. | Explain the differences between $NAD^+$ and NADH; $NADP^+$ and NADPH; FAD and $FADH_2$. |
| (77) | 16. | ATP is composed of _____, a five-carbon sugar, three _____ groups, and _____, a nitrogen-containing compound. |
| (77) | 17. | ATP directly or indirectly delivers _____ to almost all metabolic pathways. |
| (77) | 18. | Explain the functioning of the ATP/ADP cycle. |
| (77) | 19. | Adding a phosphate to a molecule is called _____. |
| (78) | 20. | Describe the components, organization, and functions of an electron transport system. |

# Integrating and Applying Key Concepts

A piece of dry ice left sitting on a table at room temperature vaporizes. As the dry ice vaporizes into $CO_2$ gas, does its entropy increase or decrease? Tell why you answered as you did.

# 5

# ENERGY-ACQUIRING PATHWAYS

---

## Interactive Exercises

### 5 - I. PHOTOSYNTHESIS: AN OVERVIEW (pp. 82 - 85)
### LIGHT-TRAPPING PIGMENTS (pp. 86 - 87)

*Selected Words:* *photo*autotroph, *chemo*autotroph, *photosynthesis,* light-dependent reactions, *light-independent* reactions, Spirogyra

### Boldfaced, Page-Referenced Terms

(83) autotrophs _____

_____

(83) heterotrophs _____

_____

(85) chloroplast _____

_____

(85) stroma _____

_____

(85) thylakoid membrane system _____

_____

(86) photons _____

_____

(86) pigments _____

_____

(87) chlorophylls _____

_____

(87) carotenoids _____

_____

(87) phycobilins _____

_____

## Fill-in-the-Blanks

(1) _____ obtain carbon and energy from the physical environment; their carbon source is
(2) _____ _____. (3) _____ autotrophs obtain energy from sunlight. (4) _____
autotrophs are represented by a few kinds of bacteria; they obtain energy by stripping (5) _____
from sulfur or other inorganic substances. (6) _____ feed on autotrophs, each other, and organic
wastes; representatives include (7) _____, fungi, many protistans, and most bacteria. Although
energy stored in organic compounds such as glucose may be released by several pathways, the
pathway known as (8) _____ _____ releases the most energy.

9. In the space below, supply the missing information to complete the summary equation for
photosynthesis:

$$12 \underline{\hspace{1cm}} + \underline{\hspace{1cm}} CO_2 \rightarrow \underline{\hspace{1cm}} O_2 + C_6H_{12}O_6 + 6\underline{\hspace{1cm}}$$

10. Supply the appropriate information to state the equation (above) for photosynthesis in words:

(a) _____ molecules of water plus six molecules of (b) _____ _____ (in the presence of
pigments, enzymes, and sunlight) yield six molecules of (c) _____ plus one molecule of (d)
_____ plus (e) _____ molecules of water.

The two major sets of reactions of photosynthesis are the (11) _____ - _____ reactions
and the (12) _____ - _____ reactions. (13) _____ _____ and (14) _____ are
the reactants of photosynthesis, and the end product is usually given as (15) _____. The internal
membranes and channels of the chloroplast are the (16) _____ membrane system and are organized
into stacks, called (17) _____. Spaces inside the thylakoid disks and channels form a continuous
compartment where (18) _____ ions accumulate to be used to produce ATP. The semifluid interior
area surrounding the grana is known as the (19) _____ and is the area where the products of
photosynthesis are produced.

The light-capturing phase of photosynthesis takes place on a system of (20) _____ membranes.
A(n) (21) _____ is a packet of light energy. Thylakoid membranes contain (22) _____, which
absorb photons of light. The principal pigments are the (23) _____, which reflect green
wavelengths but absorb (24) _____ and (25) _____ wavelengths. (26) _____ are pigments
that absorb violet and blue wavelengths but reflect yellow, orange, and red.

## Matching

Choose the most appropriate answer for each.

27. ___ chlorophylls

28. ___ chlorophyll *b* and carotenoids

29. ___ carotenoids

30. ___ violet-blue-green-yellow-red

31. ___ photons

32. ___ chlorophyll *a*

33. ___ chloroplast

34. ___ phycobilins

35. ___ grana

A. The main pigment of photosynthesis
B. Packets of energy that have an undulating motion through space
C. The two stages of photosynthesis occur here
D. Absorb violet and blue wavelengths but transmit red, orange, and yellow
E. Visible light portion of the electromagnetic spectrum
F. Pigments that transfer energy to chlorophyll *a*
G. Absorb violet-to-blue and red wavelengths; the reason leaves appear green
H. Red and blue pigments
I. The site of the first stage of photosynthesis

## 5 - II. LIGHT-DEPENDENT REACTIONS (pp. 88 - 89)
## A CLOSER LOOK AT ATP FORMATION IN CHLOROPLASTS (p. 90)

*Selected Words:* *type I* photosystem, *type II* photosystem, *chemiosmotic* theory

### Boldfaced, Page-Referenced Terms

(88) light-dependent reactions _____

_____

(88) photosystem _____

_____

(88) electron transport systems _____

_____

(88) cyclic pathway of ATP formation _____

_____

(88) noncyclic pathway of ATP formation _____

_____

(89) photolysis _____

_____

### Fill-in-the-Blanks

A cluster of 200 to 300 of these pigment proteins is a(n) (1) _____. When pigments absorb

(2) _____ energy, an (3) _____ is transferred from a photosystem to a(n) (4) _____

molecule. (5) _____ refers to the attachment of phosphate to ADP or other organic molecules. Due

to the input of light energy, electrons flow through a transport system that causes protons ($H^+$)

simultaneously to be pumped into the thylakoid compartments. Electrons then end up in (6) _____

chlorophyll at the end of this transport chain. The flow of protons from the thylakoid compartment through (7) _____ _____ drives the enzyme machinery that phosphorylates (8) _____, a sequence of events known as the (9) _____ theory of ATP formation.

### Complete the Table

10. In the table below, identify and state the role of each item given in the cyclic pathway of the light-dependent reactions.

| | |
|---|---|
| a. Photosystem I | |
| b. Electrons | |
| c. P700 | |
| d. Electron acceptor | |
| e. Electron transport system | |
| f. ADP | |

### Labeling

The diagram below illustrates noncyclic photophosphorylation. Identify each numbered part of the illustration.

11. _____ _____

12. _____ _____ _____

13. _____ _____

14. _____ _____

15. _____

16. _____

17. _____

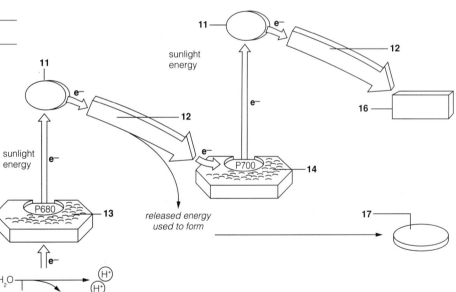

## Complete the Table

With a check mark (√) indicate for each phase of the light-dependent reactions all items from the left-hand column that are applicable.

| Light-Dependent Reactions | Cyclic Pathway | Noncyclic Pathway | Photolysis Alone |
|---|---|---|---|
| Uses $H_2O$ As A Reactant | (18) | (30) | (42) |
| Produces $H_2O$ As A Product | (19) | (31) | (43) |
| Photosystem I Involved (P700) | (20) | (32) | (44) |
| Photosystem II Involved (P680) | (21) | (33) | (45) |
| ATP Produced | (22) | (34) | (46) |
| NADPH Produced | (23) | (35) | (47) |
| Uses $CO_2$ As A Reactant | (24) | (36) | (48) |
| Causes $H^+$ to be pumped into the thylakoid compartments from the stroma | (25) | (37) | (49) |
| Produces $O_2$ As A Product | (26) | (38) | (50) |
| Produces apart $H^+$ by breaking apart $H_2O$ | (27) | (39) | (51) |
| Uses ADP and $P_i$ as Reactants | (28) | (40) | (52) |
| Uses $NADP^+$ as a Reactant | (29) | (41) | (53) |

## 5 - III. LIGHT-INDEPENDENT REACTIONS (p. 91)
### FIXING CARBON—SO NEAR, YET SO FAR (p. 92)
### *Focus on the Environment:* AUTOTROPHS, HUMANS, AND THE BIOSPHERE (p. 93)

*Selected Words:* "photorespiration"

*Boldfaced, Page-Referenced Terms*

(91) light-independent reactions _____

_____

(91) RuBP (ribulose bisphosphate) _____

_____

(91) carbon dioxide fixation _____

_____

(91) Calvin-Benson cycle _____

_____

(91) PGA (phosphoglycerate) _____

_____

(91) PGAL (phosphoglyceraldehyde) _____

_____

(92) C3 plants _____

_____

(92) C4 plants _____

_____

(92) CAM plants _____

_____

(93) chemoautotrophs _____

_____

## Label and Match

Identify each part of the illustration below. Complete the exercise by matching and entering the letter of the proper function description in the parentheses following each label.

1. _____ _____ ( )

2. _____ _____ _____ ( )

3. _____ ( )

4. _____ _____ ( )

5. _____ ( )

6. _____ ( )

7. _____ _____ ( )

8. _____ - _____ _____ ( )

9. _____ _____ ( )

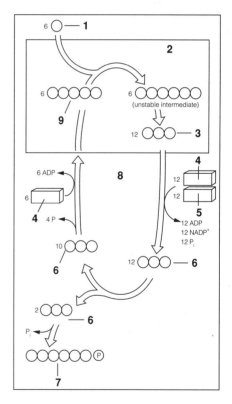

A. A three-carbon sugar, the first sugar produced; goes on to form sugar phosphate and RuBP

B. Typically used at once to form carbohydrate end products of photosynthesis

C. A five-carbon compound produced from PGALs; attaches to incoming $CO_2$

D. A compound that diffuses into leaves; attached to RuBP by enzymes in photosynthetic cells

E. Includes all the chemical reactions that "fix" carbon into an organic compound

F. Three-carbon compounds formed from the splitting of the six-carbon intermediate compound

G. A molecule that was reduced in the noncyclic pathway; furnishes hydrogen atoms to construct sugar molecules

H. A product of the light-dependent reactions; necessary in the light-independent reactions to energize molecules in metabolic pathways.

I. Includes all the chemistry that fixes $CO_2$; converts PGA to PGAL and PGAL to RuBP and sugar phosphates

## Fill-in-the-Blanks

The light-independent reactions can proceed without sunlight as long as (10) _____ and (11) _____ are available. The reactions begin when an enzyme links (12) _____ _____ to (13) _____ _____, a five-carbon compound. The resulting six-carbon compound is highly unstable and breaks apart at once into two molecules of a three-carbon compound, (14) _____. This entire reaction sequence is called carbon dioxide (15) _____. ATP gives a phosphate group to each (16) _____. This intermediate compound takes on $H^+$ and electrons from NADPH to form (17) _____. It takes (18) _____ carbon dioxide molecules to produce twelve PGAL. Most of the PGAL becomes rearranged into new (19) _____ molecules—which can be used to fix more (20) _____ _____. Two (21) _____ are joined together to form a (22) _____ _____, primed for further reactions. The Calvin-Benson cycle yields enough RuBP to replace those used in carbon dioxide (23) _____. ADP, NADP$^+$, and phosphate leftovers are sent back to the (24) _____ - _____ reaction sites, where they are again converted to (25) _____ and (26) _____. (27) _____ _____ formed in the cycle serves as a building block for the plant's main carbohydrates. When RuBP attaches to oxygen instead of carbon dioxide, (28) _____ results; this is typical of (29) _____ plants in hot, dry conditions. If less PGA is available, leaves produce a reduced amount of (30) _____. $C_4$ plants can still construct carbohydrates when the ratio of carbon dioxide to (31) _____ is unfavorable because of the attachment of carbon dioxide to (32) _____ in certain leaf cells.

(33) _____ plants open their stomata at night, capture (34)_____ _____ as part of a metabolic intermediate stored in central vacuoles and use it in photosynthesis next day when stomata are closed. (33) plants capture (34) more than once in the (35) _____ _____ at different times.

Organisms that obtain energy from oxidation of inorganic substances such as ammonium compounds, and iron or sulfur compounds, are known as (36) _____ autotrophs. Such organisms use this energy to build (37) _____ compounds. As an example, some soil bacteria use ammonia molecules as an energy source, stripping them of (38) _____ and (39) _____.

## Complete the Table

With a check mark (√) indicate for each phase of the light-independent reactions all items from the left-hand column that are applicable.

| Light-Independent Reactions | $CO_2$ Fixation Alone | Conversion of PGA to PGAL | Regeneration of RuB | Formation of Glucose and Other Organic Compounds |
|---|---|---|---|---|
| Requires RuBP as a reactant | (40) | (51) | (62) | (73) |
| Requires ATP as a reactant | (41) | (52) | (63) | (74) |
| Produces ADP as a product | (42) | (53) | (64) | (75) |
| Requires NADPH as a reactant | (43) | (54) | (65) | (76) |
| Produces $NADP^+$ as a reactant | (44) | (55) | (66) | (77) |
| Produces PGA | (45) | (56) | (67) | (78) |
| Produces PGAL | (46) | (57) | (68) | (79) |
| Requires PGAL as a reactant | (47) | (58) | (69) | (80) |
| Produces Pi as a product | (48) | (59) | (70) | (81) |
| Produces $H_2O$ as a product | (49) | (60) | (71) | (82) |
| Requires $CO_2$ as a reactant | (50) | (61) | (72) | (83) |

# Self-Quiz

___ 1. The electrons that are passed to NADPH during noncyclic photophosphorylation were obtained from _____.
a. water
b. $CO_2$
c. glucose
d. sunlight

___ 2. The cyclic pathway of the light-dependent reactions functions mainly to _____.
a. fix $CO_2$
b. make ATP
c. produce PGAL
d. regenerate ribulose biphosphate

___ 3. Chemosynthetic autotrophs obtain energy by oxidizing such inorganic substances as

_____.
a. PGA
b. PGAL
c. sulfur
d. water

___ 4. The ultimate electron and hydrogen acceptor in the noncyclic pathway is _____.
a. $NADP^+$
b. ADP
c. $O_2$
d. $H_2O$

___ 5. C4 plants have an advantage in hot, dry conditions because _____.
a. their leaves are covered with thicker wax layers than those of C3 plants
b. their stomates open wider than those of C3 plants, thus cooling their surfaces
c. $CO_2$ is fixed in the mesophyll cells, where the C3 pathway occurs, then delivered to the bundle sheath cells
d. they are also capable of capturing $CO_2$ by photorespiration

___ 6. Chlorophyll is _____.
a. on the outer chloroplast membrane
b. inside the mitochondria

c. in the stroma lamellae
d. part of the thylakoid membrane system

___ 7. Thylakoid disks are stacked in groups
called _____.
a. grana
b. stroma
c. lamellae
d. cristae

___ 8. Plant cells produce $O_2$ during photosynthesis by _____.
a. breaking apart $CO_2$ molecules
b. breaking apart water molecules
c. degradation of the stroma
d. breaking apart sugar molecules

___ 9. Plants need _____ and _____ to
carry on photosynthesis.
a. oxygen; water
b. oxygen; $CO_2$
c. $CO_2$; $H_2O$
d. sugar; water

___ 10. The two products of the light-dependent
reactions that are required for the light-
independent chemistry are _____ and
_____.
a. $CO_2$; $H_2O$
b. $O_2$; NADPH; inorganic phosphate
c. $O_2$; ATP
d. ATP; NADPH

## Chapter Objectives/Review Questions

| Page | | Objectives/Questions |
|---|---|---|
| (84) | 1. | List the major stages of photosynthesis and state what occurs in those sets of reactions. |
| (84) | 2. | Study the general equation for photosynthesis as shown on page 84 of the main text until you can remember the reactants and products. Reproduce the equation from memory on another piece of paper. |
| (84 - 85) | 3. | Describe the structural details of the green leaf. Begin with the layers of a leaf cross section and complete your description with the minute structural sites within the chloroplast where the major sets of photosynthetic reactions occur. Explain how each of the reactants needed in various phases of photosynthesis arrives at the place where they are used. Explain what happens to the products of photosynthesis. |
| (86) | 4. | Describe how the pigments found on thylakoid membranes are organized into photosystems and how they relate to photon light energy. |
| (86 - 88) | 5. | Describe the role that chlorophylls and the other chloroplast pigments play in the light-dependent reactions. After consulting Figure 5.3 of the main text, state which colors of the visible spectrum are absorbed by (a) chlorophyll $a$, (b) chlorophyll $b$, and (c) carotenoids. |
| (86) | 6. | State what T. Englemann's 1882 experiment with *Spirogyra* revealed. |
| (88 - 89) | 7. | Two energy-carrying molecules produced in the noncyclic pathways are _____ and _____; explain why these molecules are necessary for the light-independent reactions. |
| (89) | 8. | After evolution of the noncyclic pathway, _____ accumulated in the atmosphere and made _____ respiration possible. |
| (90) | 9. | Explain how the chemiosmotic theory is related to thylakoid compartments and the production of ATP. |
| (91) | 10. | Explain why the light-independent reactions are called by that name. |
| (91) | 11. | Describe the Calvin-Benson cycle as it is related to the four phases shown in the table in section 5 - III. |
| (92) | 12. | Describe the mechanism by which C4 plants thrive under hot, dry conditions; distinguish this $CO_2$-capturing mechanism from that of C3 plants. |

## Integrating and Applying Key Concepts

Suppose that humans acquired all the enzymes needed to carry out photosynthesis. Speculate about the attendant changes in human anatomy, physiology, and behavior that would be necessary for those enzymes to actually carry out photosynthetic reactions.

# 6

# ENERGY-RELEASING PATHWAYS

## Interactive Exercises

### 6 - I. HOW CELLS MAKE ATP (pp. 96 - 99)

*Selected Words:* "aerobic," "anaerobic"

*Boldfaced, Page-Referenced Terms*

(98) aerobic respiration _____

_____

(98) fermentation pathways _____

_____

(98) anaerobic electron transport _____

_____

(98) glycolysis _____

_____

(99) Krebs cycle _____

_____

(99) electron transport phosphorylation _____

_____

## Short Answer

1. Although various organisms utilize different energy sources, what is the usual form of chemical energy that will drive metabolic reactions? _____

_____

2. Describe the function of oxygen in the main degradative pathway, aerobic respiration. _____

_____

_____

3. List the most common anaerobic pathways, and describe the conditions in which they function. _____

_____

## Fill-in-the-Blanks

Virtually all forms of life depend on a molecule known as (4) _____ as their primary energy carrier. Plants produce adenosine triphosphate during (5) _____, but plants and all other organisms also can produce ATP through chemical pathways that degrade (take apart) food molecules. The main degradative pathway requires free oxygen and is called (6) _____ _____.

There are three stages of aerobic respiration. In the first stage, (7) _____, glucose is partially degraded to (8) _____. By the end of the second stage, which includes the (9) _____ cycle, glucose has been completely degraded to carbon dioxide and (10) _____. Neither of the first two stages produces much (11) _____. During both stages, protons and (12) _____ are stripped from intermediate compounds and delivered to a (13) _____ system. That system is used in the third stage of reactions, electron transport (14) _____; passage of electrons along the transport system drives the enzymatic "machinery" that phosphorylates ADP to produce a high yield of (15) _____. (16) _____ accepts "spent" electrons from the transport system and keeps the pathway clear for repeated ATP production.

Other degradative pathways are (17) _____, in that something other than oxygen serves as the final electron acceptor in energy-releasing reactions. (18) _____ and anaerobic (19) _____ _____ are the most common anaerobic pathways.

## Completion

20. Complete the equation below, which summarizes the degradative pathway known as aerobic respiration:

_____ + _____ $O_2 \rightarrow$ 6 _____ + 6 _____

21. Supply the appropriate information to state the equation (above) for aerobic respiration in words:

One molecule of glucose plus six molecules of _____ (in the presence of appropriate enzymes)

yield _____ molecules of carbon dioxide plus _____ molecules of water.

## 6 - II. GLYCOLYSIS: FIRST STAGE OF THE ENERGY-RELEASING PATHWAYS (pp. 100 - 101)

*Selected Words:* *energy-requiring* reactions, *energy-releasing* reactions, "phosphorylation," *net* energy yield

### Boldfaced, Page-Referenced Terms

(100) pyruvate _____

_____

(100) substrate-level phosphorylation _____

_____

### Fill-in-the-Blanks

(1) _____ organisms can synthesize and stockpile energy-rich carbohydrates and other food molecules from inorganic raw materials. (2) _____ is partially dismantled by the glycolytic pathway; at the end of this process some of its stored energy remains in two (3) _____ molecules. Some of the energy of glucose is released during the breakdown reactions and used in forming two molecules of the coenzyme (4) _____ and four (5) _____. These reactions take place in the cytoplasm. Glycolysis begins with two phophate groups being transferred to (6) _____ from two (7) _____ molecules. The addition of two phosphate groups to (6) energizes it and and causes it to become unstable and split apart, forming two molecules of (8) _____. Each (8) gains one (9) _____ group from the cytoplasm, then (10) _____ atoms and electrons from each PGAL are transferred to NAD$^+$, changing this coenzyme to NADH. Two (11) _____ molecules form by substrate-level phosphorylation; the cell's energy investment is paid off. One (12) _____ molecule is released from each 2-PGA as a waste product. The resulting intermediates are rather unstable; each gives up a(n) (13) _____ group to ADP. Once again, two (14) _____ molecules have formed by (15) _____-_____ phosphorylation. For each (16) _____ molecule entering glycolysis, the net energy yield is two ATP molecules that the cell can use anytime to do work. The end products of glycolysis are two molecules of (17) _____, each with a (18) _____-carbon backbone.
(number)

## Sequence

Arrange the following events of the glycolysis pathway in correct chronological sequence. Write the letter of the first step next to 1, the letter of the second step next to 2, and so on.

19. ___     A.  Two ATPs form by substrate-level phosphorylation; the cell's energy debt is paid off.

20. ___     B.  Diphosphorylated glucose (fructose 1,6-bisphosphate) molecules split to form 2 PGALs; this is the first energy-releasing step.

21. ___     C.  Two 3-carbon pyruvate molecules form as the end products of glycolysis.

22. ___     D.  Glucose is present in the cytoplasm.

23. ___     E.  Two more ATPs form by substrate-level phosphorylation, the cell gains ATP; net yield of ATP from glycolysis is 2 ATPs.

24. ___     F.  The cell invests two ATPs; one phosphate group is attached to each end of the glucose molecule (fructose 1,6-bisphosphate).

25. ___     G.  Two PGALs gain two phosphate groups from the cytoplasm.

26. ___     H.  Hydrogen atoms and electrons from each PGAL are transferred to NAD$^+$, reducing this carrier to NADH.


## 6 - III. SECOND STAGE OF THE AEROBIC PATHWAY (pp. 102 - 103)

*Selected Words:* reduced coenzyme, FAD, FADH$_2$

### Fill-in-the-Blanks

If sufficient oxygen is present, the end product of glycolysis enters a preparatory step, (1) _____

_____ formation.  This step converts pyruvate into acetyl CoA, the molecule that enters the

(2) _____ cycle, which is followed by (3) _____ _____ phosphorylation.  During these

three processes, (4) _____ (number) additional (5) _____ molecules are generated.  In the

preparatory conversions prior to the Krebs cycle and within the Krebs cycle, the food molecule

fragments are further broken down into (6) _____ _____.  During these reactions, hydrogen

atoms (with their (7) _____) are stripped from the fragments and transferred to the energy carriers

(8) _____ and (9) _____.

## Labeling

In exercises 10–14, identify the structure or location; in exercises 15–18, identify the chemical substance involved. In exercise 19, name the metabolic pathway.

10._____ _____ of mitochondrion    15._____

11._____ _____ of mitochondrion    16._____

12._____ _____ of mitochondrion    17._____

13._____ _____ of mitochondrion    18._____

14._____    19._____ _____ _____

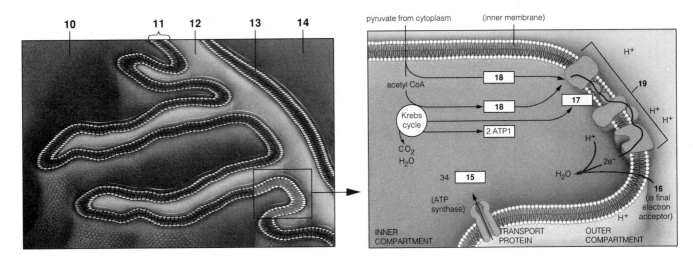

## 6 - IV. THIRD STAGE OF THE AEROBIC PATHWAY (pp. 104 - 105)

*Selected Words:* electron transport phosphorylation, chemiosmotic theory

### Fill-in-the-Blanks

NADH delivers its electrons to the highest possible point of entry into a transport system; from each NADH enough $H^+$ is pumped to produce (1) _____ (number) ATP molecules. $FADH_2$ delivers its electrons at a lower point of entry into the transport system; fewer $H^+$ are pumped, and (2) _____ (number) ATPs are produced. The electrons are then sent down highly organized (3) _____ systems located in the inner membrane of the mitochondrion; hydrogen ions are pumped into the outer mitochondrial compartment. According to the (4) _____ theory, the hydrogen ions accumulate and then follow a gradient to flow through channel proteins, called ATP (5) _____, that lead into the inner compartment. The energy of the hydrogen ion flow across the membrane is used to phosphorylate ADP to produce (6) _____. Electrons leaving the electron transport system combine with hydrogen ions and (7) _____ to form water. These reactions occur only in (8) _____. From glycolysis (in the cytoplasm) to the final reactions occurring in the mitochondria, the aerobic pathway commonly yields (9) _____ (number) ATP or (10) _____ (number) ATP for every glucose molecule degraded.

*Choice*

For questions 11–32, choose from the following. Some correct answers may require more than one letter.

a. preparatory steps to Krebs cycle    b. Krebs cycle    c. electron transport phosphorylation

_____ 11. Chemiosmosis occurs to form ATP molecules.

_____ 12. Three carbon atoms in pyruvate leave as three $CO_2$ molecules.

_____ 13. Chemical reactions occur at transport systems.

_____ 14. Coenzyme A picks up a 2-carbon acetyl group.

_____ 15. Makes two turns for each glucose molecule entering glycolysis.

_____ 16. Two NADH molecules form for each glucose entering glycolysis.

_____ 17. Oxaloacetate forms from intermediate molecules.

_____ 18. Named for a scientist who worked out its chemical details.

_____ 19. Occurs within the mitochondrion.

_____ 20. Two $FADH_2$ and six NADH form from one glucose molecule entering glycolysis.

_____ 21. Hydrogens collect in the mitochondrion's outer compartment.

_____ 22. Hydrogens and electrons are transferred to $NAD^+$ and FAD.

_____ 23. Two ATP molecules form by substrate-level phosphorylation.

_____ 24. Free oxygen withdraws electrons from the system and then combines with $H^+$ to form water molecules.

_____ 25. No ATP is produced.

_____ 26. Thirty-two ATPs are produced.

_____ 27. Delivery point of NADH and $FADH_2$.

_____ 28. Two pyruvates enter for each glucose molecule entering glycolysis.

_____ 29. The carbons in the acetyl group leave as $CO_2$.

_____ 30. One carbon in pyruvate leaves as $CO_2$.

_____ 31. An electron transport system and channel proteins are involved.

_____ 32. Two $FADH_2$ and ten NADH are sent to this stage.

# 6 - V. ANAEROBIC ROUTES (pp. 106 - 107)

*Selected Words:* *pyruvate,* Lactobacillus, Saccharomyces cerevisiae, S. ellipsoideus

## Boldfaced, Page-Referenced Terms

(106) lactate fermentation _____

_____

(107) alcoholic fermentation _____

_____

(107) anaerobic electron transport _____

_____

## Fill-in-the-Blanks

If (1) _____ is not present in sufficient amounts, the end product of glycolysis enters (2) _____ pathways; in some bacteria and muscle cells, pyruvate is converted into such products as (3) _____, or in yeast cells it is converted into (4) _____ and (5) _____ _____.

(6) _____ pathways do not use oxygen as the final (7) _____ acceptor that ultimately drives the ATP-forming machinery. Anaerobic routes must be used by many bacteria and protistans that live in an oxygen-free environment. (8) _____ precedes any of the fermentation pathways. During (8), a glucose molecule is split into two (9) _____ molecules, two energy-rich (10) _____ intermediate molecules form, and the net energy yield from one glucose molecule is two ATP.

In one kind of fermentation pathway, (11) _____ itself accepts hydrogen and electrons from NADH. (11) is then converted to a three-carbon compound, (12) _____, during this process in a few bacteria species and some animal cells. Human muscle cells can carry on (12) fermentation in times of oxygen depletion; this provides a low yield of ATP.

In yeast cells, each pyruvate molecule from glycolysis forms an intermediate called (13) _____ while a gas, (14) _____ _____, is detached from pyruvate with the help of an enzyme. This intermediate accepts hydrogen and electrons from NADH and is then converted to (15) _____, the end product of alcoholic fermentation.

In both types of fermentation pathways, the net energy yield of two ATPs is formed during (16) _____. The reactions of the fermentation chemistry regenerate the (17) _____ needed for glycolysis to occur.

Anaerobic electron transport is an energy-releasing pathway occurring among the (18) _____. For example, sulfate-reducing bacteria living in soil or water produce (19) _____ by stripping electrons from a variety of compounds and sending them through membrane transport systems. The inorganic compound (20) _____ ($SO_4^=$) serves as the final electron acceptor and is converted into foul-smelling hydrogen sulfide gas ($H_2S$). Other kinds of bacteria produce ATP by stripping electrons from nitrate ($NO_3^-$), leaving (21) _____ ($NO_2^-$) as the end product. These bacteria are important in the global cycling of (22) _____.

## Complete the Table

Include a ( √ ) in each box that correctly links an occurrence (left-hand column) with a process (or processes).

| | Glycolysis | Lactate Fermentation | Alcoholic Fermentation | Anaerobic Electron Transport |
|---|---|---|---|---|
| 6-C $\longrightarrow \begin{array}{l}3C\\3C\end{array}$ | (23) | (36) | (49) | (62) |
| NADH → NAD$^+$ | (24) | (37) | (50) | (63) |
| NAD+ → NADH | (25) | (38) | (51) | (64) |
| SO$_4$= → H$_2$S | (26) | (39) | (52) | (65) |
| NO$_3$- → NO$_2$- | (27) | (40) | (53) | (66) |
| CO$_2$ is a waste product | (28) | (41) | (54) | (67) |
| 3-C → 3-C | (29) | (42) | (55) | (68) |
| 3-C → 2-C | (30) | (43) | (56) | (69) |
| ATP is used as a reactant | (31) | (44) | (57) | (70) |
| ATP is produced | (32) | (45) | (58) | (71) |
| pyruvate $\longrightarrow$ ethanol CO$_2$ | (33) | (46) | (59) | (72) |
| Occurs in animal cells | (34) | (47) | (60) | (73) |
| Occurs in yeast cells | (35) | (48) | (61) | (74) |

## 6 - VI. ALTERNATIVE ENERGY SOURCES IN THE HUMAN BODY (pp. 108 - 109)
### Commentary: PERSPECTIVE ON LIFE (p. 110)

### True/False

If the statement is true, write a T in the blank. If the statement is false, explain why in the blank.

_____ 1. Glucose is the only carbon-containing molecule that can be fed into the glycolytic pathway.

_____ 2. Simple sugars, fatty acids, and glycerol that remain after a cell's biosynthetic and storage needs have been met are generally sent to the cell's respiratory pathways for energy extraction.

_____ 3. Carbon dioxide and water, the products of aerobic respiration, generally get into the blood and are carried to gills or lungs, kidneys, and skin, where they are expelled from the animal's body.

_____ 4. Energy is recycled along with materials.

_____ 5. The first forms of life on Earth were most probably photosynthetic eukaryotes.

## Labeling

Identify the process or substance indicated in the illustration below.

6. _____ _____

7. _____

8. _____

9. _____ _____

10. _____ _____

11. _____ _____

12. _____

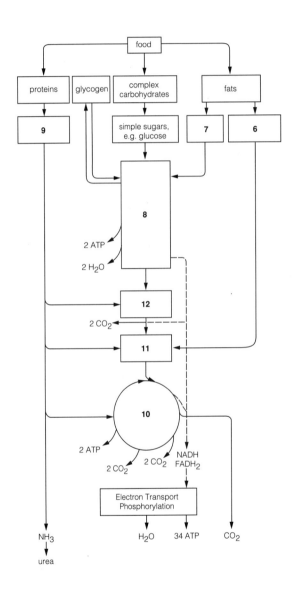

## Complete this Table

Across the top of each column are the principal phases of degradative pathways into which food molecules (in various stages of breakdown) enter or in which specific events occur. Put a check ( √ ) in each box that indicates the phase into which a specific food molecule is fed, or in which a specific event occurs. For example: if simple sugars can enter the glycolytic pathway, put a check ( √ ) in the top left hand box; if not, let the box remain blank.

| | Glycolysis (includes pyruvate) | Acetyl CoA Formation | Krebs Cycle | Electron Transport Phosphorylation | Fermentation Alcoholic | Lactate |
|---|---|---|---|---|---|---|
| Complex Carbohydrates → Simple Sugars, which enter | (13) | (28) | (43) | (58) | (73) | (88) |
| Fats → Fatty Acids which enter | (14) | (29) | (44) | (59) | (74) | (89) |
| Fats → Glycerol, which enters | (15) | (30) | (45) | (60) | (75) | (90) |
| Proteins → Amino Acids, which enter | (16) | (31) | (46) | (61) | (76) | (91) |
| Intermediate energy carriers (NADH) are produced | (17) | (32) | (47) | (62) | (77) | (92) |
| ($FADH_2$) | (18) | (33) | (48) | (63) | (78) | (93) |
| ATPs produced directly as a result of this process alone | (19) | (34) | (49) | (64) | (79) | (94) |
| $NAD^+$ produced | (20) | (35) | (50) | (65) | (80) | (95) |
| FAD produced | (21) | (36) | (51) | (66) | (81) | (96) |
| ADP produced | (22) | (37) | (52) | (67) | (82) | (97) |
| Unbound phosphate ($P_i$) required | (23) | (38) | (53) | (68) | (83) | (98) |
| $CO_2$ produced (waste product) | (24) | (39) | (54) | (69) | (84) | (99) |
| $H_2O$ produced (waste product) | (25) | (40) | (55) | (70) | (85) | (100) |
| Atoms from $O_2$ react here | (26) | (41) | (56) | (71) | (86) | (101) |
| NADH required to drive this process | (27) | (42) | (57) | (72) | (87) | (102) |

*Choice*

For questions 103 - 117, refer to the text and Figure 6.11. Choose from the following answers:

a. glucose    b. glucose-6-phosphate    c. glycogen    d. fatty acids    e. triglycerides

f. PGAL    g. acetyl-CoA    h. amino acids    i. glycerol    j. proteins

____ 103. Fats that are tapped between meals or during exercise as alternatives to glucose

____ 104. Used between meals when free glucose supply dwindles; enters glycolysis after conversion

____ 105. Its breakdown yields much more ATP than glucose

____ 106. Absorbed in large amounts immediately following a meal

____ 107. Represents only 1 percent or so of the total stored energy in the body

____ 108. Following removal of amino groups, the carbon backbones may be converted to fats or carbohydrates, or they may enter the Krebs cycle

____ 109. On average, represents 78 percent of the body's stored food

____ 110. Between meals liver cells can convert it back to free glucose and release it

____ 111. Can be stored in cells but not transported across plasma membranes

____ 112. Amino groups undergo conversions that produce urea, a nitrogen-containing waste product excreted in urine

____ 113. Converted to PGAL in the liver; a key intermediate of glycolysis

____ 114. Accumulate inside the fat cells of adipose tissues, at strategic points under the skin

____ 115. A storage polysaccharide produced from glucose-6-phosphate following food intake that exceeds cellular energy demand (and increases ATP production to inhibit glycolysis)

____ 116. Building blocks of the compounds that represent 21 percent of the body's stored food

____ 117. A product resulting from enzymes cleaving circulating fatty acids; enters the Krebs cycle

## Self-Quiz

____ 1. Glycolysis would quickly halt if the process ran out of _____, which serves as the hydrogen and electron acceptor.
a. $NADP^+$
b. ADP
c. $NAD^+$
d. $H_2O$

____ 2. The ultimate electron acceptor in aerobic respiration is _____.
a. NADH
b. carbon dioxide ($CO_2$)
c. oxygen ($\frac{1}{2} O_2$)
d. ATP

____ 3. When glucose is used as an energy source, the largest amount of ATP is generated by the _____ portion of the entire respiratory process.
a. glycolytic pathway
b. acetyl-CoA formation

c. Krebs cycle
d. electron transport phosphorylation

____ 4. The process by which about 10 percent of the energy stored in a sugar molecule is released as it is converted into two small organic-acid molecules is _____.
a. photolysis
b. glycolysis
c. fermentation
d. the dark reactions

____ 5. During which of the following phases of aerobic respiration is ATP produced directly by substrate-level phosphorylation?
a. glucose formation
b. ethyl-alcohol production
c. acetyl-CoA formation
d. glycolysis

___ 6. What is the name of the process by which reduced NADH transfers electrons along a chain of acceptors to oxygen so as to form water and in which the energy released along the way is used to generate ATP?
a. glycolysis
b. acetyl-CoA formation
c. the Krebs cycle
d. electron transport phosphorylation

___ 7. Pyruvic acid can be regarded as the end product of _____.
a. glycolysis
b. acetyl-CoA formation
c. fermentation
d. the Krebs cycle

___ 8. Which of the following is not ordinarily capable of being reduced at any time?
a. NAD
b. FAD
c. oxygen, $O_2$
d. water

___ 9. ATP production by chemiosmosis involves _____.
a. $H^+$ concentration and electric gradients across a membrane
b. ATP synthases
c. formation of ATP in the inner mitochondrial compartment
d. all of the above

___ 10. During the fermentation pathways, a net yield of two ATP molecules is produced from _____; the $NAD^+$ necessary for _____ is regenerated during the fermentation reactions.
a. the Krebs cycle; glycolysis
b. glycolysis; electron transport phosphorylation
c. the Krebs cycle; electron transport phosphorylation
d. glycolysis; glycolysis

## Matching

Match the following components of respiration to the list of words below. Some components may have more than one answer.

11. ___ lactic acid, lactate

12. ___ $NAD^+ \rightarrow NADH$

13. ___ carbon dioxide is a product

14. ___ $NADH \rightarrow NAD^+$

15. ___ pyruvic acid, pyruvate, used as a reactant

16. ___ ATP produced by substrate-level phosphorylation

17. ___ glucose

18. ___ acetyl-CoA is both a reactant and a product

19. ___ oxygen

20. ___ water is a product

A. Glycolysis
B. Preparatory conversions prior to the Krebs cycle
C. Fermentation
D. Krebs cycle
E. Electron transport

# Chapter Objectives/Review Questions

This section lists general and detailed chapter objectives that can be used as review questions. You can make maximum use of these items by writing answers on a separate sheet of paper. Fill in answers where blanks are provided. To check for accuracy, compare your answers with information given in the chapter or glossary.

Page        Objectives/Questions

(98 - 99)    1.    No matter what the source of energy might be, organisms must convert it to _____, a form of chemical energy that can drive metabolic reactions.

(98)   2.   Give the overall equation for the aerobic respiratory route; indicate where energy occurs in the equation.

(99)   3.   In the first of the three stages of aerobic respiration, _____ is partially degraded to pyruvate.

(99)   4.   By the end of the second stage of aerobic respiration, which includes the _____ cycle, _____ has been completely degraded to carbon dioxide and water.

(99)   5.   Explain, in general terms, the role of oxygen in aerobic respiration.

(100)   6.   Glycolysis occurs in the _____ of the cell.

(100)   7.   Explain the purpose served by molecules of ATP reacting first with glucose and then with fructose-6-phosphate in the early part of glycolysis (see Figure 6.3 in the text).

(100)   8.   Four ATP molecules are produced by _____ - _____ phosphorylation for every two used during glycolysis. Consult Figure 6.3 in the text.

(100)   9.   Glycolysis produces _____ (number) NADH, _____ (number) ATP (net) and _____ (number) pyruvate molecules for each glucose molecule entering the reactions.

(102)   10.   Consult Figures 6.3 and 6.5 in the text. State the events that happen during the preparatory steps and explain how the process of acetyl-CoA formation relates glycolysis to the Krebs cycle.

(103)   11.   What happens to the $CO_2$ produced during acetyl-CoA formation and the Krebs cycle?

(103)   12.   Consult Figure 6.5 in the text and predict what will happen to the NADH produced during acetyl-CoA formation and the Krebs cycle.

(104 - 105) 13.   Explain how chemiosmotic theory operates in the mitochondrion to account for the production of ATP molecules.

(104)   14.   Briefly describe the process of electron transport phosphorylation by stating what reactants are needed and what the products are. State how many ATP molecules are produced through operation of the transport system.

(104 - 105) 15.   Be able to account for the total net yield of thirty-six ATP molecules produced through aerobic respiration; that is, state how many ATPs are produced in glycolysis, the Krebs cycle, and electron transport phosphorylation.

(106)   16.   List some places where there is very little oxygen present and where anaerobic organisms might be found.

(106)   17.   Describe what happens to pyruvate in anaerobic organisms. Then explain the necessity for pyruvate to be converted to a fermentative product.

(106 - 107) 18.   State which factors determine whether the pyruvate (pyruvic acid) produced at the end of glycolysis will enter into the alcoholic fermentation pathway, the lactate fermentation pathway, or the acetyl-CoA formation pathway.

(108 - 109) 19.   You have been fasting for three days, drinking only water and eating no solid food. Tell which stored molecules your body is using to provide energy, and describe how that is occurring.

(110)   20.   After reading "Perspective on Life" in the main text, outline the supposed evolutionary sequence of energy-extraction processes.

(110)   21.   Closely scrutinize the diagram of the carbon cycle in the *Commentary*; be able to reproduce the cycle from memory.

## Integrating and Applying Key Concepts

How is the "oxygen debt" experienced by runners and sprinters related to aerobic and anaerobic respiration in humans?

# 7

# CELL DIVISION AND MITOSIS

## Interactive Exercises

### 7 -I. DIVIDING CELLS: THE BRIDGE BETWEEN GENERATIONS (pp. 114 - 116)

*Selected Words:* cell division, nuclear division mechanisms, *haploid*

In addition to the boldfaced terms, the text features other important terms essential to understanding the assigned material. "Selected Words" is a list of these terms, which appear in the text in italics, in quotation marks, and occasionally in roman type. Latin binomials found in this section are underlined and in roman type to distinguish them from other italicized words.

### Boldfaced, Page-Referenced Terms

The page-referenced terms are important; they were in boldface type in the chapter. Write a definition for each term in your own words without looking at the text. Next, compare your definition with that given in the chapter or in the text glossary. If your definition seems inaccurate, allow some time to pass and repeat this procedure until you can define each term rather quickly (how fast you can answer is a gauge of your learning effectiveness).

(115) reproduction _____

_____

(116) mitosis _____

_____

(116) meiosis _____

_____

(116) somatic cells _____

_____

(116) germ cells _____

_____

(116) chromosome _____

_____

(116) sister chromatids _____

_____

(116) centromere _____

_____

(116) diploid cell _____

_____

(116) chromosome number _____

_____

## Matching

Choose the most appropriate answer for each.

1. ___ centromere
2. ___ prokaryotic fission
3. ___ diploid cell
4. ___ chromosome
5. ___ germ cells
6. ___ reproduction
7. ___ somatic cells
8. ___ sister chromatids
9. ___ mitosis and meiosis
10. ___ chromosome number

A. Cell lineage set aside for forming gametes
B. Producing a new generation of cells or multicelled individuals
C. Any cell having two of each type of chromosome
D. The number of each type of chromosome
E. Each DNA molecule with attached proteins
F. Sort out and package parent DNA molecules into daughter cell nuclei
G. Narrowed chromosome region with attachment sites for microtubules
H. Asexual reproduction, bacterial cell division
I. The two attached DNA molecules of a duplicated chromosome
J. Body cells that reproduce by mitosis and cytoplasmic division

## 7 - II. MITOSIS AND THE CELL CYCLE (p. 117)

### Boldfaced, Page-Referenced Terms

(117) cell cycle _____

_____

(117) interphase _____

_____

## Labeling

Identify the stage in the cell cycle indicated by each number.

1. _____
2. _____
3. _____
4. _____
5. _____
6. _____
7. _____
8. _____
9. _____
10. _____

## Matching

Link each time span identified below with the most appropriate number in the preceding labeling section.

11. ___ Period after duplication of DNA during which the cell prepares for division, a second "gap"

12. ___ Period of nuclear division followed by cytoplasmic division (a separate event)

13. ___ DNA duplication occurs now; a time of "synthesis" of DNA and proteins

14. ___ Period of cell growth before DNA duplication; a "gap" of interphase

15. ___ Usually the longest part of a cell cycle

16. ___ Period of cytoplasmic division

17. ___ Period that includes $G_1$, S, $G_2$

# 7 - III STAGES OF MITOSIS (pp. 118 - 119)

## Boldfaced, Page-Referenced Terms

(118) prophase _____

_____

(118) metaphase _____

_____

(118) anaphase _____

_____

(118) telophase _____

_____

(118) spindle apparatus _____

_____

(118) centrioles _____

_____

## Label and Match

Identify each of the mitotic stages shown below by entering the correct stage in the blank beneath the sketch. Select from late prophase, transition to metaphase (prometaphase), interphase-parent cell, metaphase, early prophase, telophase, interphase-daughter cells, and anaphase. Complete the exercise by matching and entering the letter of the correct phase description in the parentheses following each label.

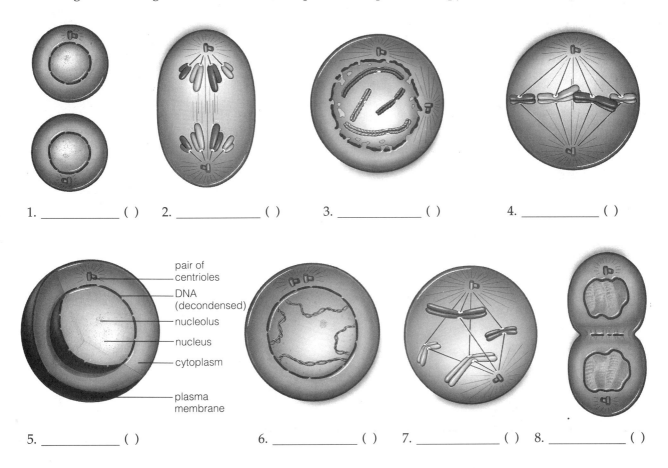

1. _____ (  )    2. _____ (  )    3. _____ (  )    4. _____ (  )

pair of centrioles
DNA (decondensed)
nucleolus
nucleus
cytoplasm
plasma membrane

5. _____ (  )    6. _____ (  )    7. _____ (  )    8. _____ (  )

A. Attachment between two sister chromatids of each chromosome breaks; the two are now separate chromosomes that move to opposite spindle poles.

B. Microtubules that form the spindle apparatus penetrate the nuclear region and form the spindle; microtubules become attached to the sister chromatids of each chromosome.

C. The DNA and its associated proteins begin to condense.

D. Chromosomes are now fully condensed and lined up at the equator of the spindle.

E. DNA is duplicated and the cell prepares for nuclear division.

F. Two daughter cells have formed, each diploid with two of each type of chromosome.

G. Chromosomes continue to condense. New microtubules are assembled, and they move one of two centriole pairs toward the opposite end of the cell. The nuclear envelope begins to break up.

H. New patches of membrane join to form nuclear envelopes around the decondensing chromosomes. Cytokinesis begins before this stage ends.

## 7 - IV. DIVISION OF THE CYTOPLASM (pp. 120 - 121)
### Focus on Science: HENRIETTA'S IMMORTAL CELLS (p. 122)

*Boldfaced, Page-Referenced Terms*

(120) cytoplasmic division _____

_____

(120) cell plate formation _____

_____

(121) cleavage _____

_____

## Choice

For questions 1 - 5, choose from the following:

                              a. plant cells     b. animal cells

___ 1. Formation of a cell plate

___ 2. Microfilaments pull the plasma membrane inward and cut the cell in two

___ 3. Cellulose deposits form a crosswall between the two daughter cells

___ 4. Possess rigid walls that cannot be pinched in two

___ 5. Cleavage furrow

# Self-Quiz

___ 1.  The replication of DNA occurs _____.
   a. between the growth phases of interphase
   b. immediately before prophase of mitosis
   c. during prophase of mitosis
   d. during prophase of meiosis

___ 2.  In the cell life cycle of a particular cell,

   _____.
   a. mitosis occurs immediately prior to S
   b. mitosis occurs immediately prior to $G_1$
   c. $G_2$ precedes S
   d. $G_1$ precedes S
   e. mitosis and S precede $G_1$

___ 3.  In eukaryotic cells, which of the following can occur during mitosis?
   a. Two mitotic divisions to maintain the parental chromosome number
   b. The replication of DNA
   c. A long growth period
   d. The disappearance of the nuclear envelope and nucleolus

___ 4.  Diploid refers to _____.
   a. having two chromosomes of each type in somatic cells
   b. twice the parental chromosome number
   c. half the parental chromosome number
   d. having one chromosome of each type in somatic cells

___ 5.  Somatic cells are _____ cells; germ cells are _____ cells.
   a. meiotic; body
   b. body; body
   c. meiotic; meiotic
   d. body; meiotic

___ 6.  If a parent cell has sixteen chromosomes and undergoes mitosis, the resulting cells will have _____ chromosomes.
   a. sixty-four
   b. thirty-two
   c. sixteen
   d. eight
   e. four

___ 7.  The correct order of the stages of mitosis is

   _____.
   a. prophase, metaphase, telophase, anaphase
   b. telophase, anaphase, metaphase, prophase
   c. telophase, prophase, metaphase, anaphase
   d. anaphase, prophase, telophase, metaphase
   e. prophase, metaphase, anaphase, telophase

8. "The nuclear envelope breaks up completely into vesicles. Microtubules are now free to interact with the chromosomes." These sentences describe the _____ of mitosis.
   a. prophase
   b. metaphase
   c. transition to metaphase
   d. anaphase
   e. telophase

9. During _____, sister chromatids of each chromosome are separated from each other, and those former partners, now chromosomes, are moved toward opposite poles.
   a. prophase
   b. metaphase

   c. anaphase
   d. telophase

10. In the process of cytokinesis, cleavage furrows are associated with _____ cell division, and cell plate formation is associated with _____ cell division.
    a. animal; animal
    b. plant; animal
    c. plant; plant
    d. animal; plant

## Chapter Objectives/Review Questions

This section lists general and detailed chapter objectives that can be used as review questions. You can make maximum use of these items by writing answers on a separate sheet of paper. Fill in answers where blanks are provided. To check for accuracy, compare your answers with information given in the chapter or glossary.

*Page*       *Objectives/Questions*

(116)    1.  Name the substance that contains the instructions for making proteins.
(116)    2.  Mitosis and meiosis refer to the division of the cell's _____.
(116)    3.  Distinguish between somatic cells and germ cells as to their location and function.
(116)    4.  The eukaryotic chromosome is composed of _____ and _____.
(116)    5.  The two attached threads of a duplicated chromosome are known as sister _____.
(116)    6.  Describe the function of the chromosome portion known as a centromere.
(116)    7.  Any cell having two of each type of chromosome is a _____ cell.
(117)    8.  Be able to list and describe, in order, the various activities occurring in the eukaryotic cell life cycle.
(117)    9.  Interphase of the cell cycle consists of $G_1$, _____, and $G_2$.
(117)   10.  S is the time in the cell cycle when _____ duplication occurs.
(118 - 119)  11.  Be able to describe the cellular events occurring in the prophase, metaphase, anaphase, and telophase of mitosis.
(118)   12.  The "_____" is a time of transition when the nuclear envelope breaks up into flattened vesicles prior to metaphase.
(120 - 121)  13.  Compare and contrast cytokinesis as it occurs in plant mitosis and animal mitosis using the following concepts: cleavage furrow, microfilaments at the cell's midsection, and cell plate formation.
(122)   14.  Explain why cells from Henrietta Lacks continue to benefit humans everywhere more than forty years after her death.

## Integrating and Applying Key Concepts

Runaway cell division is characteristic of cancer. Imagine the various points of the mitotic process that might be sabotaged in cancerous cells in order to halt their multiplication. Then try to imagine how one might discriminate between cancerous and normal cells in order to guide those methods of sabotage most effective in combating cancer.

# 8

# MEIOSIS

## Interactive Exercises

### 8 - I. COMPARISON OF ASEXUAL AND SEXUAL REPRODUCTION (pp. 124 - 126) MEIOSIS AND THE CHROMOSOME NUMBER (pp. 126 - 127)

*Boldfaced, Page-Referenced Terms*

(126) asexual reproduction _____

_____

(126) genes _____

_____

(126) sexual reproduction _____

_____

(126) allele _____

_____

(126) meiosis _____

_____

(126) gametes _____

_____

(127) diploid _____

_____

(127) homologous chromosomes _____

_____

(127) haploid _____

_____

(127) sister chromatids _____

_____

## Choice

For questions 1 - 10, choose from the following:

a. asexual reproduction        b. sexual reproduction

___ 1. Brings together new combinations of alleles in offspring

___ 2. Involves only one parent

___ 3. Both parents pass on one of each gene to their offspring.

___ 4. Commonly involves two parents

___ 5. The production of "clones"

___ 6. Involves meiosis and fertilization

___ 7. Offspring are genetically identical copies of the parent.

___ 8. Produces the variation in traits that forms the basis of evolutionary change

___ 9. The instructions in every pair of genes are identical in all individuals of a species.

___ 10. New combinations of alleles lead to variations in physical and behavioral traits.

## True/False

If the statement is true, write a T in the blank. If the statement is false, correct it by changing the under-lined word and writing the correct word(s) in the blank.

_____ 11. Each unique molecular form of the same gene is called an <u>allele</u>.

_____ 12. Sperms and eggs are cells known as <u>germ</u> cells.

_____ 13. <u>Haploid</u> germ cells produce haploid gametes.

_____ 14. <u>Diploid</u> cells possess pairs of homologous chromosomes.

_____ 15. <u>Meiosis</u> produces cells that have one member of each pair of homologous chromosomes possessed by the species.

_____ 16. Sister chromatids bear <u>identical</u> alleles.

_____ 17. The alleles on a pair of homologous chromosomes are often <u>nonidentical</u>.

# 8 - II. A VISUAL TOUR OF THE STAGES OF MEIOSIS (pp. 128 - 129)

## Matching

To review the major stages of meiosis, match the following written descriptions with the appropriate sketch. Assume that the cell in this model initially has only one pair of homologous chromosomes (one from a paternal source and one from a maternal source) and crossing over does not occur. Complete the exercise by indicating the diploid ($2n = 2$) or haploid ($n = 1$) chromosome number of the cell chosen in the parentheses following each blank.

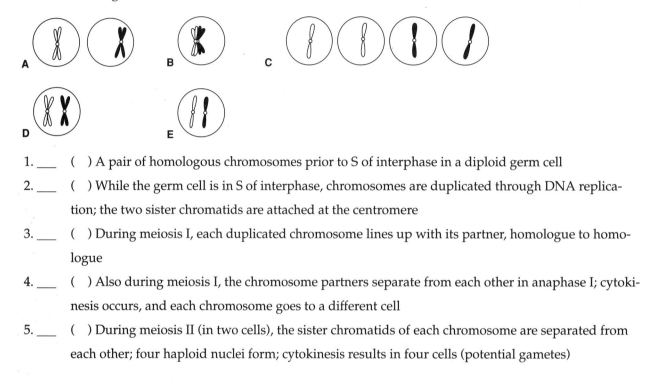

1. ___ (   ) A pair of homologous chromosomes prior to S of interphase in a diploid germ cell

2. ___ (   ) While the germ cell is in S of interphase, chromosomes are duplicated through DNA replication; the two sister chromatids are attached at the centromere

3. ___ (   ) During meiosis I, each duplicated chromosome lines up with its partner, homologue to homologue

4. ___ (   ) Also during meiosis I, the chromosome partners separate from each other in anaphase I; cytokinesis occurs, and each chromosome goes to a different cell

5. ___ (   ) During meiosis II (in two cells), the sister chromatids of each chromosome are separated from each other; four haploid nuclei form; cytokinesis results in four cells (potential gametes)

## Label and Match

Identify each of the meiotic stages shown below by entering the correct stage of either meiosis I or meiosis II in the blank beneath the sketch. Choose from prophase I, metaphase I, anaphase I, telophase I, prophase II, metaphase II, anaphase II, and telophase II. Complete the exercise by matching and entering the letter of the correct phase description in the parentheses following each label.

A. The spindle is now fully formed; all chromosomes are positioned midway between the poles of one cell.
B. Centrioles have moved to form the poles in each of two cells; a spindle forms, and microtubules attach the duplicated chromosomes to the spindle and begin moving them toward the equator of each cell.
C. Four daughter nuclei form; when the cytoplasm divides, each new cell has a haploid number of chromosomes, all in the unduplicated state. One or all cells may develop into gametes.
D. In one cell, each duplicated chromosome is pulled away from its homologue; the two are moved to opposite spindle poles.
E. Each chromosome is drawn up close to its homologue; crossing over and genetic recombination occur.
F. All of the chromosomes are now aligned at the spindle's equator; this is occurring in two haploid cells.
G. Two haploid cells form, but chromosomes are still in the duplicated state.
H. Chromatids of each chromosome separate; former "sister chromatids" are now chromosomes in their own right and are moved to opposite poles.

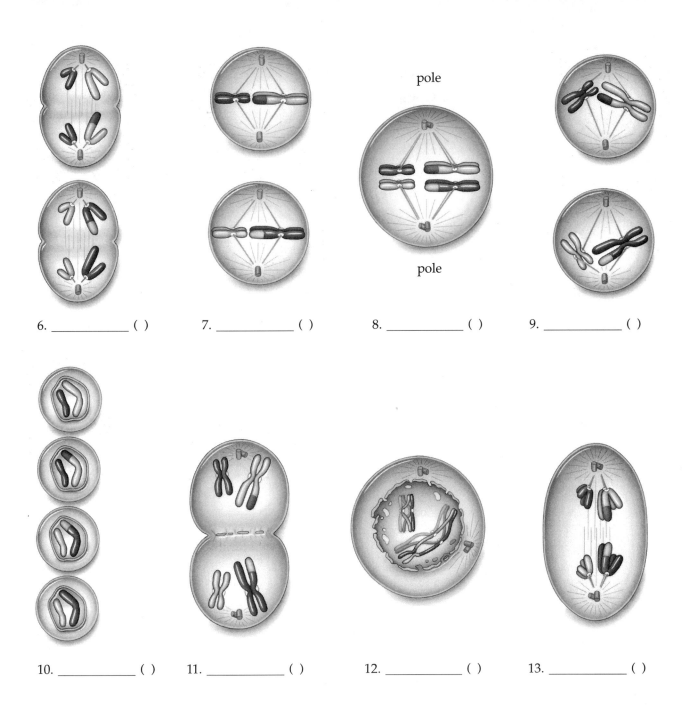

6. _____ ( )    7. _____ ( )    8. _____ ( )    9. _____ ( )

pole

pole

10. _____ ( )    11. _____ ( )    12. _____ ( )    13. _____ ( )

## 8 - III. KEY EVENTS OF MEIOSIS I (pp. 130 - 131)

*Selected Words:* synapsis, "nonsister" chromatids, "chiasmata," *maternal* chromosome, *paternal* chromosome

*Boldfaced, Page-Referenced Terms*

(130) crossing over _____

_____

## Matching

Choose the most appropriate answer for each.

1. ___ synapsis
2. ___ crossing over
3. ___ function of meiosis
4. ___ maternal chromosomes
5. ___ paternal chromosomes

A. Chromosomes inherited from your father
B. Exchange of corresponding genes between two nonsister chromatids; leads to recombination
C. Chromosomes inherited from your mother
D. A prophase I activity when each chromosome is drawn close to its homologue during prophase I
E. Reduction of the chromosome number by half for forthcoming gametes

# 8 - IV. FROM GAMETES TO OFFSPRING (pp. 132 - 133)

*Selected Words:* gamete-producing *bodies,* spore-producing *bodies*

## Boldfaced, Page-Referenced Terms

(132) spores _____

_____

(132) sperm _____

_____

(132) egg _____

_____

(132) fertilization _____

_____

## Choice

For questions 1–10, choose from the following:

a. animal life cycle      b. plant life cycle      c. both animal and plant life cycles

___  1. Meiosis results in the production of haploid spores.

___  2. A zygote divides by mitosis.

___  3. Meiosis results in the production of haploid gametes.

___  4. Haploid gametes fuse in fertilization to form a diploid zygote.

___  5. A zygote divides by mitosis to form a diploid sporophyte.

___  6. A spore divides by mitosis to produce a haploid gametophyte.

___  7. A haploid gametophyte divides by mitosis to produce haploid gametes.

___  8. A haploid spore divides by mitosis to produce a gametophyte.

___  9. A diploid body forms from mitosis of a zygote.

___ 10. A gamete-producing body and a spore-producing body develop during the life cycle.

## Sequence

Arrange the following entities in correct order of development, entering a 1 by the stage that appears first and a 5 by the stage that completes the process of spermatogenesis. Complete the exercise by indicating if each cell is n or 2n in the parentheses following each blank. Refer to Figure 8.8, page 133, in the text.

11. ___ (  )    primary spermatocyte

12. ___ (  )    sperm

13. ___ (  )    spermatid

14. ___ (  )    spermatogonium

15. ___ (  )    secondary spermatocyte

## Matching

Choose the most appropriate answer to match with each oogenesis concept. Refer to Figure 8.9, page 133, in the text.

16. ___ primary oocyte

17. ___ oogonium

18. ___ secondary oocyte

19. ___ ovum and three polar bodies

20. ___ first polar body

A. The cell in which synapsis, crossing over, and recombination occur
B. A cell that is equivalent to a diploid germ cell
C. A haploid cell formed after division of the primary oocyte that does not form an ovum at second division
D. Haploid cells, but only one of which functions as an egg
E. A haploid cell formed after division of the primary oocyte, the division of which forms a functional ovum

## Short Answer

21. List the various mechanisms that contribute to the huge number of new gene combinations that may

result from fertilization. _____

_____

_____

# 8 - V. MEIOSIS AND MITOSIS COMPARED (pp. 134 - 135)

*Selected Word:* clones

## Complete the Table

1. Complete the table below by entering the word "mitosis" or "meiosis" in the blank adjacent to the statement describing one of these processes.

| Description | Mitosis/Meiosis |
|---|---|
| a. Involves one division cycle | |
| b. Functions in growth and repair | |
| c. Daughter cells are haploid. | |
| d. Initiated in germ cells | |
| e. Involves two division cycles | |
| f. Daughter cells have one chromosome from each homologous pair. | |
| g. Produces spores in plant life cycles | |
| h. Daughter cells have the diploid chromosome number. | |
| i. Completed when four daughter cells are formed | |

## Matching

The cell model used in this exercise has two pairs of homologous chromosomes, one long pair and one short pair. Match the descriptions to the numbers of chromosomes shown in the sketches below.

2. ___ Beginning of meiosis II following interkinesis

3. ___ A daughter cell at the end of meiosis II

4. ___ Metaphase I of meiosis

5. ___ Metaphase of mitosis

6. ___ G$_1$ in a daughter cell following mitosis

7. ___ Prophase of mitosis

A      B      C      D      E      F

The following questions refer to the sketches above; enter answers in the blanks after each question.

8. How many chromosomes are present in cell E? ___

9. How many chromatids are present in cell E? ___

10. How many chromatids are present in cell C? ___

11. How many chromatids are present in cell D? ___

12. How many chromosomes are present in cell F? ___

# Self-Quiz

___ 1. Which of the following does not occur in prophase I of meiosis?
   a. a cytoplasmic division
   b. a cluster of four chromatids
   c. homologues pairing tightly
   d. crossing over

___ 2. Crossing over is one of the most important events in meiosis because _____.
   a. it produces new combinations of alleles on chromosomes
   b. homologous chromosomes must be separated into different daughter cells
   c. the number of chromosomes allotted to each daughter cell must be halved
   d. homologous chromatids must be separated into different daughter cells

___ 3. Crossing over _____.
   a. generally results in pairing of homologues and binary fission
   b. is accompanied by gene-copying events
   c. involves breakages and exchanges between sister chromatids
   d. alters the composition of chromosomes and results in new combinations of alleles being channeled into the daughter cells

___ 4. The appearance of chromosome ends lapped over each other in meiotic prophase I provides evidence of _____.
   a. meiosis
   b. crossing over
   c. chromosomal aberration
   d. fertilization
   e. spindle fiber formation

___ 5. Which of the following does not increase genetic variation?
   a. Crossing over
   b. Random fertilization
   c. Prophase of mitosis
   d. Random homologue alignments at metaphase I

___ 6. Which of the following is the most correct sequence of events in animal life cycles?
   a. meiosis ———> fertilization ———> gametes ———> diploid organism
   b. diploid organism ———> meiosis ———> gametes ———> fertilization
   c. fertilization ———> gametes ———> diploid organism ———> meiosis
   d. diploid organism ———> fertilization ———> meiosis ———> gametes

___ 7. In sexually reproducing organisms, the zygote is _____.
   a. an exact genetic copy of the female parent
   b. an exact genetic copy of the male parent
   c. unlike either parent genetically
   d. a genetic mixture of male parent and female parent

___ 8. Which of the following is the most correct sequence of events in plant life cycles?
   a. fertilization ———> zygote ———> sporophyte ———> meiosis ———> spores ———> gametophytes ———> gametes
   b. fertilization ———> sporophyte ———> zygote ———> meiosis ———> spores ———> gametophytes ———> gametes
   c. fertilization ———> zygote ———> sporophyte ———> meiosis ———> gametes ———> gametophyte ———> spores
   d. fertilization ———> zygote ———> gametophyte ———> meiosis ———> gametes ———> sporophyte ———> spores

___ 9. The cell in the diagram is a diploid that has three pairs of chromosomes. From the number and pattern of chromosomes, the cell _____.
   a. could be in the first division of meiosis
   b. could be in the second division of meiosis
   c. could be in mitosis
   d. could not be in mitosis or meiosis, because this stage is not possible in a cell with three pairs of chromosomes

____ 10. You are looking at a cell from the same organism as in the previous question. Now the cell _____.

a. could be in the first division of meiosis
b. could be in the second division of meiosis

c. could be in mitosis
d. could not be in mitosis or meiosis, because this stage is not possible in this organism

# Chapter Objectives/Review Questions

| Page | | Objectives/Questions |
|---|---|---|
| (126) | 1. | "One parent always passes on a duplicate of all its genes to offspring" describes _____ reproduction. |
| (126) | 2. | Sexual reproduction puts together new combinations of _____ in offspring. |
| (127) | 3. | Describe the relationship between the following terms: *homologous chromosomes, diploid,* and *haploid.* |
| (127) | 4. | During interphase a germ cell duplicates its DNA; a duplicated chromosome consists of two DNA molecules that remain attached to a constriction called the _____. |
| (127) | 5. | As long as the two DNA molecules remain attached they are referred to as _____ of the chromosome. |
| (127) | 6. | During meiosis I, homologous chromosomes pair; each homologue consists of _____ chromatids. |
| (127) | 7. | During meiosis II, the two sister _____ of each _____ are separated from each other. |
| (127) | 8. | If the diploid chromosome number for a particular plant species is 18, the haploid number is _____. |
| (130) | 9. | _____ _____ breaks up old combinations of alleles and puts new ones together in pairs of homologous chromosomes. |
| (131) | 10. | The _____ attachment and subsequent positioning of each pair of maternal and paternal chromosomes at metaphase I leads to different _____ of maternal and paternal traits in offspring. |
| (132 - 133) | 11. | Using the special terms for the cells at the various stages, describe spermatogenesis in male animals and oogenesis in female animals. |
| (132) | 12. | Meiosis in the animal life cycle results in haploid _____; meiosis in the plant life cycle results in haploid _____. |
| (134) | 13. | Crossing over, the distribution of random mixes of homologous chromosomes into gametes, and fertilization contribute to _____ in traits of offspring. |

# Integrating and Applying Key Concepts

A few years ago, it was claimed that the actual cloning of a human being had been accomplished. Later, this claim was admitted to be fraudulent. If sometime in the future cloning of humans becomes possible, speculate about the effects of reproduction without sex on human populations.

# 9

# OBSERVABLE PATTERNS OF INHERITANCE

## Interactive Exercises

### 9 - I. MENDEL'S INSIGHT INTO PATTERNS OF INHERITANCE (pp. 138 - 141)
### MENDEL'S THEORY OF SEGREGATION (pp. 142 - 143)
### INDEPENDENT ASSORTMENT (pp. 144 - 145)

*Selected Words:* Pisum sativum, *hybrid, homozygous, heterozygous, dominant* allele, *recessive* allelle

*Boldfaced, Page-Referenced Terms*

(141) true-breeding _____

_____

(141) genes _____

_____

(141) alleles _____

_____

(141) homozygous dominant _____

_____

(141) homozygous recessive _____

_____

(141) heterozygous _____

_____

(141) genotype _____

_____

(141) phenotype _____

_____

(142) monohybrid crosses _____

_____

(142) probability _____

_____

(143) Punnett-square method _____

_____

(143) testcross _____

_____

(144) dihybrid crosses _____

_____

(145) independent assortment _____

_____

## Matching

Choose the most appropriate answer for each.

1. ___ genotype
2. ___ alleles
3. ___ heterozygous
4. ___ dominant allele
5. ___ phenotype
6. ___ genes
7. ___ recessive allele
8. ___ homozygous
9. ___ diploid cell
10. ___ locus

A. All the different molecular forms of a gene that exist
B. Particular location of a gene on a chromosome
C. Describes an individual having a pair of nonidentical alleles
D. Gene whose effect is masked by its partner
E. Refers to an individual's observable traits
F. Refers to the genes present in an individual organism
G. Gene whose effect "masks" the effect of its partner
H. Describes an individual for which two alleles of a pair are the same
I. Units of information about specific traits; passed from parents to offspring
J. Has a pair of genes for each trait, one on each of two homologous chromosomes

## Fill-in-the-Blanks

Offspring of (11) _____ crosses are heterozygous for the one trait being studied. Because

fertilization is a chance event, the rules of (12) _____ apply to genetics crosses. The separation of

A and a as members of a pair of homologous chromosomes move to different gametes during meiosis is known as Mendel's theory of (13) _____ . Crossing F$_1$ hybrids (possibly of unknown genotype) back to a plant known to be a true-breeding recessive plant is known as Mendel's (14) _____ . From this cross, a ratio of (15) _____ is expected. When F$_1$ offspring inherit two gene pairs, each consisting of two nonidentical alleles, the cross is known as a (16) _____ cross. "Gene pairs assorting into gametes independently of other gene pairs located on nonhomologous chromosomes" describes Mendel's theory of (17) _____ _____ .

## Problems

18. In garden pea plants, Tall (*T*) is dominant over dwarf (*t*). In the cross *Tt* × *tt*, the *Tt* parent would produce a gamete carrying *T* (tall) and a gamete carrying *t* (dwarf) through segregation; the *tt* parent could only produce gametes carrying the *t* (dwarf) gene. Use the Punnett-square method (refer to Figures 9.4, 9.6, and 9.7 in the text) to determine the genotype and phenotype probabilities of offspring from the above cross, *Tt* × *tt*:

Although the Punnett-square (checkerboard) method is a favored method for solving single-factor genetics problems, there is a quicker way. Only six different outcomes are possible from single-factor crosses. Studying the following relationships allows one to obtain the result of any such cross by inspection.

1. *AA* × *AA* = all *AA*
   (Each of the four blocks of the Punnett square would be *AA*.)
2. *aa* × *aa* = all *aa*
3. *AA* × *aa* = all *Aa*
4. *AA* × *Aa* = ½ *AA*; ½ *Aa* or *Aa* × *AA*
   (Two blocks of the Punnett square are *AA*, and two blocks are *Aa*.)
5. *aa* × *Aa* = ½ *aa*; ½ *Aa* or *Aa* × *aa*
6. *Aa* × *Aa* = ¼ *AA*; ½ *Aa*; ¼ *aa*
   (One block in the Punnett square is *AA*, two blocks are *Aa*, and one block is *aa*.)

## Complete the Table

19. Using the gene symbols (tall and dwarf pea plants) in exercise 18, apply the six Mendelian ratios listed above to complete the following table of single-factor crosses by inspection. State results as phenotype and genotype ratios.

| Cross | Phenotype Ratio | Genotype Ratio |
|---|---|---|
| a. *Tt* × *tt* | | |
| b. *TT* × *Tt* | | |
| c. *tt* × *tt* | | |
| d. *Tt* × *Tt* | | |
| e. *tt* × *Tt* | | |
| f. *TT* × *tt* | | |
| g. *TT* × *TT* | | |
| h. *Tt* × *TT* | | |

*Problems*

When working genetics problems dealing with two gene pairs, you can visualize the independent assortment of gene pairs located on nonhomologous chromosomes into gametes by use of a fork-line device. Assume that in humans, pigmented eyes (*B*) are dominant (an eye color other than blue) over blue (*b*), and right-handedness (*R*) is dominant over left-handedness (*r*). To learn to solve a problem, cross the parents *BbRr* × *BbRr*. A sixteen-block Punnett square is required with gametes from each parent arrayed on two sides of the Punnett square (refer to Figures 9.8 and 9.9 in the text). The gametes receive genes through independent assortment using a fork-line method:

B b R r          B b R r
BR, Br, bR, br    ×    BR, Br, bR, br

20. Array the gametes above on two sides of the Punnett square; combine these haploid gametes to form diploid zygotes within the squares. In the blank spaces below, enter the probability ratios derived within the Punnett square for the phenotypes listed:

   a. _____ pigmented eyes, right-handed
   b. _____ pigmented eyes, left-handed
   c. _____ blue-eyed, right-handed
   d. _____ blue-eyed, left-handed

21. Albinos cannot form the pigments that normally produce skin, hair, and eye color, so albinos have white hair and pink eyes and skin (because the blood shows through). To be an albino, one must be homozygous recessive (*aa*) for the pair of genes that code for the key enzyme in pigment production. Suppose a woman of normal pigmentation (*A __* ) with an albino mother marries an albino man. State the possible kinds of pigmentation possible for this couple's children, and specify the ratio of each kind of child the couple is likely to have. Show the genotype(s) and state the phenotype(s).

_____

_____

22. In horses, black coat color is influenced by the dominant allele (*B*), and chestnut coat color is influenced by the recessive allele (*b*). Trotting gait is due to a dominant gene (*T*), pacing gait to the recessive allele (*t*). A homozygous black trotter is crossed to a chestnut pacer.

   a. What will be the appearance of the $F_1$ and $F_2$ generations? _____

_____

_____

   b. Which phenotype will be most common? _____

   c. Which genotype will be most common? _____

   d. Which of the potential offspring will be certain to breed true? _____

# 9 - II. DOMINANCE RELATIONS (p. 146)
## MULTIPLE EFFECTS OF SINGLE GENES (p. 147)
## INTERACTIONS BETWEEN GENE PAIRS (pp. 148 - 149)

*Selected Words:* transfusions, sickle-cell anemia, albinism

## Boldfaced, Page-Referenced Terms

(146) incomplete dominance _____

_____

(146) codominance _____

_____

(146) ABO blood typing _____

_____

(146) multiple allele system _____

_____

(147) pleiotropy _____

_____

(148) epistasis _____

_____

## Complete the Table

1. Complete the following table by supplying the type of inheritance illustrated by each example. Choose from the gene interaction inheritance of pleiotropy, multiple alleles, incomplete dominance, codominance, and epistasis.

| Type of Inheritance | Example |
|---|---|
| a. | Pink-flowered snapdragons produced from red- and white-flowered parents |
| b. | AB type blood from a gene system of three alleles, *A, B,* and *O* |
| c. | A gene with three or more alleles such as the *ABO* blood-typing alleles |
| d. | Black, brown, or yellow fur of Labrador retrievers and comb shape in poultry |
| e. | The multiple phenotypic effects of the gene causing human sickle-cell anemia |

## Problems

2. Genes that are not always dominant or recessive may blend to produce a phenotype of a different appearance. This is termed *incomplete dominance.* In four o'clock plants, red flower color is determined by gene R and white flower color by *R'*, while the heterozygous condition, *RR'*, is pink. Complete the table below by determining the phenotypes and genotypes of the offspring of the following crosses:

| Cross | Phenotype | Genotype |
|---|---|---|
| a. $RR \times R'R'$ = | | |
| b. $R'R' \times R'R'$ = | | |
| c. $RR \times RR'$ = | | |
| d. $RR \times RR$ = | | |

3. Sickle-cell anemia is a genetic disease in which children that are homozygous for a defective gene produce defective hemoglobin ($Hb^SHb^S$). The genotype of normal persons is $Hb^AHb^A$. If the level of blood oxygen drops below a certain level in a person with the $Hb^SHb^S$ genotype, the hemoglobin chains stiffen and cause the red blood cells to form sickle, or crescent shapes. These cells clog and rupture capillaries, which results in oxygen-deficient tissues where metabolic wastes collect. Several body functions are badly damaged. Severe anemia and other symptoms develop, and death nearly always occurs before adulthood. The sickle-cell gene is considered *pleiotropic*. Persons that are heterozygous ($Hb^AHb^S$) are said to possess *sickle-cell trait*. They are able to produce enough normal hemoglobin molecules to appear normal but their red blood cells will sickle if they encounter oxygen tension (such as at high altitudes).

   a. A man whose sister died of sickle-cell anemia married a woman whose blood is found to be normal. What advice would you give this couple about the inheritance of this disease as they plan their family?

   _____

   _____

   _____

   b. If a man and a woman, each with sickle-cell trait, planned to marry, what information could you provide them regarding the genotypes and phenotypes of their future children? _____

   _____

   _____

4. In one example of a *multiple allele system* with *codominance*, the three genes $I^Ai$, $I^Bi$, and $i$ produce proteins found on the surfaces of red blood cells that determine the four blood types in the ABO system: A, B, AB, and O. Genes $I^A$ and $I^B$ are both dominant over $i$ but not over each other. They are codominant. Recognize that blood types A and B may be heterozygous or homozygous ($I^AI^A$, $I^Ai$ or $I^BI^B$, $I^Bi$) while blood type O is homozygous ($ii$). Indicate the genotypes and phenotypes of the offspring and their probabilities from the parental combinations in exercises a–e.

   a. $I^Ai \times I^AI^B =$ _____

   b. $I^Bi \times I^Ai =$ _____

   c. $I^AI^A \times ii =$ _____

   d. $ii \times ii =$ _____

   e. $I^AI^B \times I^AI^B =$ _____

5. In one type of gene interaction, two alleles of a gene mask the expression of alleles of another gene, and some expected phenotypes never appear. *Epistasis* is the term given such interactions. Work the following problems on scratch paper to understand epistatic interactions.

   In sweet peas, genes $C$ and $P$ are necessary for colored flowers. In the absence of either (__ $pp$ or $cc$ __), or both ($ccpp$), the flowers are white. What will be the color of the offspring of the following crosses, and in what proportions will they appear?

a. $CcPp \times ccpp$ = _____

b. $CcPP \times Ccpp$ = _____

c. $Ccpp \times ccPp$ = _____

     In poultry, an epistatic interaction occurs in which two genes produce a phenotype that neither gene can produce alone. The two interacting genes ($R$ and $P$) produce comb shape in chickens. The possible genotypes and phenotypes are:

| Genotypes | Phenotypes |
|-----------|------------|
| $R\_P\_$ | walnut comb |
| $R\_pp$ | rose comb |
| $rrP\_$ | pea comb |
| $rrpp$ | single comb |

d. What are the genotype and phenotype ratios of the offspring of a heterozygous walnut-combed male

and a single-combed female? _____

_____

e. Cross a homozygous rose-combed rooster with a homozygous single-combed hen and list the geno-

type and phenotype ratios of their offspring. _____

_____

## 9 - III. LESS PREDICTABLE VARIATIONS IN TRAITS (pp. 150 - 151)
##        EXAMPLES OF ENVIRONMENTAL EFFECTS ON PHENOTYPE (p. 152)

*Selected Words:* camptodactyly, "bell-shaped" curve

*Boldfaced, Page-Referenced Terms*

(150) continuous variation _____

_____

## Choice

For questions 1 - 5, choose from the following primary contributing factors:

        a. environment     b. gene interaction     c. a number of genes affecting a trait

____ 1. Height of human beings

____ 2. Campodactyly, a human genetic disorder

____ 3. Development of different water buttercup leaf shapes under and above water level

____ 4. The range of eye colors in the human population

____ 5. Heat-sensitive version of one of the enzyme required for melanin Production in Himalayan rabbits

## Self-Quiz

____ 1. The best statement of Mendel's principle of independent assortment is that _____.
- a. one allele is always dominant to another
- b. hereditary units from the male and female parents are blended in the off-spring
- c. the two hereditary units that influence a certain trait separate during gamete formation
- d. each hereditary unit is inherited separately from other hereditary units

____ 2. One of two or more alternative forms of a gene for a single trait is a(n) _____.
- a. chiasma
- b. allele
- c. autosome
- d. locus

____ 3. In the $F_2$ generation of a monohybrid cross involving complete dominance, the expected phenotypic ratio is _____.
- a. 3:1
- b. 1:1:1:1
- c. 1:2:1
- d. 1:1

____ 4. In the $F_2$ generation of a cross between a red-flowered four o'clock (homozygous) and a white-flowered four o'clock, the expected phenotypic ratio of the offspring is _____.
- a. ¾ red, ¼ white
- b. 100 percent red
- c. ¼ red, ½ pink, ¼ white
- d. 100 percent pink

____ 5. In a testcross, $F_1$ hybrids are crossed to an individual known to be _____ for the trait.
- a. heterozygous
- b. homozygous dominant
- c. homozygous
- d. homozygous recessive

____ 6. A man with type A blood could be the father of _____.
- a. a child with type A blood
- b. a child with type B blood
- c. a child with type O blood
- d. a child with type AB blood
- e. all of the above

____ 7. A single gene that affects several seemingly unrelated aspects of an individual's phenotype is said to be _____.
- a. pleiotropic
- b. epistatic
- c. mosaic
- d. continuous

____ 8. Suppose two individuals, each heterozygous for the same characteristic, are crossed. The characteristic involves complete dominance. The expected genotypic ratio of their progeny is_____.
- a. 1:2:1
- b. 1:1
- c. 100 percent of one genotype
- d. 3:1

____ 9. If the two homozygous classes in the $F_1$ generation of the cross in exercise 8 are

allowed to mate, the observed genotypic ratio of the offspring will be _____.
a. 1:1
b. 1:2:1
c. 100 percent of one genotype
d. 3:1

___ 10. Applying the types of inheritance learned in this chapter in the text, the skin color trait in humans exhibits _____.
a. pleiotropy
b. epistasis
c. mosaicism
d. continuous variation

## Chapter Objectives/Review Questions

| Page | | Objectives/Questions |
|---|---|---|
| (140) | 1. | What was the prevailing method of explaining the inheritance of traits before Mendel's work with pea plants? |
| (141) | 2. | Garden pea plants are naturally _____-fertilizing, but Mendel took steps to _____-fertilize them for his experiments. |
| (141) | 3. | _____ are units of information about specific traits; they are passed from parents to offspring. |
| (141) | 4. | What is the general term applied to the location of a gene on a chromosome? |
| (141) | 5. | Define *allele*; how many alleles are present in the genotype *Tt*? *tt*? *TT*? |
| (141) | 6. | When two alleles of a pair are identical, it is a _____ condition; if the two alleles are different, this is a _____ condition. |
| (141) | 7. | Distinguish a dominant allele from a recessive allele. |
| (141) | 8. | _____ refers to the genes present in an individual; _____ refers to an individual's observable traits. |
| (142) | 9. | Offspring of _____ crosses are heterozygous for the one trait being studied. |
| (142 - 143) | 10. | Explain why probability is useful to genetics. |
| (143) | 11. | Be able to use the Punnett-square method of solving genetics problems. |
| (143) | 12. | Define *testcross*, and cite an example. |
| (145) | 13. | Mendel's theory of _____ _____ states that gene pairs on homologous chromosomes tend to be sorted into one gamete or another independently of how gene pairs on other chromosomes are sorted out. |
| (145) | 14. | Mendel's theory of _____ states that during meiosis, the two genes of each pair separate from each other and end up in different gametes. |
| (146) | 15. | Distinguish between complete dominance, incomplete dominance, and codominance. |
| (146) | 16. | Define *multiple allele system* and cite an example. |
| (147) | 17. | Explain why sickle-cell anemia is a good example of pleiotropy. |
| (148) | 18. | Gene interaction involving two alleles of a gene that mask alleles of another gene is called _____. |
| (152) | 19. | Himalayan rabbits and water buttercups are good examples of environmental effects on _____ _____. |
| (152) | 20. | List possible explanations for less predictable trait variations that are observed. |

## Integrating and Applying Key Concepts

Solve the following genetics problem:

In garden peas, one pair of alleles controls the height of the plant and a second pair of alleles controls flower color. The allele for tall (*D*) is dominant to the allele for dwarf (*d*), and the allele for purple (*P*) is dominant to the allele for white (*p*). A tall plant with purple flowers crossed with a tall plant with white flowers produces ⅜ tall purple, ⅜ tall white, ⅛ dwarf purple, and ⅛ dwarf white. What are the genotypes of the parents?

# 10

# CHROMOSOMES AND HUMAN GENETICS

## Interactive Exercises

### 10 - I. THE CHROMOSOMAL BASIS OF INHERITANCE—AN OVERVIEW (pp. 156 - 158)
### *Focus on Science:* PREPARING A KARYOTYPE DIAGRAM (pp. 159 - 160)

*Selected Words:* in vitro, <u>Colchicum</u> <u>autumnale</u>, *centrifugation*

### *Boldfaced, Page-Referenced Terms*

(158) genes _____

_____

(158) homologous chromosomes _____

_____

(158) alleles _____

_____

(158) crossing over _____

(158) X chromosome _____

(158) Y chromosome _____

(158) sex chromosomes _____

(158) autosomes _____

(158) karyotype _____

## Fill-in-the-Blanks

Heritable traits arise from units of molecular information (located on chromosomes) that are known as (1) _____. A pair of chromosomes that have the same length, shape, gene sequence, and interact during meiosis, are known as (2) _____ chromosomes. (3) _____ are slightly different forms of the same gene that arose through mutation. An event known as (4) _____ _____ occurring early in the meiotic process functions to exchange gene segments between the members of a pair and provide new combinations of genes. Human X and Y chromosomes that determine gender are examples of (5) _____ chromosomes. All other chromosomes are designated (6) _____. The (7) _____ of an individual (or species) is the number of metaphase chromosomes and their characteristics. To culture cells for the preparation of an organized chromosome diagram summarizing this information, a chemical compound, colchicine, is placed in a cell medium to arrest all nuclear divisions at the (8) _____ stage of mitosis.

## 10 - II. SEX DETERMINATION IN HUMANS (pp. 160 - 161)
## EARLY QUESTIONS ABOUT GENE LOCATIONS (pp. 162 - 163)

*Selected Words:* *nonsexual* traits, *Y chromosome,* <u>Drosophila</u> <u>melanogaster</u>, <u>Zea</u> <u>mays</u>, *wild-type, reciprocal* crosses

## Boldfaced, Page-Referenced Terms

(162) X-linked genes _____

(162) Y-linked genes _____

(162) linkage groups _____

## Matching

Choose the most appropriate answer for each term.

1. ___ crossing over
2. ___ X chromosome
3. ___ Y chromosome
4. ___ sex chromosomes
5. ___ autosomes
6. ___ karyotype
7. ___ X-linked genes
8. ___ Y-linked genes
9. ___ linkage groups
10. ___ SRY gene

A. Examples are the X and Y chromosomes of humans.
B. The number of metaphase chromosomes and their defining characteristics
C. Found only on the Y chromosome
D. Male-determining gene found on the Y chromosome; expression leads to testes formation
E. Any chromosome not concerned with sex determination; identical in males and females
F. Only one is found in a male karyotype; meiotic synapse is with an X chromosome.
G. The block of genes located on each type of chromosome; these genes tend to travel together in inheritance
H. Found only on the X chromosome
I. Interrupts gene linkage; brings about the recombination of genes on each homologue from mother and the matching homologue from father
J. Human females possess two; genes are mostly concerned with nonsexual traits.

## Complete the Table

11. Complete the Punnett square table at the right, which brings Y-bearing and X-bearing sperm together randomly in fertilization.

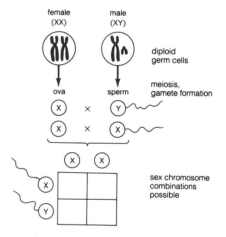

## Dichotomous Choice

Answer the following questions related to the Punnett Square table just completed.

12. Male humans transmit their Y chromosome only to their (sons/daughters).
13. Male humans receive their X chromosome only from their (mothers/fathers).
14. Human mothers and fathers each provide an X chromosome for their (sons/daughters).
15. If genes A and B are twice as far apart on a chromosome as genes C and D, we would expect that crossing over occurs (twice/half) as often between genes A and B as between genes C and D.
16. Morgan's experiments with fruit flies indicated a correlation between sex and eye color; eye color must be carried on the (X/Y) chromosome.

## Problems

Early investigators realized that the frequency of crossing over (data from actual genetic crosses) could be used as a tool to map genes on chromosomes. Genetic maps are graphic representations of the relative positions and distances between genes on each chromosome (linkage group). Geneticists arbitrarily equate 1 percent of crossover (recombination) with 1 genetic map unit. In the following questions, a line is used to represent a chromosome with its linked genes.

17. Which of the following represents a chromosome that has undergone crossover and recombination? It is assumed that the organism involved is heterozygous with the genotype: $\begin{array}{c}A\\B\end{array}\Big| \begin{array}{c}a\\b\end{array}\Big|$. _____

   **a** $\begin{array}{c}A\\B\end{array}\Big|$     **b** $\begin{array}{c}a\\b\end{array}\Big|$     **c** $\begin{array}{c}B\\A\end{array}\Big|$     **d** $\begin{array}{c}a\\B\end{array}\Big|$

18. After breeding experiments and a study of crossovers, the following map distances were recorded: gene C to gene B = 35 genetic map units; B to A = 10 map units, and A to C = 45 map units. Which of the following maps is correct?

    A B  C           B  A  C          B  C A         B A  C
  **a** ⌊_⌊__⌋      **b** ⌊__⌊__⌋      **c** ⌊___⌊_⌋      **d** ⌊_⌊__⌋

19. In *Drosophila* white eyes are determined by a recessive X-linked gene and the wild-type or normal brick-red eyes are due to its dominant allele. Use symbols of the following types: $X^wY$ = a white-eyed male; $X^WX^W$ = a homozygous normal red female.
    a. What offspring can be expected from a cross of a white-eyed male and a homozygous normal female?
    b. In addition, show the genotypes and phenotypes of the $F_2$ offspring.

## 10 - III. HUMAN GENETIC ANALYSIS (pp. 164 - 165)
##           PATTERNS OF INHERITANCE (pp. 166 - 167)

*Selected Words: polydactyly, "diseases," galactosemia, Huntington disorder, achondroplasia, hemophilia A, Duchenne muscular dystrophy, faulty enamel trait*

### Boldfaced, Page-Referenced Terms

(164) pedigrees _____

_____

(165) genetic abnormality _____

_____

(165) genetic disorder _____

_____

## Choice

For questions 1 - 18, choose from the following patterns of inheritance; some items may require more than one letter.

a. autosomal recessive    b. autosomal dominant    c. X-linked recessive    d. X-linked dominant

1. ___ The trait is expressed in heterozygous females.

2. ___ Heterozygotes can remain undetected.

3. ___ The trait appears in each generation.

4. ___ The recessive phenotype shows up far more often in males than in females.

5. ___ Both parents may be heterozygous normal.

6. ___ If one parent is heterozygous and the other homozygous recessive, there is a 50 percent chance any child of theirs will be heterozygous.

7. ___ The allele is usually expressed, even in heterozygotes.

8. ___ The trait is expressed in heterozygous females.

9. ___ Heterozygous normal parents can expect that one-fourth of their children will be affected by the disorder.

10. ___ A son cannot inherit the recessive allele from his father, but his daughter can.

11. ___ Females can mask this gene, males cannot.

12. ___ The trait is expressed in heterozygotes of either sex.

13. ___ Heterozygous women transmit the allele to half their offspring, regardless of sex.

14. ___ Individuals displaying this type of disorder will always be homozygous for the trait.

15. ___ The trait is expressed in both the homozygote and the heterozygote.

16. ___ Heterozygous females will transmit the recessive gene to half their sons and half their daughters.

17. ___ If both parents are heterozygous, there is a 50 percent chance that each child will be heterozygous.

18. ___ A son cannot inherit the allele responsible for the trait from an affected father, but all his daughters will.

## Problems

19. The autosomal allele that causes albinism (*a*) is recessive to the allele for normal pigmentation (*A*). A normally pigmented woman whose father is an albino marries an albino man whose parents are normal. They have three children, two normal and one albino. Give the genotypes for each person listed.

_____

_____

_____

20. Huntington disorder is a rare form of autosomal dominant inheritance, *H*; the normal gene is *h*. The disease causes progressive degeneration of the nervous system with onset exhibited near middle age. An apparently normal man in his early twenties learns that his father has recently been diagnosed as having Huntington disorder. What are the chances that the son will develop this disorder?

_____

_____

21. A color-blind man and a woman with normal vision whose father was color-blind have a son. Color blindness, in this case, is caused by an X-linked recessive gene. If only the male offspring are considered, what is the probability that their son is color-blind?

_____

_____

_____

22. Hemophilia A is caused by an X-linked recessive gene. A woman who is seemingly normal but whose father was a hemophiliac marries a normal man. What proportion of their sons will have hemophilia? What proportion of their daughters will have hemophilia? What proportion of their daughters will be carriers?

_____

_____

_____

23. Refer to Figure 10.8, page 164, in the text. The following pedigree shows the pattern of inheritance of color blindness in a family (persons with the trait are indicated by black circles). What is the chance that the third-generation female ( indicated by the arrow) will have a color-blind son if she marries a normal male? If she marries a color-blind male?

_____

_____

## Complete the Table

24. Complete the table below by indicating whether the genetic disorder listed is due to inheritance that is autosomal recessive, autosomal dominant, X-linked recessive, or X-linked dominant. Refer to Table 10.1, page 165, in the text.

| Genetic Disorder | Inheritance Pattern |
|---|---|
| a. Galactosemia | |
| b. Achondroplasia | |
| c. Hemophilia A | |
| d. Huntington's disorder | |
| e. Duchenne muscular dystrophy | |
| f. Faulty enamel trait | |
| g. Red-green color blindness | |
| h. Progeria | |
| i. Phenylketonuria | |

## 10 - IV. CHANGES IN CHROMOSOME NUMBER (pp. 168 - 169)
## CHANGES IN CHROMOSOME STRUCTURE (p. 170)

*Selected Words:* "tetraploid," *Down syndrome,* "syndrome," *Turner syndrome, Klinefelter syndrome, double-blind studies, cri-du-chat, fragile X syndrome*

### Boldfaced, Page-Referenced Terms

(168) aneuploidy _____

_____

(168) polyploidy _____

_____

(168) nondisjunction _____

_____

(170) deletion _____

_____

(170) inversion _____

_____

(170) translocation _____

_____

(170) duplications _____

_____

### Short Answer

1.  If a nondisjunction occurs at anaphase I of the first meiotic division, what will be the proportion of abnormal gametes (for the chromosomes involved in the nondisjunction)? (p. 168) _____

_____

_____

2.  If a nondisjunction occurs at anaphase II of the second meiotic division, what will be the proportion of abnormal gametes (for the chromosomes involved in the nondisjunction)? (p. 168) _____

_____

_____

3.  Describe the effects of polyploidy in humans. (p. 168) _____

_____

_____

## Complete the Table

4. Complete the table below to summarize two major types of chromosome number change in organisms.

| Category of Change | Description |
|---|---|
| a. Trisomy | |
| b. Monosomy | |

## Choice

For questions 5 - 14, choose from the following:

    a. Down syndrome    b. Turner syndrome    c. Klinefelter syndrome    d. XYY condition

5. ___ XXY male

6. ___ Ovaries nonfunctional and secondary sexual traits fail to develop at puberty

7. ___ Testes smaller than normal, sparse body hair, and some breast enlargement

8. ___ Could be caused only by a nondisjunction in males

9. ___ Older children smaller than normal with distinctive facial features; small skin fold over the inner corner of the eyelid

10. ___ X0 female; often abort early; distorted female phenotype

11. ___ Males that tend to be taller than average; some mildly retarded but most are phenotypically normal

12. ___ Injections of testosterone reverse feminized traits but not the mental retardation.

13. ___ Trisomy 21; skeleton develops more slowly than normal with slack muscles

14. ___ At one time these males were thought to be genetically predisposed to become criminals.

## Label and Match

On rare occasions, chromosome structure becomes abnormally rearranged. Such changes may have profound effects on the phenotype of an organism. Label the following diagrams of abnormal chromosome structure as a deletion, a duplication, an inversion, or a translocation. Complete the exercise by matching and entering the letter of the proper description in the parentheses following each label.

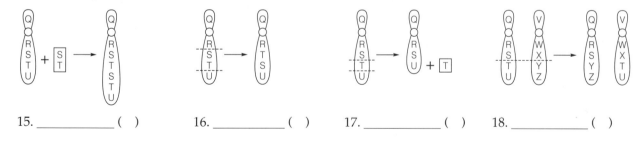

15. _____ ( )    16. _____ ( )    17. _____ ( )    18. _____ ( )

A. The loss of a chromosome segment; an example is cri-du-chat disorder
B. A gene sequence in excess of its normal amount in a chromosome; an example is the fragile X syndrome
C. A chromosome segment that separated from the chromosome and then was inserted at the same place, but in reverse; this alters the position and order of the chromosome's genes
D. The transfer of part of one chromosome to a nonhomologous chromosome; an example is when chromosome 14 ends up with a segment of chromosome 8

*Selected Words:* cleft lip, preimplantation diagnosis

## Boldfaced, Page-Referenced Terms

(171) amniocentesis _____

_____

(171) chorionic villi sampling, or CVS _____

_____

(171) abortion _____

_____

(171) in-vitro fertilization _____

_____

## Complete the Table

1.  Complete the table below, which summarizes methods of dealing with the problems of human genetics. Choose from phenotypic treatments, genetic screening, genetic counseling, and prenatal diagnosis.

| Method | Description |
|---|---|
| a. | Detects genetic disorders before birth; may use karyotypes, biochemical tests, amniocentesis, CVS, in-vitro fertilization, and possibly abortion; an example is a pregnancy at risk in a mother forty-five years old |
| b. | Parents at risk requesting emotional support, advice, and risk predictions from clinical psychologists, geneticists, and social workers |
| c. | Suppressing or minimizing symptoms of genetic disorders by surgical intervention, controlling diet or environment, or chemically modifying genes; PKU and cleft lip are examples |
| d. | Large-scale programs to detect affected persons or carriers in a population; early detection may allow introduction of preventive measures before symptoms develop; PKU screening for newborns is an example |

# Self-Quiz

____ 1.  All the genes located on a given chromosome compose a _____.
  a. karyotype
  b. bridging cross
  c. wild-type allele
  d. linkage group

____ 2.  Chromosomes other than those involved in sex determination are known as _____.
  a. nucleosomes
  b. heterosomes

  c. alleles
  d. autosomes

____ 3.  The farther apart two genes are on a chromosome, _____.
  a. the less likely that crossing over and recombination will occur between them
  b. the greater will be the frequency of crossing over and recombination between them

c. the more likely they are to be in two different linkage groups
d. the more likely they are to be segregated into different gametes when meiosis occurs

___ 4. Karyotype analysis is _____ .
a. a means of detecting and reducing mutagenic agents
b. is a surgical technique that separates chromosomes that have failed to segregate properly during meiosis II
c. used in prenatal diagnosis to detect chromosomal mutations and metabolic disorders in embryos
d. a process that substitutes defective alleles with normal ones

___ 5. Which of the following did Morgan and his research group not do?
a. They isolated and kept under culture fruit flies with the sex-linked recessive white-eyed trait.
b. They developed the technique of amniocentesis.
c. They discovered X-linked genes.
d. Their work reinforced the concept that each gene is located on a specific chromosome.

___ 6. Red-green color blindness is a sex-linked recessive trait in humans. A color-blind woman and a man with normal vision have a son. What are the chances that the son is color-blind? If the parents ever have a daughter, what is the chance for each birth that the daughter will be color-blind? (Consider only the female offspring).
a. 100 percent, 0 percent
b. 50 percent, 0 percent
c. 100 percent, 100 percent
d. 50 percent, 100 percent
e. none of the above

___ 7. Suppose that a hemophilic male (X-linked recessive allele) and a female carrier for the hemophilic trait have a nonhemophilic daughter with Turner syndrome. Nondisjunction could have occurred in _____ .
a. both parents
b. neither parent
c. the father only
d. the mother only

___ 8. Nondisjunction involving the X chromosome occurs during oogenesis and produces two kinds of eggs, XX and O (no X chromosome). If normal Y sperm fertilize the two types, which genotypes are possible?
a. XX and XY
b. XXY and YO
c. XYY and XO
d. XYY and YO

___ 9. Of all phenotypically normal males in prisons, the type once thought to be genetically predisposed to becoming criminals was the group with _____ .
a. XXY disorder
b. XYY disorder
c. Turner syndrome
d. Down syndrome

___ 10. Amniocentesis is _____ .
a. a surgical means of repairing deformities
b. a form of chemotherapy that modifies or inhibits gene expression or the function of gene products
c. used in prenatal diagnosis to detect chromosomal mutations and metabolic disorders in embryos
d. a form of gene-replacement therapy

## Chapter Objectives/Review Questions

| *Page* | | *Objectives/Questions* |
|---|---|---|
| (158) | 1. | The units of information about heritable traits are known as _____ . |
| (158) | 2. | Diploid (2*n*) cells have pairs of _____ chromosomes. |
| (158) | 3. | _____ are different molecular forms of the same gene that arise through mutation. |
| (158) | 4. | State the circumstances required for crossing over and describe the results. |
| (158) | 5. | Name and describe the sex chromosomes in human males and females. |

(158 - 159) 6. Define *karyotype;* briefly describe its preparation and value.

(158) 7. Distinguish between chromosomes and autosomes.

(160) 8. A newly identified region of the Y chromosome called _____ appears to be the master gene for sex determination.

(160) 9. Explain meiotic segregation of sex chromosomes to gametes and the subsequent random fertilization that determines sex in many organisms.

(162) 10. In whose laboratory was sex linkage in fruit flies discovered? When?

(162) 11. A specific chromosome with all its genes is known as a _____ group.

(163) 12. State the relationship between crossover frequency and the location of genes on a chromosome.

(164) 13. A _____ chart or diagram is used to study genetic connections between individuals.

(165) 14. A genetic _____ is a rare, uncommon version of a trait whereas a genetic _____ causes mild to severe medical problems.

(166 - 167) 15. As viewed on a pedigree chart, describe the characteristics of autosomal recessive inheritance, autosomal dominant inheritance, as well as recessive and dominant X-linked inheritance; cite one example of each.

(168) 16. When gametes or cells of an affected individual end up with one extra or one less than the parental number of chromosomes, it is known as _____; relate this to monosomy and trisomy.

(168) 17. Having three or more complete sets of chromosomes is called _____.

(168) 18. _____ is the failure of the chromosomes to separate at either meiosis I or meiosis II.

(168) 19. Trisomy 21 is known as _____ syndrome; Turner syndrome has the chromosome constitution, _____; XXY chromosome constitution is _____ syndrome; taller than average males with sometimes slightly depressed IQs have the _____ condition.

(169) 20. Explain what is meant by double-blind studies.

(170) 21. A(n) _____ is a loss of a chromosome segment; a(n) _____ is a gene segment that separated from a chromosome and then was inserted at the same place, but in reverse; a(n) _____ is a repeat of several gene sequences on the same chromosome; a(n) _____ is the transfer of part of one chromosome to a nonhomologous chromosome.

(171) 22. Define phenotypic treatment, and describe one example.

(171) 23. Explain the procedures used in two types of prenatal diagnosis, amniocentesis and chorionic villi sampling; compare the risks.

(171) 24. List some benefits of genetic screening and genetic counseling to society.

(171) 25. Discuss some of the ethical considerations that might be associated with a decision of induced abortion .

(171) 26. _____ fertilization is the fertilizing of eggs in a petri dish.

---

## Integrating and Applying Key Concepts

1. The parents of a young boy bring him to their doctor. They explain that the boy does not seem to be going through the same vocal developmental stages as his older brother. The doctor orders a common cytogenetics test to be done, and it reveals that the young boy's cells contain two X chromosomes and one Y chromosome. Describe the test that the doctor ordered and explain how and when such a genetic result, XXY, most logically occurred.

2. Solve the following genetics problem. Show rationale, genotypes, and phenotypes. A husband sues his wife for divorce, arguing that she has been unfaithful. His wife gave birth to a girl with a fissure in the iris of her eye, an X-linked recessive trait. Both parents have normal eye structure. Can the genetic facts be used to argue for the husband's suit? Explain your answer.

# 11

## DNA STRUCTURE AND FUNCTION

**DISCOVERY OF DNA FUNCTION**
    Early Clues
    Confirmation of DNA Function

**DNA STRUCTURE**
    Components of DNA
    Patterns of Base Pairing

**A CLOSER LOOK AT DNA**
    DNA Replication and Repair
    Organization of DNA in Chromosomes

*Focus on Health:* **WHEN DNA CAN'T BE FIXED**

---

## Interactive Exercises

### 11 - I. DISCOVERY OF DNA FUNCTION (pp. 174 - 177)

*Selected Words:* <u>Streptococcus pneumoniae</u>, <u>Escherichia coli</u>

*Boldfaced, Page-Referenced Terms*

(174) deoxyribonucleic acid (DNA) _____

_____

(176) bacteriophages _____

_____

*Complete the Table*

1. Complete the table below that traces the discovery of DNA function.

| Investigators | Year(s) | Contribution |
|---|---|---|
| a. Miescher | 1868 | |
| b. | 1928 | Discovered the transforming principle in *Streptococcus pneumoniae;* live, harmless R cells were mixed with dead S cells, R cells became S cells |
| c. Avery (also MacLeod and McCarty) | 1944 | |
| d. Hershey and Chase | 1952 | |

# 11 - II. DNA STRUCTURE (pp. 178 - 179)

## Boldfaced, Page-Referenced Terms

(178) nucleotide _____

_____

(178) adenine (A) _____

_____

(178) guanine (G) _____

_____

(178) thymine (T) _____

_____

(178) cytosine (C) _____

_____

(178) x-ray diffraction images _____

_____

## Short Answer

1. List the three parts of a nucleotide._____

_____

## Labeling

Four nucleotides are illustrated below. In the blank, label each nitrogen-containing base correctly as guanine, thymine, cytosine, or adenine. In the parentheses following each blank, indicate whether that nucleotide base is a purine (pu) or a pyrimidine (py). (See p. 115 for definitions.)

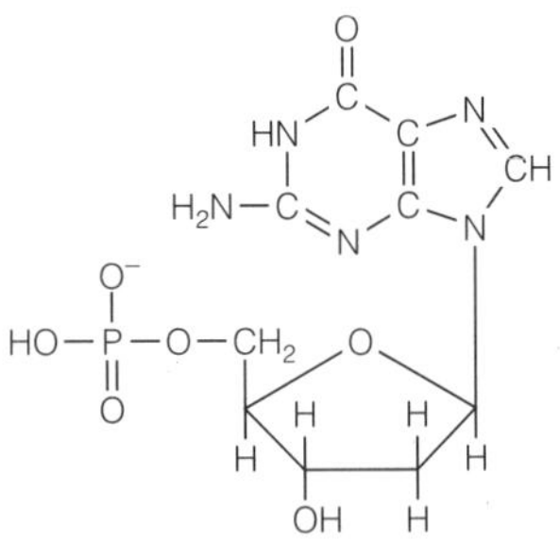

2._____ 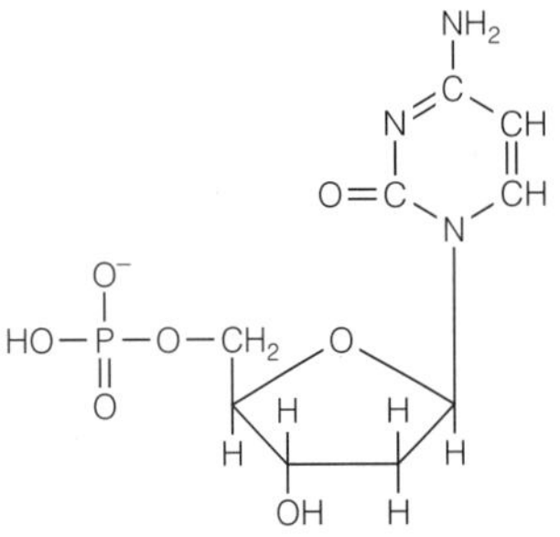 ( )     3._____ 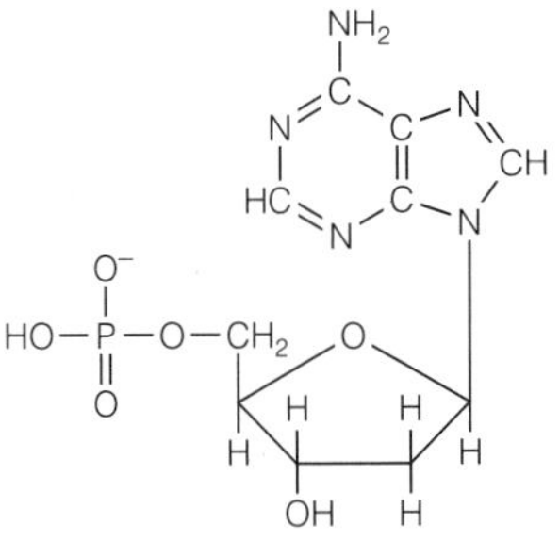 ( )     4._____ 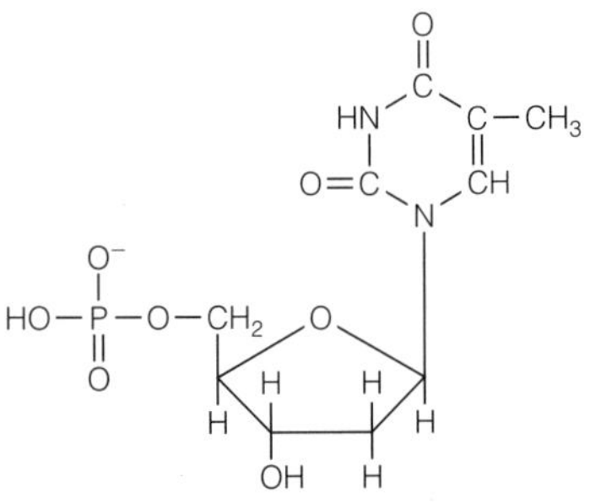 ( )     5._____ ( )

## Label and Match

Identify each indicated part of the DNA illustration below. Choose from these answers: phosphate group, purine, pyrimidine, nucleotide, and deoxyribose. Complete the exercise by matching and entering the letter of the proper structure description in the parentheses following each label.

The following DNA memory devices may be helpful: Use pyrCUT to remember that the single-ring pyrimidines are cytosine, uracil, and thymine; use purAG to remember that the double-ring purines are adenine and guanine; pyrimidine is a long name for a narrow molecule; purine is a short name for a wide molecule; to recall the number of hydrogen bonds between DNA bases, remember that AT = 2 and CG = 3.

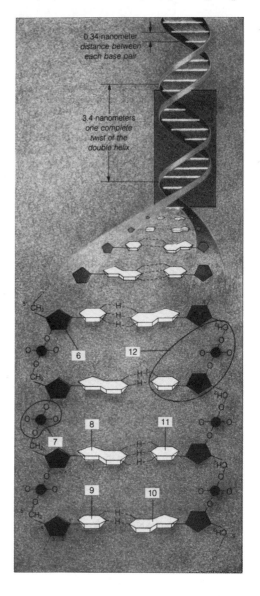

6. _____ ( )

7. _____   _____ ( )

8. _____ ( )

9. _____ ( )

10. _____ ( )

11. _____ ( )

12. _____ ( )

A. The pyrimidine is thymine because it has two hydrogen bonds.
B. A five-carbon sugar joined to two phosphate groups in the upright portion of the DNA ladder
C. The purine is guanine because it has three hydrogen bonds.
D. The pyrimidine is cytosine because it has three hydrogen bonds.
E. The purine is adenine because it has two hydrogen bonds.
F. Composed of three smaller molecules: a phosphate group, five-carbon deoxyribose sugar, and a nitrogenous base (in this case, a pyrimidine)
G. A chemical group that joins two sugars in the upright portion of the DNA ladder

## True/False

If the statement is true, write a T in the blank. If the statement is false, correct it by changing the underlined word(s) and writing the correct word(s) in the blank.

_____ 13. DNA is composed of <u>four</u> different types of nucleotides.

_____ 14. In the DNA of every species, the amount of adenine present always equals the amount of <u>thymine</u>, and the amount of cytosine always equals the amount of <u>guanine</u> (A = T and C = G).

_____ 15. In a nucleotide, the phosphate group is attached to the <u>nitrogen-containing base</u>, which is attached to the five-carbon sugar.

_____ 16. Watson and Crick built their model of DNA in the early <u>1950s</u>.

_____ 17. Guanine pairs with <u>cytosine</u> and adenine pairs with <u>thymine</u> by forming hydrogen bonds between them.

## Fill-in-the-Blanks

Base (18) _____ between the two nucleotide strands in DNA is (19) _____ for all species (A-T; G-C). The base (20) _____ (determining which base follows the next in a nucleotide strand) is (21) _____ from species to species.

## Short Answer

22. Explain why understanding the structure of DNA helps scientists understand how living organisms can have so much in common at the molecular level and yet be so diverse at the whole organism level (p. 179).

_____

_____

## 11 - III. A CLOSER LOOK AT DNA (pp. 180–181)
### *Focus on Health:* WHEN DNA CAN'T BE FIXED (p. 182)

*Selected Words:* *semiconservative* replication, "supercoiled" arrangement, structural "scaffold," chromosomal "domains," "decondensed" DNA

## Boldfaced, Page-Referenced Terms

(180) DNA replication _____

_____

(180) DNA polymerases _____

_____

(180) DNA ligases _____

_____

(180) DNA repair _____

_____

(180) histones _____

_____

(181) nucleosome _____

_____

## Labeling

1. The term *semiconservative replication* refers to the fact that each new DNA molecule resulting from the replication process is "half-old, half-new." In the illustration below, complete the replication required in the middle of the molecule by adding the required letters representing the missing nucleotide bases. Recall that ATP energy and the appropriate enzymes are actually required in order to complete this process.

T-_____    _____-A

G-_____    _____-C

A-_____    _____-T

C-_____    _____-G

C-_____    _____-G

C-_____    _____-G

old new    new old

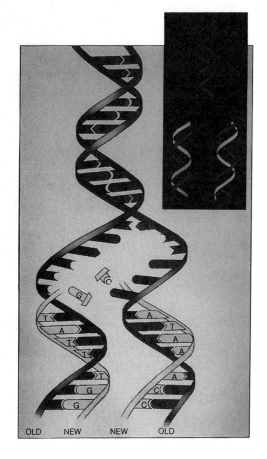

## True/False

If the statement is true, write a T in the blank. If the statement is false, correct it by changing the underlined word(s) and writing the correct word(s) in the blank.

_____ 2. The hydrogen bonding of adenine to <u>guanine</u> is an example of complementary base pairing.

_____ 3. The replication of DNA is considered a <u>semiconservative</u> process because <u>the same four nucleotides are used again and again during replication</u>.

_____ 4. Each parent strand <u>remains intact</u> during replication, and a new companion strand is assembled on each of those parent strands.

_____ 5. Some of the enzymes associated with DNA assembly repair <u>errors</u> during the replication process.

## Fill-in-the-Blanks

Each chromosome has one (6) _____ molecule coursing through it. Eukaryotic DNA is complexed tightly with many (7) _____. Some (8) _____ proteins act as spools to wind up small pieces of (9) _____. A (10) _____ is a histone-DNA spool. The way the chromosome is packed is known to influence the activity of different (11) _____.

## Labeling

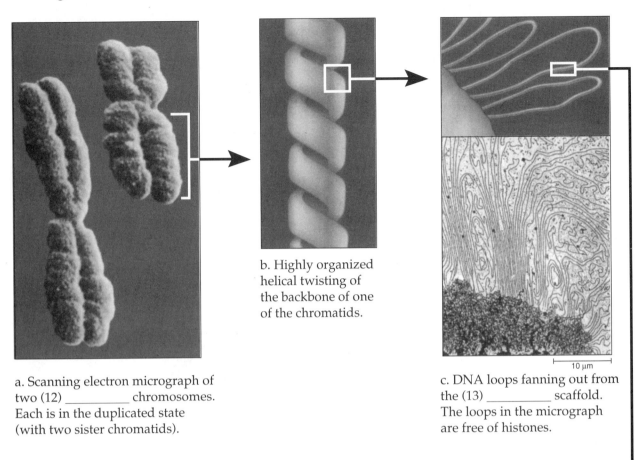

a. Scanning electron micrograph of two (12) _____ chromosomes. Each is in the duplicated state (with two sister chromatids).

b. Highly organized helical twisting of the backbone of one of the chromatids.

c. DNA loops fanning out from the (13) _____ scaffold. The loops in the micrograph are free of histones.

10 µm

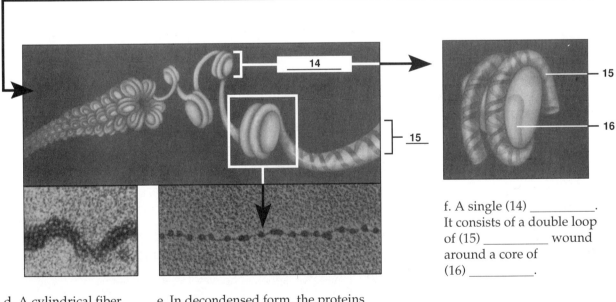

d. A cylindrical fiber (solenoid), 30 nm in diameter. It results from DNA-protein interactions that lead to repeated coiling.

e. In decondensed form, the proteins look like beads on a string (the "string" being DNA). Each "bead" is a (14) _____.

f. A single (14) _____. It consists of a double loop of (15) _____ wound around a core of (16) _____.

# Self-Quiz

___ 1. Each DNA strand has a backbone that consists of alternating _____.
a. purines and pyrimidines
b. nitrogen-containing bases
c. hydrogen bonds
d. sugar and phosphate molecules

___ 2. In DNA, complementary base pairing occurs between _____.
a. cytosine and uracil
b. adenine and guanine
c. adenine and uracil
d. adenine and thymine

___ 3. Adenine and guanine are _____.
a. double-ringed purines
b. single-ringed purines
c. double-ringed pyrimidines
d. single-ringed pyrimidines

___ 4. Franklin used the technique known as _____ to determine many of the physical characteristics of DNA.
a. transformation
b. transmission electron microscopy
c. density-gradient centrifugation
d. x-ray diffraction

___ 5. The significance of Griffith's experiment that used two strains of pneumonia-causing bacteria is that _____.
a. the conserving nature of DNA replication was finally demonstrated
b. it demonstrated that harmless cells had become permanently transformed through a change in the bacterial hereditary system
c. it established that pure DNA extracted from disease-causing bacteria transformed harmless strains into "pathogenic strains"
d. it demonstrated that radioactively labeled bacteriophages transfer their DNA but not their protein coats to their host bacteria

___ 6. The significance of the experiments in which $^{32}P$ and $^{35}S$ were used is that _____.
a. the semiconservative nature of DNA replication was finally demonstrated

b. it demonstrated that harmless cells had become permanently transformed through a change in the bacterial hereditary system
c. it established that pure DNA extracted from disease-causing bacteria transformed harmless strains into "killer strains"
d. it demonstrated that radioactively labeled bacteriophages transfer their DNA but not their protein coats to their host bacteria

___ 7. Franklin's research contribution was essential in _____.
a. establishing the double-stranded nature of DNA
b. establishing the principle of base pairing
c. establishing most of the principal structural features of DNA
d. all of the above

___ 8. When Griffith injected mice with a mixture of dead pathogenic cells—encapsulated S cells and living, unencapsulated R cells of pneumonia bacteria—he discovered that _____.
a. the previously harmless strain had permanently inherited the capacity to build protective capsules
b. the dead mice teemed with living pathogenic (R) cells
c. the killer strain R was encased in a protective capsule
d. all of the above

___ 9. A single strand of DNA with the base-pairing sequence C-G-A-T-T-G is compatible only with the sequence _____.
a. C-G-A-T-T-G
b. G-C-T-A-A-G
c. T-A-G-C-C-T
d. G-C-T-A-A-C

___ 10. The nucleosome is a _____.
a. subunit of a nucleolus
b. coiled bead of histone-DNA
c. DNA packing arrangement within a chromosome
d. term synonymous with gene
e. both (b) and (c)

# Chapter Objectives/Review Questions

*Page*        *Objectives/Questions*

(174)       1.  Before 1952, _____ molecules and _____ molecules were suspected of housing the genetic code.

(174 - 176) 2.  Summarize the research carried out by Miescher, Griffith, Avery and colleagues, and Hershey and Chase; state the specific advances made by each in the understanding of genetics.

(176)       3.  Viruses called _____ were used in early research efforts to discover the genetic material.

(177)       4.  Summarize the specific research that demonstrated that DNA, not protein, governed inheritance.

( * )       5.  DNA is composed of double-ring nucleotides known as _____ and single-ring nucleotides known as _____; the two purines are _____ and _____, while the two pyrimidines are _____ and _____.

(179)       6.  Draw the basic shape of a deoxyribose molecule, and show how a phosphate group is joined to it when forming a nucleotide.

(179)       7.  Show how each nucleotide base would be joined to the sugar-phosphate combination drawn in objective 6.

(179)       8.  List the pieces of information about DNA structure that Rosalind Franklin discovered through her x-ray diffraction research.

(179)       9.  The two scientists who assembled the clues to DNA structure and produced the first model were _____ and _____.

(178 - 179) 10. Explain what is meant by the pairing of nitrogen-containing bases (base pairing), and explain the mechanism that causes bases of one DNA strand to join with bases of the other strand.

(179)       11. Assume that the two parent strands of DNA have been separated and that the base sequence on one parent strand is A-T-T-C-G-C; the base sequence that will complement that parent strand is _____.

(180)       12. Describe how double-stranded DNA replicates from stockpiles of nucleotides.

(180)       13. Explain what is meant by "each parent strand is conserved in each new DNA molecule."

(180)       14. During DNA replication, enzymes called DNA _____ assemble new DNA strands.

(181)       15. The basic histone-DNA packing unit of the chromosome is the _____.

(181)       16. List possible reasons for the highly organized packing of nucleoprotein into chromosomes.

# Integrating and Applying Key Concepts

Review the stages of mitosis and meiosis, as well as the process of fertilization. Include what has now been learned about DNA replication and the relationship of DNA to a chromosome. As you cover the stages, be sure each cell receives the proper number of DNA threads.

*Page 114 of this study guide.

# 12

# FROM DNA TO PROTEINS

## Interactive Exercises

## 12 - I. TRANSCRIPTION OF DNA INTO RNA (pp. 184 - 187)

*Selected Words:* genetic "code words," 5' cap of pre-mRNA, "poly-A tail," to "pace" access

*Boldfaced, Page-Referenced Terms*

(185) base sequence _____

_____

(185) transcription _____

_____

(185) translation _____

_____

(185) ribonucleic acid (RNA) _____

_____

(186) messenger RNA, mRNA _____

_____

(186) ribosomal RNA, rRNA _____

_____

(186) transfer RNA, tRNA _____

_____

(186) uracil _____

_____

(186) RNA polymerases _____

_____

(186) promoter _____

_____

(187) introns _____

_____

(187) exons _____

_____

## Fill-in-the-Blanks

Which base follows the next in a strand of DNA is referred to as the base (1) _____. A region of DNA that calls for the assembly of specific amino acids into a polypeptide chain is a (2) _____. The two steps from genes to proteins are called (3) _____ and (4) _____. In (5) _____, single-stranded molecules of RNA are assembled on DNA templates in the nucleus. In (6) _____, the RNA molecules are shipped from the nucleus into the cytoplasm, where they are used as templates for assembling (7) _____ chains. Following translation, one or more chains become (8) _____ into the three-dimensional shape of protein molecules. Proteins have (9) _____ and (10) _____ roles in cells, including control of DNA.

## Complete the Table

11. Three types of RNA are transcribed from DNA in the nucleus (from genes that code only for RNA). Complete the following table, which summarizes information about these molecules.

| RNA Molecule | Abbreviation | Description/Function |
|---|---|---|
| a. Ribosomal RNA | | |
| b. Messenger RNA | | |
| c. Transfer RNA | | |

## Short Answer

12. List three ways in which a molecule of RNA is structurally different from a molecule of DNA._____

_____

_____

13. Cite two similarities in DNA replication and transcription. _____

_____

14. What are the three key ways in which transcription differs from DNA replication? _____

_____

## Sequence

Arrange the steps of transcription in correct chronological sequence. Write the letter of the first step next to 15, the letter of the second step next to 16, and so on.

15. ____

16. ____

17. ____

18. ____

19. ____

A. The RNA strand grows along exposed bases until RNA polymerase meets a DNA base sequence that signals "stop."

B. RNA polymerase binds with the DNA promoter region to open up a local region of the DNA double helix.

C. An RNA polymerase enzyme locates the DNA bases of the promoter region of one DNA strand by recognizing DNA-associated proteins near a promoter.

D. RNA is released from the DNA template as a free, single-stranded transcript.

E. RNA polymerase moves stepwise along exposed nucleotides of one DNA strand; as it moves, the DNA double helix keeps unwinding.

## Completion

20. Suppose the line below represents the DNA strand that will act as a template for the production of mRNA through the process of transcription. Fill in the blanks below the DNA strand with the sequence of complementary bases that will represent the message carried from DNA to the ribosome in the cytoplasm.

DpDpDpDpDpDpDpDpDpDpDpDpDpDpDpDpDpDpDpDpDpDpDpDpDpDpDpDpDpDp

T A C A A G A T A A C A T T A T T T C C T A C C G T C A T C

ᴚd ᴚd ᴚd ᴚd ᴚd ᴚd ᴚd ᴚd ᴚd ᴚd ᴚd ᴚd ᴚd ᴚd ᴚd ᴚd ᴚd ᴚd ᴚd ᴚd ᴚd ᴚd ᴚd ᴚd ᴚd ᴚd ᴚd ᴚd ᴚd ᴚd

(transcribed single-strand of mRNA)

## Label and Match

Newly transcribed mRNA contains more genetic information than is necessary to code for a chain of amino acids. Before the mRNA leaves the nucleus for its ribosome destination, an editing process occurs as certain portions of nonessential information are snipped out. Identify each indicated part of the illustration below; use abbreviations for the nucleic acids. Complete the exercise by matching and entering the letter of the description in the parentheses following each label.

21. _____ (   )

22. _____ (   )

23. _____ (   )

24. _____ (   )

25. _____ (   )

26. _____   _____

   _____ (   )

A. The actual coding portions of mRNA
B. Noncoding portions of the newly transcribed mRNA
C. Presence of cap and tail, introns snipped out and exons spliced together
D. Acquiring of a poly-A tail by the modified mRNA transcript
E. The region of the DNA template strand to be copied
F. Reception of a nucleotide cap by the 5′ end of mRNA (the first synthesized)

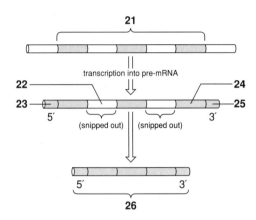

# 12 - II. DECIPHERING mRNA TRANSCRIPTS (pp. 188 - 189)
## STAGES OF TRANSLATION (pp. 190 - 191)

*Selected Words:* "three-bases-at-a-time," "start" signal, tRNA "hook," "wobble effect," *initiation, elongation, termination,* polysomes

### Boldfaced, Page-Referenced Terms

(188) codons _____

_____

(188) genetic code _____

_____

(189) ribosomes _____

_____

(189) anticodon _____

_____

## Matching

Choose the most appropriate answer for each.

1. ___ codon
2. ___ three at a time
3. ___ sixty-one
4. ___ the genetic code
5. ___ release factors
6. ___ ribosome
7. ___ anticodon
8. ___ the "stop" codons

A. Composed of two subunits, the small subunit with P and A amino acid binding sites as well as a binding site for mRNA
B. Reading frame of the nucleotide bases in mRNA
C. Detach protein and mRNA from the ribosome
D. UAA, UAG, UGA
E. A sequence of three nucleotide bases that can pair with a specific mRNA codon
F. Name for each base triplet in mRNA
G. The number of codons that actually specify amino acids
H. Term for how the nucleotide sequences of DNA and then mRNA correspond to the amino acid sequence of a polypeptide chain

## Complete the Table

9. Complete the following table, which distinguishes the stages of translation.

| Translation Stage | Description |
|---|---|
| a. | Special initiator tRNA loads onto small ribosomal subunit and recognizes AUG; small subunit binds with mRNA, and large ribosomal subunit joins small one. |
| b. | Amino acids are strung together in sequence dictated by mRNA codons as the mRNA strand passes through the two ribosomal subunits; two tRNAs interact at P and A sites. |
| c. | mRNA "stop" codon signals the end of the polypeptide chain; release factors detach the ribosome and polypeptide chain from the mRNA. |

## Completion

10. Given the following DNA sequence, deduce the composition of the mRNA transcript:

TAC  AAG  ATA  ACA  TTA  TTT  CCT  ACC  GTC  ATC

___  ___  ___  ___  ___  ___  ___  ___  ___  ___
(mRNA transcript)

11. Deduce the composition of the tRNA anticodons that would pair with the above specific mRNA codons as these tRNAs deliver the amino acids (identified below) to the P and A binding sites of the small ribosomal subunit.

___  ___  ___  ___  ___  ___  ___  ___  ___  ___
(tRNA anticodons)

12. From the mRNA transcript in exercise 10, use Figure 12.7 of the text to deduce the composition of the amino acids of the polypeptide sequence.

___  ___  ___  ___  ___  ___  ___  ___  ___  ___
(amino acids)

## Fill-in-the-Blanks

The order of (13) _____ _____ in a protein is specified by a sequence of nucleotide bases. The genetic code is read in units of (14) _____ nucleotides; each unit of three codes for (15) _____ amino acid(s). In the table that showed which triplet specified a particular amino acid, the triplet code was incorporated in (16) _____ molecules. Each of these triplets is referred to as a(n) (17) _____. (18) _____ alone carries the instructions for assembling a particular sequence of amino acids from the DNA to the ribosomes in the cytoplasm, where (19) _____ of the polypeptide occurs. (20) _____ RNA acts as a shuttle molecule as each type brings its particular (21) _____ _____ to the ribosome where it is to be incorporated into the growing (22) _____. A(n) (23) _____ is a triplet on mRNA that forms hydrogen bonds with a(n) (24) _____, which is a triplet on tRNA.

## 12 - III. HOW MUTATIONS AFFECT PROTEIN SYNTHESIS (p. 192)
## SUMMARY OF PROTEIN SYNTHESIS (p. 193)

*Selected Words:* "fix" the wrong base, "base-pair substitution," "frameshift mutation," "jumping genes"

### Boldfaced, Page-Referenced Terms

(192) gene mutations _____

_____

(192) mutagens _____

_____

(193) mature mRNA _____

_____

(193) ribosomal subunits _____

_____

(193) mature tRNA _____

_____

## Short Answer

1. Cite several examples of mutagens. _____

_____

## Fill-in-the-Blanks

In addition to changes in chromosomes (crossing over, recombination, deletion, addition, translocation, and inversion), changes can also occur in the structure of DNA; these modifications are referred to as gene mutations. Complete the following exercise on types of spontaneous gene mutations.

Viruses, ultraviolet radiation, and certain chemicals are examples of environmental agents called (2) _____ that may enter cells and damage strands of DNA. If A becomes paired with C instead of T during DNA replication, this spontaneous mutation is a base-pair (3) _____. Sickle-cell anemia is a genetic disease whose cause has been traced to a single DNA base pair; the result is that one (4) _____ _____ is substituted for another in the beta chain of (5) _____. Some DNA regions "jump" to new DNA locations and often inactivate the genes in their new environment; such (6) _____ elements may give rise to observable changes in the phenotype of an organism.

## Label and Match

A summary of the flow of genetic information in protein synthesis is useful as an overview. Identify the indicated parts of the illustration below by filling in the blanks with the names of the appropriate structures or functions. Choose from the following: DNA, mRNA, tRNA, polypeptide, rRNA subunits, intron, exon, mature mRNA transcript, new mRNA transcript, anticodon, amino acids, ribosome-mRNA complex. Complete the exercise by matching and entering the letter of the description in the parentheses following each label.

7. _____ ( )

8. _____

   _____

   _____ ( )

9. _____ ( )

10. _____ ( )

11. _____

   _____

   _____ ( )

12. _____ ( )

13. _____

   _____ ( )

14. _____ ( )

15. _____ ( )

16. _____

   _____ ( )

17. _____ ( )

18. _____ -

   _____

   _____ ( )

19. _____ ( )

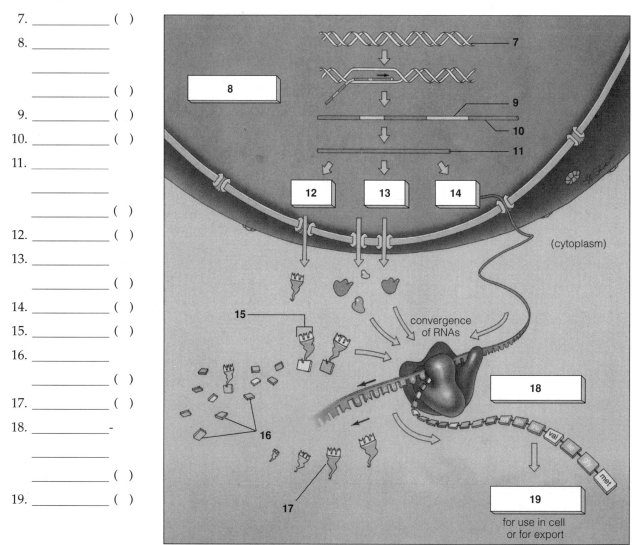

A. Coding portion of mRNA that will translate into proteins
B. Carries a modified form of the genetic code from DNA in the nucleus to the cytoplasm
C. Transports amino acids to the ribosome and mRNA
D. The building blocks of polypeptides
E. Noncoding portions of newly transcribed mRNA
F. tRNA after delivering its amino acid to the ribosome-mRNA complex
G. Join when translation is initiation
H. Holds the genetic code for protein production
I. Place where translation occurs
J. Includes introns and exons
K. A sequence of three bases that can pair with a specific mRNA codon
L. Snipping out of introns, only exons remaining
M. May serve as a functional protein (enzyme) or a structural protein

# 12 - IV. THE NATURE OF CONTROLS OVER GENE EXPRESSION (p. 194)
## EXAMPLES OF GENE CONTROL IN PROKARYOTIC CELLS (pp. 194–195)

*Selected Words:* *block, promote,* <u>Escherichia</u> <u>coli</u>, CAP, cAMP

## Boldfaced, Page-Referenced Terms

(194) regulatory proteins _____

_____

(194) negative control systems _____

_____

(194) positive control systems _____

_____

(194) operator _____

_____

(194) repressor _____

_____

(194) operon _____

_____

(195) activator protein _____

_____

## Complete the Table

1.  All of the diploid cells in an organism possess the same genes, and every cell utilizes most of the same genes; yet specialized cells must activate only certain genes. Some agents of gene control have been discovered. Transcriptional controls are the most common. Complete the following table to summarize the agents of gene control.

| Agents of Gene Control | Method of Gene Control |
|---|---|
| a. Repressor proteins | |
| b. Hormones | Major agents of vertebrate gene control; signaling molecules that move through the bloodstream to affect gene expression in target cells |
| c. Promoters | |
| d. | Short DNA base sequences between a promoter and the start of a gene; binding sites for control agents |

## Label and Match

*Escherichia coli*, a bacterial cell living in mammalian digestive tracts, is able to exert a negative type of gene control over lactose metabolism. Use the numbered blanks to identify each part of the illustration below. Use abbreviations for nucleic acids. Complete the exercise by matching and entering the letter of the proper function description in the parentheses following each label. Choose from the following:

lactose    regulator gene    promoter    mRNA transcript    genes that code for synthesizing ions

lactose operon    lactose-metabolizing enzymes    repressor-lactose complex

repressor protein    RNA polymerase    operator

2. _____ _____ ( )

3. _____ ( )

4. _____ _____ ( )

5. _____ ( )

6. _____ ( )

7. _____ _____ ( )

8. _____ _____ ( )

9. _____-_____ _____ ( )

10. _____ ( )

11. _____-_____ _____ ( )

12. _____ _____ ( )

A. Includes promoter, operator, and the genes that code for lactose-metabolizing enzymes
B. Short DNA base sequence between promoter and the beginning of a gene
C. The nutrient molecule in the lactose operon
D. Major enzyme that catalyzes transcription
E. Binds to operator and overlaps promoter; this prevents RNA polymerase from binding to DNA and initiating transcription
F. Prevents repressor from binding to the operator
G. Genes that produce lactose-metabolizing enzymes
H. Catalyze the digestion of lactose
I. Carries genetic instructions to ribosomes for production of lactose enzymes
J. Specific base sequence that signals the beginning of a gene
K. Gene that contains coding for production of repressor protein

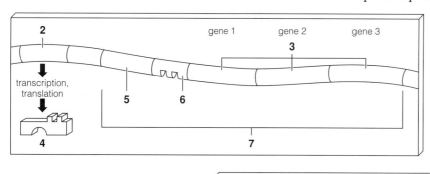

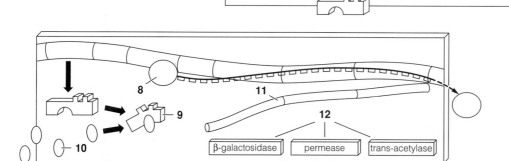

## Fill-in-the-Blanks

(13) _____ _____ is a species of bacterium that lives in mammalian digestive tracts and provided some of the first clues about gene control. A(n) (14) _____ is any group of genes together with its promoter and operator sequence. Promoter and operator provide (15) _____ _____. The (16) _____ codes for the formation of mRNA, which assembles a repressor protein. The affinity of the (17) _____ for RNA polymerase dictates the rate at which a particular operon will be transcribed. Repressor protein allows (18) _____ _____ over the lactose operon. Repressor binds with operator and overlaps promoter when lactose concentrations are (19) _____. This blocks (20) _____ _____ from the genes that will process lactose. This (21) (choose one) ❑ blocks ❑ promotes production of lactose-processing enzymes. When lactose is present, lactose molecules bind with the (22) _____ _____. Thus, the repressor cannot bind to the (23) _____, and RNA polymerase has access to the lactose-processing genes. This gene control works well because lactose-degrading enzymes are not produced unless they are (24) _____.

## 12 - V. GENE CONTROL IN EUKARYOTIC CELLS (pp. 196–197)
### Focus on Science: GENES, PROTEINS, AND CANCER (pp. 198–199)

*Selected Words:* "internal environment," inactivated chromosome, genetic "mosaic," *anhidrotic ectodermal dysplasia,* "calico" cat, "spotting gene," *benign, malignant, myc* gene, *Burkitt's lymphoma*

### Boldfaced, Page-Referenced Terms

(196) cell differentiation _____

_____

(196) Barr body _____

_____

(197) phytochrome _____

_____

(198) tumor _____

_____

(198) metastasis _____

_____

(198) cancer _____

_____

(198) oncogene _____

_____

(198) proto-oncogenes _____

_____

(198) carcinogens _____

_____

## Short Answer

1. Although a complex organism such as a human being arises from a single cell (the zygote), differentiation occurs in development. Define *differentiation,* and relate it to a definition of selective gene expression.

_____

_____

## Fill-in-the-Blanks

All diploid cells in our bodies contain copies of the same (2) _____. These genetically identical

cells become structurally and functionally distinct from one another through a process called

(3) _____, which arises through (4) _____ gene expression in different cells. Cells depend on

(5) _____, which govern transcription, translation, and enzyme activity. Controls that operate

during transcription and transcript processing utilize (6) _____ proteins, especially (7) _____

that are turned on and off by the addition and removal of cAMP. A (8) _____ body is a condensed

X chromosome. X chromosome inactivation produces adult human females who are

(9) _____ for X-linked traits. This effect is shown in human females with patches of skin that lack

normal sweat glands, a disorder known as (10) _____ _____ _____ and provides

evidence for (11) _____ gene expression.

## True/False

If the statement is true, write a T in the blank. If the statement is false, correct it by changing the underlined word(s) and writing the correct word(s) in the blank.

_____ 12. When cells become cancerous, cell populations <u>decrease</u> to very <u>low</u> densities and <u>stop</u> dividing.

_____ 13. <u>All</u> abnormal growths and massings of new tissue in any region of the body are called tumors.

_____ 14. Malignant tumors have cells that <u>migrate and divide</u> in other organs.

_____ 15. Oncogenes are genes that <u>combat</u> cancerous transformations.

_____ 16. Proto-oncogenes <u>rarely</u> trigger cancer.

_____ 17. The normal expression of proto-oncogenes is vital, even though their <u>normal</u> expression may be lethal.

# Self-Quiz

___ 1. Transcription _____.
   a. occurs on the surface of the ribosome
   b. is the final process in the assembly of a protein
   c. occurs during the synthesis of any type of RNA by use of a DNA template
   d. is catalyzed by DNA polymerase

___ 2. _____ carry(ies) amino acids to ribosomes, where amino acids are linked into the primary structure of a polypeptide.
   a. mRNA
   b. tRNA
   c. Introns
   d. rRNA

___ 3. Transfer RNA differs from other types of RNA because it _____.
   a. transfers genetic instructions from cell nucleus to cytoplasm
   b. specifies the amino acid sequence of a particular protein
   c. carries an amino acid at one end
   d. contains codons

___ 4. _____ dominates the process of transcription.
   a. RNA polymerase
   b. DNA polymerase
   c. Phenylketonuria
   d. Transfer RNA

___ 5. _____ and _____ are found in RNA but not in DNA.
   a. Deoxyribose; thymine
   b. Deoxyribose; uracil
   c. Uracil; ribose
   d. Thymine; ribose

___ 6. Each "word" in the mRNA language consists of _____ letters.
   a. three
   b. four
   c. five
   d. more than five

___ 7. If each kind of nucleotide is coded for only one kind of amino acid, how many different types of amino acids could be selected?
   a. four
   b. sixteen
   c. twenty
   d. sixty-four

___ 8. The genetic code is composed of _____ codons.
   a. three
   b. twenty
   c. sixteen
   d. sixty-four

___ 9. _____ binds to operator whenever lactose concentrations are low.
   a. Operon
   b. Repressor

   c. Promoter
   d. Operator

___ 10. Any gene or group of genes together with its promoter and operator sequence is a(n) _____.
   a. repressor
   b. operator
   c. promoter
   d. operon

___ 11. The operon model explains the regulation of _____ in prokaryotes.
   a. replication
   b. transcription
   c. induction
   d. Lyonization

___ 12. In multicelled eukaryotes, cell differentiation occurs as a result of _____.
   a. growth
   b. selective gene expression
   c. repressor molecules
   d. the death of certain cells

___ 13. _____ controls govern the rates at which mRNA transcripts that reach the cytoplasm will be translated into polypeptide chains at the ribosomes.
   a. Transport
   b. Transcript processing
   c. Translational
   d. Transcriptional

___ 14. The cause of sickle-cell anemia has been traced to _____.
   a. a mosquito-transmitted virus
   b. two DNA mutations that result in two incorrect amino acids in a hemoglobin chain
   c. three DNA mutations that result in three incorrect amino acids in a hemoglobin chain
   d. one DNA mutation that results in one incorrect amino acid in a hemoglobin chain

# Chapter Objectives/Review Questions

| Page | | Objectives/Questions |
|---|---|---|
| (185 - 186) | 1. | State how RNA differs from DNA in structure and function, and indicate what features RNA has in common with DNA. |
| (186) | 2. | _____ RNA combines with certain proteins to form the ribosome; _____ RNA carries genetic information for protein construction from the nucleus to the cytoplasm; _____ RNA picks up specific amino acids and moves them to the area of mRNA and the ribosome. |
| (186 - 187) | 3. | Describe the process of transcription, and indicate three ways in which it differs from replication. |
| (186) | 4. | Transcription starts at a _____, a specific sequence of bases on one of the two DNA strands that signals the start of a gene. |
| (186 - 187) | 5. | The first end of the mRNA to be synthesized is the _____ end; at the opposite end, the most mature transcripts acquire a _____ tail. |
| (188 - 189) | 6. | What RNA code would be formed from the following DNA code: TAC-CTC-GTT-CCC-GAA? |
| (188) | 7. | Each base triplet in mRNA is called a _____. |
| (187 - 188) | 8. | State the relationship between the DNA genetic code and the order of amino acids in a protein chain. |
| (188) | 9. | Scrutinize Figure 12.7 in the text, and decide whether the genetic code in this instance applies to DNA, mRNA, or tRNA. |
| (188) | 10. | Explain how the DNA message TAC-CTC-GTT-CCC-GAA would be used to code for a segment of protein, and state what its amino acid sequence would be. |
| (189 - 191) | 11. | Describe how the three types of RNA participate in the process of translation. |
| (192) | 12. | List some of the environmental agents, or _____, that can cause mutations. |
| (192) | 13. | Briefly describe the spontaneous DNA mutations known as base-pair substitution, frameshift mutation, and transposable element. |
| (192) | 14. | Cite an example of a change in one DNA base pair that has profound effects on the human phenotype. |
| (193) | 15. | Using a diagram, summarize the steps involved in the transformation of genetic messages into proteins (see Figure 12.13 in the text). |
| (194 - 195) | 16. | The negative control of _____ protein prevents the enzymes of transcription from binding to DNA; the positive control of _____ protein enhances the binding of RNA polymerases to DNA. |
| (194 - 195) | 17. | The cells of *E. coli* manage to produce enzymes to degrade lactose when those molecules are _____ and to stop production of lactose-degrading enzymes when lactose is _____. |
| (196) | 18. | Explain how selective gene expression relates to cell differentiation in multicelled eukaryotes. |
| (196 - 197) | 19. | Explain how X chromosome inactivation provides evidence for selective gene expression; use the example of anhidrotic ectodermal dysplasia. |
| (198 - 199) | 20. | Describe the relationship of proto-oncogenes, environmental irritants, and oncogenes. |

## Integrating and Applying Key Concepts

Genes code for specific polypeptide sequences. Not every substance in living cells is a polypeptide. Explain how genes might be involved in the production of a storage starch (such as glycogen) that is constructed from simple sugars.

# 13

## RECOMBINANT DNA AND GENETIC ENGINEERING

## Interactive Exercises

### 13 - I. RECOMBINATION IN NATURE—AND IN THE LABORATORY (pp. 202 - 205)
### WORKING WITH DNA FRAGMENTS (pp. 206 - 207)
### *Focus on Science:* RIFF-LIPS AND DNA FINGERPRINTS (p. 208)

*Selected Words: familial cholesterolemia, antibiotics,* <u>Escherichia</u> <u>coli</u>, *staggered* cuts, "sticky" ends, "recombinant plasmids," *amplified,* "cloned," *Southern blot method*

### *Boldfaced, Page-Referenced Terms*

(203) gene therapy _____

_____

(204) recombinant DNA technology _____

_____

(204) genetic engineering _____

_____

(204) plasmids _____

_____

(205) restriction enzymes _____

_____

(205) DNA ligase _____

_____

(205) DNA library _____

_____

(206) polymerase chain reaction (PCR) _____

_____

(206) DNA polymerase _____

_____

(206) gel electrophoresis _____

_____

(208) RFLPs (restriction fragment length polymorphisms) _____

_____

(208) DNA fingerprint _____

_____

## Short Answer

1. Describe the bacterial chromosome and plasmids present in a bacterial cell, and distinguish between
   them. _____

_____

_____

## True/False

If the statement is true, write T in the blank. If the statement is false, make it correct by changing the underlined words and writing the correct word(s) in the blank.

_____ 2. Plasmids are <u>organelles on the surfaces of which amino acids are assembled into
              polypeptides.</u>

_____ 3. <u>Gene transfer</u> and <u>recombination</u> are common in nature.

## Fill-in-the-Blanks

Genetic experiments have been occurring in nature for billions of years as a result of gene
(4) _____, crossing over and recombination, and other events. Humans now are causing genetic
change by using (5) _____ _____ technology in which researchers cut out, and splice together,
gene regions from different (6) _____, then greatly (7) _____ the number of copies of the
genes that interest them. The genes, and in some cases their (8) _____ products, are produced in
quantities that are large enough for (9) _____ and for practical applications. (10) _____

_____ involves isolating, modifying, and inserting particular genes back into the same organism or into a different one.

Many bacteria can transfer (11) _____ genes to a bacterial neighbor that may integrate them into the recipient's chromosome, forming a recombinant DNA molecule naturally. In nature, (12) _____ as well as bacteria dabble in gene transfers and recombination of genes, and so do eukaryotic organisms.

## Complete the Table

13. Complete the table below, which summarizes some of the basic tools and procedures used in recombinant DNA technology.

| Tool/Procedure | Definition and Role in Recombinant DNA Technology |
| --- | --- |
| a. Restriction enzymes | |
| b. DNA ligase | |
| c. DNA library | |
| d. Cloned DNA | |
| e. PCR | |
| f. DNA sequencing | |
| g. RFLP | |
| h. DNA fingerprint | |

## Matching

Match the steps in the formation of a DNA library with the parts of the illustration below.

14. ___
15. ___
16. ___
17. ___
18. ___
19. ___

A. Joining of chromosomal and plasmid DNA using DNA ligase
B. Restriction enzyme cuts chromosomal DNA at specific recognition sites
C. Cut plasmid DNA
D. Recombinant plasmids containing cloned library
E. Fragments of chromosomal DNA
F. Same restriction enzyme is used to cut plasmids

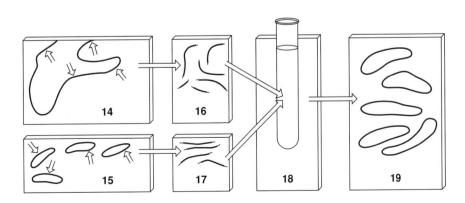

## True/False

A genetic engineer used restriction enzymes to prepare fragments of DNA from two different species that were then mixed. Four of these fragments are illustrated below. Fragments (a) and (c) are from one species, (b) and (d) from the other species. Answer exercises 20-24 with a T for true or an F for false.

| _____ TACA | _____ TTCA | _____ CGTA | _____ ATGT |
|---|---|---|---|
| a. | b. | c. | d. |

___ 20. Some of the fragments represent sticky ends.

___ 21. The same restriction enzyme was used to cut fragments (b), (c), and (d).

___ 22. Different restriction enzymes were used to cut fragments (a) and (d).

___ 23. Fragment (a) will base-pair with fragment (d) but not with fragment (c).

___ 24. The same restriction enzyme was used to cut the different locations in the DNA of the two species shown.

## Matching

Choose the most appropriate answer for each.

25. ___ polymerase chain reaction (PCR)

26. ___ DNA ligase

27. ___ DNA library

28. ___ cloned DNA

29. ___ DNA fingerprint

30. ___ recombinant DNA technology

31. ___ plasmids

32. ___ restriction enzymes

33. ___ genome

A. All the DNA in a haploid set of chromosomes
B. Process by which a gene is split into two strands and then copied over and over (most common type of gene amplification) by enzymes
C. Caused by a unique array of RFLPs inherited from each parent
D. Connects DNA fragments
E. Small circular DNA molecules that carry only a few genes
F. Cuts DNA molecules; produces "sticky ends"
G. A collection of DNA fragments produced by restriction enzymes and incorporated into plasmids
H. Multiple, identical copies of DNA fragments from an original chromosome
I. Method of genetic engineering

## Fill-in-the-Blanks

Each person has a genetic (34) _____: a unique pattern of RFLPs that can be used to map the human genome, apprehend criminals, and resolve cases of disputed paternity and maternity. Bacterial host cells lack the proper (35) _____ _____ to translate cloned genes unless the (36) _____ have been cut out. Identification of the order and identity of nucleotides in DNA is called (37) _____.

*Short Answer*

38. Outline a gene therapy procedure that has been used somewhat successfully to reverse the traumatic effects of familial cholesterolemia. _____

_____

_____

_____

*Fill-in-the-Blanks*

39. From the migration results shown, determine the sequence of nucleotides in the code molecule.

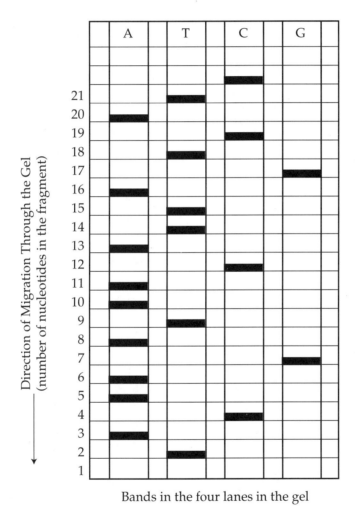

Bands in the four lanes in the gel

⇑ First Letter

⇑ Last Letter

# 13 - II. MODIFIED HOST CELLS (p. 209)
## BACTERIA, PLANTS, AND THE NEW TECHNOLOGY (pp. 210 - 211)
## GENETIC ENGINEERING OF ANIMALS (p. 212)
### *Focus on Bioethics:* REGARDING HUMAN GENE THERAPY (p. 213)

**Selected Words:** "fail-safe" genes, "ice-minus bacteria," *potato blight*, <u>Agrobacterium</u> <u>tumefaciens</u>, "gene gun," "super-mouse," "biotech barnyards," *eugenic engineering*

## Boldfaced, Page-Referenced Terms

(209) DNA probes _____

_____

(209) nucleic acid hybridization _____

_____

(209) cDNA ("copied" DNA) _____

_____

(209) reverse transcription _____

_____

(212) human genome project _____

_____

## Complete the Table

1.  Complete the table below, which summarizes some of the basic tools and procedures used in recombinant DNA technology.

| Tool/Procedure | Definition and Role in Recombinant DNA Technology |
|---|---|
| a. DNA probe | |
| b. Nucleic acid hybridization | |
| c. cDNA | |
| d. Reverse transcription | |

a.

b.

c.

d.

*Fill-in-the-Blanks*

Host cells that take up modified genes may be identified by procedures involving DNA (2) _____ . Gene (3) _____ does not automatically follow the successful taking-in of a modified gene by a host cell; genes must be suitably (4) _____ first, which involves getting rid of the (5) _____ from the mRNA transcripts, then using the enzyme reverse (6) _____ to produce cDNA. The process goes like this:

a. A mature (7) _____ transcript of a desired gene is used as a template for assembling a DNA strand. An enzyme (6) does the assembling. b. An mRNA-(8) _____ hybrid molecule results. c. (9) _____ action removes the mRNA and assembles a second strand of (10) _____ on the first strand. d. The result is double-stranded (11) _____ , "copied" from an mRNA (12) _____ .

*Short Answer*

13. Explain the goal of the human genome project. _____

_____

In exercises 14 - 19, summarize the results of the given experimentation dealing with genetic modifications of plants and animals.

14. Cattle may soon be producing human collagen: _____

_____

15. The bacterium, *Agrobacterium tumefaciens:* _____

_____

16. Cotton plants: _____

_____

17. Introduction of the rat and human somatotropin gene into fertilized mouse eggs:_____

_____

18. "Ice-minus" bacteria and strawberry plants: _____

_____

19. Bacteria that can degrade oil spills: _____

_____

## Fill-in-the-Blanks

The harmful gene is called the "ice-forming" gene, and bacteria that lack it are known as

(20) _____-_____ bacteria; genetic engineers were able to remove the harmful gene and test

the modified bacterium on strawberry plants with no adverse effects. Inserting one or more genes into

the (21) _____ _____ of an organism for the purpose of correcting genetic defects is known

as (22) _____ _____. Attempting to modify a human trait by inserting genes into sperm

or eggs is called (23) _____ _____.

# Self-Quiz

___ 1. Small circular molecules of DNA in bacteria are called _____.
   a. plasmids
   b. desmids
   c. pili
   d. F particles
   e. transferins

___ 2. Base-pairing between nucleotide sequences from different sources is called _____.
   a. nucleic acid hybridization
   b. reverse replication
   c. heterocloning
   d. plasmid formation

___ 3. Enzymes used to cut genes in recombinant DNA research are _____.
   a. ligases
   b. restriction enzymes
   c. transcriptases
   d. DNA polymerases
   e. replicases

___ 4. The total DNA in a haploid set of chromosomes of a species is its _____.
   a. plasmid
   b. enzyme potential
   c. genome
   d. DNA library
   e. none of the above

___ 5. An enzyme that heals random base-pairing of chromosomal fragments and plasmids is _____.
   a. reverse transcriptase
   b. DNA polymerase
   c. cDNA
   d. DNA ligase

___ 6. A DNA library is _____.
   a. a collection of DNA fragments produced by restriction enzymes and incorporated into plasmids
   b. cDNA plus the required restriction enzymes
   c. mRNA-cDNA
   d. composed of mature mRNA transcripts

___ 7. Amplification results in _____.
   a. plasmid integration
   b. bacterial conjugation
   c. cloned DNA
   d. production of DNA ligase

___ 8. Any DNA molecule that is copied from mRNA is known as _____.
   a. cloned DNA
   b. cDNA
   c. DNA ligase
   d. hybrid DNA

___ 9. The most commonly used method of DNA amplification is _____.
   a. polymerase chain reaction
   b. gene expression
   c. genome mapping
   d. RFLPs

___ 10. Restriction fragment length polymorphisms are valuable because _____.
   a. they reduce the risks of genetic engineering
   b. they provide an easy way to sequence the human genome
   c. they allow fragmenting DNA without enzymes
   d. they provide DNA fragment sizes unique to each person

# Chapter Objectives/Review Questions

| Page | | Objectives/Questions |
|---|---|---|
| (204) | 1. | List the means by which natural genetic recombination occurs. |
| (204) | 2. | Define *recombinant DNA technology.* |
| (204) | 3. | _____ are small, circular, self-replicating molecules of DNA or RNA within a bacterial cell. |
| (205) | 4. | Some bacteria produce _____ enzymes that cut apart DNA molecules injected into the cell by viruses; such DNA fragments or "_____ ends" often have staggered cuts capable of base-pairing with other DNA molecules cut by the same _____ enzymes. |
| (205) | 5. | Base-pairing between chromosomal fragments and cut plasmids is made permanent by DNA _____. |
| (205) | 6. | Be able to explain what a DNA library is; review the steps used in creating such a library. |
| (206) | 7. | List and define the two major methods of DNA amplification. |
| (206) | 8. | Polymerase chain reaction is the most commonly used method of DNA _____. |
| (206) | 9. | Multiple, identical copies of DNA fragments produced by restriction enzymes are known as _____ DNA. |
| (206 - 207) | 10. | Explain how gel electrophoresis is used to sequence DNA. |
| (208) | 11. | List some practical genetic uses of RFLPs. |
| (209) | 12. | How is a cDNA probe used to identify a desired gene carried by a modified host cell? |
| (209) | 13. | A special viral enzyme, _____ _____, presides over the process by which mRNA is transcribed into DNA. |
| (209) | 14. | Define *cDNA.* |
| (209) | 15. | Why do researchers prefer to work with cDNA when working with human genes? |
| (210 - 211) | 16. | Explain why you believe that human "tinkering" with genes in different organisms is primarily a benefit, or a disaster about to happen. Use examples from your text or from recent publications. |
| (212 - 213) | 17. | Tell about the human genome project and its implications. |
| (212 - 213) | 18. | Define *gene therapy* and *eugenic engineering.* |

## Integrating and Applying Key Concepts

How can scientists guarantee that *Escherichia coli,* the human intestinal bacterium, will not be transformed into a severely pathogenic form and released into the environment if researchers use the bacterium in recombinant DNA experiments?

# 14

# MICROEVOLUTION

## Interactive Exercises

### 14 - I. EARLY BELIEFS, CONFOUNDING DISCOVERIES (pp. 216 - 219)
### A FLURRY OF NEW THEORIES (pp. 220 - 221)
### DARWIN'S THEORY TAKES FORM (pp. 222 - 223)

*Selected Words:* microevolution, "species," "fluida," H.M.S. *Beagle, Principles of Geology, artificial* selection, *natural* selection, *descent with modification,* "missing links," Archaeopteryx

In addition to the boldfaced terms, the text features other important terms essential to understanding the assigned material. "Selected Words" is a list of these terms, which appear in the text in italics, in quotation marks, and occasionally in roman type. Latin binomials found in this section are underlined and in roman type to distinguish them from other italicized words.

## Boldfaced, Page-Referenced Terms

The page-referenced terms are important; they were in boldface type in the chapter. Write a definition for each term in your own words without looking at the text. Next, compare your definition with that given in the chapter or in the text glossary. If your definition seems inaccurate, allow some time to pass and repeat this procedure until you can define each term rather quickly (how fast you can answer is a gauge of your learning effectiveness).

(218) biogeography _____

_____

(218) comparative anatomy _____

_____

(219) sedimentary beds _____

_____

(219) fossils _____

_____

(219) evolution _____

_____

(220) catastrophism _____

_____

(220) theory of inheritance of acquired characteristics _____

_____

(221) theory of uniformity _____

_____

## Matching

Choose the most appropriate answer for each.

1. ___ comparative anatomy
2. ___ biogeography
3. ___ fossils
4. ___ school of Hippocrates
5. ___ evolution
6. ___ Great Chain of Being
7. ___ Buffon
8. ___ species
9. ___ Aristotle
10. ___ sedimentary beds

A. Rock layers deposited at different times; may contain fossils
B. Came to view nature as a continuum of organization, from lifeless matter through complex forms of plant and animal life
C. Each kind of being; represent links in the great Chain of Being
D. Modification of species over time
E. Extended from the lowest forms of life to humans and on to spiritual beings
F. Suggested that perhaps species had originated in more than one place and perhaps had been modified over time
G. Studies of body structure comparisons and patterning
H. Studies of the world distribution of plants and animals
I. Suggested that the gods were not the cause of the sacred disease
J. One type of direct evidence that organisms lived in the past

## Choice

For questions 11 - 22, choose from the following:

a. Georges Cuvier      b. Jean-Baptiste Lamarck

11. ___ Stretching directed "fluida" to the necks of giraffes, which lengthened them permanently

12. ___ Acknowledged there were abrupt changes in the fossil record that corresponded to discontinuities between certain layers of sedimentary beds; thought this was evidence of change in populations of ancient organisms

13. ___ The force for change in organisms is the drive for perfection.

14. ___ There was only one time of creation that populated the world with all species.

15. ___ Catastrophism

16. ___ Permanently stretched giraffe necks were bestowed on offspring

17. ___ When a global catastrophe destroyed many organisms, a few survivors repopulated the world.

18. ___ Theory of inheritance of acquired characteristics

19. ___ The survivors of global catastrophe were not new species (they may have differed from related fossils, but naturalists had not discovered them yet).

20. ___ During a lifetime, environmental pressures and internal "desires" bring about permanent changes.

21. ___ The force for change is a drive for perfection, up the Chain of Being.

22. ___ The drive to change is centered in nerves that direct an unknown "fluida" to body parts in need of change.

## Complete the Table

23. Several key players and events in the life of Charles Darwin led him to his conclusions about natural selection and evolution. Summarize these influences by completing the table below.

| Event/Person | Importance to Synthesis of Evolutionary Theory |
|---|---|
| a. | Botanist at Cambridge University who perceived Darwin's real interests and arranged for Darwin to become a ship's naturalist |
| b. | Where Darwin earned a degree in theology but also developed his love for natural history |
| c. | British ship that carried Darwin (as a naturalist) on a five-year voyage around the world |
| d. | Wrote Principles of Geology; advanced the theory of uniformity; suggested the Earth was much older than 6,000 years |
| e. | Wrote an influential essay (read by Darwin) on human populations asserting that people tend to produce children faster than food supplies, living space, and other resources can be sustained |
| f. | Volcanic islands 900 kilometers from the South American coast where Darwin correlated differences in various species of finches with their environmental challenges |
| g. | The key point in Darwin's theory of evolution; involves reproductive capacity, heritable variations, and adaptive traits |
| h. | English naturalist contemporary with Darwin; independently developed Darwin's theory of evolution before Darwin published |
| i. | Unearthed in 1861; the first transitional fossil (between reptiles and birds); provided evidence for Darwin's theory |

# 14 - II. INDIVIDUALS DON'T EVOLVE; POPULATIONS DO (pp. 224 - 225)
## *Focus on Science:* WHEN IS A POPULATION NOT EVOLVING? (pp. 226 - 227)

*Selected Words:* *morphological* traits, *physiological* traits, *behavioral* traits, *phenotypes,* "Hardy-Weinberg formula," *natural selection, gene flow, genetic drift,* beneficial mutations

## Boldfaced, Page-Referenced Terms

(224) population _____

_____

(224) polymorphism _____

_____

(224) gene pool _____

_____

(224) alleles _____

_____

(225) allele frequencies _____

_____

(225) genetic equilibrium _____

_____

(225) microevolution _____

_____

(225) mutation rate _____

_____

(225) lethal mutations _____

_____

(225) neutral mutations _____

_____

## Labeling

The traits of individuals in a population are often classified as being morphological, physiological, or behavioral traits. In the list of traits below, enter M if the trait seems to be morphological, P if the trait is physiological, and B if the trait is behavioral.

____ 1.  Frogs have a three-chambered heart.

____ 2.  The active transport of $Na^+$ and $K^+$ ions is unequal.

____ 3.  Humans and orangutans possess an opposable thumb.

____ 4.  An organism periodically seeking food

____ 5.  Some animals have a body temperature that fluctuates with the environmental temperature.

____ 6.  The platypus is a strange mammal. It has a bill like a duck and lays eggs.

____ 7.  Some vertebrates exhibit greater parental protection for offspring than others.

___ 8. During short photoperiods, the pituitary gland releases small quantities of gonadotropins.

___ 9. Red grouse defend large multipurpose territories where they forage, mate, nest, and rear young.

___ 10. Lampreys have an elongated cylindrical body without scales.

## Matching

Choose the most appropriate answer to match the sources of genetic variation.

11. ___ fertilization

12. ___ changes in chromosome structure or number

13. ___ crossing over at meiosis

14. ___ gene mutation

15. ___ independent assortment at meiosis

A. Leads to mixes of paternal and maternal chromosomes in gametes
B. Produces new alleles
C. Leads to the loss, duplication, or alteration of alleles
D. Brings together combinations of alleles from two parents
E. Leads to new combinations of alleles in chromosomes

## Short Answer

16. Briefly discuss the role of the environment and genetics in the production of an organism's phenotype.

_____

_____

_____

17. List the conditions (in any order) that must be met before genetic equilibrium (or nonevolution) will

occur. _____

_____

## Problems

18. For the following situation, assume that the conditions listed in question 17 do exist; therefore, there should be no change in gene frequency, generation after generation. Consider a population of hamsters in which dominant gene $B$ produces black coat color and recessive gene $b$ produces gray coat color (two alleles are responsible for color). The dominant gene has a frequency of 80 percent (or .80). It would follow that the frequency of the recessive gene is 20 percent (or .20). From this, the assumption is made that 80 percent of all sperm and eggs have gene $B$. Also, 20 percent of all sperm and eggs carry gene $b$. (See pages 226 - 227 in the text.)

a. Calculate the probabilities of all possible matings in the Punnett square.
b. Summarize the genotype and phenotype frequencies of the $F_1$ generation.

| Genotypes | Phenotypes |
| --- | --- |
| _____ BB | |
| _____ Bb | _____ % black |
| _____ bb | _____ % gray |

Sperm

|  | 0.80 B | 0.20 b |
| --- | --- | --- |
| Eggs 0.80 B | BB | Bb |
| 0.20 b | Bb | bb |

c. Further assume that the individuals of the $F_1$ generation produce another generation and the assumptions of the Hardy-Weinberg rule still hold. What are the frequencies of the sperm produced?

| Parents ($F_1$) | B sperm | b sperm |
|---|---|---|
| _____BB | _____ | _____ |
| _____Bb | _____ | _____ |
| _____bb | _____ | _____ |
| Totals = | _____ | _____ |

The egg frequencies may be similarly calculated. Note that the gamete frequencies of the $F_2$ are the same as the gamete frequencies of the last generation. Phenotype percentage also remains the same. Thus, the gene frequencies did not change between the $F_1$ and the $F_2$ generation. Again, given the assumptions of the Hardy-Weinberg equilibrium, gene frequencies do not change generation after generation.

19. In a population, 81 percent of the organisms are homozygous dominant, and 1 percent are homozygous recessive. Find the following:

a. the percentage of heterozygotes_____

b. the frequency of the dominant allele_____

c. the frequency of the recessive allele_____

20. In a population of 200 individuals, determine the following for a particular locus if $p = 0.80$.

a. the number of homozygous dominant individuals_____

b. the number of homozygous recessive individuals _____

c. the number of heterozygous individuals if $p = 0.80$ _____

21. If the percentage of gene $D$ is 70 percent in a gene pool, find the percentage of gene $d$. _____

_____

22. If the frequency of gene $R$ in a population is 0.60, what percentage of the individuals are heterozygous

$Rr$? _____

_____

## Matching

Choose the most appropriate answer for each.

23. ___ gene flow
24. ___ allele frequencies
25. ___ neutral mutations
26. ___ microevolution
27. ___ lethal mutation
28. ___ alleles
29. ___ mutation
30. ___ genetic drift
31. ___ genetic equilibrium
32. ___ natural selection

A. A heritable change in DNA
B. Zero evolution
C. Different molecular forms of a gene
D. Change in allele frequencies as individuals leave or enter a population
E. Change or stabilization of allele frequencies due to differences in survival and reproduction among variant members of a population
F. Random fluctuation in allele frequencies over time due to chance
G. Change in which expression of a mutated gene always leads to the death of the individual
H. The abundance of each kind of allele in the entire population
I. Neither harmful nor helpful to the individual
J. Changes in allele frequencies brought about by mutation, genetic drift, gene flow, and natural selection

## Fill-in-the-Blanks

A(n) (33) _____ is a group of individuals of the same species that occupy a given area at a specific time. The (34) _____-_____ principle allows researchers to establish a theoretical reference point (baseline) against which changes in allele frequency can be measured. Variation can be expressed in terms of (35) _____ _____, which means the relative abundance of different alleles carried by the individuals in that population. The stability of allele ratios that would occur if all individuals had equal probability of surviving and reproducing is called (36) _____ _____. Over time, allele frequencies tend to change through infrequent but inevitable (37) _____, which are the original source of genetic variation. Random fluctuation in allele frequencies over time due to chance occurrence alone is called (38) _____ _____; it is more pronounced in small populations than in large ones. (39) _____ flow associated with immigration and/or emigration also changes allele frequencies. (40) _____ _____ is the differential survival and reproduction of individuals of a population that differ in one or more traits. (41) _____ _____ is the most important microevolutionary process.

## 14 - III. A CLOSER LOOK AT NATURAL SELECTION (p. 227)
### DIRECTIONAL CHANGE IN THE RANGE OF VARIATION (pp. 228 - 229)
### SELECTION AGAINST OR IN FAVOR OF EXTREME PHENOTYPES (pp. 230 - 231)
### SPECIAL OUTCOMES OF SELECTION (pp. 232 - 233)

*Selected Words:* Biston betularia, *mark-release-recapture,* "superbugs," Eurosta, *HbA allele, HbS allele, sickle-cell anemia, malaria*

## Boldfaced, Page-Referenced Terms

(228) directional selection _____

_____

(229) pest resurgence _____

_____

(229) biological control _____

_____

(229) antibiotic _____

_____

(230) stabilizing selection _____

_____

(231) disruptive selection _____

_____

(232) sexual dimorphism _____

_____

(232) sexual selection _____

_____

(232) balanced polymorphism _____

_____

## Complete the Table

1.  Complete the following table, which defines three types of natural selection effects.

| Effect | Characteristics | Example |
|---|---|---|
| a. Stabilizing selection | | |
| b. Directional selection | | |
| c. Disruptive selection | | |

## Labeling

2.  Identify the three curves below as stabilizing selection, directional selection, or disruptive selection.

a. _____     b. _____     c. _____

## Fill-in-the-Blanks

When individuals of different phenotypes in a population differ in their ability to survive and reproduce, their alleles are subject to (3) _____ selection. (4) _____ selection favors the most common phenotypes in the population. (5) _____ selection occurs when a specific change in the environment causes a heritable trait to occur with increasing frequency and the whole population tends to shift in a parallel direction. (6) _____ selection favors the development of two or more distinct polymorphic varieties such that they become increasingly represented in a population and the population splits into different phenotypic variations. (7) _____ provide an excellent example of stabilizing selection, because they have existed essentially unchanged for hundreds of millions of years. (8) _____-_____ _____, a genetic disorder that produces an abnormal form of hemoglobin, is an example of (9) _____ selection that maintains a high frequency of the harmful (10) _____ along with the normal gene. Such a genetic juggling act is called a (11) _____ _____. Differences in appearance between males and females of a species are known as (12) _____ _____. Among birds and mammals, females act as agents of (13) _____ when they choose their mates. (14) _____ selection is based on any trait that gives the individual a competitive edge in mating and producing offspring. The more colorful, showier, and larger appearance of male pheasants when compared with females is the result of (15) _____ selection.

## Choice

For questions 16 - 29, choose from the following categories of natural selection; in some cases, two letters may be correct.

> a. directional selection    b. stabilizing selection    c. disruptive selection
>
> d. balanced polymorphism    e. sexual selection

16. ___ May result even when heterozygotes are selected against; unstable Rh markers in the human population provide an example

17. ___ Sexual dimorphism

18. ___ Phenotypic forms at both ends of the variation range are favored, and intermediate forms are selected against.

19. ___ May account for the persistence of certain phenotypes over time; horsetail plants still bear strong resemblance to their ancient relatives

20. ___ Selection maintains two or more alleles for the same trait in steady fashion, generation after generation.

21. ___ The most common forms of a trait in a population are favored; over time, alleles for uncommon forms are eliminated.

22. ___ Allele frequencies shift in a steady, consistent direction; this may be due to a change in the environment.

23. ___ Often results when environmental conditions favor homozygotes over heterozygotes

24. ___ Counters the effects of mutation, genetic drift, and gene flow

25. ___ Human newborns weighing an average of 7 pounds are favored.

26. ___ In West Africa, finches known as black-bellied seedcrackers have either large or small bills.

27. ___ When a trait gives an individual an advantage in reproductive success

28. ___ The most frequent wing color of peppered moths shifted from a light form to a dark form as tree trunks became soot-darkened due to coal used for fuel during the Industrial Revolution.

29. ___ Females of a species choosing mates to directly affect reproductive success

## 14 - IV. GENE FLOW (p. 233)
### GENETIC DRIFT (pp. 234 - 235)

*Selected Words:* *feline infectious peritonitis*

### Boldfaced, Page-Referenced Terms

(233) gene flow _____

_____

(234) genetic drift _____

_____

(234) sampling error _____

_____

(234) fixation _____

_____

(234) bottleneck _____

_____

(235) inbreeding _____

_____

(235) founder effect _____

_____

(235) endangered species _____

_____

## Choice

For questions 1 - 10, choose from the following microevolutionary forces that can change gene frequency:

a. genetic drift    b. gene flow

1. ___ A random change in allele frequencies over the generations, brought about by chance alone

2. ___ Emigration

3. ___ Prior to the turn of the century, hunters killed all but twenty of a large population of northern elephant seals.

4. ___ When allele frequencies change due to individuals leaving a population or new individuals entering it

5. ___ Long ago, seabirds, winds, or ocean currents carried a few seeds from the Pacific Northwest to the Hawaiian Islands to establish plant populations.

6. ___ Sometimes result in endangered species

7. ___ Founder effects and bottlenecks are two extreme cases.

8. ___ Immigration

9. ___ Only 20,000 cheetahs have survived to the present.

10. ___ Scrubjays make hundreds of round trips carrying acorns from oak trees as much as a mile away to soil in their home territories for winter storage; this introduces new alleles to oak stands.

## Short Answer

11. Distinguish between the two extreme cases of genetic drift: the founder effect and bottlenecks. _____

_____

# Self-Quiz

___ 1. An acceptable definition of evolution is _____.
a. changes in organisms that are extinct
b. changes in organisms since the flood
c. changes in organisms over time
d. changes in organisms in only one place

___ 2. The two scientists most closely associated with the concept of evolution are _____.
a. Lyell and Malthus
b. Henslow and Cuvier
c. Henslow and Malthus
d. Darwin and Wallace

___ 3. The distribution of organisms on Earth is known as _____.
a. the Great Chain of Being
b. stratification
c. geological evolution
d. biogeography

___ 4. Buffon suggested that _____.
a. tail bones in a human have no place in a perfectly designed body
b. perhaps species originated in more than one place and have been modified over time
c. the force for change in organisms was a built-in drive for perfection, up the Chain of Being
d. gradual processes now molding Earth's surface had also been at work in the past

___ 5. In Lyell's book Principles of Geology, he suggested that _____.
a. tail bones in a human have no place in a perfectly designed body
b. perhaps species originated in more than one place and have been modified over time

c. the force for change in organisms was a built-in drive for perfection, up the Chain of Being

d. gradual processes now molding Earth's surface had also been at work in the past

___ 6. One of the central ideas of Lamarck's theory of desired evolution was that

_____.

a. tail bones in a human have no place in a perfectly designed body

b. perhaps species originated in more than one place and have been modified over time

c. the force for change in organisms was a built-in drive for perfection, up the Chain of Being

d. gradual processes now molding Earth's surface had also been at work in the past

___ 7. The theory of catastrophism is associated with _____.

a. Darwin
b. Cuvier
c. Buffon
d. Lamarck

___ 8. The idea that any population tends to out-grow its resources and that its members must compete for what is available belonged to _____.

a. Malthus
b. Darwin
c. Lyell
d. Henslow

___ 9. The ideas that all natural populations have the reproductive capacity to exceed the resources required to sustain them and that members of a natural population show great variation in their traits are associated with _____.

a. catastrophism
b. desired evolution
c. natural selection
d. Malthus's idea of survival

___ 10. *Archaeopteryx* is evidence for _____.

a. the theory that birds descended from reptiles
b. evolution
c. an evolutionary "link" between two major groups of organisms
d. the existence of fossils
e. all of the above

## Chapter Objectives/Review Questions

This section lists general and detailed chapter objectives that can be used as review questions. You can make maximum use of these items by writing answers on a separate sheet of paper. Fill in answers where blanks are provided. To check for accuracy, compare your answers with information given in the chapter or glossary.

| Page | | Objectives/Questions |
|---|---|---|
| (218) | 1. | It was the study of _____ _____ that raised the following question: If all species were created at the same time in the same place, why were certain species found in only some parts of the world and not others? |
| (218 - 219) | 2. | Give examples of the type of evidence found by comparative anatomists that suggested living things may have changed with time. |
| (220 - 223) | 3. | Be able to state the significance of the following: Henslow, H.M.S. *Beagle*, Lyell, Darwin, Malthus, Wallace, and *Archaeopteryx*. |
| (222 - 223) | 4. | What conclusion was Darwin led to when he considered the various species of finches living on the separate islands of the Galápagos? |
| (224) | 5. | A _____ is a group of individuals occupying a given area and belonging to the same species. |
| (224) | 6. | Distinguish between morphological, physiological, and behavioral traits. |
| (224) | 7. | All of the genes of an entire population belong to a _____ _____. |
| (224) | 8. | Each kind of gene usually exists in one or more molecular forms, called _____. |
| (224) | 9. | Review the five categories through which genetic variation occurs among individuals. |
| (225) | 10. | The abundance of each kind of allele in the whole population is referred to as allele _____. |

| (225) | 11. | Be able to list the five conditions that must exist before conditions for the Hardy-Weinberg principle are met. |
|---|---|---|
| (225) | 12. | A point at which allele frequencies for a trait remain stable through the generations is called genetic _____. |
| (225) | 13. | Changes in allele frequencies brought about by mutation, genetic drift, gene flow, and natural selection are called _____. |
| (225) | 14. | A _____ is a random, heritable change in DNA. |
| (225) | 15. | Distinguish a lethal mutation from a neutral mutation. |
| (226) | 16. | Be able to calculate allele and genotype frequencies when provided with the genotype frequency of the recessive allele. |
| (228 - 231) | 17. | Be able to define and provide an example of directional selection, stabilizing selection, and disruptive selection. |
| (232) | 18. | Define *balanced polymorphism.* |
| (232) | 19. | The occurrence of phenotypic differences between males and females of a species is called _____ _____. |
| (232) | 20. | _____ selection is based on any trait that gives an individual a competitive edge in mating and producing offspring. |
| (233) | 21. | Allele frequencies change as individuals leave or enter a population; this is gene _____. |
| (234) | 22. | Random fluctuations in allele frequencies over time, due to chance, is called _____ _____. |
| (234 - 235) | 23. | Distinguish the founder effect from a bottleneck. |
| (235) | 24. | Relate bottlenecks to endangered species. |

## Integrating and Applying Key Concepts

Can you imagine any way in which directional selection may have occurred or may be occurring in humans? Which factors do you suppose are the driving forces that sustain the trend? Do you think the trend could be reversed? If so, by what factor(s)?

# 15

# SPECIATION

## Interactive Exercises

### 15 - I. ON THE ROAD TO SPECIATION (pp. 238 - 241)

*Selected Words:* <u>Helix</u> <u>aspersa</u>, hybrid, *temporal* isolation, *ecological* isolation, *behavioral* isolation, *mechanical* isolation, *gametic* isolation

### Boldfaced, Page-Referenced Terms

(239) speciation _____

_____

(240) species _____

_____

(240) biological species concept _____

_____

(240) gene flow_____

_____

(240) genetic divergence_____

_____

(240) isolating mechanisms _____

_____

(241) prezygotic isolation _____

_____

(241) postzygotic isolation _____

_____

## Choice

For questions 1 - 10, choose from the following:

a. Mayr's biological species    b. genetic divergence    c. isolating mechanisms

1. ___ Heritable aspect of body form, physiology, or behavior that prevents gene flow between genetically divergent populations

2. ___ Groups of interbreeding natural populations

3. ___ May occur as a result of stopping gene flow

4. ___ Phenotypically very different but can interbreed and produce fertile offspring

5. ___ Defined as the buildup of differences in the separated pools of alleles of populations

6. ___ A "kind" of living thing

7. ___ Factors that prevent horses and zebras from mating in the wild

8. ___ Prezygotic

9. ___ Does not work well with organisms that reproduce solely by asexual means or those known only from the fossil record

10. ___ Postzygotic

## Matching

Choose the most appropriate answer to match the isolating mechanisms; complete the exercise by entering "pre" in the parentheses if the mechanism is prezygotic and "post" if the mechanism is postzygotic.

11. ___ ecological (   )

12. ___ temporal (   )

13. ___ hybrid inviability (   )

14. ___ mechanical (   )

15. ___ zygote mortality (   )

16. ___ gametic mortality (   )

17. ___ behavioral (   )

18. ___ hybrid offspring (   )

A. Potential mates occupy overlapping ranges but reproduce at different times.
B. The first-generation hybrid forms but shows very low fitness.
C. Potential mates occupy different local habitats within the same area.
D. Potential mates meet but cannot figure out what to do about it.
E. Sperm is transferred but the egg is not fertilized (gametes die or gametes are incompatible).
F. The hybrid is sterile or partially so.
G. Potential mates attempt engagement, but sperm cannot be successfully transferred.
H. The egg is fertilized, but the zygote or embryo dies.

## Choice

For questions 19 - 25, choose from the following isolating mechanisms:

   a. temporal   b. behavioral   c. mechanical   d. gametic   e. ecological   f. postzygotic

19. ___ Sterile zebroids

20. ___ Two sage species, each has its flower petals arranged as a "landing platform" for a different pollinator

21. ___ Two species of cicada, one matures, emerges, and reproduces every thirteen years, the other every seventeen years

22. ___ Populations of the grass *Agrostis tenuis* demonstrate differing tolerances to copper at varying distances from copper mines; speciation may be underway.

23. ___ Molecularly mismatched gametes of different sea urchin species

24. ___ Prior to copulation, male and female birds engage in complex courtship rituals recognized only by birds of their own species.

25. ___ Sterile mules

## Identification

26. The illustration below depicts the divergence of one species as time passes. Each horizontal line represents a population of that species and each vertical line, a point in time. Answer the questions below the illustration.

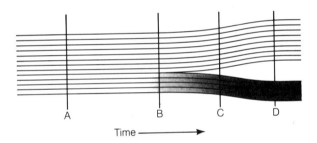

Time ⟶

a. How many species are represented at time A?_____

b. How many species exist at time D? _____

c. Between what times (letters) does divergence begin?_____

d. What letter represents the time that complete divergence is reached?_____

e. Between what two times (letters) is divergence clearly underway but not complete?_____

## 15 - II. MODELS OF SPECIATION (pp. 242 - 243)
## PATTERNS OF SPECIATION (pp. 244 - 245)

*Selected Words: Allo-, patria, sym-, para-, ana-, physical* access, *evolutionary* access, *ecological* access

*Boldfaced, Page-Referenced Terms*

(242) allopatric speciation _____

_____

(242) sympatric speciation _____

_____

(243) polyploidy _____

(243) parapatric speciation _____

_____

(243) hybrid zone _____

(244) cladogenesis _____

_____

(244) anagenesis _____

_____

(244) evolutionary tree diagrams _____

_____

(244) gradual model of speciation _____

_____

(244) punctuation model of speciation _____

_____

(245) adaptive radiation _____

_____

(245) adaptive zones _____

_____

(245) background extinction _____

_____

(245) mass extinction _____

_____

## Complete the Table

1.  Complete the following table, which defines three speciation models.

| Speciation Model | Description |
| --- | --- |
| a. | Suggests that species can form within the range of an existing species, in the absence of physical or ecological barriers; a controversial model; evidence comes from crater lake fishes |
| b. | Emphasizes disruption of gene flow; perhaps the main speciation route; through divergence and evolution of isolating mechanisms, separated populations become reproductively incompatible and cannot interbreed |
| c. | Occurs where populations share a common border; the borders may be permeable to gene flow; in some, gene exchange is confined to a common border, the hybrid zone |

## Dichotomous Choice

Circle one of two possible answers given between parentheses in each statement.

2. Instant speciation by polyploidy in many flowering plant species represents (parapatric/sympatric) speciation.
3. In the past, hybrid orioles existed in a hybrid zone between eastern Baltimore orioles and western Bullock orioles; today, hybrid orioles are becoming less frequent in this example of (allopatric/parapatric) speciation.
4. In the absence of gene flow between geographically separate populations, genetic divergence may become sufficient to bring about (allopatric/parapatric) speciation.
5. The uplifted ridge of land we now call the Isthmus of Panama divides the Atlantic Ocean from the Pacific Ocean and forms a natural laboratory for biologists to study (allopatric/sympatric) speciation.
6. In the 1800s, a major earthquake changed the course of the Mississippi River; this isolated some populations of insects that could not swim or fly, and set the stage for (parapatric/allopatric) speciation.

## Short Answer

7. Study the illustration below, which shows the possible course of wheat evolution; similar genomes (a genome is a complete set of chromosomes) are designated by the same letter. Answer the questions below the diagram.

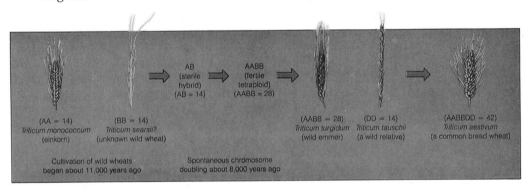

a. How many pairs of chromosomes are found in *Triticum monococcum*? _____

b. How are the *T. monococcum* chromosomes designated?_____

c. How many pairs of chromosomes are found in *T. searsii*? _____

d. How are the *T. searsii* chromosomes designated?_____

e. Why is the hybrid designated AB sterile? _____

f. What cellular event must have occurred to create the *T. turgidum* genome? _____

g. Describe the genome of the plant arising from the cross of *T. turgidum* and *T. tauschii* (not shown on the diagram). _____

h. What cellular event must have occurred to create the *T. aestivum* genome? _____

i. What is the source of the A genome in *T. aestivum*? The source of the B genome in *T. aestivum*? The source of the D genome in *T. aestivum*? _____

## Complete the Table

8. Complete the table below to summarize information for interpreting evolutionary tree diagrams. Refer to the text, Figure 15.8. Ask yourself, "How would the interpretation be illustrated on an evolutionary tree diagram?"

| Branch Form | Interpretation |
|---|---|
| a. | Traits changed rapidly around the time of speciation |
| b. | An adaptive radiation occurred |
| c. | Extinction |
| d. | Speciation occurred through gradual changes in traits over geologic time |
| e. | Traits of the new species did not change much thereafter |
| f. | Evidence of this presumed evolutionary relationship is only sketchy |

## Matching

Choose the most appropriate answer for each.

9. ___ cladogenesis

10. ___ evolutionary tree diagrams

11. ___ gradual model of speciation

12. ___ punctuation model of speciation

13. ___ adaptive radiation

14. ___ adaptive zones

15. ___ background extinction

16. ___ mass extinction

17. ___ physical, evolutionary, and ecological access

18. ___ anagenesis

A. Expected rate of species disappearance as local conditions change

B. Conditions allowing a lineage to radiate into an adaptive zone

C. Speciation in which most morphological changes are compressed into a brief period of hundreds or thousands of years when populations first start to diverge; evolutionary tree branches make abrupt ninety-degree turns

D. Ways of life such as "burrowing in the seafloor"

E. Speciation with slight angles on the evolutionary tree, which conveys many small changes in form over long time spans

F. An abrupt rise in the rates of species disappearance above background level

G. Speciation occurring within a single, unbranched line of descent

H. A method of summarizing information about the continuities of relationships among species

I. A branching speciation pattern; isolated populations diverge

J. A burst of microevolutionary activity within a lineage; results in the formation of new species in a wide range of habitats

## Fill-in-the-Blanks

(19) _____ is the process whereby species are formed. A (20) _____ is composed of one or more populations of individuals who can interbreed under natural conditions and produce fertile, reproductively isolated offspring. (21) _____ can be described as a process in which differences in alleles accumulate between populations. Any aspect of structure, function, or behavior that prevents

interbreeding is a (22) _____ _____ mechanism. Physical barriers that prevent gene flow as in the case of large rivers changing course or forests giving way to grasslands are (23) _____ barriers. Wheat is an example of a species formed by (24) _____ and (25) _____ .

## Self-Quiz

___ 1. Of the following, _____ is (are) not a major consideration of the biological species concept.
   a. groups of populations
   b. reproductive isolation
   c. asexual reproduction
   d. interbreeding natural populations
   e. production of fertile offspring

___ 2. When something prevents gene flow between two populations or subpopulations, _____ may occur.
   a. genetic drift, natural selection, and mutation
   b. genetic divergence
   c. a buildup of differences in the separated gene pools of alleles
   d. all of the above

___ 3. When potential mates occupy different local habitats within the same area, it is termed _____ isolation.
   a. temporal
   b. behavioral
   c. mechanical
   d. ecological

___ 4. "Potential mates occupy overlapping ranges but reproduce at different times" is a description of _____ isolation.
   a. temporal
   b. behavioral
   c. mechanical
   d. ecological

___ 5. Of the following, _____ is a postzygotic isolating mechanism.
   a. behavioral isolation
   b. gametic mortality
   c. hybrid inviability
   d. mechanical isolation

For questions 6 - 8, choose from the following answers:
   a. parapatric speciation
   b. sympatric speciation
   c. allopatric speciation

___ 6. When the Isthmus of Panama was formed, it provided contemporary scientists with an ideal natural laboratory to study _____ .

___ 7. Polyploidy and cichlid fishes of crater lakes provide evidence for _____ .

___ 8. The identification of a hybrid zone between the ranges of eastern Baltimore orioles and western Bullock orioles is an example of _____ .

___ 9. The branching speciation pattern revealed by the fossil record is called _____ .
   a. anagenesis
   b. the gradual model
   c. cladogenesis
   d. the punctuation model

___ 10. Species formation by many small changes over long time spans fit the _____ model of speciation; rapid speciation with morphological changes compressed into a brief period of population divergence describes the _____ model of speciation.
   a. anagenesis; punctuation
   b. punctuation; gradual
   c. gradual; cladogenesis
   d. gradual; punctuation

___ 11. Adaptive radiation is _____ .
   a. a burst of microevolutionary activity within a lineage
   b. the formation of new species in a wide range of habitats
   c. the spreading of lineages into unfilled adaptive zones
   d. a lineage radiating into an adaptive zone when it has physical, ecological, or evolutionary access to it
   e. all of the above

___ 12. The expected rate of species disappearance over time is called _____.
   a. mass extinction
   b. reverse background radiation
   c. background extinction
   d. speciation extinction

For questions 13 - 14, choose from the following answers:
   a. stabilizing selection
   b. divergence

   c. a reproductive isolating mechanism
   d. polyploidy

___ 13. Two species of sage plants with differently shaped floral parts that prevent pollination by the same pollinator serve as an example of _____.

___ 14. _____ occurs when isolated populations accumulate allele frequency differences between them over time.

## Chapter Objectives/Review Questions

*Page Objectives/Questions*

| | | |
|---|---|---|
| (240) | 1. | Be able to give the biological species concept as stated by Ernst Mayr. |
| (240) | 2. | _____ _____ is a buildup of differences in separated pools of alleles. |
| (240) | 3. | The process by which species are formed is known as _____. |
| (240) | 4. | The evolution of reproductive _____ paves the way for genetic divergence and speciation. |
| (241) | 5. | Be able to briefly define the major categories of prezygotic and postzygotic isolating mechanisms (see text, Table 15.2). |
| (242) | 6. | _____ speciation occurs when daughter species form gradually by divergence in the absence of gene flow between geographically separate populations. |
| (242) | 7. | In _____ speciation, daughter species arise, sometimes rapidly, from a small proportion of individuals within an existing population. |
| (243) | 8. | When daughter species form from a small proportion of individuals along a common border between two populations, it is called _____ speciation. |
| (243) | 9. | Explain why sympatric speciation by polyploidy is a rapid method of speciation. |
| (243) | 10. | In some cases of parapatric speciation, gene exchange between two species is confined to a _____ zone. |
| (244) | 11. | Distinguish cladogenesis from anagenesis. |
| (244) | 12. | Be able to explain the use of evolutionary tree diagrams and the symbolism used. |
| (244) | 13. | Explain why branches on an evolutionary tree that have slight angles indicate the _____ model of speciation. |
| (244) | 14. | Branches on an evolutionary tree that make abrupt 90-degree turns are consistent with the _____ model of speciation. |
| (245) | 15. | An adaptive radiation is a burst of _____ activity within a lineage that results in the formation of new species in a variety of habitats. |
| (245) | 16. | Cite examples of adaptive zones. |
| (245) | 17. | How does background extinction differ from mass extinction? |

## Integrating and Applying Key Concepts

*Systematics* is defined as the practice of describing, naming, and classifying living things; this includes the comparative study of organisms and all relationships among them. Plant systematists who work with flowering plants readily accept Ernst Mayr's definition of a "biological species" as described in this chapter. However, in real practice, systematists often must also work with a concept known as the "morphological species." For example, the statement is sometimes made that "two plant specimens belong to the same morphological species but not to the same biological species." Explain the meaning of that statement. What type of experimental evidence would be necessary as a basis for this statement? From your study of this chapter, can you suggest a reason that application of the term "morphological species" is sometimes necessary? What difficulties and inaccuracies, in terms of identifying a species, might the use of both species concepts present?

# 16

# THE MACROEVOLUTIONARY PUZZLE

FOSSILS—EVIDENCE OF ANCIENT LIFE
    Fossilization
    Interpreting the Geologic Tombs
    Interpreting the Fossil Record

EVIDENCE FROM COMPARATIVE EMBRYOLOGY
    Developmental Program of Larkspurs
    Developmental Program of Vertebrates

EVIDENCE OF MORPHOLOGICAL DIVERGENCE
    Homologous Structures
    Potential Confusion from Analogous Structures

EVIDENCE FROM COMPARATIVE BIOCHEMISTRY
    Molecular Clocks
    Protein Comparisons
    Nucleic Acid Comparisons

ORGANIZING THE EVIDENCE—CLASSIFICATION SCHEMES
    Assigning Names to Species
    A Five-Kingdom, Phylogenetic Scheme

## Interactive Exercises

### 16 - I. FOSSILS—EVIDENCE OF ANCIENT LIFE (pp. 248 - 251)
### EVIDENCE FROM COMPARATIVE EMBRYOLOGY (pp. 252 - 253)

*Selected Words:* *species, trace* fossils, <u>Cooksonia</u>, *"geologic time scale,"* <u>Delphinium</u> <u>decorum</u>, <u>D.</u> <u>nudicaule</u>, allometric changes

### *Boldfaced, Page-Referenced Terms*

(249) macroevolution _____

_____

(250) fossils _____

_____

(251) fossilization _____

_____

(251) stratification _____

_____

(252) comparative morphology _____

_____

## Matching

Choose the most appropriate answer for each.

1. ___ fossilization
2. ___ stratification
3. ___ trace fossils
4. ___ continuity of relationship
5. ___ macroevolution
6. ___ fossils
7. ___ biased fossil record
8. ___ formation of sedimentary deposits
9. ___ oldest fossils
10. ___ rupture or tilt of sedimentary layers

A. Evolutionary connectedness between organisms, past to present
B. Found in the deepest rock layers
C. Affected by movements of Earth's crust, hardness of organisms, population sizes, and certain environments
D. Produced by gradual deposition of silt, volcanic ash, and other materials
E. Refers to large-scale patterns, trends, and rates of change among groups of species
F. A process favored by rapid burial in the absence of oxygen
G. Evidence of later geologic disturbance
H. Recognizable, physical evidence of ancient life; the most common are bones, teeth, shells, spore capsules, seeds, and other hard parts
I. Includes the indirect evidence of coprolites and imprints of leaves, stems, tracks, trails, and burrows
J. A layering of sedimentary deposits

## 16 - II. EVIDENCE OF MORPHOLOGICAL DIVERGENCE (pp. 254 - 255)
## EVIDENCE FROM COMPARATIVE BIOCHEMISTRY (pp. 256 - 257)

*Selected Words:* morpho-, homo-, homologous

### Boldfaced, Page-Referenced Terms

(254) morphological divergence _____

_____

(254) homology _____

_____

(255) morphological convergence _____

_____

(255) analogy _____

_____

(256) neutral mutations _____

_____

(256) molecular clock _____

_____

(256) DNA-DNA hybridization _____

_____

## True/False

If the statement is true, write a T in the blank. If the statement is false, correct it by changing the underlined word(s) and writing the correct word(s) in the blank.

_____ 1. The extent to which a strand of DNA or RNA isolated from individuals of one species will base-pair with a comparable strand from individuals of a different species is a rough measure of the evolutionary <u>distance</u> between them.

_____ 2. The amino acid sequence of cytochrome c is very similar in such distantly related organisms as (1) yeast, (2) wheat, and (3) humans; the probability that this striking resemblance resulted from chance alone is very <u>high.</u>

_____ 3. Analogous structures in two different lineages indicate strong evidence of morphological <u>divergence.</u>

_____ 4. Body parts that have similar form and function in different lineages yet develop from similar tissues in their embryos are homologous structures that provide very strong evidence of morphological <u>divergence.</u>

_____ 5. Biochemical <u>similarities</u> are fewest among the most closely related species and most numerous among the most distantly related.

## Choice

For questions 6 - 10, choose from the following:

a. analogy    b. homology    c. an example of a lineage

d. a specific example of morphological convergence    e. a specific example of morphological divergence

6. ___ Sharks, penguins, and porpoises

7. ___ Ancestral ape-human, <u>Homo erectus,</u> and modern humans (*Homo sapiens*)

8. ___ Body parts that were once quite different among distantly related lineages but are now quite similar in function and in apparent structure

9. ___ The human arm and hand, the porpoise's front flipper, and the bird's wing

10. ___ Body parts of different organisms that develop from similar embryonic parts yet might result in quite differently appearing adult structures that would probably not have similar functions

## Labeling

Evidence for macroevolution comes from comparative morphology and comparative biochemistry. For each of the following listed items, write an M in the blank if the evidence comes from comparative morphology or a B if the evidence comes from comparative biochemistry.

11. ___ Early in the vertebrate developmental program, embryos of different lineages proceed through strikingly similar stages.

12. ___ Comparison of the developmental rates of changes in a chimpanzee skull and a human skull

13. ___ The evolution of different rates of flower development in two larkspur species also led to the coevolution of two different pollinators, honeybees and hummingbirds.

14. ___ Using accumulated neutral mutations to date the divergence of two species from a common ancestor

15. ___ The amino acid sequence in the cytochrome c of humans precisely matches the sequence in chimpanzees.

16. ___ Establishing a rough measure of evolutionary distance between two organisms by use of DNA hybridization studies that demonstrate the extent to which the DNA from one species base-pairs with another

17. ___ Regulatory genes control the rate of growth of different body parts and can produce large differences in the development of two very similar embryos.

18. ___ Results from DNA-DNA hybridization studies supported an earlier view that giant pandas, bears, and red pandas are related by common descent.

19. ___ Distantly related vertebrates, such as sharks, penguins, and porpoises, show a similarity to one another in their proportion, position, and function of body parts.

20. ___ Finding similarities in vertebrate forelimbs when comparing the wings of pterosaurs, birds, bats, and porpoise flippers

## Short Answer

21. List three methods employed by systematics to elucidate the patterns of diversity in an evolutionary context. _____

_____

_____

_____

_____

_____

# 16 - III. ORGANIZING THE EVIDENCE—CLASSIFICATION SCHEMES (pp. 258 - 259)

**Selected Words:** *relative degrees* of relationship

## Boldfaced, Page-Referenced Terms

(258) higher taxa _____

_____

(258) classification schemes _____

_____

(258) binomial system _____

_____

(258) genus _____

_____

(258) phylogeny _____

_____

(258) Monera _____

_____

(259) Protista _____

_____

(259) Fungi _____

(259) Plantae _____

(259) Animalia _____

## Matching

Choose the most appropriate answer for each.

1. ___ Monera     A. Multicelled heterotrophs that feed by extracellular digestion and absorption
2. ___ Protista     B. Single-celled prokaryotes, some are autotrophs, others are heterotrophs
    C. Diverse multicelled heterotrophs, including predators and parasites
3. ___ Fungi     D. Multicelled photosynthetic autotrophs
4. ___ Plantae     E. Diverse single-celled eukaryotes, some photosynthetic autotrophs, many heterotrophs
5. ___ Animalia

## Short Answer

6. Arrange the following jumbled taxa in proper order, with the most inclusive first: family, class, species, kingdom, genus, phylum (or division), and order. _____

_____

_____

_____

7. Arrange the following jumbled taxa and their categories in proper order, the most inclusive first: genus: *Archibaccharis*; kingdom: Plantae; order: Asterales; species: *lineariloba*; class: Dicotyledonae; division: Anthophyta; family: Asteraceae. _____

_____

_____

_____

*Matching*

Choose the most appropriate answer for each.

8. ___ black bear

9. ___ phylogeny

10. ___ binomial system

11. ___ *Pinus strobus* and *Pinus banksiana*

12. ___ species

13. ___ genus

14. ___ family

15. ___ taxa

16. ___ *Ursus americanus* (black bear) and

   *Homarus americanus* (Atlantic lobster)

17. ___ Linnaeus

A. A group of similar species
B. The taxon most often studied
C. Evolutionary relationships among species, reflected in classification schemes
D. The organizing units of classification schemes
E. Classification level that includes similar genera
F. Originated the modern practice of providing two scientific names for organisms
G. An example of a common name
H. Two species belonging to one genus
I. An example of two very different organisms having the same specific epithet (species name)
J. System of assigning a two-part Latin name to a species

## Self-Quiz

___ 1. Phylogeny is _____.
   a. the identification of organisms and assigning names to them
   b. producing a "retrieval system" of several levels
   c. an evolutionary history of organism(s), both living and extinct
   d. a study of the adaptive responses of various organisms

___ 2. The significant contribution by Linnaeus was to develop _____.
   a. the idea that organisms should be placed in two kingdoms
   b. the binomial system
   c. the theory of evolution
   d. the idea that a species could have several common names

___ 3. The process of grouping organisms according to similarities derived from a common ancestor is known as _____.
   a. taxonomy
   b. classification
   c. a binomial system
   d. macroevolution

___ 4. Early embryos of vertebrates strongly resemble one another *because* _____.
   a. the genes that guide early embryonic development are the same (or similar) in all vertebrates

   b. the analogous structures they each contain are evidence of morphological convergence
   c. the homologous structures they each contain provide very strong evidence of morphological divergence
   d. DNA-DNA hybridization techniques reveal many structural alterations that have resulted from gene mutations in early embryos

___ 5. All living organisms have eukaryotic cell structure except the _____.
   a. Animalia
   b. Plantae
   c. Fungi
   d. Monera
   e. Protista

For questions 6 - 7, choose from the following answers:
   a. convergence
   b. divergence

___ 6. The wings of pterosaurs, birds, and bats serve as examples of _____.

___ 7. Penguins and porpoises serve as examples of _____.

For questions 8 - 12, choose from the following answers:
- a. DNA-DNA hybridization
- b. homologous
- c. macroevolution
- d. analogous
- e. regulatory genes

___ 8. _____ structures resemble one another due to common descent.

___ 9. A rough measure of the evolutionary distance between two species is provided by _____.

___ 10. _____ control the rate of growth in different body parts.

___ 11. Examples of _____ structures are the shark fin and the penguin wing.

___ 12. _____ refers to changes in groups of species.

---

# Chapter Objectives/Review Questions

| Page | Objectives/Questions |
|---|---|
| (249) | 1. _____ refers to large-scale patterns, trends, and rates of change among groups of species. |
| (250) | 2. Give a generalized definition of a "fossil"; cite examples. |
| (251) | 3. _____ begins with rapid burial in sediments or volcanic ash. |
| (251) | 4. What is the name given to the layering of sedimentary deposits? |
| (251) | 5. Are the oldest fossils found in the deepest layers of sedimentary rocks or nearer the surface? |
| (251) | 6. The fossil record is heavily _____ toward certain environments and certain types of organisms. |
| (252) | 7. Early in the vertebrate developmental program, _____ of different lineages proceed through strikingly similar stages. |
| (252) | 8. Explain why two species of larkspurs have flowers with different shapes that attract different pollinators. |
| (253) | 9. Mutation of _____ genes explains the differences in the timing of growth sequences of chimps and humans (even though they have nearly identical genes). |
| (254) | 10. The macroevolutionary pattern of change from a common ancestor is known as morphological _____. |
| (255) | 11. Distinguish homologies from analogies as applied to organisms. |
| (255) | 12. A pattern of long-term change in similar directions among remotely related lineages is called morphological _____. |
| (256) | 13. Define *neutral mutations*, and discuss their use as molecular clocks. |
| (256) | 14. Some nucleic acid comparisons are based on _____ hybridization. |
| (258) | 15. Describe the binomial system as developed by Linnaeus; why is Latin used as the language of binomials? |
| (258 - 259) | 16. Be able to arrange the classification groupings correctly, from most to least inclusive. |
| (258) | 17. Modern classification schemes reflect _____, the evolutionary relationships among species, starting with the most ancestral and including all the branches leading to their descendants. |
| (259) | 18. List the five kingdoms of life in the Whittaker system, and cite examples of the organisms in each. |

---

# Integrating and Applying Key Concepts

Water crowfoot plants (family Ranunculaceae) often have two or more distinct leaf types within the same species and on the same plant. One type is capillary (hairlike), found submerged in the water or growing well above the water. The other type is laminate (flat and expanded), and is found floating on the water or submerged. In some *Ranunculus* species three different leaf shapes form on a plant in a sequence. Suppose two taxonomists describe the same crowfoot plant as being two different and distinct species due to these two different leaf shapes. Can you think of difficulties that might arise when other researchers are required to refer to these "two species" in the course of their work?

# 17

# THE ORIGIN AND EVOLUTION OF LIFE

## Interactive Exercises

### 17 - I. CONDITIONS ON THE EARLY EARTH (pp. 262 - 265)
### EMERGENCE OF THE FIRST LIVING CELLS (pp. 266 - 267)
### LIFE ON A CHANGING GEOLOGIC STAGE (pp. 268 - 269)

*Selected Words:* "big bang," Stanley Miller, "proto-cells," *seafloor spreading, superplumes*

In addition to the boldfaced terms, the text features other important terms essential to understanding the assigned material. "Selected Words" is a list of these terms, which appear in the text in italics, in quotation marks, and occasionally in roman type. Latin binomials found in this section are underlined and in roman type to distinguish them from other italicized words.

### Boldfaced, Page-Referenced Terms

The page-referenced terms are important; they were in boldface type in the chapter. Write a definition for each term in your own words without looking at the text. Next, compare your definition with that given in the chapter or in the text glossary. If your definition seems inaccurate, allow some time to pass and repeat this procedure until you can define each term rather quickly (how fast you can answer is a gauge of your learning effectiveness).

(267) RNA world _____

_____

(268) plate tectonic theory _____

_____

(269) Proterozoic _____

_____

(269) Paleozoic _____

_____

(269) Mesozoic _____

_____

(269) Cenozoic _____

_____

(269) geologic time scale _____

_____

(269) mass extinctions _____

_____

(269) Archean _____

_____

## Choice

Answer questions 1 - 20 by choosing from the following probable stages of the physical and chemical evolution of life:

a. Formation of Earth     b. Early Earth and the first atmosphere

c. The synthesis of organic compounds     d. Origin of agents of metabolism

e. Origin of self-replicating systems     f. Origin of the first plasma membranes

1. ___ Most likely, the proto-cells were little more than membrane sacs protecting information-storing templates and various metabolic agents from the environment.

2. ___ Simple systems (of this type) of RNA, enzymes, and coenzymes have been created in the laboratory.

3. ___ Four billion years ago, Earth was a thin-crusted inferno.

4. ___ Sunlight, lightning, or heat escaping from Earth's crust could have supplied the energy to drive their condensation into complex organic molecules.

5. ___ During the first 600 million years of Earth history, enzymes, ATP, and other molecules could have assembled spontaneously at the same locations.

6. ___ Were gaseous oxygen ($O_2$) and water also present? Probably not.

7. ___ Sidney Fox heated amino acids under dry conditions to form protein chains, which he placed in hot water. The cooled chains self-assembled into small, stable spheres. The spheres were selectively permeable.

8. ___ Stanley Miller mixed hydrogen, methane, ammonia, and water in a reaction chamber that recirculated the mixture and bombarded it with a spark discharge. Within a week, amino acids and other small organic compounds had been formed.

9. ___ Without an oxygen-free atmosphere, the organic compounds that started the story of life never would have formed on their own. Free oxygen would have attacked them.

10. ___ Much of Earth's inner rocky material melted.

11. ___ Imagine an ancient sunlit estuary, rich in clay deposits. Countless aggregations of organic molecules stick to the clay.

12. ___ In other experiments, fatty acids and glycerol combined to form long-tail lipids under conditions that simulated evaporating tidepools. The lipids self-assembled into small, water-filled sacs, which in many were like cell membranes.

13. ___ We still don't know how DNA entered the picture.

14. ___ Rocks collected from Mars, meteorites, and the moon contain chemical carbon-based precursors that must have been present on early Earth.

15. ___ This differentiation resulted in the formation of a crust of basalt, granite, and other types of low-density rock, a rocky region of intermediate density (the mantle), and a high-density, partially molten core of nickel and iron.

16. ___ When the crust finally cooled and solidified, water condensed into clouds and the rains began. For millions of years, runoff from rains stripped mineral salts and other compounds from Earth's parched rocks.

17. ___ The close association of enzymes, ATP, and other molecules would have promoted chemical interactions. Sunlight energy alone could have driven the spontaneous formation of RNA molecules.

18. ___ Even if amino acids did form in the early seas, they wouldn't have lasted long. In water, the favored direction of most spontaneous reactions is toward hydrolysis, not condensation.

19. ___ Between 4.6 and 4.5 billion years ago, the outer regions of the cloud cooled.

20. ___ At first, it probably consisted of gaseous hydrogen ($H_2$), nitrogen ($N_2$), carbon monoxide (CO), and carbon dioxide ($CO_2$).

## Sequence

Earth's history has been divided into four great eras that are based on four abrupt transitions in the fossil record. The oldest era has been subdivided. Arrange the eras in correct chronological sequence from the *oldest* to the *youngest*.

21. ___     A. Mesozoic

22. ___     B. Cenozoic

23. ___     C. Proterozoic

24. ___     D. Archean

25. ___     E. Paleozoic

## Matching

Choose the most appropriate answer for each.

26. ___ Archean
27. ___ Mesozoic
28. ___ Pangea
29. ___ Proterozoic
30. ___ plate tectonic theory
31. ___ Paleozoic
32. ___ Gondwana
33. ___ Cenozoic
34. ___ influenced the evolution of life
35. ___ geologic time scale

A. The "modern era" of geologic time
B. Source of the most ancient fossils prior to subdivision of this era
C. An ancient continent that drifted southward from the tropics, across the south polar region, then northward
D. The boundaries mark the times of mass extinctions
E. The first era of the geologic time scale; a recent subdivision of the Proterozoic
F. An era that follows the Paleozoic
G. Plumes of molten rock from Earth's mantle spread out beneath the crust or break through it
H. An era whose fossils followed those of the Proterozoic
I. Formed from fusion with Gondwana and other land masses; a single world continent that extended from pole to pole
J. Changes in land masses, the oceans, and the atmosphere

## 17 - II. ORIGIN OF PROKARYOTIC AND EUKARYOTIC CELLS (pp. 270 - 271)
### Focus on Science: WHERE DID ORGANELLES COME FROM? (pp. 272 - 273)
### LIFE IN THE PALEOZOIC ERA (pp. 274 - 275)

*Selected Words: hydrothermal* vents, "Precambrian" times, Lynn Margulis, Euglena, Nitrobacter, Cyanophora, Psilophyton, Dimetrodon, Pangea

### Boldfaced, Page-Referenced Terms

(270) prokaryotic cells _____

_____

(270) archaebacteria _____

_____

(270) eubacteria _____

_____

(270) eukaryotic cells _____

_____

(270) stromatolites _____

_____

(272) theory of endosymbiosis _____

_____

(273) protistans _____

_____

## Choice

For questions 1 - 15, choose from the following:

a. Archean     b. Proterozoic

1. ___ By 2.5 billion years ago, the noncyclic pathway of photosynthesis had evolved in some eubacterial species.

2. ___ By 1.2 billion years ago, eukaryotes had originated.

3. ___ The first prokaryotic cells emerged.

4. ___ Oxygen, one of the by-products of photosynthesis began to accumulate.

5. ___ There was an absence of free oxygen.

6. ___ There was a divergence of the original prokaryotic lineage into three evolutionary directions.

7. ___ About 800 million years ago, stromatolites began a dramatic decline.

8. ___ Between 3.5 and 3.2 billion years ago, the cyclic pathway of photosynthesis evolved in some species of eubacteria.

9. ___ Fermentation pathways were the most likely sources of energy.

10. ___ An oxygen-rich atmosphere stopped the further chemical origin of living cells.

11. ___ Food was available, predators were absent, and biological molecules were free from oxygen attacks.

12. ___ Aerobic respiration became the dominant energy-releasing pathway.

13. ___ Archaebacteria and eubacteria arose.

14. ___ The first cells may have originated in tidal flats or on the seafloor, in muddy sediments warmed by heat from volcanic vents.

15. ___ Near the shores of the supercontinent Laurentia, small, soft-bodied animals were leaving tracks and burrows on the seafloor.

## Fill-in-the Blanks

The most important feature of eukaryotic cells is the abundance of membrane-bound (16) _____ in the cytoplasm. The origin of these structures remains a mystery. Some probably evolved gradually, through (17) _____ mutations and natural selection. Some extant prokaryotic cells do possess infoldings of the (18) _____ membrane, a site where enzymes and other metabolic agents are embedded. In prokaryotic cells that were ancestral to eukaryotic cells, similar membranous infoldings may have served as a (19) _____ from the surface to deep within the cell. These membranes may have evolved into (20) _____ channels and into an envelope around the DNA. The advantage of such membranous enclosures may have been to protect (21) _____ and their products from foreign invader cells. Evidence for this is that yeasts, simple nucleated eukaryotic cells, have an abundance of (22) _____, which they transfer from cell to cell much as prokaryotic cells do.

    As the evolution of eukaryotes proceeded, accidental partnerships between different (23) _____ species must have formed countless times. Some partnerships resulted in the origin of mitochondria, chloroplasts, and other organelles. This is the theory of (24) _____, developed in greatest detail by Lynn Margulis. According to this theory, (25) _____ arose after the noncyclic pathway of photosynthesis emerged and oxygen had accumulated to significant levels.

(26) _____ transport systems had been expanded in some bacteria to include extra cytochromes to donate electrons to (27) _____ in a system of aerobic respiration. By 1.2 billion years ago or earlier, amoebalike ancestors of eukaryotes were engulfing aerobic (28) _____ and perhaps forming endocytic vesicles around the food for delivery to the cytoplasm for digestion. Some aerobic bacteria resisted digestion and thrived in a new, protected, nutrient-rich environment. In time they released extra (29) _____, which the hosts came to depend on for growth, greater activity, and assembly of other structures. The guest aerobic bacteria came to depend on (30) _____ metabolic functions. The aerobic and anaerobic cells were now (31) _____ of independent existence. The guests had become (32) _____, supreme suppliers of (33) _____.

Mitochondria are bacteria-size and replicate their own (34) _____, dividing independently of the host cell's division process. In addition, (35) _____ may be descended from aerobic eubacteria that engaged in oxygen-producing photosynthesis. Such cells may have been engulfed, resisted digestion, and provided their respiring (36) _____ cells with needed oxygen. Chloroplasts resemble some (37) _____ in metabolism and overall nucleic acid sequence. Chloroplasts have self-replicating DNA, and they divide independently of the host cell's division.

New cells appeared on the evolutionary stage that were equipped with a nucleus, ER membranes, and mitochondria or chloroplasts (or both); they were the first eukaryotic cells, the first (38) _____.

## Sequence

Refer to Figure 17.22 in the text. Study of the geologic record reveals that as the major events in the evolution of the earth and its organisms occurred, there were periodic major *extinctions* of organisms followed by major *radiations* of organisms. Using the list below, arrange the letters of the extinctions and radiations in the approximate order in which they occurred, from *youngest* to *oldest*.

39. ___    A. Pangea, worldwide ocean forms; shallow seas squeezed out. Major radiations of reptiles, gymnosperms.

40. ___    B. Origin of animals with hard parts. Simple marine communities.

41. ___    C. Mass extinction of many marine invertebrates, most fishes.

42. ___    D. Gondwana moves south. Major radiations of marine invertebrates, early fishes.

43. ___    E. Laurasia forms. Gondwana moves north. Vast swamplands, early vascular plants. Radiation of fishes continues. Origin of amphibians.

44. ___    F. Mass extinction. Nearly all species in seas and on land perish.

45. ___    G. Tethys sea forms. Recurring glaciations. Major radiations of insects, amphibians. Spore-bearing plants dominant; gymnosperms present; origin of reptiles.

## Chronology of Events—Geologic Time

Refer to Figure 17.8 in the text. From the list of evolutionary events above (A-G), select the letters occurring in a particular era of time by circling the appropriate letters.

46. Border of Mesozoic-Paleozoic:    A-B-C-D-E-F-G
47. Paleozoic:                       A-B-C-D-E-F-G

# 17 - III. LIFE IN THE MESOZOIC ERA (pp. 276 - 277)
## *Focus on the Environment:* HORRENDOUS END TO DOMINANCE (p. 278)
## LIFE IN THE CENOZOIC ERA (p. 279)
## SUMMARY OF EARTH AND LIFE HISTORY (pp. 280 - 281)

*Selected Words:* Lystrosaurus, Triceratops, Velociraptor, *global broiling theory,* Smilodon, Laurentia, Gondwana, Laurasia

## Boldfaced, Page-Referenced Terms

(278) asteroids _____

_____

(281) mass extinction _____

_____

(281) radiation _____

_____

## Sequence

Refer to Figure 17.22 in the text. Study of the geologic record reveals that as the major events in the evolution of Earth and its organisms occurred, there were periodic major *extinctions* of organisms followed by major *radiations* of organisms. Using the list below, arrange the letters of the extinctions and radiations in the approximate order in which they occurred, from *youngest* to *oldest.*

1. ___    A. Pangea breakup begins. Rich marine communities. Major radiations of dinosaurs.

2. ___    B. Asteroid impact? Mass extinction of all dinosaurs and many marine organisms.

3. ___    C. Recovery, radiations of marine invertebrates, fishes, dinosaurs. Gymnosperms the dominant land plants. Origin of mammals.

4. ___    D. Major glaciations. Modern humans emerge and begin what may be the greatest mass extinction of all time on land, starting with Ice Age hunters.

5. ___    E. Unprecedented mountain building as continents rupture, drift, collide. Major climatic shifts; vast grasslands emerge. Major radiations of flowering plants, insects, birds, mammals. Origin of earliest human forms.

6. ___    F. Pangea breakup continues, broad inland seas form. Major radiations of marine invertebrates, fishes, insects, dinosaurs, origin of angiosperms (flowering plants).

7. ___    G. Asteroid impact? Mass extinction of many organisms in seas, some on land; dinosaurs, mammals survive.

## Chronology of Events—Geologic Time

Refer to Figure 17.8 in the text. From the list of evolutionary events above (A-G), select the letters occurring in a particular era of time by circling the appropriate letters.

8. Cenozoic:                      A-B-C-D-E-F-G
9. Border of Cenozoic-Mesozoic:   A-B-C-D-E-F-G
10. Mesozoic:                     A-B-C-D-E-F-G

## Complete the Table

11. Refer to Figure 17.8 in the text. To review some of the events that occurred in the geologic past, complete the table below by entering the geologic era (Archean, Proterozoic, Paleozoic, Mesozoic, and Cenozoic) and the *approximate* time in millions of years since the time of the events.

| Era | Time | Events |
|-----|------|--------|
| a. | | A few reptile lineages give rise to mammals and the dinosaurs. |
| b. | | Formation of Earth's crust, early atmosphere, oceans; chemical evolution leading to the origin of life. |
| c. | | Origin of amphibians. |
| d. | | Origin of animals with hard parts. |
| e. | | Rocks 3.5 billion years old contain fossils of well-developed prokaryotic cells that probably lived in tidal mud flats. |
| f. | | Flowering plants emerge, gymnosperms begin their decline. |
| g. | | In the Carboniferous, there were major radiations of insects and amphibians; gymnosperms present; origin of reptiles. |
| h. | | Oxygen accumulated in the atmosphere. |
| i. | | Before the close of this era, the first photosynthetic bacteria had evolved. |
| j. | | Insects, amphibians, and early reptiles flourished in the swamp forests of the Permian. |
| k. | | Most of the major animal phyla evolved in rather short order. |
| l. | | Dinosaurs ruled. |
| m. | | Origin of aerobic metabolism; origin of protistans algae, fungi, animals. |
| n. | | Grasslands emerge and serve as new adaptive zones for plant-eating mammals and their predators. |
| o. | | Humans destroy habitats and many species. |
| p. | | The first ice age initiated the first global mass extinction; reef life everywhere collapsed. |
| q. | | The invasion of land begins; small stalked plants establish themselves along muddy margins, and the lobe-finned fishes ancestral to amphibians move onto land. |

## Self-Quiz

___ 1. Between 4.6 and 4.5 billion years ago, _____.
  a. Earth was a thin-crusted inferno
  b. the outer regions of the cloud from which our solar system formed was cooling
  c. life originated
  d. the atmosphere was laden with oxygen

___ 2. More than 3.8 billion years ago, _____.
  a. Earth was a thin-crusted inferno
  b. the outer regions of the cloud from which our solar system formed was cooling
  c. life originated
  d. the atmosphere was laden with oxygen

For questions 3 - 10, choose from these answers:
  a. Archean
  b. Cenozoic
  c. Mesozoic
  d. Paleozoic
  e. Proterozoic

___ 3. Dinosaurs and gymnosperms were the dominant forms of life during the _____ era.

___ 4. The Alps, Andes, Himalayas, and Cascade Range were born during major reorganization of land masses early in the _____ era.

___ 5. The composition of Earth's atmosphere changed during the _____ era from one that was anaerobic to one that was aerobic.

___ 6. Invertebrates, primitive plants, and primitive vertebrates were the principle groups of organisms on Earth during the _____ era.

___ 7. Before the close of the _____ era, the first photosynthetic bacteria had evolved.

___ 8. The _____ era ended with the greatest of all extinctions, the Permian extinction.

___ 9. Late in the _____ era, flowering plants arose and underwent a major radiation.

___ 10. The _____ era included adaptive zones into which plant-eating mammals and their predators radiated.

## Chapter Objectives/Review Questions

This section lists general and detailed chapter objectives that can be used as review questions. You can make maximum use of these items by writing answers on a separate sheet of paper. Fill in answers where blanks are provided. To check for accuracy, compare your answers with information given in the chapter or glossary.

| Page | | Objectives/Questions |
|---|---|---|
| (264) | 1. | Cloudlike remnants of stars are mostly _____, but they also contain water, iron, silicates, hydrogen cyanide, methane, ammonia, formaldehyde, and many other simple inorganic and organic substances. |
| (264) | 2. | Describe the formation of early Earth prior to the formation of the first atmosphere. |
| (264) | 3. | Be able to list the probable chemical constituents of Earth's first atmosphere. |
| (264) | 4. | _____ and water were probably not present in early Earth's atmosphere. |
| (265) | 5. | Describe experimental evidence provided by Stanley Miller (and others) that the formation of biological molecules from simple precursor molecules might have occurred on early Earth. |
| (265) | 6. | _____ might have been the medium on which condensation reactions yielding complex organic compounds occurred. |

(266)    7.  During the first _____ years of Earth's history, enzymes, ATP, and other molecules could have assembled at the same location; their close association would have promoted chemical interactions—and the beginning of _____ pathways.

(267)    8.  Although simple, self-replicating systems of RNA, enzymes, and coenzymes have been created in the laboratory, the chemical ancestors of _____ and DNA remain unknown.

(267)    9.  Describe the experiments performed by Sidney Fox that aided understanding of the the origin of plasma membranes.

(268)   10.  Describe plate tectonic theory and its application to the evolution of Earth's geology as we understand it today.

(269)   11.  Be able to list the five geologic eras in proper order, from oldest to youngest.

(269)   12.  The first era of the geologic time scale is now called the _____, "the beginning."

(270)   13.  List the three evolutionary directions of the original prokaryotic lineage during the early Archean era.

(270)   14.  Populations of some early anaerobic eubacteria utilized the cyclic photosynthetic pathway; their populations formed huge mats called _____.

(273)   15.  Describe how the endosymbiosis theory may help to explain the origin of eukaryotic cells.

(280–281) 16.  Be able to generally discuss the important geological and biological events occurring throughout the Archean, Proterozoic, Paleozoic, Mesozoic, and Cenozoic eras.

## Integrating and Applying Key Concepts

As Earth becomes increasingly loaded with carbon dioxide and various industrial waste products, how do you think living forms on Earth will evolve to cope with these changes?

# 18

# BACTERIA, VIRUSES, AND PROTISTANS

## Interactive Exercises

### 18 - I. CHARACTERISTICS OF BACTERIA (pp. 284 - 287)
### BACTERIAL REPRODUCTION (p. 288)

*Selected Words:* Escherichia coli, *photoautotrophic, photoheterotrophic, chemoautotrophic, chemoheterotrophic,* Staphylococcus aureus, *Gram-positive, Gram-negative,* Bacillus cereus

*Boldfaced, Page-Referenced Terms*

(284) microorganisms _____

_____

(284) pathogens _____

_____

(287) prokaryotic cells _____

_____

(287) cell wall _____

_____

(287) bacterial flagella (sing., flagellum) _____

_____

(288) binary fission _____

_____

(288) plasmid _____

_____

(288) bacterial conjugation _____

_____

## Choice

For questions 1 - 5, choose from the following:

      a. archaebacteria    b. chemoautotrophic eubacteria    c. chemoheterotrophic eubacteria

          d. photoautotrophic eubacteria    e. photoheterotrophic eubacteria

1. ___ Use $CO_2$ from the environment as a source of carbon atoms *and* use electrons, hydrogen, and energy released from reactions between various inorganic substances to assemble chains of carbon (food storage)

2. ___ *Use $CO_2$ and $H_2O$ from the environment* as sources of carbon, hydrogen, and oxygen atoms, *and use sunlight* to power the assembly of food storage molecules

3. ___ *Cannot* use $CO_2$ from the environment to construct own carbon chains; instead obtain nutrients from the products, wastes, or remains of other organisms; can break down glucose to pyruvate and follow it with fermentation of some sort or another

4. ___ *Cannot* use $CO_2$ from the environment to construct own cellular molecules, but *can* absorb sunlight and transfer some of that energy to the bonds of ATP; must obtain food molecules (carbon chains) produced by other organisms to construct their own molecules

5. ___ Cell walls lack peptidoglycan.

## Fill-in-the-Blanks

Bacteria are microscopic, (6) _____ cells having one bacterial chromosome and, often, a number of smaller (7) _____ . The cells of nearly all bacterial species have a(n) (8) _____ around the plasma membrane, and a(n) (9) _____ or slime layer surrounding the cell wall. Typically, the width or length of these cells falls between 1 and 10 (10) (choose one) ❏ millimeters ❏ nanometers ❏ centimeters ❏ micrometers. Most bacteria reproduce by (11) _____ _____. Spherical bacteria are (12) _____, rod-shaped bacteria are (13) _____, and helical bacteria are (14) _____. (15) Gram-_____ bacteria retain the purple stain when washed with alcohol.

# 18 - II. BACTERIAL CLASSIFICATION (p. 289)
## MAJOR GROUPS OF BACTERIA (pp. 290 - 291)

*Selected Words:* <u>Anabaena</u>, <u>Lactobacillus</u>, <u>Azospirillum</u>, <u>Rhizobium</u>, <u>Clostridium</u> botulinum, *botulism,* *tetanus,* <u>Borrelia</u> burgdorferi, *Lyme disease,* <u>Myxococcus</u> xanthus

## *Boldfaced, Page-Referenced Terms*

(289) photoheterotrophic eubacteria _____

_____

(290) Archaebacteria _____

_____

(290) methanogens _____

_____

(290) halophiles _____

_____

(290) extreme thermophiles _____

_____

(290) Eubacteria _____

_____

(2909) photoautotrophic eubacteria _____

_____

(290) heterocysts _____

_____

(290) chemoautotrophic eubacteria _____

_____

(290) chemoheterotrophic eubacteria _____

_____

(291) endospore _____

_____

(291) fruiting bodies _____

_____

## Fill-in-the-Blanks

In many respects, the cell structure, metabolism, and nucleic acid sequences are unique to the

(1) _____ ; for example, none has a cell wall that contains (2) _____ . On the other hand,

(3) _____ are far more common than the three rather unusual types of (1). The most common

photoautotrophic bacteria are the (4) _____ (also called the blue-green algae). *Anabaena* and others

of the (4) group produce oxygen during photosynthesis. Heterocysts are cells in *Anabaena* that carry out

(5) _____ _____ . Many species of chemoautotrophic eubacteria affect the global cycling of

nitrogen, sulfur, (6) _____ and other nutrients. Sugarcane and corn plants benefit from a

(7) _____ -fixing spirochete, *Azospirillum*. Beans and other legumes benefit from the nitrogen-fixing

activities of (8) _____ , which dwells in their roots.

When environmental conditions become adverse, many bacteria form (9) _____ , which resist

moisture loss, irradiation, disinfectants, and even acids. Endospore formation by bacteria can kill

humans if they enter the food supply and are not killed by high temperature and high (10) _____ .

Anaerobic bacteria can live in canned food, reproduce, and produce deadly toxins. Two examples of

pathogenic (disease-causing) bacteria that form endospores harmful to humans are (11) _____

_____ and (12) _____ _____ . (13) _____ _____ may be the most common tick-

borne disease in the United States by now; tick bites deliver the (14) _____ from one host to

another. Bacterial behavior depends on (15) _____ _____ , which change shape when they

absorb or connect with chemical compounds. Cyanobacteria require (16) _____ as their source of

energy that drives their metabolic activities. Most of the world's bacteria are (17) (choose one)

❐ producers   ❐ consumers   ❐ decomposers, so we think of them as "good" heterotrophs. Antibiotics

such as (18) _____ block protein synthesis in their target cells; other antibiotics such as

(19) _____ disrupt the formation of cell walls in their target cells. *Escherichia coli*, which dwell in

our gut, synthesize vitamin (20) ____ and substances useful in digesting fats.

## Matching

Match each of the items in the left-hand column with a lowercase letter designating its principal bacterial group and an uppercase letter denoting its best descriptor from the right-hand column.

a. Archaebacteria   b. Chemoautotrophic eubacteria   c. Chemoheterotrophic eubacteria

d. Photoautotrophic eubacteria   e. Photoheterotrophic eubacteria

21. ___, ___ *Anabaena*

22. ___, ___ *Bacillus, Clostridium*

23. ___, ___ *Escherichia coli*

24. ___, ___ *Halobacterium*

25. ___, ___ *Lactobacillus*

26. ___, ___ *Methanobacterium*

27. ___, ___ *Nitrobacter, Nitrosomonas*

28. ___, ___ *Rhizobium, Agrobacterium*

29. ___, ___ *Rhodospirillum*

30. ___, ___ *Salmonella*

31. ___, ___ *Spirochaeta, Treponema*

32. ___, ___ *Staphylococcus, Streptococcus*

33. ___, ___ *Streptomyces, Actinomyces*

34. ___, ___ *Thermoplasma, Sulfolobus*

A. Lives in anaerobic sediments of lakes and in animal gut; chemosynthetic; used in sewage treatment facilities

B. Purple; generally in anaerobic sediments of lakes or ponds; do not produce oxygen, do not use water as a source of electrons

C. Endospore-forming rods and cocci that live in the soil and in the animal gut; some major pathogens

D. Gram-positive cocci that live in the soil and in the skin and mucous membranes of animals; some major pathogens

E. Gram-positive nonsporulating rods that ferment plant and animal material; some are important in dairy industry; others contaminate milk, cheese

F. In acidic soil, hot springs, hydrothermal vents on seafloor; may use sulfur as a source of electrons for ATP formation

G. Live in extremely salty water; have a unique form of photosynthesis

H. Gram-negative aerobic rods and cocci that live in soil or aquatic habitats or are parasites of animals and/or plants; some fix nitrogen

I. Nitrifying bacteria that live in the soil, fresh water, and marine habitats; play a major role in the nitrogen cycle

J. Gram-negative anaerobic rod that inhabits the human colon where it produces vitamin K

K. Major gram-negative pathogens of the human gut that cause specific types of food poisoning

L. Mostly in lakes and ponds; cyanobacteria; produce $O_2$ from water as an electron donor

M. Major producer of antibiotics; an actinomycete that lives in soil and some aquatic habitats

N. Helically coiled, motile parasites of animals; some are major pathogens

## 18 - III. THE VIRUSES (pp. 292 - 293)
### VIRAL MULTIPLICATION CYCLES (pp. 294 - 295)
### *Focus on Health:* EBOLA AND OTHER EMERGING PATHOGENS (p. 295)

**Selected Words:** *helical, polyhedral, enveloped or complex* viruses, *lytic, lysogenic,* Herpes simplex, *cold sores,* Ebola, *sporadic, endemic*

### Boldfaced, Page-Referenced Terms

(292) virus _____

_____

(292) bacteriophages _____

_____

(293) prions _____

_____

(293) viroids _____

_____

(294) lysis _____

_____

(295) infection _____

_____

(295) disease _____

_____

(295) emergent pathogens _____

_____

(295) epidemic _____

_____

(295) pandemic _____

_____

## Short Answer

1.  a. State the principal characteristics of viruses (p. 292). _____

_____

   b. Describe the structure of viruses (pp. 292 - 293). _____

_____

   c. Distinguish between the ways viruses replicate themselves (pp. 294 - 295). _____

_____

2.  a. List five specific viruses that cause human illness (pp. 293 - 295). _____

_____

   b. Describe how each virus in (a) does its dirty work (pp. 294  - 295). _____

_____

_____

_____

_____

_____

## Fill-in-the-Blanks

A(n) (3) _____ is a noncellular, nonliving infectious agent, each of which consists of a central
(4) _____ _____ core surrounded by a protective (5) _____ _____. (6) _____
contain the blueprints for making more of themselves but cannot carry on metabolic activities.
Chickenpox and shingles are two infections caused by DNA viruses from the (7) _____ category.
Naked strands or circles of RNA that lack a protein coat are called (8) _____. (9) _____ are
RNA viruses that infect animal cells, cause AIDS, and follow (10) _____ pathways of replication.
(11) _____ are the usual units of measurement with which to measure viruses, while
microbiologists measure bacteria and protistans in terms of (12) _____. A bacterium 86
micrometers in length is (13) _____ nanometers long. During a period of (14) _____, viral
genes remain inactive inside the host cell and any of its descendants. Pathogenic protein particles are
called (15) _____.

## Identification

Identify the virus that causes the following illnesses by writing the name of the virus in the blank preceding the disease. Then, tell whether it is a DNA virus or an RNA virus.

_____, ____     16. Common colds

_____, ____     17. AIDS, leukemia

_____, ____     18. Cold sores, chickenpox

## Matching

Match each item below with the correct lettered description.

19. ___ antibiotic

20. ___ antiviral drug

21. ___ bacteriophage

22. ___ endemic

23. ___ epidemic

24. ___ lysogenic pathway

25. ___ lytic pathway

26. ___ microorganism

27. ___ pathogen

28. ___ sporadic

29. ___ vector

30. ___ virus

A. An agent that transports to other creatures or temporarily houses pathogens coming from an infected organism
B. Disease that breaks out irregularly, affects few organisms
C. Chemical substance that interferes with gene expression or other normal functions of bacteria
D. Disease abruptly spreads through large portions of a population
E. Disease that occurs continuously, but is localized to a relatively small portion of the population
F. Acyclovir and AZT, for example
G. Any organism too small to be seen without a microscope
H. A virus that infects a bacterium
I. Damage and destruction to host cells occurs quickly
J. Noncellular infectious agent that must take over a living cell in order to reproduce itself
K. Viral nucleic acid is integrated into the nucleic acid system of the host cell and replicated during this time
L. Any disease-causing organism or agent

# 18 - IV. PROTISTAN CLASSIFICATION (p. 296)
## PREDATORY AND PARASITIC MOLDS (pp. 296 - 297)

***Selected Words:*** *protistans, late blight,* cellular *slime molds,* plasmodial *slime molds,* <u>Saprolegnia</u>, <u>Phytophthora infestans</u>, <u>Dictyostelium discoideum</u>

## *Boldfaced, Page-Referenced Terms*

(296) protistans _____

_____

(296) heterotrophs _____

_____

(296) autotrophs _____

_____

(296) chytrids _____

_____

(296) water molds _____

_____

(296) slime molds _____

_____

## *Choice*

1. For each structure, write a P if it is a prokaryotic (bacterial) characteristic, and write an E if it is a eukaryotic characteristic.

| | |
|---|---|
| a. Double-membraned nucleus | |
| b. Mitochondria present | |
| c. Reproduce by binary fission | |
| d. Engage in mitosis | |
| e. Circular chromosome present | |
| f. Endoplasmic reticulum present | |
| g. Cilia or flagella with 9+2 core | |

## More Choice

2. For each group below, write a "+" if it has chloroplasts, and a "−" if members of that group lack chloroplasts and the ability to do photosynthesis.

| | |
|---|---|
| a. Brown algae | f. Protozoans |
| b. Chytrids | g. Red algae |
| c. Chrysophytes | h. Slime molds |
| d. Dinoflagellates | i. Sporozoans |
| e. Green algae | j. Water molds |

## Fill-in-the-Blanks

(3) _____ and (4) _____ _____ are the only fungi that produce motile spores; this is a primitive trait that may resemble ancestral fungi that lived several hundred million years ago in watery habitats. Water molds are only distantly related to other fungi and are major (5) _____ of aquatic habitats. The cells of some (6) _____ _____ differentiate and form (7) _____ _____, stalked structures which produce gametes or (8) _____. The cytoplasm of a (9) _____ slime mold flows throughout the entire mass, distributing nutrients and oxygen within the organism, which may grow to become several square (10)_____ in size .

(units)

Several parasitic water molds cause difficulties for their hosts or for humans. *Phytophthora infestans*, known as (11) _____, rots potato and tomato plants. *Saprolegnia* destroys tissues of aquarium (12) _____. *Plasmopara viticola* causes (13) _____ in grapes, threatening large vineyards in France and North America. When marsh grasses die, break off, and travel out to sea on the tides, bacteria and (14) _____ decompose the dead grasses by secreting (15) _____ that digest the organic matter to a form that other detritis feeders can use.

## 18 - V. ANIMAL-LIKE PROTISTANS (pp. 298 - 299)
### *Focus on Health:* MALARIA AND THE NIGHT-FEEDING MOSQUITOES (p. 300)

**Selected Words:** Amoeba proteus, Entamoeba histolytica, *amoebic dysentery,* Paramecium, Trichomonas vaginalis, Giardia lamblia, Trypanosoma brucei, *African sleeping sickness,* T. cruzi, *Chagas disease,* Plasmodium, Toxoplasma, *toxoplasmosis, malaria,* Anopheles

## Boldfaced, Page-Referenced Terms

(298) protozoans _____

_____

(298) cysts _____

_____

(298) amoeboid protozoans _____

_____

(298) ciliated protozoans _____

_____

(298) contractile vacuoles _____

_____

(299) protozoan conjugation _____

_____

(299) flagellated protozoans _____

_____

(299) sporozoan _____

_____

## *Fill-in-the-Blanks*

Amoebas move by sending out (1) _____, which surround food and engulf it. (2) _____ secrete a hard exterior covering of calcareous material that is peppered with tiny holes through which sticky, food-trapping pseudopods extend. Needlelike (3) _____ often support the pseudopods. Accumulated shells of (4) _____, which generally have a skeleton of silica (glass), and foraminiferans are key components of many oceanic sediments.

Examples of flagellated protozoans that are parasitic include the (5) _____, two species of which cause African sleeping sickness and Chagas disease.

*Paramecium* is a ciliate that lives in (6) _____ environments and depends on (7) _____ _____ for eliminating the excess water constantly flowing into the cell. *Paramecium* has a (8) _____, a cavity that opens to the external watery world. Once inside the cavity, food particles become enclosed in (9) _____ - _____ _____, where digestion takes place.

(10) _____ is a famous sporozoan that causes malaria. When a particular (11) _____ draws blood from an infected individual, (12) _____ of the parasite fuse to form zygotes, which eventually develop within the mosquito.

## Matching

Choose as many appropriate answers as are applicable for each.

13. ___ *Entamoeba histolytica*

14. ___ foraminiferans

15. ___ *Giardia lamblia*

16. ___ *Paramecium*

17. ___ *Plasmodium*

18. ___ *Trichomonas vaginalis*

19. ___ *Trypanosoma brucei*

A. Ciliophora
B. Mastigophora
C. Sarcodina
D. Sporozoans
E. Amoeboid protozoans
F. Flagellated protozoans
G. African sleeping sickness
H. Malaria
I. Amoebic dysentery
J. Causes intestinal disorders in humans
K. Primary component of many ocean sediments

## Matching

Match the pictures below with the names below.

20. ___

21. ___

22. ___

23. ___

24. ___

A. *Didinium,* a ciliate
B. *Euglena*
C. Flagellated protozoans
D. Foraminiferans
E. *Paramecium*

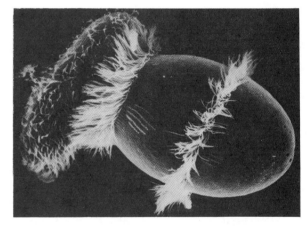

20. _____

21. _____

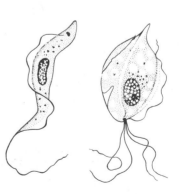

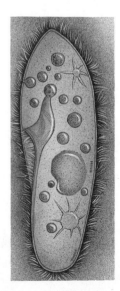

22. _____

23. _____

24. _____

# 18 - VI. THE (MOSTLY) PHOTOSYNTHETIC SINGLE-CELLED PROTISTANS (p. 301)
## THE (MOSTLY) MULTICELLED PHOTOSYNTHETIC PROTISTANS (pp. 302 - 303)

*Selected Words:* Euglena, Gymnodinium breve, Macrocystis, Laminaria, Sargassum, Chlamydomonas, Spirogyra, Volvox, Halimeda, Codium magnum

## *Boldfaced, Page-Referenced Terms*

(301) phytoplankton _____

_____

(301) euglenoids _____

_____

(301) chrysophytes _____

_____

(301) golden algae _____

_____

(301) diatoms _____

_____

(301) dinoflagellates _____

_____

(301) red tides _____

_____

302) red algae _____

_____

(302) brown algae _____

_____

(303) green algae _____

_____

## *Fill-in-the-Blanks*

Euglenoids contain (1) _____, which enable them to carry out photosynthesis.  A(n) (2) _____ of carotenoid pigment granules partly shields a light-sensitive receptor and enables *Euglena* to remain where light is optimal for its activities.  Some strains of *Euglena* can be converted from photosynthetic, chloroplast-containing forms to strains that are (3) _____.

The term (4) "_____" no longer has formal classification significance, because organisms once lumped under that term are now assigned to different kingdoms.  (5) _____ include about 500 species of golden algae and more than 5,600 existing species of golden-brown (6) _____. Photosynthetic chrysophytes contain xanthophylls and (7) _____; those pigments mask the green color of chlorophyll in golden algae and diatoms. Diatom cells have external thin, overlapping "shells" of (8) _____ that fit together like a pill box. Two hundred seventy thousand metric tons of (9) _____ _____ are extracted annually from a quarry near Lompoc, California, and are used to make abrasives, (10) _____ and insulating materials. Dinoflagellates are photosynthetic members of marine (11) _____ and freshwater ecosystems; some forms are also heterotrophic. Some (12) _____ undergo explosive population growth and color the seas red or brown, causing a red tide that may kill hundreds or thousands of fish and, occasionally, people.

Several species of red algae secrete (13) _____ (used in culture media) as part of their cell walls. Most red algae live in (14) _____ habitats. Some red algae have stonelike cell (15) _____, participate in coral reef building, and are major producers. The (16) _____ algae live offshore or in intertidal zones and have many representatives with large sporophytes known as kelps; some species produce (17) _____, a valuable thickening agent. Green algae are thought to be ancestral to more complex plants, because they have the same types and proportions of (18) _____ pigments, have (19) _____ in their cell walls, and store their carbohydrates as (20) _____.

## Complete the Table

21. Complete the table below.

| Type of Alga | Typical Pigments | Probably Evolved From | Uses by Humans | Representatives |
|---|---|---|---|---|
| Red algae (Rhodophyta) | a. | b. | c. | d. |
| Brown algae (Phaeophyta) | e. | f. | g. | h. |
| Green algae (Chlorophyta) | i. | j. | k. | l. |

## Label and Match

Identify each indicated part of the illustration below by entering its name in the appropriate numbered blank. Choose from the following terms: cytoplasmic fusion, asexual reproduction, resistant zygote, fertilization, zygote, meiosis and germination, spore mitosis, gametes meet. Complete the exercise by matching from the list below, entering the correct letter in the parentheses following each label.

22. _____ ( )

23. _____ _____ ( )

24. _____ and _____ ( )

25. _____ _____ ( )

26. _____ _____ ( )

27. _____ _____ ( )

28. _____ _____ ( )

29. _____ ( )

A. Fusion of two gametes of different mating types
B. A device to survive unfavorable environmental conditions
C. More spore copies are produced.
D. Fusion of two haploid nuclei
E. Haploid cells form smaller haploid gametes when nitrogen levels are low.
F. Formed after fertilization
G. Two haploid gametes coming together
H. Reduction of the chromosome number

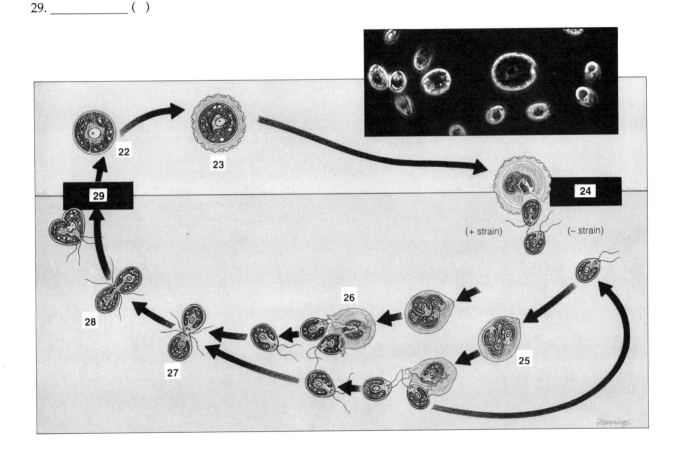

---

# Self-Quiz

___1. Which of the following diseases is not caused by a virus?
   a. smallpox
   b. polio
   c. influenza
   d. syphilis

___2. Bacteriophages are _____.
   a. viruses that parasitize bacteria
   b. bacteria that parasitize viruses
   c. bacteria that phagocytize viruses
   d. composed of a protein core surrounded by a nucleic acid coat

___3. Many biologists believe that chloroplasts are descendants of _____ that were able to live symbiotically within a predatory host cell.
a. aerobic bacteria with "extra" cytochromes
b. endosymbiont prokaryotic autotrophs
c. dinoflagellates
d. bacterial heterotrophs

___4. Which of the following specialized structures is not correctly paired with a function?
a. gullet—ingestion
b. cilia—food gathering
c. contractile vacuole—digestion
d. anal pore—waste elimination

___5. Population "blooms" of _____ cause red tides and extensive fish kills.
a. *Euglena*
b. specific dinoflagellates
c. diatoms
d. *Plasmodium*

___6. Which of the following protists does not cause great misery to humans?
a. *Dictyostelium discoideum*
b. *Entamoeba histolytica*
c. *Plasmodium*
d. *Trypanosoma brucei*

___7. For an organism to be considered truly multicellular, _____.
a. its cells must be heterotrophic
b. there must be division of labor and cellular specialization

c. the organisms cannot be parasitic
d. the organisms must at least be motile

___8. Red, brown, and green algae are found in the kingdom _____.
a. Plantae
b. Monera
c. Protista
d. all of the above

___9. Red algae _____.
a. are primarily marine organisms
b. are thought to have developed from green algae
c. contain xanthophyll as their main accessory pigments
d. all of the above

___10. Stemlike structure, leaflike blades, and gas-filled floats are found in the species of

_____.
a. red algae
b. brown algae
c. bryophytes
d. green algae

___11. Because of pigmentation, cellulose walls, and starch storage similarities, the _____ algae are thought to be ancestral to more complex plants.
a. red
b. brown
c. blue-green
d. green

## Matching

Match all applicable letters with the appropriate terms. A letter may be used more than once, and a blank may contain more than one letter.

12. _____ *Amoeba proteus*

13. _____ diatoms

14. _____ *Dictyostelium*

15. _____ foraminifera

16. _____ *Gymnodinium breve* (red tide)

17. _____ *Paramecium*

18. _____ *Plasmodium*

19. _____ *Volvox*

A. Protista
B. Slime mold
C. Photosynthetic flagellates
D. Dinoflagellates
E. Obtain food by using pseudopodia
F. Causes malaria
G. A sporozoan
H. A ciliate
I. Live in "glass" houses
J. Live in hardened shells that have thousands of tiny holes, through which pseudopodia protrude

*Matching*

20. ___    A.  Amoeba proteus
21. ___    B.  Diatom
           C.  Paramecium
22. ___    D.  foraminiferans
23. ___

20. _____

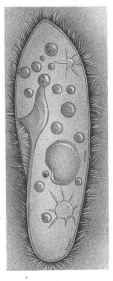

21. _____

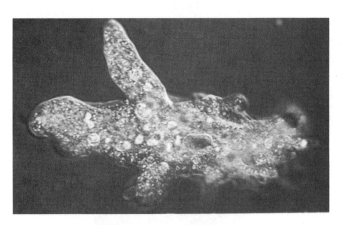

22. _____

23. _____

## Matching

Match pictures 24 - 29 with the terms below.

24. ___
25. ___
26. ___
27. ___
28. ___
29. ___

A. *Bacillus*
B. bacteriophage
C. *Clostridium tetani*
D. *Cyanobacterium*
E. *Herpes simplex*
F. HIV

25. _____

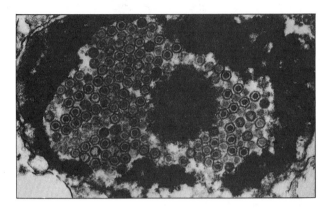

24. _____

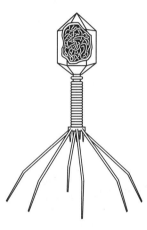

27. _____

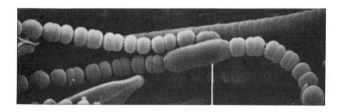

26. _____

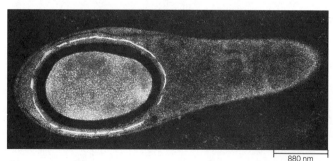

880 nm

28. _____

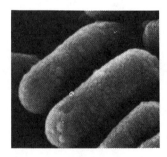

29. _____

## Matching

Match all applicable letters with the appropriate terms. A letter may be used more than once, and a blank may contain more than one letter.

30. _____ *Anabaena, Nostoc*
31. _____ *Clostridium botulinum*
32. _____ *Escherichia coli*
33. _____ *Herpes simplex*
34. _____ HIV
35. _____ *Lactobacillus*
36. _____ *Staphylococcus*

A. Bacteria
B. Virus
C. Cyanobacteria
D. Gram-positive eubacteria
E. Cause cold sores and a type of venereal disease
F. Associated with AIDS, ARC

---

# Chapter Objectives/Review Questions

| Page | | Objectives/Questions |
|---|---|---|
| (286) | 1. | Distinguish chemoautotrophs from photoautotrophs. |
| (286 - 287) | 2. | Describe the principal body forms of bacteria (kingdom: Monera). |
| (288) | 3. | Explain how, with no nucleus, or few if any, membrane-bound organelles, bacteria reproduce themselves and obtain energy to carry on metabolism. |
| (289 - 290) | 4. | State the ways in which archaebacteria differ from eubacteria. |
| (292 - 293) | 5. | Describe the general structure of viruses, and tell how they are classified into two main groups. |
| (293) | 6. | Name ten different diseases caused by viruses. |
| (294 - 295) | 7. | Describe and distinguish between the two pathways used by viruses to replicate themselves. |
| (296) | 8. | _____ protistans generally make their own food by photosynthesis, but _____ protistans act as decomposers, predators, or parasites in order to obtain the energy that fuels life. |
| (298 - 299) | 9. | State the principal characteristics of the amoebas, radiolarians, and foraminiferans. Indicate how they generally move from one place to another and how they obtain food. |
| (298 - 299) | 10. | List the features common to most ciliated protozoans. |
| (299) | 11. | Two flagellated protozoans that cause human misery are _____ and _____ . |
| (299 - 300) | 12. | Characterize the sporozoan group, identify the group's most prominent representative, and describe the life cycle of that organism. |
| (301) | 13. | How do golden algae resemble diatoms? |
| (301) | 14. | Explain what causes red tides. |
| (302 - 303) | 15. | State the outstanding characteristics of organisms of the red, brown, and green algae divisions. |

---

# Integrating and Applying Key Concepts

The textbook (Figure 39.15) identifies natural gas as a nonrenewable fuel resource, yet there is a group of archaebacteria that produce methane, the burning of which can serve as a fuel for heating and/or cooking. Recall or imagine how these bacteria could be incorporated into a system that could serve human societies by generating methane in a cycle that is renewable. Why did your text categorize natural gas as a nonrenewable resource? Is methane a constituent of natural gas? Why or why not?

# 19

# PLANTS AND FUNGI

## Interactive Exercises

### 19 - I. EVOLUTIONARY TRENDS AMONG PLANTS (pp. 306 - 309)

*Selected Words:* heterospory, homospory

*Boldfaced, Page-Referenced Terms*

(308) vascular plants _____

_____

(308) bryophytes _____

_____

(308) gymnosperms _____

_____

(308) angiosperms _____

(308) root systems _____

(308) shoot systems _____

(308) lignin _____

(308) xylem _____

(308) phloem _____

(308) cuticle _____

(308) stomata _____

(308) sporophyte _____

(308) spores _____

(309) gametophytes _____

(309) pollen grains _____

(309) seed _____

## Matching

Choose the most appropriate answer for each.

1. ___ seedless vascular plants
2. ___ angiosperms
3. ___ vascular plants
4. ___ bryophytes
5. ___ gymnosperms

A. Liverworts, hornworts, and mosses
B. Seed-bearing plants that include cycads, ginkgo, and conifers
C. Whisk ferns, lycophytes, horsetails, and ferns
D. A group of plants producing seeds by means of flowers
E. In general, a large number of plants having internal conducting tissues

## Complete the Table

6. As plants evolved, several key evolutionary events occurred that solved the problems of living in new land environments. Complete the following table to summarize these events. Choose from Phloem, Sporophyte (diploid) dominance of the life cycle, Well-developed shoot systems, Cuticle, Heterospory, Lignin production, Xylem, Seeds, Well-developed root systems, Interaction of young roots and mycorrhizal fungi, Stomata, Pollen grains.

| Evolutionary Trends | Survival Problem Solved |
| --- | --- |
| a. | Provides a large surface area for rapidly taking up soil water and scarce mineral ions; often anchors the plant |
| b. | Consist of stems and leaves that function in the absorption of sunlight energy and $CO_2$ from the air |
| c. | Provides cellular pipelines to distribute water and dissolved ions through plant parts |
| d. | Provides cellular pipelines to distribute sugars and other photosynthetic products |
| e. | Allows extensive growth of stems and branches; a very hard substance that strengthens cell walls, thus allowing erect plant parts to display leaves to sunlight |
| f. | Allows water conservation for plants living on land |
| g. | Provides the main route for plant absorption of carbon dioxide while restricting evaporative water loss |
| h. | A symbiotic relationship that provides many sporophytes with water and scarce nutrients, even in seasonally dry habitats |
| i. | Development of a multicelled plant body that forms through mitotic cell divisions of a new zygote |
| j. | Seed-bearing plants produce two types of spores. |
| k. | One spore type becomes mature, sperm-bearing male gametophytes. |
| l. | Embryos became packaged with protective and nutritive tissues for survival on dry land. |

## Matching

Match each of the following life-cycle sketches with the appropriate description. The order illustrates an evolutionary trend from haploid to diploid dominance during the colonization of land.

7. ____            A. Typical life cycle for some algae (haploid dominance)

8. ____            B. Typical life cycle for bryophytes (haploid dominance)

9. ____            C. Typical life cycle for vascular plants (diploid dominance)

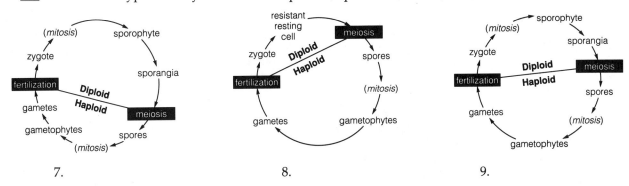

7.            8.            9.

## 19 - II. BRYOPHYTES (pp. 310 - 311)

*Selected Word:* Sphagnum

*Boldfaced, Page-Referenced Terms*

(310) mosses _____

_____

(310) liverworts _____

_____

(310) hornworts _____

_____

*True/False*

If the statement is true, write a T in the blank. If the statement is false, make it correct by changing the underlined word(s) and writing the correct word(s) in the blank.

_____ 1.    Mosses are highly sensitive to <u>water</u> pollution.

_____ 2.    Bryophytes <u>have</u> leaflike, stemlike, and rootlike parts although they do not contain xylem or phloem.

_____ 3.    Most bryophytes have <u>rhizomes</u>, elongated cells or threads that attach gametophytes to soil and serve as absorptive structures.

_____ 4.    Bryophytes are the simplest plants to exhibit a cuticle, cellular jackets around gamete-producing parts, and large gametophytes that retain nutritionally <u>dependent</u> sporophytes.

_____ 5.    The true <u>liverworts</u> are the most common bryophytes.

_____ 6.    Following fertilization, zygotes give rise to <u>gametophytes</u>.

_____ 7.  Each <u>sporophyte</u> consists of a stalk and a sporangium.

_____ 8.  <u>Club</u> moss is a bog moss whose large, dead cells in their leaflike parts soak up five times as much water as cotton.

_____ 9.  Bryophyte sperm reach eggs by movement through <u>air</u>.

_____ 10. Eggs and sperm of moss plants develop at gametophyte shoot tips, in <u>tiny jacketed vessels</u>.

## Fill-in-the-Blanks

The numbered items in the illustration below represent missing information about a typical moss life cycle; complete the numbered blanks in the narrative to supply this information.

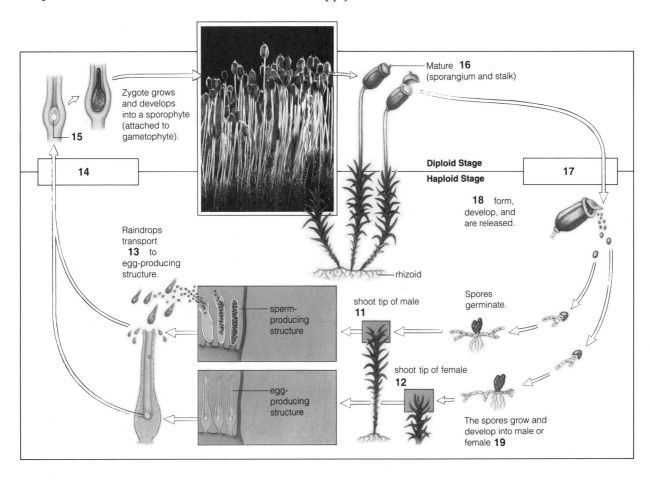

The gametophytes are the green leafy "moss plants."  Sperms develop in jacketed structures at the shoot tip of the male (11) _____ and eggs develop in jacketed structures at the shoot tip of the female (12) _____. Raindrops transport (13) _____ to the egg-producing structure.  (14) _____ occurs within the egg-producing structure.  The (15) _____ grows and develops into a mature (16) _____ (with sporangium and stalk) while attached to the gametophyte.  (17) _____ occurs within the sporangium of the sporophyte where haploid (18) _____ form, develop, and are released. The released spores grow and develop into male or female (19) _____.

**19 - III.** *Focus on the Environment:* **ANCIENT PLANTS, CARBON TREASURES** (p. 311)
**EXISTING SEEDLESS VASCULAR PLANTS** (pp. 312 - 313)

*Selected Words:* Psilotum, Selaginella, Equisetum

*Boldfaced, Page-Referenced Terms*

(311) peat _____

_____

(311) coal _____

_____

(312) whisk ferns _____

_____

(312) lycophytes _____

_____

(312) horsetails _____

_____

(312) ferns _____

_____

(312) rhizomes _____

_____

(312) strobilus _____

_____

*Choice*

For questions 1 - 20 , choose from the following:

    a. whisk ferns    b. lycophytes    c. horsetails    d. ferns    e. applies to a, b, c, and d

1. ___ A group in which only the genus *Equisetum* survives

2. ___ Familiar club mosses growing as mats on forest floors

3. ___ Seedless vascular plants

4. ___ *Psilotum*

5. ___ Rust-colored patches, the sori, are on the lower surface of their fronds.

6. ___ Some tropical species are the size of trees.

7. ___ Ancestral plants living in the Carboniferous; through time, heat, and pressure these plants became peat and coal, the "ancient carbon treasures"

8. ___ Stems were used by pioneers of the American West to scrub cooking pots.

9. ___ The sporophytes have no roots or leaves.

10. ___ Mature leaves are usually divided into leaflets.

11. ___ The sporophyte has vascular tissues.

12. ___ When the spore chamber snaps open, spores catapult through the air.

13. ___ Grow in mud soil of streambanks and in disturbed habitats, such as roadsides and railroad beds

14. ___ Sporophytes have rhizomes and a hollow, photosynthetic, aboveground stem with scalelike leaves.

15. ___ The sporophyte is the larger, longer lived phase of the life cycle.

16. ___ *Lycopodium*

17. ___ The young leaves are coiled into the shape of a fiddlehead.

18. ___ A spore develops into a small, green, heart-shaped gametophyte.

19. ___ *Selaginella* produces two spore types, qualifying it as a heterosporous genus.

20. ___ "Amphibians" of the plant kingdom; life cycles require water

## Fill-in-the-Blanks

The numbered items in the illustration of a generalized fern life cycle below represent missing information; complete the numbered blanks in the narrative to supply this information.

Fern leaves (fronds) of the sporophyte are usually divided into leaflets. The underground stem of the sporophyte is termed a (21) _____. On the undersides of many fern fronds, rust-colored patches (the sori) of spore chambers (sporangia) occur. (22) _____ of diploid cells within each sporangium produces haploid (23) _____. The (24) _____ are catapulted into the air when each spore chamber snaps open. A spore may germinate and grow into a (25) _____ that is small, green, and heart-shaped. Jacketed structures develop on the underside of the mature (26) _____. Each male jacketed structure produces many (27) _____ while each female jacketed structure produces a single (28) _____. These gametes meet in (29) _____. The diploid (30) _____ is first formed inside the female jacketed structure; it divides to form the developing (31) _____, still attached to the gametophyte.

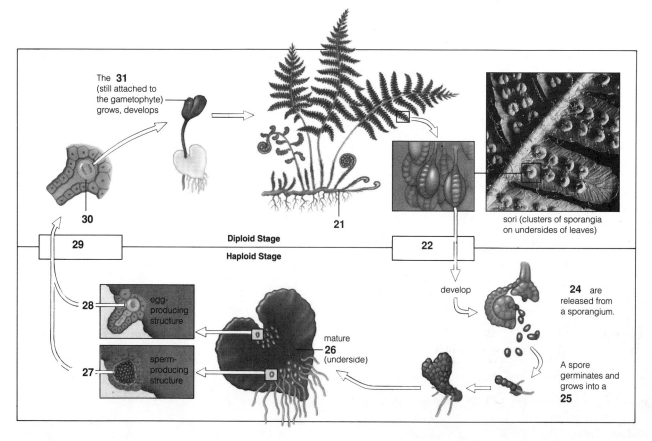

# 19 - IV. GYMNOSPERMS—PLANTS WITH "NAKED" SEEDS (pp. 314 - 315)

*Selected Words:* *evergreen, deciduous,* Ginkgo biloba, Gnetum, Ephedra, Welwitschia

## Boldfaced, Page-Referenced Terms

(314) gymnosperms _____

_____

(314) ovules _____

_____

(315) conifers _____

_____

(315) cones _____

_____

(315) microspores _____

_____

(315) megaspores _____

_____

(315) pollination _____

_____

(315) deforestation _____

_____

(315) cycads _____

_____

(315) ginkgos _____

_____

(315) gnetophytes _____

_____

## Choice

For questions 1 - 12, choose from the following:

    a. cycads    b. ginkgos    c. gnetophytes    d. conifers    e. gymnosperms (includes a, b, c, d)

1. ___ Fleshy-coated seeds of female trees produce an awful stench when stepped on.

2. ___ Includes pines, spruces, redwoods, firs, and junipers

3. ___ Seeds and a flour made from the trunk are edible following removal of poisonous alkaloids.

4. ___ Only a single species survives, the maidenhair tree.

5. ___ Includes *Welwitschia* of hot deserts of south and west Africa, *Gnetum* of humid tropical regions, and *Ephedra* of deserts and other arid regions

6. ___ Have massive, cone-shaped structures that bear either pollen or ovules; superficially resemble palm trees

7. ___ Seeds are mature ovules.

8. ___ The favored male trees are now planted in cities; they have attractive, fan-shaped leaves and are resistant to insects, disease, and air pollutants.

9. ___ Their ovules and seeds are not covered; they are borne on surfaces of spore-producing reproductive structures.

10. ___ Some sporophyte plants in this group mainly have a deep-reaching taproot; the exposed part is a woody disk-shaped stem bearing cone-shaped strobili and one or two strap-shaped leaves that split lengthwise repeatedly as the plant ages.

11. ___ Most species are evergreen trees and shrubs with needlelike or scalelike leaves.

12. ___ Includes conifers, cycads, ginkgos, and gnetophytes

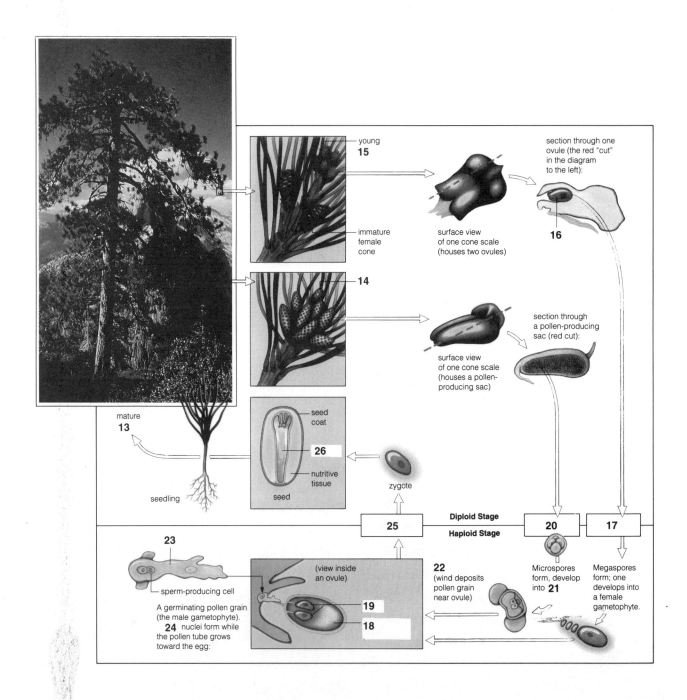

## Fill-in-the-Blanks

The numbered items in the illustration on page 208 represent missing information; complete the numbered blanks in the narrative to supply this information.

The familiar pine tree, a conifer, represents the mature (13) _____. Pine trees produce two kinds of spores in two kinds of cones. Pollen grains are produced in male (14) _____. Ovules are produced in young female (15) _____. Inside each (16) _____, (17) _____ occurs to produce haploid megaspores; one develops into a many-celled female (18) _____ that contains haploid (19) _____. Diploid cells within pollen sacs of male cones undergo (20) _____ to produce haploid microspores. Microspores develop into (21) _____ grains. (22) _____ occurs when spring air currents deposit pollen grains near ovules of female cones. A pollen (23) _____ representing a male gametophyte grows toward the female gametophyte. (24) _____ nuclei form within the pollen tube as it grows toward the egg. (25) _____ follows and the ovule becomes a seed that is composed of an outer seed coat, the (26) _____ diploid sporophyte plant, and nutritive tissue.

## 19 - V. ANGIOSPERMS—FLOWERING, SEED-BEARING PLANTS (p. 316)

*Selected Word:* Eucalyptus

### Boldfaced, Page-Referenced Terms

(316) angiosperms _____

_____

(316) pollinators _____

_____

(316) fruits _____

_____

## Matching

Choose the most appropriate answer for each.

1.___ examples of monocot plants
2.___ flower
3.___ pollinators
4.___ fruits
5.___ endosperm
6.___ examples of dicot plants
7.___ seeds

A. Nutritive seed tissue
B. Palms, lilies, orchids, wheat, corn, rice, rye, and sugarcane
C. Unique angiosperm reproductive structure
D. Mature ovaries
E. Insects, bats, birds, and other animals that coevolved with the life cycles of flowering plants
F. Packaged in fruits
G. Most shrubs and trees, most nonwoody plants, cacti, and water lilies

*Complete the Table*

8. Complete the table below to compare the plant groups studied in this chapter.

| Plant Group | Dominant Generation | Vascular Tissue Present? | Seeds Present? |
|---|---|---|---|
| a. Bryophytes | | | |
| b. Lycophytes | | | |
| c. Horsetails | | | |
| d. Ferns | | | |
| e. Gymnosperms | | | |
| f. Angiosperms | | | |

# 19 - VI. CHARACTERISTICS OF FUNGI (p. 318)

*Boldfaced, Page-Referenced Terms*

(318) saprobes _____

_____

(318) parasites _____

_____

(318) mycelium (pl., mycelia) _____

_____

(318) hypha (pl., hyphae) _____

_____

*Matching*

Choose the most appropriate answer for each.

1.___ saprobes
2.___ parasites
3.___ mycelium
4.___ hypha
5.___ spores

A. Nonmotile reproductive cells or multicelled structures; often walled and germinate following dispersal from the parent body
B. A mesh of branching fungal filaments that grows over and into organic matter, secretes digestive enzymes and functions in food absorption
C. Fungi that obtain nutrients from nonliving organic matter and so cause its decay
D. Fungi that extract nutrients from tissues of a living host
E. Each filament in a mycelium; consists of tube-shaped cells with chitin-reinforced walls

# 19 - VII. CONSIDER THE CLUB FUNGI (pp. 318 - 319)

*Selected Words:* <u>Armillaria</u> <u>bulbosa,</u> <u>Agaricus</u> <u>brunnescens,</u> *dikaryotic*

### Boldfaced, Page-Referenced Terms

(318) club fungi _____

_____

### Fill-in-the-Blanks

The numbered items in the illustration of a club fungus life cycle below represent missing information; complete the numbered blanks in the narrative below to supply this information.

The mature mushroom is actually a short-lived (1) _____-bearing body; each consists of a cap and a stalk. (2) _____-shaped structures that bear spores on their outer surface form at the sides of each gill. Each bears two haploid ($n + n$) nuclei. (3) _____ fusion occurs within the club-shaped structures, which yields a (4) _____ stage. (5) _____ occurs within the club-shaped structures and four haploid (6) _____ emerge at the tip of each. The spores are released and each may germinate into a haploid ($n$) mycelium. When hyphae of two compatible mating strains meet, (7) _____ fusion occurs. Following this, a (8) "_____" ($n + n$) mycelium gives rise to the spore-bearing mushrooms.

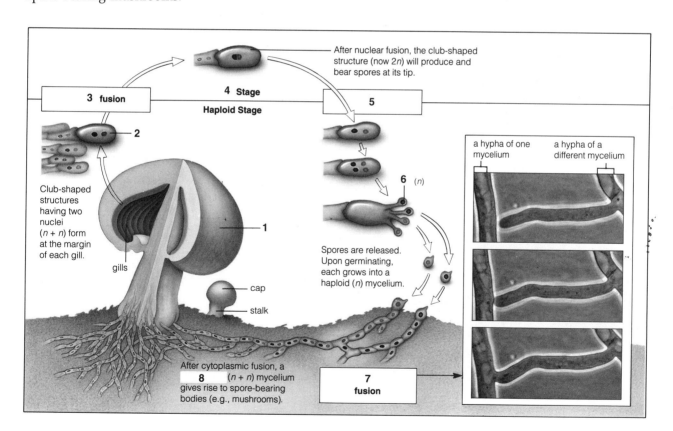

*Matching*

Choose the most appropriate answer for each.

9. ___ rust and smut fungi

10. ___ *Agaricus brunnescens*

11. ___ *Polyporous*

12. ___ *Armillaria bulbosa*

13. ___ *Amanita muscaria* (p. 318, Figure 19.13*b*)

A. Fly agaric mushroom, causes hallucinations when eaten; ritualistic use
B. Among the oldest and largest of all organisms
C. Destroys entire fields of wheat, corn, and other major crops
D. Common cultivated mushroom
E. Shelf fungus on rotting logs

## 19 - VIII. SPORES AND MORE SPORES (pp. 320 - 321)

*Selected Words:* Rhizopus stolonifer, Penicillium, Aspergillus, Neurospora sitophila, N. crassa, Saccharomyces cerevisiae, Candida albicans

*Boldfaced, Page-Referenced Terms*

(320) zygomycetes _____

_____

(320) sac fungi _____

_____

## Fill-in-the-Blanks

The numbered items in the illustration below (*Rhizopus* life cycle) represent missing information; complete the corresponding numbered blanks in the narrative to supply this information about zygomycetes.

The sexual phase begins when haploid (1) _____ of two different mating strains grow into each other and fuse. Two (2) _____ form between two hyphae, and several haploid nuclei are produced inside each. Later, their nuclei fuse, forming a zygote with a thick protective wall called a (3) _____ (the key defining feature of the zygomycetes). Meiosis proceeds and haploid (4) _____ are produced when this structure germinates. Each gives rise to stalked structures that can produce many spores, each of which can be the start of an extensive (5) _____.

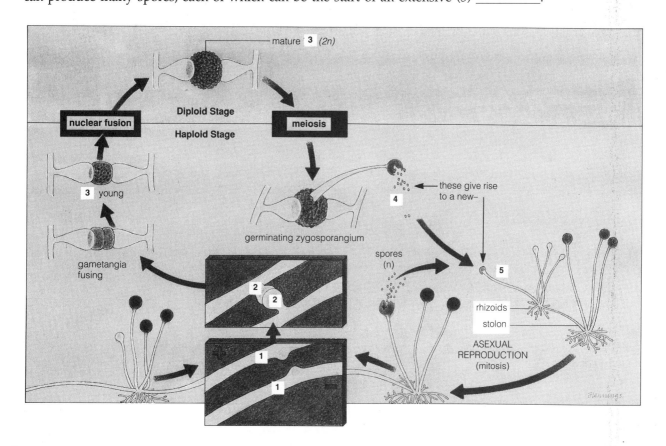

## True/False

If the statement is true, write a T in the blank. If the statement is false, make it correct by changing the underlined word(s) and writing the correct word(s) in the blank.

_____ 6. Most of the sac fungi species are <u>single-celled</u>.

_____ 7. <u>Fermentation</u> carried on by *Saccharomyces cerevisiae* furnishes carbon dioxide for leavening bread and ethanol for alcoholic beverages.

_____ 8. <u>*Neurospora sitophila*</u> is a sac fungus important in genetic research.

_____ 9. Trained pigs and dogs are used to snuffle out edible <u>morels.</u>

_____ 10. _Penicillium_ produces citric acid for candies and soft drinks.

_____ 11. _Aspergillis_ is a multicelled sac fungus that flavors cheeses.

_____ 12. Multicelled sac fungi species form specialized <u>asexual</u> spores called conidiospores.

_____ 13. <u>Ascospores</u> can be shaped like flasks, globes, and shallow cups.

_____ 14. Saclike structures that produce <u>sexual</u> spores usually form on the inner surface of ascocarps.

_____ 15. _Candida albicans_ is a yeast associated with infections of the vagina, mouth, intestines, and skin.

## Short Answer

16. What criterion is used to determine if a particular fungal species is an "imperfect" fungus? Cite one example of an "imperfect" fungus. _____

_____

_____

_____

_____

_____

_____

## 19 - IX. _Commentary:_ A LOOK AT THE UNLOVED FEW (p. 322) BENEFICIAL ASSOCIATIONS BETWEEN FUNGI AND PLANTS (pp. 322 - 323)

_**Selected Words:**_ <u>Histoplasma capsulatum</u>, _histoplasmosis_, <u>Claviceps purpurea</u>, _ergotism, exo_mycorrhiza, _endo_mycorrhiza

## _Boldfaced, Page-Referenced Terms_

(322) symbiosis _____

_____

(322) mutualism _____

_____

(322) lichen _____

_____

(322) mycorrhizae _____

_____

## Complete the Table

1. After reading the *Commentary,* "A Look at the Unloved Few" (p. 322), complete the following table, which deals with a few pathogenic and toxic fungi.

| Fungus Latin Name | Disease | Host and Symptoms |
|---|---|---|
| a. | Histoplasmosis | |
| b. | Ergotism | |
| c. | Athlete's foot | |
| d. | Apple scab | |

## Fill-in-the-Blanks

(2) _____ refers to species that live in close association. In cases of symbiosis called (3) _____, both partners benefit. A (4) _____ is commonly called a mutualistic interaction between a fungus and a photosynthetic species. However, it may well be a form of controlled (5) _____, in which the fungus holds a cyanobacterium, green alga, or both captive. The captive organism(s) may benefit some from sheltering.

Lichens can colonize places that are too (6) _____ for other organisms. Lichens absorb minerals from rocks and nitrogen from the (7) _____. Metabolic activities of lichens slowly change their substrate composition; they can also contribute to (8) _____ formation. Lichens also serve as early warnings of deteriorating (9) _____ conditions in that their death around cities signals that air pollution is getting bad.

Mutually beneficial associations of fungi with the young roots of most vascular plants is known as (10) _____, or "fungus-roots." The fungus benefits by absorbing (11) _____ from the plant, which benefits by absorbing (12) _____ and water from the fungus. In an (13) _____, hyphae form a dense net around living root cells but do not penetrate them; they are common in (14) _____ forests where they help trees withstand adverse seasonal changes in temperature and rainfall. A mycorrhizal association that is more common is the (15) _____ form that occurs in about 80 percent of all vascular plants; these fungal hyphae penetrate plant cells, as they do in lichens. They also extend for several centimeters into the surrounding (16) _____. Mycorrhizae are highly susceptible to various types of (17) _____ _____, and this adversely affects the world's forests.

# Self-Quiz

## *Label and Match*

In the blank beneath each illustration below (1 - 7), identify the organism by common name (or scientific name if a common name is unavailable). Then match each organism with the appropriate item (some may be used more than once) from the list below the illustrations by entering the letter in the parentheses.

1. _____ ( )

2. _____ ( )

3. _____ ( )

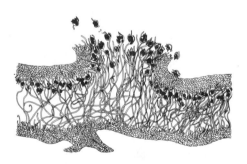

4. _____ ( )

5. _____ ( )

6. _____ ( )

A. Sac fungi

B. Zygomycetes

C. Club fungi

D. Imperfect fungi

E. Algae and fungi

7. _____ ( )

___8.  The _____ is *not* a trend in the evolution of plants.
   a. evolution of complex sporophytes
   b. shift from homospory to heterospory
   c. shift from diploid to haploid dominance
   d. development of xylem and phloem

___9.  Existing nonvascular plants do *not* include _____.
   a. horsetails
   b. mosses
   c. liverworts
   d. hornworts

___10. Plants possessing xylem and phloem are called _____ plants.
   a. gametophyte
   b. nonvascular
   c. vascular
   d. seedless

___11. The principle function of xylem in a plant is to _____; the principle function of phloem in a plant is to _____.
   a. distribute water and dissolved minerals; distribute photosynthetic products
   b. distribute water and dissolved minerals; distribute water and dissolved minerals
   c. distribute photosynthetic products; distribute water and dissolved minerals
   d. distribute photosynthetic products; distribute photosynthetic products

___12. Bryophytes _____.
   a. have vascular systems that enable them to live on land
   b. include lycopods, horsetails, and ferns
   c. have true roots but not stems
   d. include mosses, liverworts, and hornworts

___13.    _____ are not seedless vascular plants.
a. Lycophytes
b. Gymnosperms
c. Horsetails
d. Whisk Ferns
e. Ferns

___14.   In horsetails, lycopods, and ferns, _____.
a. spores give rise to gametophytes
b. the main plant body is a gametophyte
c. the sporophyte bears sperm- and egg-producing structures
d. all of the above

___15.    _____ are seed plants.
a. Cycads and ginkgos
b. Conifers
c. Angiosperms
d. all of the above

___16.   In complex land plants, the diploid stage is resistant to adverse environmental conditions such as dwindling water supplies and cold weather. The diploid stage progresses through this sequence: _____.
a. gametophyte → male and female gametes
b. spores → sporophyte
c. zygote → sporophyte
d. zygote → gametophyte

___17:   Monocots and dicots are groups of _____.
a. gymnosperms
b. club mosses
c. angiosperms
d. horsetails

___18.   Most true fungi send out cellular filaments called _____.
a. mycelia
b. hyphae
c. mycorrhizae
d. asci

___19.   Heterotrophic species of fungi can be _____.
a. saprobic
b. parasitic
c. mutualistic
d. all of the above

For questions 20 - 29, choose from the following:
a. club fungi
b. imperfect fungi
c. sac fungi
d. zygomycetes

___20.   The group that includes *Rhizopus stolonifer*, the notorious black bread mold is _____.

___21.   The group that includes delectable morels and truffles but also includes bakers' and brewers' yeasts is _____.

___22.   The group that includes shelf fungi, which decompose dead and dying trees, and mycorrhizal symbionts, which help trees extract mineral ions from the soil, is _____.

___23.   The group that includes the commercial mushroom *Agaricus brunnescens,* as well as the fly agaric mushroom, *Amanita muscaria* is _____.

___24.   The group that includes *Penicillium,* which has a variety of species that produce penicillin and substances that flavor Camembert and Roquefort cheeses, is _____.

___25.   The group whose spore-producing structures (ascocarps) are shaped like flasks, globes, and cups is _____.

___26.   The group that forms a thick wall around the zygote to produce a zygosporangium is _____.

___27.   **The group whose spore-producing structures are club-shaped is _____.**

___28.   The groups that are symbiotic with young roots of shrubs and trees in mycorrhizal associations is _____ and _____.

___29.   A group of fungi whose members were assigned to it because their sexual phase was undetected or absent.

# Chapter Objectives/Review Questions

| Page | | Objectives/Questions |
|---|---|---|
| (308) | 1. | Most of the members of the plant kingdom are _____ plants, with internal tissues that conduct and distribute water and solutes. |
| (308) | 2. | _____ are called the nonvascular plants. |
| (308) | 3. | Distinguish between the vascular seed plants known as gymnosperms and angiosperms. |
| (308) | 4. | State the general functions of the root systems and shoot systems of vascular plants. |
| (308) | 5. | _____ tissue distributes water and dissolved ions through plant parts; _____ tissue distributes sugars and other photosynthetic products. |
| (308) | 6. | Explain the structure and function of a cuticle and stomata. |
| (308 - 309) | 7. | Give the reasons that diploid dominance allowed plants to successfully exploit the land environment. |
| (309) | 8. | _____ means spore-producing body. |
| (309) | 9. | As plants evolved two spore types (heterospory), one spore type developed into pollen grains that become sperm-bearing male _____. |
| (309) | 10. | The combination of a plant embryo, nutritive tissues, and protective tissues constitutes a _____. |
| (310) | 11. | What group of plants first displayed cuticles, cellular jackets around gametangia, and large gametophytes that retain sporophytes? |
| (311) | 12. | In mosses and their relatives, a _____ consists of a jacketed structure and a stalk. |
| (311) | 13. | Define *peat moss, peat,* and *coal.* |
| (312) | 14. | The _____ is the dominant phase in the seedless vascular plants. |
| (312 - 313) | 15. | Be able to describe structural characteristics of *Psilotum, Lycopodium, Equisetum,* and a fern; be generally familiar with their life cycles. |
| (313) | 16. | A cluster of fertile leaves is known as a _____. |
| (314) | 17. | An _____ contains the female gametophyte, surrounded by nutritive tissue and a jacket of cell layers; a mature, coat-enclosed ovule is a _____. |
| (315) | 18. | Cycads, conifers, ginkgos, and gnetophytes are all members of the _____ lineage. |
| (315) | 19. | Define *conifers, cones, microspores, megaspores,* and *pollination.* |
| (314 - 315) | 20. | Be generally familiar with the gymnosperm life cycle. |
| (315) | 21. | State the immediate threat to the coniferous forests. |
| (315) | 22. | Briefly characterize plants known as cycads, ginkgos, and gnetophytes. |
| (315) | 23. | Explain why *Ginkgo biloba* is a unique plant. |
| (316) | 24. | A group of plants known as _____ alone produce flowers. |
| (316) | 25. | Name and cite examples of the two classes of flowering plants. |
| (316) | 26. | Unlike gymnosperms, flowering plants provide their seeds with a unique, protective package of stored food, the _____. |
| (316) | 27. | List some familiar flower pollinators. |
| (318) | 28. | Fungi are heterotrophs; most are _____ and obtain nutrients from nonliving organic matter and so cause its decay. |
| (318) | 29. | Other fungi are _____; they extract nutrients from tissues of a living host. |
| (318) | 30. | Distinguish between the meanings of the following terms: *hypha, hyphae, mycelium,* and *mycelia.* |
| (318) | 31. | What is the most used reproductive mode in the fungi? |
| (318) | 32. | Describe the diverse appearances of the fungi classified as club fungi. |
| (318) | 33. | What is the importance of the club fungi found in the genus *Amanita*? |
| (319) | 34. | Relate the generalized life cycle for many club fungi (mushroom). |
| (319) | 35. | A mushroom has a cap and a _____. |
| (319) | 36. | A _____ mycelium is one in which the hyphae have undergone cytoplasmic fusion but not nuclear fusion. |
| (320) | 37. | Be able to relate the life cycle of *Rhizopus stolonifer.* |

(320)  38. Only sac fungi produce sexual spores known as _____ during the sexual phase of their life cycle.
(321)  39. What are some typical shapes of ascocarps?
(321)  40. Explain why a classification group, the "imperfect fungi," is necessary.
(322)  41. Give the Latin name of the fungus that causes the disease known as ergotism; list the symptoms of ergotism.
(322)  42. Define *histoplasmosis;* name the causative organism.
(322)  43. Define *mutualism* and explain why a lichen seems to fit that definition; explain why this relationship may be a form of controlled parasitism.
(323)  44. Describe the fungus–plant root association known as mycorrhizae.
(323)  45. Distinguish exomycorrhizae from endomycorrhiza.
(323)  46. Explain why the effect of pollution on mycorrhizae could have far-reaching effects.

## Integrating and Applying Key Concepts

1. Explain why totally submerged aquatic plants that live in deep water never developed heterosporous life cycles.

2. Suppose humans acquired a few well-placed fungal genes that caused them to reproduce in the manner of a "typical" fungus (p. 319, Figure 19.14). Try to imagine the behavioral changes that humans would likely undergo. Would their food supplies necessarily be different? Table manners? Stages of their life cycle? Courtship patterns? Habitat? Would the natural limits to population increase be the same? Would their body structure change? Would there necessarily have to be separate sexes? Compose a descriptive science-fiction tale about two mutants who find each other and set up "housekeeping" together.

# 20

# ANIMALS: THE INVERTEBRATES

## Interactive Exercises

### 20 - I. OVERVIEW OF THE ANIMAL KINGDOM (pp. 326 - 329)
### PUZZLES ABOUT ORIGINS (p. 330)
### SPONGES—SUCCESS IN SIMPLICITY (pp. 330 - 331)

*Selected Words:* *thoracic* cavity, *abdominal* cavity, <u>Paramecium</u>, <u>Volvox</u>, <u>Trichoplax</u> adhaerens

*Boldfaced, Page-Referenced Terms*

(326) invertebrates _____

_____

(328) animals _____

_____

(328) ectoderm _____

_____

(328) endoderm _____

_____

(328) mesoderm _____

_____

(328) vertebrates _____

_____

(328) invertebrates _____

_____

(328) radial symmetry _____

_____

(328) bilateral symmetry _____

_____

(329) cephalization _____

_____

(329) gut _____

_____

(329) coelom _____

_____

(330) placozoan _____

_____

(330) sponges _____

_____

(330) collar cells _____

_____

(331) larva _____

_____

## Matching

Choose the most appropriate answer for each term.

1. ___ animals
2. ___ ectoderm, endoderm, mesoderm
3. ___ vertebrates
4. ___ invertebrates
5. ___ radial symmetry
6. ___ bilateral symmetry
7. ___ gut
8. ___ coelom
9. ___ thoracic cavity
10. ___ abdominal cavity
11. ___ segmentation
12. ___ cephalization

A. All animals whose ancestors evolved before backbones did
B. An evolutionary process whereby sensory structures and nerve cells became concentrated in a head
C. Animal body cavity lined with a peritoneum—found in most bilateral animals; some worms lack this cavity, other worms have a false cavity
D. Animals having body parts arranged regularly around a central axis, like spokes of a bike wheel
E. Upper coelom cavity holding a heart and lungs
F. Primary tissue layers that give rise to all adult animal tissues and organs
G. Region inside animal body in which food is digested
H. Animals having right and left halves that are mirror images of each other
I. Series of animal body units that may or may not be similar to one another
J. Lower coelom cavity holding a stomach, intestines, and other organs
K. Multicellular organisms with tissues forming organs and organ systems, diploid body cells, heterotrophic, aerobic respiration, sexual reproduction, sometimes asexual, most are motile in some part of the life cycle, and the life cycle shows embryonic development
L. Animals with a "backbone"

## Complete the Table

13. Complete the table below by filling in the appropriate phylum or representative group name.

| Phylum | Some Representative(s) | Number of Known Species |
|---|---|---|
| a. | Trichoplax; simplest animal | 1 |
| b. Porifera | | 8,000 |
| c. | Hydrozoans, jellyfishes, corals, sea anemones | 11,000 |
| d. Platyhelminthes | | 15,000 |
| e. | Pinworms, hookworms | 20,000 |
| f. | Tiny body with crown of cilia, great internal complexity; "wheel animals" | 1,800 |
| g. Mollusca | | 110,000 |
| h. | Leeches, earthworms, polychaetes | 15,000 |
| i. Arthropoda | | 1,000,000 |
| j. | Sea stars, sea urchins | 6,000 |
| k. | Invertebrate chordates: tunicates, lancelets | 2,100 |
| l. Chordata | | 45,000 |

## Choice

For questions 14 - 23, answer questions about animal origins by choosing from the following:

a. *Paramecium*    b. *Volvox*    c. *Trichoplax adhaerens*

d. different animal lineages arose from more than one group of protistanlike ancestors

14. ___ The only known placozoan

15. ___ Similar ciliate forerunners may have had multiple nuclei within a single cell.

16. ___ Similar to colonies that became flattened and crept on the seafloor

17. ___ The answer to animal origins might actually be more than one answer.

18. ___ As simple as an animal can get

19. ___ By another hypothesis, multicelled animals arose from flagellated cells that live in similar hollow, spherical colonies.

20. ___ A soft-bodied marine animal, shaped a bit like a plate

21. ___ By one hypothesis, the animal forerunners were ciliates, much like this organism.

22. ___ Has no symmetry, no tissues, and no mouth

23. ___ In a similar organism, the division of labor characterizing multicellularity might have begun.

## Labeling

Identify the groups in the evolutionary tree shown by writing the group name in the appropriate blank.

24. _____

25. _____

26. _____

27. _____

28. _____

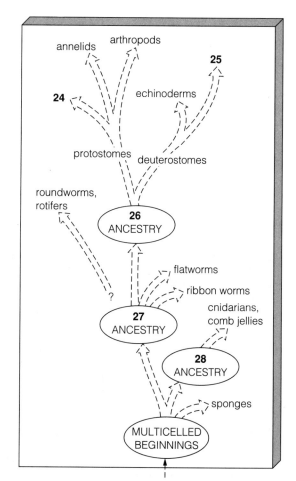

## Matching

Choose the most appropriate answer for each.

29. ___ osculum, oscula

30. ___ fragmentation

31. ___ sponge phylum

32. ___ larva

33. ___ amoeboid cells

34. ___ *Trichoplax*

35. ___ sponge skeletal elements

36. ___ microvilli

37. ___ collar cells

38. ___ water entering a sponge body

39. ___ gemmules

A. Flagellated cells that absorb and move water through a sponge

B. Reside in a gelatinlike substance between inner and outer cell linings

C. An organism whose two cell layers resemble those of a sponge

D. Form the "collars" of collar cells

E. Clusters of sponge cells capable of establishing new colonies

F. Microscopic pores and chambers

G. Random chunks of sponge tissue break off and grow into more sponges

H. Spongin fibers, glasslike spicules of silica or calcium carbonate or both

I. Sexually immature form preceding the adult stage

J. Openings where water leaves a sponge

K. Porifera

## 20 - II. CNIDARIANS—TISSUES EMERGE (pp. 332 - 333)

*Selected Words:* Hydra, Trichoplax, Physalia, Obelia

### Boldfaced, Page-Referenced Terms

(332) cnidarians _____

_____

(332) nematocysts _____

_____

(332) medusa _____

_____

(332) polyp _____

_____

(332) epithelium _____

_____

(333) nerve cells _____

_____

(333) sensory cells _____

_____

(333) contractile cells _____

_____

(333) hydrostatic skeleton _____

_____

(333) gonads _____

(333) planulas _____

## Fill-in-the-Blanks

All (1) _____ are tentacled, radial animals; they include jellyfishes, sea anemones, corals, and animals like *Hydra*. Most of these animals live in the sea, and they alone produce (2) _____, which are capsules capable of discharging threads that entangle prey or fend off predators.

Cnidarians have two common body plans, the (3) _____ that look like bells or upside-down saucers, and the (4) _____ that has a tubelike body with a tentacle-fringed mouth at one end. The saclike cnidarian gut processes food with its (5) _____, a sheetlike lining with glandular cells that secrete digestive enzymes. The (6) _____ covers the rest of the body's surfaces. Each of these linings is referred to as an (7) _____. (8) _____ cells form a "nerve net" that coordinates responses to stimulation. Working together with (9) _____ cells and (10) _____ cells, the nerve net controls movement and shape changes.

The (11) _____ is a layer of secreted material that lies between the epidermis and gastrodermis. Most polyps have little mesoglea and use water in their gut as a (12) _____ skeleton, a fluid-filled cavity or cell mass. Reef-forming (13) _____ consist of interconnected polyps that secrete calcium-reinforced external skeletons that, over time, build reefs. Many cnidarians have only a polyp or a medusa stage in the life cycle with the medusa being the sexual form. They have simple (14) _____. They rupture and release gametes. Zygotes formed at fertilization develop into (15) _____, a type of swimming or creeping larva that usually has ciliated epidermal cells. Eventually a mouth opens at one end, transforming the larva into a polyp or a medusa, and the cycle begins anew.

## Labeling

Identify each numbered part of the illustration on p. 227.

16. _____ _____

17. _____ _____

18. _____ _____

19. _____

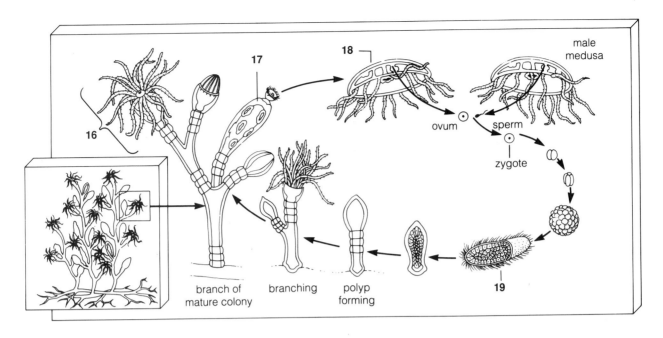

16

17

18

male medusa

ovum    sperm

zygote

19

branch of mature colony    branching    polyp forming

## 20 - III. FLATWORMS, ROUNDWORMS, ROTIFERS—AND SIMPLE ORGAN SYSTEMS (pp. 334 - 335)
### *Focus on Health:* A ROGUE'S GALLERY OF PARASITIC WORMS (p. 336)

**Selected Words:** *primary* host, *intermediate* host, *schistosomiasis,* <u>Schistosoma,</u> <u>Enterobius,</u> <u>Trichinella,</u> <u>Wuchereria,</u> *elephantiasis*

### *Boldfaced, Page-Referenced Terms*

(334) organ _____

_____

(334) organ system _____

_____

(334) flatworms _____

_____

(334) hermaphrodite _____

_____

(335) roundworms _____

_____

(335) rotifers _____

_____

## Choice

For questions 1 - 25, choose from the following:

a. flatworms    b. roundworms    c. rotifers

1. ___ They live nearly everywhere; a cupful of rich soil has thousands of them.

2. ___ The group includes parasitic flukes.

3. ___ They are abundant in plankton.

4. ___ Sexual maturity is reached in a vertebrate primary host; larval stages develop or become encysted in an intermediate host.

5. ___ Two "toes" exude substances that attach these animals to a substrate at feeding time.

6. ___ All of them have a bilateral, cylindrical body, usually tapered at both ends.

7. ___ All of them have a crown of cilia used in swimming and wafting food to the mouth.

8. ___ They are the simplest animals that have a complete digestive system.

9. ___ Hookworms, pinworms, and other types parasitize plants and animals.

10. ___ Planarians and a few others live in freshwater.

11. ___ Its motions reminded early microscopists of a turning wheel.

12. ___ Some resemble cnidarian planula larvae.

13. ___ They possess a tough, flexible cuticle.

14. ___ Tapeworms thrive in predigested food in vertebrate intestines.

15. ___ Each is a hermaphrodite.

16. ___ They eat bacteria and tiny algae.

17. ___ Most species are less than a millimeter long, yet they have a pharynx, an esophagus, digestive glands, a stomach, usually an intestine and anus, and protonephridia.

18. ___ These parasites attach to the intestinal wall by a scolex.

19. ___ Planarians regenerate lost parts.

20. ___ Protonephridia and flame cells carry on water regulation.

21. ___ *Enterobius vermicularis* parasitizes humans in temperate regions.

22. ___ *Schistosoma japonicum* causes schistosomiasis.

23. ___ Adult worms become lodged in the body's lymph nodes; elephantiasis occurs, an enlargement of legs and other body regions.

24. ___ Humans become infected with *Trichinella spiralis* mostly by eating insufficiently cooked meat from pigs or certain game animals.

25. ___ Humans become infected by walking barefoot where a juvenile may penetrate the skin; the parasite then travels in the bloodstream to the lungs.

## Labeling

Identify the parts of the animal shown dissected in the accompanying drawings.

26. _____ _____

27. _____

28. _____

29. _____ _____

30. _____

31. _____

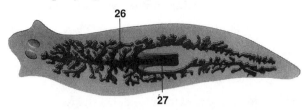

Answer exercises 32 - 35 about the drawing of the accompanying dissected animal.

32. What is the common name (or genus) of the animal dissected? _____

33. Is the animal parasitic? _____

34. Is the animal hermaphroditic? _____

35. Name the coelom type exhibited by this animal.
_____

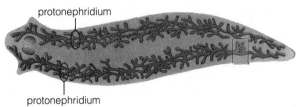

protonephridium

protonephridium

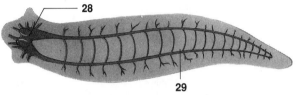

28

29

Answer questions 36 - 38 about the drawing of the dissected animal below.

36. What is the common name of the animal dissected? _____

37. Is the animal hermaphroditic? _____

38. Name the coelom type exhibited by this animal. _____

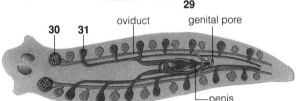

30    31    oviduct    genital pore

penis

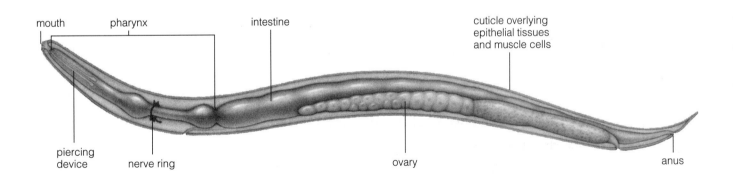

mouth    pharynx    intestine    cuticle overlying epithelial tissues and muscle cells

piercing device    nerve ring    ovary    anus

## Fill-in-the-Blanks

The numbered items in the illustration below (blood fluke life cycle) represent missing information; complete the corresponding numbered blanks in the narrative to supply this information.

The life cycle of the Southeast Asian blood fluke (*Schistosoma japonicum*) requires a human primary host standing in water in which the fluke larvae can swim. This life cycle also requires an aquatic snail as an intermediate host. Flukes reproduce (39) _____, and eggs mature in a human body. Eggs leave the body in feces, then hatch into ciliated, swimming (40) _____ that burrow into a (41) _____ and multiply asexually. In time, many fork-tailed (42) _____ develop. These leave the snail and swim until they contact (43) _____ skin. They bore inward and migrate to thin-walled intestinal veins, and the cycle begins anew. In infected humans, white blood cells that defend the body attack the masses of fluke eggs, and grainy masses form in tissues. In time, the liver, spleen, bladder, and kidneys deteriorate.

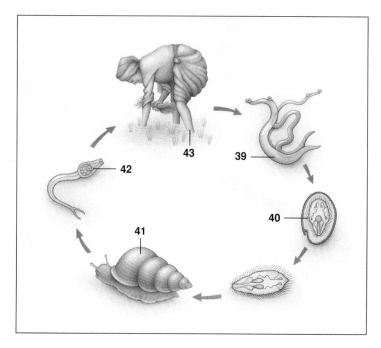

## Fill-in-the-Blanks

The numbered items in the illustration below (beef tapeworm life cycle) represent missing information; complete the corresponding numbered blanks in the narrative to supply this information.

(44) _____, each with an inverted scolex of a future tapeworm, become encysted in (45) _____ host tissues (for example, skeletal muscle). A (46) _____, a definitive host, eats infected and undercooked beef containing tapeworm cysts. The scolex of a larva turns inside out, attaches to the wall of the host's (47) _____ _____, and begins to absorb host nutrients. Many (48) _____ form by budding; each of these segments becomes sexually mature and has both male and female

reproductive (49) _____. Ripe (50) _____ containing fertilized eggs leave the host in

(51) _____, which may contaminate water and vegetation. Inside each fertilized egg, an embryonic

(52) _____ form develops. Cattle may ingest embryonated eggs or ripe proglottids and so become

(53) _____ hosts.

**b.** A **46**, a definitive host eats infected, undercooked beef (mainly skeletal muscle).

**a. 44** each with inverted scolex of future tapeworm become encysted in **45** host tissues (e.g. skeletal muscle).

**c.** Scolex of larva turns inside out, attaches to small **47** wall. Larva absorbes host nutrients.

**d.** Many **48** form by budding.

**f.** Inside each fertilized egg, an embryonic **52** form develops. Cattle may ingest embryonated eggs or ripe proglottids. and so become **53** hosts.

**e.** Each sexually mature proglottid has female and male **49**. Ripe **50** containing fertilized eggs leave host in **51** which may contaminate water and vegetation.

## 20 - IV. A MAJOR DIVERGENCE (p. 337)
## MOLLUSKS—A WINNING BODY PLAN (pp. 338 - 339)

### *Boldfaced, Page-Referenced Terms*

(337) protostomes _____

_____

(337) deuterostomes _____

_____

(337) spiral cleavage _____

_____

(337) radial cleavage _____

_____

(338) mollusks _____

_____

(338) gastropods _____

_____

(339) bivalves _____

_____

(339) cephalopods _____

_____

## Choice

For questions 1 - 10, choose from the following:

a. protostomes     b. deuterostomes

1. ___ The first external opening in these embryos becomes the anus; the second becomes the mouth.

2. ___ Animals having a developmental pattern in which the early cell divisions are parallel and perpendicular to the axis.

3. ___ A coelom arises from spaces in the mesoderm.

4. ___ Radial cleavage

5. ___ The first external opening in these embryos becomes the mouth.

6. ___ A coelom forms from outpouchings of the gut wall.

7. ___ Spiral cleavage

8. ___ Animal having a developmental pattern in which early cell divisions are at oblique angles relative to the original body axis

9. ___ Echinoderms and chordates

10. ___ Mollusks, annelids, and arthropods

## Matching

Identify the animals pictured in the next exercise by matching each with the appropriate description.

I. Bivalve     II. Cephalopod     III. Gastropod

11. ____ Animal A

12. ____ Animal B

13. ____ Animal C

## Labeling

Identify each numbered part in the drawings below by writing its name in the appropriate blank.

14. _____     21. _____     28. _____ _____

15. _____     22. _____     29. _____

16. _____     23. _____     30. _____

17. _____     24. _____     31. _____

18. _____     25. _____     32. _____ _____

19. _____     26. _____     33. _____

20. _____     27. _____

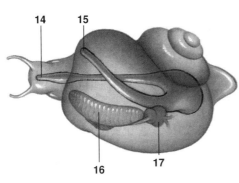

**Animal A**

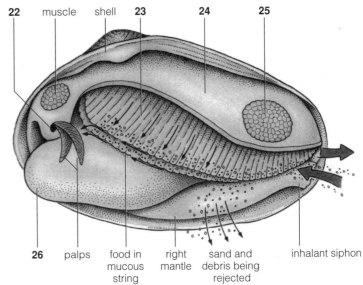

**22** muscle shell **23** **24** **25**

**26** palps food in right sand and inhalant siphon
mucous mantle debris being
string rejected

**Animal B**

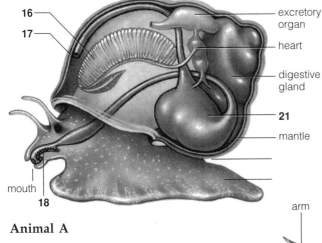

**16**
**17**

mouth
**18**

**Animal A**

excretory organ
heart
digestive gland
**21**
mantle

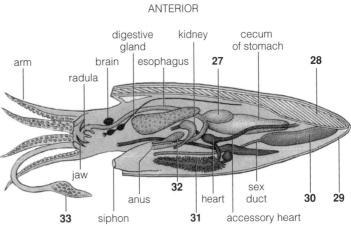

ANTERIOR

digestive gland   kidney   cecum of stomach
brain   esophagus   **27**   **28**
arm   radula
jaw
anus   **32**   heart   sex duct
**33**   siphon   **31**   accessory heart   **30** **29**

POSTERIOR

**Animal C**

## Fill-in-the-Blanks

The name of the phylum, Mollusca, means (34) _____ bodied. All mollusks have (35) _____ symmetry and they alone have a (36) _____, a tissue fold that hangs skirtlike over the body. Most have a (37) _____ of calcium carbonate. They have (38) _____ that are the respiratory organs constructed of thin-walled leaflets for gas exchange. Mollusk embryos undergo (39) _____ while they develop. The vast majority of mollusks have a fleshy (40) _____. Those with a well-developed (41) _____ have eyes and tentacles. Many have a (42) _____, a tonguelike, toothed organ that preshreds food.

The three classes of mollusks are the gastropods, bivalves, and the (43) _____. With 90,000 species, (44) _____, or "belly foots," are the largest class. These are the shelled snails and (45) _____ that have no shell or only a remnant of one. Clams, scallops, oysters, and mussels are well-known (46) _____. (47) _____ have been eating one type or another since prehistoric times. A few bivalves produce pearls. The bivalve head is greatly reduced but the (48) _____ is usually large and good for burrowing. In nearly all bivalves the gills function in collecting food and in (49) _____. As water moves through the mantle cavity, (50) _____ on the gills traps bits of food. Cilia move the mucous to (51) _____, which sort out the food before acceptable bits are driven to the mouth. Bivalves hunkered in mud or sand have a pair of (52) _____; water is drawn into the mantle cavity by one and it leaves through the other carrying (53) _____ from the anus and kidneys.

(54) _____ include squids, cuttlefish, octopuses, and nautiluses. Some are the largest and (55) _____ of the known invertebrates. They have separate sexes, are footless, and are active marine predators. They have prey-capturing (56) _____, a (57) _____ to draw prey into the mouth, and a beaklike pair of jaws to bite or crush it. They move by (58) _____ propulsion by forcing water out of the mantle cavity through a funnel-shaped siphon. Active cephalopods demand (59) _____, and they alone among the mollusks have a (60) _____ circulatory system with two (61) _____. Cephalopods have a well-developed (62) _____ system. Compared to other mollusks, they have the largest (63) _____ relative to body size. Cephalopod eyes resemble those of (64) _____, although they form in a different way. In terms of memory and learning, the cephalopods are the most complex (65) _____.

## 20 - V. ANNELIDS—SEGMENTS GALORE (pp. 340 - 341)

*Selected Words: poly-, hydrostatic* skeleton

### Boldfaced, Page-Referenced Terms

(340) annelids _____

_____

(341) nephridia _____

_____

(341) nerve cord _____

_____

(341) ganglion _____

_____

## Matching

Choose the most appropriate answer for each.

1. ___ cuticle
2. ___ annelids
3. ___ earthworms
4. ___ annelid lineage
5. ___ marine polychaetes
6. ___ nephridia
7. ___ nerve cord
8. ___ advantage of segmentation
9. ___ hydrostatic skeleton
10. ___ leeches
11. ___ earthworm locomotion
12. ___ ganglion

A. Fluid-cushioned coelomic chambers
B. Oligochaete scavengers with a closed circulatory system
C. A bundle of slender extensions of nerve cell bodies
D. Possess many bristles per segment
E. Muscle contraction with protraction and retraction of segment bristles
F. Secreted wrapping around the body surface of most annelids that permits respiratory exchange
G. Lack bristles
H. A group that diverged early from protostome ancestors
I. Different body parts can evolve separately and specialize in different tasks.
J. The term means "ringed forms."
K. An enlargement of the nerve cord in each segment of an earthworm
L. Regulates volume and composition of body fluids; begins with a funnel-shaped structure in each segment

## Labeling

Identify each numbered part of the illustration in questions 13 - 19, then answer questions 20 ·

13. _____
14. _____
15. _____ _____
16. _____
17. _____ _____
18. _____
19. _____
20. Name this animal. _____
21. Name this animal's phylum. _____
22. Name two distinguishing characteristics of this group. _____

23. Protostome ( ) or deuterostome ( )
24. Is this animal segmented? ( ) yes ( ) no
25. Symmetry of adult: ( ) radial ( ) bilateral
26. Does this animal have a true coelom? ( ) yes ( ) no

***Selected Words:*** larval *feeding and growth,* adult *dispersal and reproduction*

## Boldfaced, Page-Referenced Terms

(342) arthropods _____

_____

(342) exoskeleton _____

_____

(342) molting _____

_____

(342) tracheas _____

_____

(342) metamorphosis _____

_____

## Choice

For questions 1 - 15, choose from the following six adaptations that contributed to the success of arthropods:

a. a hardened exoskeleton    b. specialized segments    c. jointed appendages

d. specialized respiratory structures    e. efficient nervous system and sensory organs

f. division of labor in the life cycle

1. ___ Intricate eyes and other sensory organs

2. ___ Sexually immature larvae that molt and change as they grow

3. ___ Hardened, light, and flexible due to protein and chitin

4. ___ Became fewer in number, grouped in various ways and became more specialized

5. ___ Metamorphosis from larvae into adult forms

6. ___ Probably evolved as defenses against predators

7. ___ Larval stages specialize in feeding and growth; the adult is concerned with dispersal and reproduction.

8. ___ Gills of aquatic arthropods

9. ___ In insects, different segments became combined into a head, a thorax, and an abdomen.

10. ___ Some paired appendages are used in feeding, detecting stimuli, locomotion, transferring sperm to females, or spinning silk.

11. ___ Many species have a wide angle of vision and can process visual information from many directions.

12. ___ Tracheas, the air-conducting tubes of land-dwelling arthropods

13. ___ In spiders, segments fused as a forebody and hindbody

14. ___ Restrict evaporative water loss and can support a body deprived of water's buoyancy

15. ___ Crabs and lobsters are equipped with a calcium-stiffened armor plating.

## Short Answer

16. Why are the arthropods said to be the most biologically successful organisms on Earth? _____

_____

_____

_____

_____

_____

## 20 - VII. A LOOK AT SPIDERS AND THEIR KIN (p. 343)
##      A LOOK AT THE CRUSTACEANS (pp. 344 - 345)
##      VARIABLE NUMBERS OF MANY LEGS (p. 345)

*Selected Words:* arachnids, *open* circulatory system, <u>Scutigera</u> <u>coleoptrata</u>

### Boldfaced, Page-Referenced Terms

(345) millipedes _____

_____

(345) centipedes _____

_____

### Matching

Choose the most appropriate answer for each.

1. ___ ticks
2. ___ arachnid forebody appendages
3. ___ arachnids
4. ___ arachnid hindbody appendages
5. ___ chelicerates
6. ___ spider, internal organs
7. ___ predatory arachnids

A. Scorpions and spiders that sting, bite, and sometimes dispense venom
B. An open circulatory system and book lungs
C. A group including mites, horseshoe crabs, sea spiders, spiders, ticks, and chigger mites
D. Spin out silk thread for webs and egg cases
E. Some transmit bacterial agents of Rocky Mountain spotted fever or Lyme disease to humans.
F. A familiar group including scorpions, spiders, ticks, and chigger mites
G. Four pairs of legs, a pair of chelicereae (with pincers or fangs), and a pair of pedipalps (inflict wounds and discharge venom)

### Dichotomous Choice

Circle one of two possible answers given between parentheses in each statement.

8. Nearly all arthropods possess (strong claws/an exoskeleton).
9. The giant crustaceans are (lobsters and crabs/barnacles and pillbugs).
10. The simplest crustaceans have (different/similar) appendages along their length.
11. (Barnacles/Lobsters and crabs) have strong claws that collect food, intimidate other animals, and sometimes dig burrows.

12. (Barnacles/Lobsters and crabs) have feathery appendages that comb microscopic bits of food from the water.
13. (Barnacles/Copepods) are the most numerous animals in aquatic habitats and are premier consumers of phytoplankton.
14. Of all arthropods, only (lobsters and crabs/barnacles) have a calcified "shell."
15. Adult (barnacles/copepods) cement themselves to wharf pilings, rocks, and similar surfaces.
16. As is true of other arthropods, crustaceans undergo a series of (rapid feedings/molts) and so shed the exoskeleton during their life cycle.
17. Adult (millipedes/centipedes) have a rounded body, two pairs of legs per segment and move slowly.
18. Adult (millipedes/centipedes) have a flattened body, are fast-moving, and have a pair of legs on every segment except one.
19. (Millipedes/Centipedes) mainly scavenge for decaying vegetation.
20. (Millipedes/Centipedes) are aggressive predators of insects, earthworms, and snails (with fangs and venom glands).

## Labeling

Identify each numbered part of the animal pictured here, then answer questions 27 - 33.

21. _____

22. _____

23. _____

24. _____

25. _____

26. _____

27. Name the animal pictured. _____

28. Name the subgroup of arthropods to which this animal belongs. _____

29. Name two distinguishing characteristics of this group. _____

30. Protostome ( ) or deuterostome ( )

31. Is this animal segmented? ( ) yes ( ) no

32. Symmetry of adult: ( ) radial ( ) bilateral

33. Does this animal have a true coelom? ( ) yes ( ) no

Identify each numbered body part in this illustration, then answer question 39.

34. _____ _____

35. _____

36. _____

37. _____

38. _____ _____

39. Name the subgroup of arthropods to which the animal belongs. _____

## 20 - VIII. A LOOK AT INSECT DIVERSITY (pp. 346 - 347)

### Boldfaced, Page-Referenced Terms

(346) Malpighian tubules _____

_____

(346) adult _____

_____

(346) larvae _____

_____

(346) nymphs _____

_____

(346) pupae _____

_____

(346) metamorphosis _____

_____

### Matching

Choose the most appropriate answer for each.

1. ___ the most successful species of insects

2. ___ Malpighian tubules

3. ___ metamorphosis

4. ___ insect life-cycle stages

5. ___ shared insect adaptations

A. Post-embryonic resumption of growth and transformation into an adult form
B. Head, thorax, and abdomen, paired antennae and mouthparts, three pairs of legs and two pairs of wings
C. Winged insects, also the only winged invertebrates
D. Larva - nymph - pupa - adult
E. Structures that collect nitrogen-containing wastes from blood and convert them to harmless uric acid

## 20 - IX. THE PUZZLING ECHINODERMS (pp. 348 - 349)

### Boldfaced, Page-Referenced Terms

(348) tube feet _____

_____

(348) water-vascular system _____

_____

### Fill-in-the Blanks

The second lineage of coelomate animals is referred to as the (1) _____. The major invertebrate members of this lineage include (2) _____. The body wall of all echinoderms has protective spines, spicules, or plates stiffened with (3) _____. Oddly, adult echinoderms have (4) _____ symmetry but some produce larvae with (5) _____ symmetry. Adult echinoderms lack a (6) _____, but a decentralized nervous system provides essential environmental information. The

(7) _____ feet of sea stars are used for walking, burrowing, clinging to a rock, or gripping a meal of clam or snail. These "feet" are part of a (8) _____ vascular system unique to echinoderms. Each tube foot contains an (9) _____ that acts like a rubber bulb on a medicine dropper as it contracts and forces fluid into a foot that then lengthens. Tube feet change shape constantly as (10) _____ action redistributes fluid through the water vascular system. Some sea stars are able to swallow their prey (11) _____; some can push part of their stomach outside the mouth and around their prey, then start (12) _____ the prey even before swallowing it. Undigestible remnants are regurgitated back through the mouth.

## Labeling

Identify each numbered part of the illustration below.

13. _____ _____          17. _____
14. _____ _____          18. _____ _____
15. _____                      19. _____
16. _____                      20. _____ _____

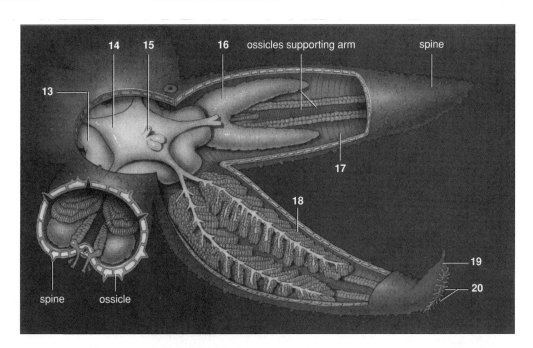

Answer questions 21 - 23 for the animal pictured above.

21. Name the animal shown. _____

22. Does this animal have a true coelom? ( ) yes ( ) no

23. Protostome ( ) or deuterostome ( )

## Identifying

24. Name the system shown at the right. _____

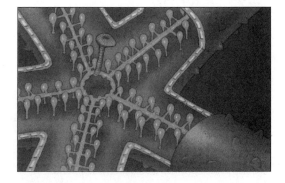

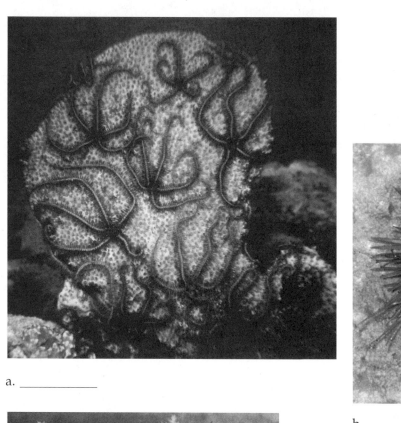

a. _____

b. _____

c. _____

d. _____

25. Identify each creature above by its common name.
26. Name the phylum of the animals pictured above. _____
27. Name two distinguishing characteristics of this group. _____

_____

28. Symmetry of adults represented above: ( ) radial ( ) bilateral

# Self-Quiz

___ 1. Which of the following is *not* true of sponges? They have no _____.
   a. distinct cell types
   b. nerve cells
   c. muscles
   d. gut

___ 2. Which of the following is *not* a protostome?
   a. earthworm
   b. crayfish or lobster
   c. sea star
   d. squid

___ 3. Deuterostomes undergo _____ cleavage; protostomes undergo _____ cleavage.
   a. radial; spiral
   b. radial; radial
   c. spiral; radial
   d. spiral; spiral

___ 4. Bilateral symmetry is characteristic of _____.
   a. cnidarians
   b. sponges
   c. jellyfish
   d. flatworms

___ 5. Flukes and tapeworms are parasitic _____.
   a. leeches
   b. flatworms
   c. jellyfish
   d. flatworms

___ 6. Insects include _____.
   a. spiders, mites, ticks
   b. centipedes and millipedes
   c. termites, aphids, and beetles
   d. all of the above

___ 7. Creeping behavior and a mouth located toward the "head" end of flatworms, may have led, in some evolutionary lines, to _____.
   a. development of a circulatory system with blood
   b. sexual reproduction
   c. feeding on nutrients suspended in the water (filter feeding)
   d. concentration of sense organs in the head region

___ 8. Which of the following is associated with the shift from radial to bilateral body form?
   a. a circulatory system
   b. a one-way gut
   c. paired organs
   d. the development of a water-vascular system

___ 9. The _____ body plan is characterized by bilateral symmetry, simple gas-exchange mechanisms, two-way traffic through a relatively unspecialized gut, and a thin cephalized body with all cells fairly close to the gut.
   a. annelid
   b. nematode
   c. echinoderm
   d. flatworm

___ 10. The _____ have bilateral symmetry, a tough cuticle, longitudinal muscles, and the simplest example of a complete digestive system.
   a. nematodes
   b. cnidarians
   c. flatworms
   d. echinoderms

___ 11. _____ with a peritoneum separates the gut and body wall of most bilateral animals.
   a. A coelom
   b. Mesoderm
   c. A mantle
   d. A water-vascular system

___ 12. The annelid _____ is composed of cells similar to flatworm flame cells.
   a. trachea
   b. nephridium
   c. mantle
   d. parapodium

## Matching

Match each phylum below with the corresponding characteristics (a - j) and representatives (A - P). A phylum may match with more than one letter from the group of representatives.

___, ___ 13. Annelida

___, ___ 14. Arthropoda

___, ___ 15. Chordata

___, ___ 16. Cnidaria

___, ___ 17. Echinodermata

___, ___ 18. Mollusca

___, ___ 19. Nematoda

___, ___ 20. Platyhelminthes

___, ___ 21. Porifera

___, ___ 22. Rotifera

a. choanocytes (= collar cells) + spicules
b. jointed legs + an exoskeleton
c. gill slits in pharynx + dorsal, tubular nerve cord + notochord
d. pseudocoelomate + wheel organ + soft body
e. soft body + mantle; may or may not have radula or shell
f. bilateral symmetry + blind-sac gut
g. radial symmetry + blind-sac gut; stinging cells
h. body compartmentalized into repetitive segments; coelom containing nephridia (= primitive kidneys)
i. tube feet + calcium carbonate structures in skin
j. complete gut + bilateral symmetry + cuticle; includes many parasitic species, some of which are harmful to humans

A. Dinosaurs
B. Corals, sea anemones, and *Hydra*
C. Salamanders and toads
D. Whales and opossums
E. Tapeworms and *Planaria*
F. Insects
G. Jellyfish and the Portuguese man-of-war
H. Sand dollars and starfishes
I. Earthworms and leeches
J. Lobsters, shrimp, and crayfish
K. Organisms with spicules and choanocytes
L. Scorpions and millipedes
M. Octopuses and oysters
N. Flukes
O. Hookworm, trichina worm
P. Small animals with a crown of cilia and two "exuding toes"

---

# Chapter Objectives/Review Questions

*Page*          *Objectives/Questions*

(328)      1.  Be able to list the general characteristics that define an animal.

(328)      2.  _____ cells are the forerunners of the primary tissue layers—the ectoderm, endoderm, and, in most species, mesoderm.

(328)      3.  List the primary characteristic that separates vertebrates from invertebrates.

(328)      4.  Distinguish radial symmetry from bilateral symmetry, and generally describe various animal gut types.

(329)      5.  List two benefits that the development of a coelom brings to an animal.

(329)      6.  Describe the general structure of a segmented animal.

(330 - 331) 7.  List two characteristics that distinguish sponges from other animal groups.

(332)      8.  State what nematocysts are used for, and explain how they operate.

(332)      9.  Two cnidarian body types are the _____ and the _____.

(334)     10.  List the three main types of flatworms.

(335)     11.  Describe the body plan of a roundworm, comparing its various systems with those of the flatworm.

(335)     12.  Describe the size, structure, and environment of rotifers.

(336 - 337) 13. Southeast Asian blood flukes, tapeworms, and pinworms are examples of _____ parasites.

(337) 14. Be able to discuss the evolutionary significance of animal embryos exhibiting spiral or radial cleavage.

(337) 15. Define *protostome* and *deuterostome* and give examples of each group.

(339) 16. Define *mantle*, and tell what role it plays in the molluscan body.

(339) 17. Explain why you think cephalopods came to have such well-developed sensory and motor systems and why they are able to learn.

(340) 18. Describe the advantages of segmentation and tell how this relates to the development of specialized internal organs.

(342) 19. Be able to list the six arthropod adaptations that led to their success.

(342) 20. List five different groups of arthropods.

(342 - 343) 21. What criteria would you use to distinguish an insect from an arachnid?

(344) 22. Name some common types of crustaceans.

(344) 23. Name the most obvious characteristic shared by most crustaceans.

(345) 24. How is the hard exoskeleton of arthropods able to change size to accommodate growth of the organism within?

(346) 25. The developmental process of many insects proceeds through very different post-embryonic stages, and then to an adult form by a process known as _____.

(346 - 347) 26. Name the characteristics that all insects share, even though they may appear very dissimilar.

(348) 27. List five examples of echinoderms; describe how locomotion occurs in echinoderms.

## Integrating and Applying Key Concepts

Most highly evolved animals have a complete gut, a closed blood vascular system, both central and peripheral nervous systems, and are dioecious. Why do you suppose having two sexes in separate individuals is considered to be more highly evolved than the monoecious condition utilized by many flatworms? Wouldn't it be more efficient if all individuals in a population could produce both kinds of gametes? Cross-fertilization would then result in both individuals being able to produce offspring.

# 21

# ANIMALS: THE VERTEBRATES

THE CHORDATE HERITAGE
    Characteristics of Chordates
    Chordate Classification

INVERTEBRATE CHORDATES
    Tunicates
    Lancelets

EVOLUTIONARY TRENDS AMONG THE VERTEBRATES
    Puzzling Origins, Portentous Trends
    The First Vertebrates

EXISTING JAWLESS FISHES

EXISTING JAWED FISHES
    Cartilaginous Fishes
    Bony Fishes

AMPHIBIANS
    Origin of Amphibians
    Salamanders
    Frogs and Toads
    Caecilians

REPTILES
    The Rise of Reptiles
    Crocodilians

    Turtles
    Tuataras
    Lizards and Snakes

BIRDS

MAMMALS
    Egg-Laying Mammals
    Pouched Mammals
    Placental Mammals

EVOLUTIONARY TRENDS AMONG THE PRIMATES
    Primate Classification
    Key Evolutionary Trends

FROM EARLY PRIMATES TO HOMINIDS
    Origins and Early Divergences
    The First Hominids

EMERGENCE OF HUMANS

---

## Interactive Exercises

**21 - I. THE CHORDATE HERITAGE** (pp. 352 - 354)
    **INVERTEBRATE CHORDATES** (pp. 354 - 355)
    **EVOLUTIONARY TRENDS AMONG THE VERTEBRATES** (pp. 356 - 357)
    **EXISTING JAWLESS FISHES** (p. 357)
    **EXISTING JAWED FISHES** (pp. 358 - 359)

*Selected Words:* "vertebrate chordates," "invertebrate chordates," *jawless fishes,* "sea squirts," larva

*Boldfaced, Page-Referenced Terms*

(354) chordates _____

(354) notochord _____

_____

(354) nerve cord _____

_____

(354) pharynx _____

_____

(354) tunicates _____

_____

(354) filter feeders _____

_____

(354) gill slits _____

_____

(354) metamorphosis _____

_____

(354) lancelets _____

_____

(356) vertebrae (sing., -bra) _____

_____

(356) jaws _____

_____

(356) fins _____

_____

(356) gills _____

_____

(356) lungs _____

_____

(357) ostracoderms _____

_____

(357) placoderms _____

_____

(357) hagfishes _____

_____

(357) lampreys _____

_____

(358) swim bladder _____

_____

(358) cartilaginous fishes _____

_____

(358) scales _____

_____

(358) bony fishes _____

_____

(359) lobe-finned fish _____

_____


## Fill-in-the-Blanks

Four major features distinguish the embryos of chordates from those of all other animals: a hollow dorsal (1) _____ _____, a(n) (2) _____ with slits in its wall, a(n) (3) _____, and a tail that extends past the anus at least during part of their lives. In some chordates, the (4) _____ chordates, the notochord is *not* divided into a skeletal column of separate, hard segments; in others, the (5) _____, it is.

Invertebrate chordates living today are represented by tunicates and (6) _____, which obtain their food by (7) _____ - _____; they draw in plankton-laden water through the mouth and pass it over sheets of mucus, which trap the particulate food before the water exits through the (8) _____ _____ in the pharynx. (9) _____ are among the most primitive of all living chordates; when they are larvae, they look and swim like (10) _____. A rod of stiffened tissue, the (11) _____, runs much of the length of the larval body; it functions like a(n) (12) _____ _____. Most adults remain attached to rocks and hard substrates.

As the ancestors of land vertebrates began spending less time immersed and more of their lives exposed to air, use of gills declined and (13) _____ evolved. More elaborate and efficient lungs and (14) _____ systems also evolved.

Even though the adult forms of hemichordates, echinoderms, and tunicates look very different, they have similar embryonic developmental patterns. Chordates may have developed from a mutated ancestral deuterostome (15) _____ of an attached, filter-feeding adult. A larva is an immature, motile form of an organism, but if a (16) _____ occurred that caused (17) _____ _____ to become functional in the larval body, then a motile larva that could reproduce would have been more successful in finding food and avoiding (18) _____ than an attached adult. In time, the attached stages in the species would be eliminated.

The ancestors of the vertebrate line may have been mutated forms of their closest relatives, the (19) _____, in which the notochord became segmented and the segments became hardened (20) _____. The vertebral column was the foundation for fast-moving (21) _____, some of which were ancestral to all other vertebrates. The evolution of (22) _____ intensified the competition for prey and the competition to avoid being preyed upon; animals in which mutations expanded the nerve cord into a (23) _____ that enabled the animal to compete effectively survived more frequently than their duller-witted fellows and passed along their genes into the next generations. Fins became (24) _____; in some fishes, those became (25) _____ and equipped with skeletal supports. These forms set the stage for the development of legs, arms and wings in later groups.

Exercises 26 - 27 refer to the illustrations below.

26. Name the creature in illustration B. _____
27. Name the creature in illustration C. _____

## Labeling

Name the structures numbered in illustrations A - C.

28. _____
29. _____ with _____ _____
30. _____
31. _____ _____
32. _____
33. _____
34. _____

35. _____ _____
36. _____ _____
37. _____ _____ _____
38. _____
39. _____
40. _____ _____ _____

A

mantle

31

gonad

B  B

genital duct
34
atrial cavity
33
tunic
32
circulatory system (with heart)

40    39

tentacles

37    atrial cavity    38    pore of atrial cavity    segmental muscles

C

D

42

41

43

41. Name the creature whose head is shown in illustration D. _____

_____

42. Name the structures. _____ _____

43. Name the structures. _____ _____

44. Name the creature shown in illustration E. _____

45. What structures occupy the front of its head space?

_____

44 _____

45

E

## Complete the Table

Provide common names wherever (*) appears, and provide the class name wherever ❒ appears.

*Phylum: Hemichordata*    *Phylum: Chordata*

| 46 * | *Subphylum: Urochordata* | *Subphylum: Cephalochordata* | *Subphylum: Vertebrata* | |
|---|---|---|---|---|
| | 47 * | 48 * | Class: *Agnatha* | 49 * |
| | | | Class: *Placodermi* | 50 * |
| | | | Class: *Chondrichthyes* | 51* |
| | | | Class: *Osteichthyes* | 52 * |
| | | | Class: 53 ❒ | amphibians |
| | | | Class: 54 ❒ | reptiles |
| | | | Class: 55 ❒ | birds |
| | | | Class: 56 ❒ | mammals |

## Fill-in-the-Blanks

Cartilaginous fishes include about 850 species of rays, skates, (57) _____, and chimaeras. They have conspicuous fins and five to seven (58) _____ _____ on both sides of the pharynx. Ninety-six percent of the existing species of fishes are (59) _____. Their ancestors arose during the Silurian period perhaps as early as 450 million years ago and soon gave rise to three lineages: the (60) _____-_____ fishes, the lobe-finned fishes, and lungfishes.

## Matching

Match the numbered item with its letter. One letter is used twice.

61. ___
62. ___
63. ___
64. ___
65. ___
66. ___
67. ___
68. ___
69. ___
70. ___
71. ___
72. ___
73. ___
74. ___
75. ___
76. ___

A. Anal fin
B. Anus
C. Brain
D. Caudal fin
E. Dorsal fin
F. Gallbladder
G. Heart
H. Intestine
I. Kidney
J. Pectoral fin (paired)
K. Pectoral fin (paired)
L. Pelvic fin (paired)
M. Stomach
N. Swim bladder
O. Urinary bladder

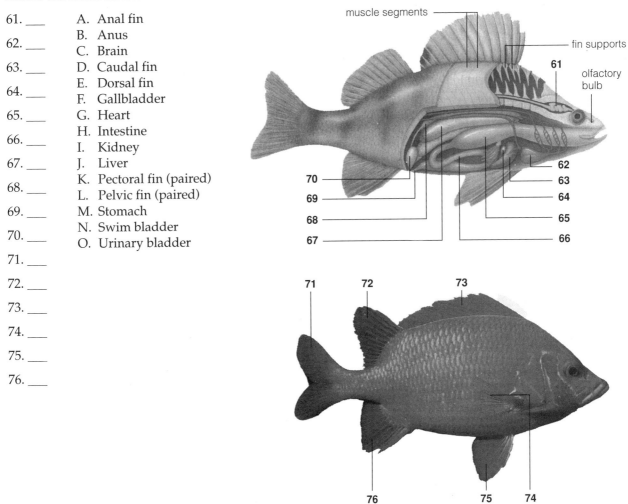

## Analysis and Short Answer

Figure 21.11 mentions features that distinguish the amphibian–lungfish lineage from the lineage that leads to sturgeons and other bony fishes. At least one of those features may have developed near ⑤ in this evolutionary tree.

Exercises 77 - 87 refer to the evolutionary diagram illustrated on page 251.

77. What single feature do the lampreys, hagfishes, and extinct ostracoderms have in common that is different from the placoderms?_____

78. How did ostracoderms feed?_____

79. A mutation in ostracoderm stock led to the development of what feature in all organisms that descended from ① ?_____

80. Mutation at ② led to the development of an endoskeleton made of what?_____

81. Mutations at ③ led to an endoskeleton of what?_____

82. Mutations at ④ led to which spectacularly diverse fishes that have delicate fins originating from the dermis? _____

83. Mutations at ⑤ led to which fishes whose fins incorporate fleshy extensions from the body?

_____

84. Which branch, ④ or ⑤ , gave rise to the amphibians? _____

85. Which branch gave rise to the modern bony fishes? _____

86. In which period did three distinctly different lineages of *bony* fishes appear in the fossil record?

_____

87. Approximately how many million years ago did the fork in the evolutionary path that led to the

amphibians occur? _____

## 21 - II. AMPHIBIANS (pp. 360 - 361)
### REPTILES (pp. 362 - 363)
### BIRDS (pp. 364 - 365)
### MAMMALS (pp. 366 - 367)

*Selected Words:* Sphenodon, Archaeopteryx, *animal migration,* marsupial

*Boldfaced, Page-Referenced Terms*

(360) amphibian _____

_____

(361) salamanders _____

_____

(361) frogs _____

_____

(361) caecilians _____

_____

(362) reptiles _____

_____

(362) amniote egg _____

_____

(362) crocodilians _____

_____

(362) turtles _____

_____

(362) tuataras _____

_____

(362) lizards _____

_____

(362) snakes _____

_____

(364) birds _____

_____

(364) feathers _____

_____

(366) mammals _____

_____

(366) behavioral flexibility _____

_____

(366) dentition _____

_____

(367) placenta _____

_____

## Fill-in-the-Blanks

Natural selection acting on lobe-finned fishes during the Devonian period favored the evolution of ever more efficient (1) _____ used in gas exchange and stronger (2) _____ used in locomotion. Without the buoyancy of water, an animal traveling over land must support its own weight against the pull of gravity. [(3) _____ _____ inside the lobed fins of certain fishes evolved over long periods of time into the limb bones of early amphibians.] The (4) _____ of early amphibians underwent dramatic modifications that involved evaluating incoming signals related to vision, hearing, and (5) _____. Changes in the (6) _____ system converted the two-chambered heart of fishes into a three-chambered one, and more efficient changes in (7) _____ distribution patterns delivered oxygen to cells throughout the body. The Devonian period brought humid, forested swamps with an abundance of aquatic invertebrates and (8) _____—ideal prey for amphibians.

There are three groups of existing amphibians: (9) _____, frogs and toads, and caecilians. Amphibians require free-standing (10) _____ or at least a moist habitat to (11) _____. Amphibian skin generally lacks scales but contains many glands, some of which produce (12) _____.

In the late Carboniferous, (13) _____ began a major adaptive radiation into the lush habitats on land, and only amphibians that mutated and developed certain (14) _____ features were able to follow them and exploit an abundant food supply. Several features helped: modification of (15) _____ bones favored swiftness, modification of teeth and jaws enabled them to feed efficiently on a variety of prey items, and the development of a (16) _____ egg protected the embryo inside from drying out, even in dry habitats.

(For questions 17 - 28, consult Figure 21.13c of the main text, and the time line and figure for the Analysis and Short Answer exercise on page 254 of this study guide.)

Today's reptiles include (17) _____, crocodilians, snakes, and (18) _____. All rely on (19) _____ fertilization, and most lay leathery-shelled eggs. Although amphibians originated during Devonian times, ancestral "stem" reptiles appeared during the (20) _____ period, about 340 million years ago. Reptilian groups living today that have existed on Earth longest are the (21) _____; their ancestral path diverged from that of the "stem" reptiles during the (22) _____ period. Crocodilian ancestors appeared in the early (23) _____ period, about 220 million years ago. Snake ancestry diverged from lizard stocks during the late (24) _____ period, about 140 million years ago. (25) _____ are more closely related to extinct dinosaurs and crocodiles than to any other existing vertebrates; they, too, have a (26) _____-chambered heart. Mammals have descended from therapsids, which in turn are descended from the (27) _____ group of reptiles, which diverged earlier from the stem reptile group during the (28) _____ period, approximately 320 million years ago.

In blanks 29 - 35, arrange the following groups in sequence, from earliest to latest, according to their appearance in the fossil record:

A. Birds     B. Crocodilians     C. Dinosaurs     D. Early ancestors of mammals (= synapsid reptiles)

E. Early ancestors of turtles (= anapsid reptiles)     F. Lizards and snakes     G. "Stem" reptiles.

29. ____   30. ____   31. ____   32. ____   33. ____   34. ____   35. ____

In blanks 36 - 42, select from the choices immediately below the geologic period in which each group (29 - 35 on p. 253) first appeared and write it in the correct space. For example: The group in blank 29 first appeared when? Write the answer in blank 36.

| *Paleozoic Era* | | *Mesozoic Era* | | |
|---|---|---|---|---|
| A. Carboniferous | B. Permian | C. Triassic | D. Jurassic | E. Cretaceous |
| 360-290 mya | 290-240 mya | 240-205 mya | 205-138 mya | 138-65 mya |

36. _____  37. _____  38. _____  39. _____  40. _____  41. _____  42. _____

## Analysis and Short Answer

43. The text, page 366, mentions the features that distinguish, say, the mammals from all of the remaining groups on the right-hand side of the diagram; therefore, at some time during the evolution of mammals, mutations that produced those features appeared. Consult the evolutionary diagram below and imagine what sort of mutation(s) occurred at the numbered places.

a. What may have occurred at ① ?_____

b. What may have occurred at ② ?_____

c. What may have occurred at ③ ?_____

d. What may have occurred at ④ ?_____

e. What may have occurred at ⑤ ?_____

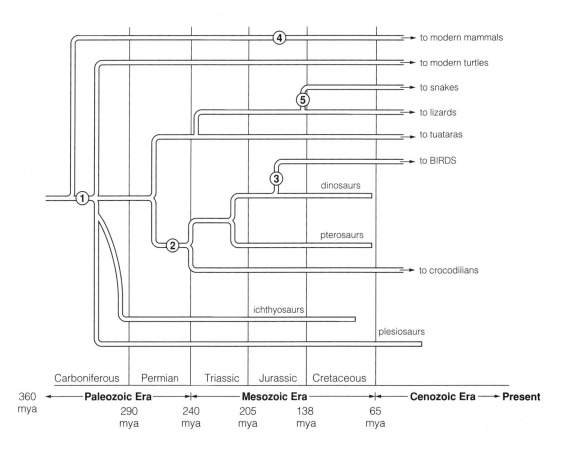

## Labeling

Label the structures pictured at the right.

44. _____

45. _____

46. _____ _____

allantois    **44**    amnion    chorion

**45**

air space

**46**

shell

## Fill-in-the-Blanks

Birds descended from (47) _____, which ran around on two legs some 160 million years ago. All birds have (48) _____, which insulate them and help get them aloft. Generally, birds have a greatly enlarged (49) _____ to which flight muscles are attached. Bird bones contain (50) _____ _____, and air flows through sacs, not into and out of them, for gas exchange in the lungs. Birds also lay (51) _____ _____ eggs, have complex courtship behaviors, and generally nurture their offspring. Birds are able to regulate their body (52) _____, which is generally higher than that of mammals. Existing birds do not have socketed (53) _____ in their beaks.

There are three groups of existing mammals: those that lay eggs (examples are the (54) _____ and the spiny anteater), those that are (55) _____ (examples are the opossum and the kangaroo), and those that are (56) _____ mammals (there are more than 4,500 species of these). Mammals regulate their body temperature, have a (57) _____-chambered heart, and show a high degree of parental nurture. Most mammals have (58) _____ as a means of insulation, and mammalian mothers generally suckle their young with milk.

## 21 - III. EVOLUTIONARY TRENDS AMONG THE PRIMATES (pp. 368 - 369)
## FROM EARLY PRIMATES TO HOMINIDS (pp. 370 - 371)
## EMERGENCE OF HUMANS (pp. 372 - 373)

*Selected Words:* prosimian, tarsioid, *prehensile, opposable,* Plesiadapis, Aegyptopithecus, dryopiths, early Homo, H. erectus, H. sapiens

### Boldfaced, Page-Referenced Terms

(368) Primates _____

_____

(368) anthropoids _____

_____

(368) hominoids _____

_____

(368) hominids _____

_____

(368) savannas _____

_____

(369) bipedal _____

_____

(369) culture _____

_____

(371) australopiths _____

_____

## Fill-in-the-Blanks

During the Cenozoic Era (65 million years ago to the present), birds, (1) _____, and flowering plants evolved to dominate Earth's assemblage of organisms. (1) are warm-blooded vertebrates with (2) _____ that began their evolution more than 200 million years ago. There are many groups, or (3) _____, within the class Mammalia; each order has its distinctive array of characteristics. Humans, apes, monkeys, and prosimians are all (4) _____; members of this order have excellent (5) _____ perception as a result of their forward-directed eyes. They also have hands that are (6) _____; that is they are adapted for (7) _____ instead of running. Primates rely less on their sense of (8) _____ and more on daytime vision. Their (9) _____ became larger and more complex; this trend was accompanied by refined technologies and the development of (10) _____: the collection of behavior patterns of a social group, passed from generation to generation by learning and by symbolic behavior (language). Modification in the (11) _____ led to increased dexterity and manipulative skills. Changes in primate (12) _____ indicate that there was a shift from eating insects to fruit and leaves and then to a mixed diet. Primates began to evolve from ancestral mammals more than (13) _____ million years ago, during the Paleocene. The first primates resembled small (14) _____ or tree shrews; they foraged at night for (15) _____, seeds, buds, and eggs on the forest floor, and they could clumsily climb trees searching for safety and sleep.

During the (16) _____ (25 million to 5 million years ago), the continents began to assume their current positions, and climates became cooler and (17) _____. Under these conditions, forests began to give way to mixed woodlands and (18) _____; an adaptive radiation of apelike forms— the first hominoids—took place as subpopulations of apes became reproductively isolated within the shrinking stands of trees. The chimpanzee-sized "tree apes," or (19) _____, originated during this time; eventually, they ranged throughout Africa, Europe, and Southern (20) _____. Between

(21) _____ million and (22) _____ million years ago, three divergences occurred that gave rise to the ancestors of modern gorillas, chimpanzees, and humans.

## Choice

Choose from the choices below for 23 - 32. Choose the single *smallest* group to which each belongs.

a. anthropoids    b. hominids    c. hominoids    d. prosimians    e. tarsioids

23. ___australopiths    27. ___gorillas    31. ___Old World monkeys

24. ___chimpanzees    28. ___humans    32. ___orangutans

25. ___dryopiths    29. ___lemurs

26. ___gibbons    30. ___New World monkeys

Choose from these choices for 33 - 40 in order to fill in the blanks correctly:

a. anthropoids    b. apes    c. hominids    d. hominoids    e. monkeys

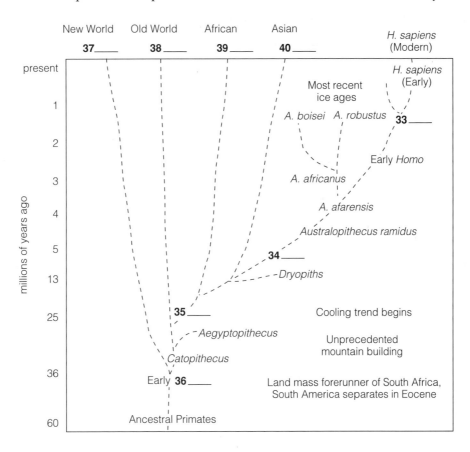

## Fill-in-the-Blanks

The family Hominidae (hominids) includes all species on the genetic path leading to humans since the time when that path diverged from the path leading to the (41) _____. Hominids emerged between (42) _____ million and (43) _____ million years ago, during the late Miocene. (44) _____-million-year-old fossils of humanlike forms have been discovered in Africa, and they all were bipedal and omnivorous and had an expanded brain. Lucy was one of the earliest (45) _____,

a collection of forms that combined ape and human features; they were fully two-legged, or
(46) _____, with essentially human bodies and ape-shaped heads.

The oldest fossils of the genus *Homo*, makers of stone tools, date from approximately
(47) _____ million years ago. Between 1.8 million and 400,000 years ago, during the Pleistocene
periods of glaciation, there were also intermittent periods of warming; during these interglacial times, a
larger- brained human species, (48) _____ _____, migrated out of Africa and into China,
Southeast Asia, and Europe. Over time, (48) became better at toolmaking and learned how to control
(49) _____ as members of the species became adapted to a wide range of habitats and were
associated with abundant (50) _____ artifacts. The (51) _____ were a distinct hominid
population that appeared 130,000 years ago in Southern France, Central Europe, and the Near East;
their cranial capacity was indistinguishable from our own, and they had a complex culture.

Anatomically modern humans evolved from (52) _____ _____. Analysis of
(53) _____ and (54) _____ comparisons and specimens from the fossil record support the
model of human origins that hypothesizes that *Homo erectus* migrated out of Africa, then formed
distinctive subpopulations of modern humans as an outcome of genetic divergence in different
geographic regions. From (55) _____ years ago to the present, human evolution has been almost
entirely cultural rather than biological. By 30,000 years ago, there was only one remaining hominid
species: (56) _____ _____.

## Matching

Suppose you are a student who wants to be chosen to accompany a paleontologist who has spent forty years teaching and roaming the world in search of human ancestors. Above the desk in her office, she keeps reconstructions of many skull types. Six are shown on this page, and you have heard that she chooses the graduate students who accompany her on her summer safaris on the basis of their ability on a short matching quiz. The quiz is presented below. *Without consulting your text (or any other)* while you are taking the quiz, match up all six skulls with all applicable letters. Eighteen correctly placed letters win you a place on the expedition. Sixteen correct answers put you on a waiting list. Fifteen or fewer and she suggests that you may wish to investigate dinosaur fossils instead of human ancestry. Which will it be for you?

57.

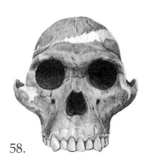
58.

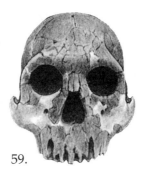
59.

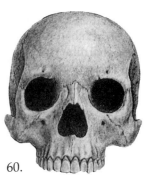
60.

61.

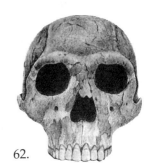
62.

57. _____

58. _____

59. _____

60. _____

61. _____

62. _____

A. Gracile form(s) of australopiths
B. Robust form(s) of australopiths
C. Fashioned stone tools and used them first
D. Has a chin
E. First controlled use of fire
F. Lived 4 million years ago
G. From a lineage that lived some time during the time span from approximately 3 million years ago until about 1.25 million years ago
H. Lived 2 million years ago
I. Lived from 1.8 million years ago until at least 100,000 years ago
J. The most recent hominid to appear
K. A species of *Australopithecus*
L. early *Homo*
M. *Homo erectus*
N. *Homo sapiens sapiens*

## Self-Quiz

___1. Filter-feeding chordates rely on _____, which have cilia that create water currents and mucous sheets that capture nutrients suspended in the water.
a. notochords
b. differentially permeable membranes
c. filiform tongues
d. gill slits

___2. In true fishes, the gills serve primarily _____ function.
a. a gas-exchange
b. a feeding
c. a water-elimination
d. a feeding and a gas-exchange

___3. The heart in amphibians _____.
a. pumps blood more rapidly than the heart of fish
b. is efficient enough for amphibians but would not be efficient for birds and mammals
c. has three chambers (ventricle and two atria)
d. all of the above

___4. The feeding behavior of true fishes selected for highly developed _____.
a. parapodia
b. notochords
c. sense organs
d. gill slits

## Identifying

Provide the common name and the major chordate group to which each creature pictured below and on pages 261 - 262 belongs.

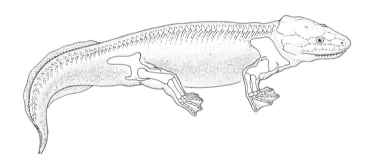

5. _____, _____

6. _____, _____

7. _____, _____

8. _____, _____

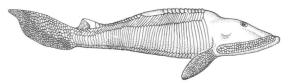

9. _____, _____

10. _____, _____

11. _____, _____

12. _____, _____

13. _____, _____

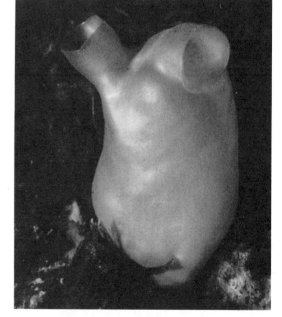

14. _____, _____

15. _____, _____

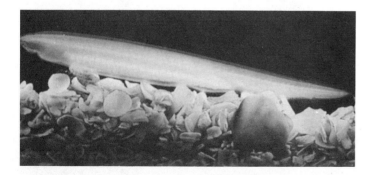

## Matching

Match the following groups and classes with the corresponding characteristics (a - i) and representatives (A - I).

16. __, __ Amphibians
17. __, __ Birds
18. __, __ Bony fishes
19. __, __ Cartilaginous fishes
20. __, __ Cephalochordates
21. __, __ Jawless fishes
22. __, __ Mammals
23. __, __ Reptiles
24. __, __ Urochordates

a. hair + vertebrae
b. feathers + hollow bones
c. jawless + cartilaginous skeleton (in existing species)
d. two pairs of limbs (usually) + glandular skin + "jelly"-covered eggs
e. amniote eggs + scaly skin + bony skeleton
f. invertebrate + an attached adult that cannot swim
g. jaws + cartilaginous skeleton + vertebrae
h. in adult, notochord stretches from head to tail; mostly burrowed-in, adult can swim
i. bony skeleton + skin covered with scales and mucus

A. lancelet
B. loons, penguins, and eagles
C. tunicates, sea squirts
D. sharks and manta rays
E. lampreys and hagfishes (and ostracoderms)
F. true eels and sea horses
G. lizards and turtles
H. caecilians and salamanders
I. platypuses and opossums

## Multiple Choice

___25. Which of the following is *not* considered to have been a key character in early primate evolution?
   a. Eyes adapted for discerning color and shape in a three-dimensional field
   b. Body and limbs adapted for tree climbing
   c. Bipedalism and increased cranial capacity
   d. Eyes adapted for discerning movement in a three-dimensional field

___26. Primitive primates generally live
   _____.
   a. in tropical and subtropical forest canopies
   b. in temperate savanna and grassland habitats
   c. near rivers, lakes, and streams in the East African Rift Valley
   d. in caves where there are abundant supplies of insects

___27. All the ancestral placental mammals apparently arose from ancestral forms of a group _____.
   a. that includes omnivorous shrews and moles
   b. that includes dogs, cats, and seals
   c. that includes mice and beavers
   d. that includes the koala ("teddy bear") and flying phalanger (resembles the flying squirrel)

___28. The hominid evolutionary line stems from a divergence from the ape line that apparently occurred _____.
   a. somewhere between 10 million and 5 million years ago
   b. about 3 million years ago
   c. during the Pliocene epoch
   d. less than 2 million years ago

___29. _____ was an Oligocene anthropoid that probably predated the divergence leading to Old World monkeys and the apes, with dentition more like that of dryopiths and less like that of the Paleocene primates with rodentlike teeth.
   a. *Aegyptopithecus*
   b. *Australopithecus*
   c. *Homo erectus*
   d. *Plesiadapis*

___30. Donald Johanson, from the University of California, discovered Lucy (named for the Beatles tune), who was a(n)

_____.

a. dryopith
b. australopith
c. member of *Homo*
d. prosimian

___31. A hominid of Europe and Asia that became extinct nearly 30,000 years ago was

_____.

a. a dryopith
b. *Australopithecus*
c. *Homo erectus*
d. Neandertal

## Matching

Choose the most appropriate answer for each.

32. ___ anthropoids

33. ___ australopiths

34. ___ Cenozoic

35. ___ hominids

36. ___ hominoids

37. ___ Miocene

38. ___ primates

39. ___ prosimians

A. A group that includes apes and humans
B. Organisms in a suborder that includes New World and Old World monkeys, apes, and humans
C. An era that began 65 million to 63 million years ago; characterized by the evolution of birds, mammals, and flowering plants
D. A group that includes humans and their most recent ancestors
E. An epoch of the Cenozoic era lasting from 25 million to 5 million years ago; characterized by the appearance of primitive apes, whales, and grazing animals of the grasslands
F. Organisms in a suborder that includes tree shrews, lemurs, and others
G. A group that includes prosimians, tarsioids, and anthropoids
H. Bipedal organisms living from about 4 million to 1 million years ago, with essentially human bodies and ape-shaped heads; brains no larger than those of chimpanzees

# Chapter Objectives/Review Questions

| Page | | Objectives/Questions |
|---|---|---|
| (353 - 354) | 1. | List three characteristics found only in chordates. |
| (354 - 355) | 2. | Describe the adaptations that sustain the sessile or sedentary life-style seen in primitive chordates such as tunicates and lancelets. |
| (356 - 357) | 3. | State what sort of changes occurred in the primitive chordate body plan that could have encouraged the emergence of vertebrates. |
| (357 - 359) | 4. | Describe the differences between primitive and advanced fishes in terms of skeleton, jaws, special senses, and brain. |
| (360 - 361) | 5. | Describe the changes that enabled aquatic fishes to give rise to swamp dwellers. |
| (360) | 6. | List the principal skin structures that each kind of four-limbed vertebrate produces. |
| (368) | 7. | Where do tarsier survivors dwell today on Earth? |
| (368, 371) | 8. | Beginning with the primates most closely related to humans, list the main groups of primates in order by decreasing closeness of relationship to humans. |
| (368 - 369) | 9. | Five key characters of primate evolution are _____, _____, _____, _____, and _____. |
| (370) | 10. | Describe the general physical features and behavioral patterns attributed to early primates. |
| (370 - 373) | 11. | Trace primate evolutionary development through the Cenozoic era. Describe how Earth's climates were changing as primates changed and adapted. Be specific about times of major divergence. |

(370 - 372) 12. State which anatomical features underwent the greatest changes along the evolutionary line from early anthropoids to humans.

(372 - 373) 13. Explain how you think *Homo sapiens sapiens* arose. Make sure your theory incorporates existing paleontological (fossil), biochemical, and morphological data.

## Integrating and Applying Key Concepts

Suppose someone told you that some time between 12 million and 6 million years ago dryopiths were forced by larger predatory members of the cat family to flee the forests and take up residence in estuarine, riverine, and sea coastal habitats where they could take refuge in the nearby water to evade the tigers. Those that, through mutations, became naked, developed an upright stance, developed subcutaneous fat deposits as insulation, and developed a bridged nose that had advantages in watery habitats (features that other dryopiths remaining inland never developed) survived and expanded their populations.

As time went on, predation by the big cats and competition with other animals for available food caused most of the terrestrial dryopiths to become extinct, but the water-habitat varieties survived as scattered remnant populations, adapting to easily available shellfish and fish, wild rice and oats, and various tubers, nuts, and fruits. It was in these aquatic habitats that the first food-getting tools (baskets, nets, and pebble tools) were developed, as well as the first words that signified different kinds of food. How does such a story fit with current speculations about and evidence of human origins? How could such a story be shown to be true or false?

# 22

# PLANT TISSUES

## Interactive Exercises

### 22 - I. OVERVIEW OF THE PLANT BODY (pp. 376 - 379)

*Selected Words:* *primary* tissues, lateral meristems, *secondary* tissues, *primary* growth, *secondary* growth

In addition to the boldfaced terms, the text features other important terms essential to understanding the assigned material. "Selected Words" is a list of these terms, which appear in the text in italics, in quotation marks, and occasionally in roman type. Latin binomials found in this section are underlined and in roman type to distinguish them from other italicized words.

## Boldfaced, Page-Referenced Terms

The page-referenced terms are important; they were in boldface type in the chapter. Write a definition for each term in your own words without looking at the text. Next, compare your definition with that given in the chapter or in the text glossary. If your definition seems inaccurate, allow some time to pass and repeat this procedure until you can define each term rather quickly (how fast you can answer is a gauge of your learning effectiveness).

(378) angiosperms _____

_____

(378) shoots _____

_____

(378) roots _____

_____

(378) ground tissue system _____

_____

(378) vascular tissue system _____

_____

(378) dermal tissue system _____

_____

(379) meristems _____

_____

(379) apical meristems _____

_____

(379) vascular cambium _____

_____

(379) cork cambium _____

_____

## Label and Match

Identify each part of the accompanying illustration. Choose from dermal tissues, root system, ground tissues, shoot system, and vascular tissues. Complete the exercise by matching and entering the letter of the proper description in the parentheses following each label.

1. _____ _____ ( )

2. _____ _____ ( )

3. _____ _____ ( )

4. _____ _____ ( )

5. _____ _____ ( )

A. Typically consists of stems, leaves, and reproductive structures

B. Usually grows below ground, absorbs soil water and minerals, stores food, and anchors the aboveground parts

C. Tissues that make up the bulk of the plant body

D. Protective covering for the plant body

E. Two conducting tissues that distribute water and solutes through the plant body

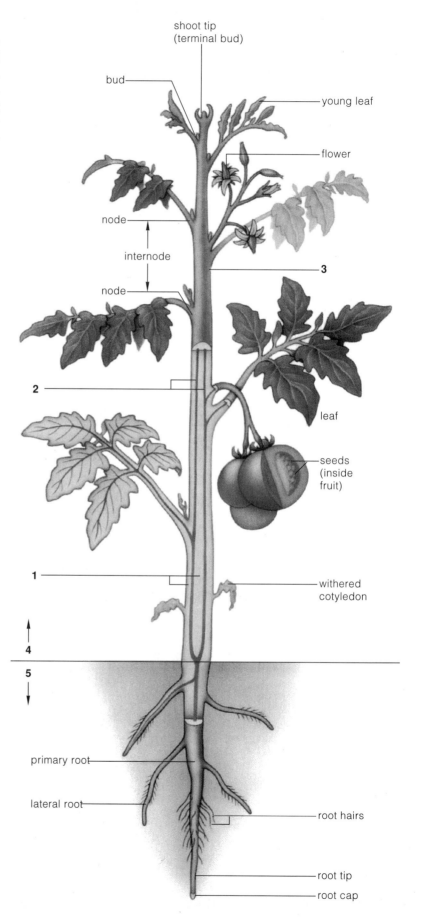

## Matching

Choose the most appropriate answer for each term.

6. ___ secondary tissues

7. ___ shoots

8. ___ vascular cambium and cork cambium

9. ___ meristems

10. ___ ground tissue system

11. ___ apical meristems

12. ___ primary tissues

13. ___ vascular tissue system

14. ___ roots

15. ___ dermal tissue system

A. Produced by growth at apical meristems and at tissues derived from them
B. Located inside dome-shaped tips of stems and roots; cell divisions and enlargements give rise to three primary meristems
C. Descending parts that penetrate soil; specialized for absorbing water and dissolved nutrients
D. Contains two kinds of conducting tissues that distribute water and solutes throughout the plant body
E. Two types of lateral meristems; growth there produces secondary tissues of the plant body
F. Tissues that make up the bulk of the plant body
G. Localized regions of self-perpetuating embryonic cells
H. Produces thickening of stems and roots, originates at lateral meristems
I. Covers and protects the plant's surfaces
J. Consist of stems and their branchings, leaves, flowers, or other structures

## Labeling

Identify each numbered part of the accompanying illustration. Choose from the following and enter in the numbered blanks: vascular cambium, apical meristem, cork cambium, primary tissues derived from the apical meristem.

16. _____

17. _____

18. _____

19. _____

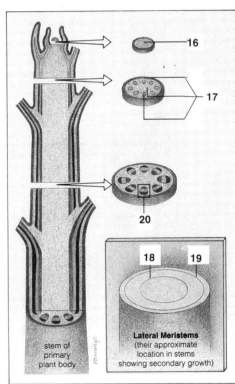

*Selected Words:* *fibers, sclereids, "companion" cells, "cotyledons"*

## *Boldfaced, Page-Referenced Terms*

(380) parenchyma _____

_____

(380) collenchyma _____

_____

(380) sclerenchyma _____

_____

(380) lignin _____

_____

(380) xylem _____

_____

(381) phloem _____

_____

(381) epidermis _____

_____

(381) cuticle _____

_____

(381) stoma _____

_____

(381) periderm _____

_____

(381) dicots _____

_____

(381) monocots _____

_____

## Labeling

Place one of these three labels under each of the following illustrations: collenchyma, parenchyma, or sclerenchyma.

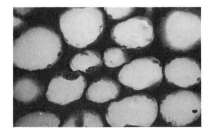

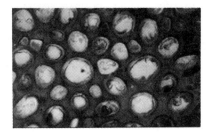

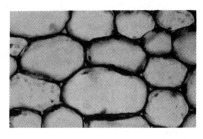

1. _____

2. _____

3. _____

## Choice

For questions 4 - 14, choose from the following:

a. parenchyma     b. collenchyma     c. sclerenchyma

4. ___ Patches or cylinders of this tissue are found near the surface of lengthening stems.

5. ___ The cells have thick, lignin-impregnated walls.

6. ___ Cells are alive at maturity and retain the capacity to divide.

7. ___ Some types specialize in storage, secretion, and other tasks.

8. ___ Cells are mostly elongated, with unevenly thickened walls.

9. ___ Provides flexible support for primary tissues.

10. ___ Abundant air spaces are found around the cells.

11. ___ Supports mature plant parts and often protects seeds.

12. ___ Cells are thin-walled, pliable, and many-sided.

13. ___ Mesophyll is specialized for photosynthesis.

14. ___ Fibers and sclereids.

## Labeling

Identify the cell types and cellular structures in the illustrations below by entering the correct name in the blanks. Complete the exercise by entering the name of the complex tissue (xylem or phloem) in which that cell or structure is found.

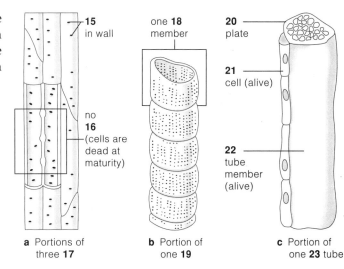

a Portions of three **17**

b Portion of one **19**

c Portion of one **23** tube

15. _____ (_____)

16. _____ (_____)

17. _____ (_____)

18. _____ (_____)

19. _____ (_____)

20. _____ (_____)

21. _____ (_____)

22. _____ (_____)

23. _____ (_____)

## Complete the Table

24. The two types of vascular tissues, xylem and phloem, occur in strands called vascular bundles within the plant body (each of these complex tissues also has fibers and parenchyma). Complete the table below, which summarizes important information about these vascular tissues.

| Vascular Tissue | Major Conducting Cells | Cells Alive | Function(s) |
|---|---|---|---|
| a. Xylem | | | |
| b. Phloem | | | |

25. Two types of dermal tissues cover the plant body. Complete the table below, which summarizes information about the dermal tissues.

| Dermal Tissue | Primary/Secondary Plant Body | Function(s) |
|---|---|---|
| a. Epidermis | | |
| b. Periderm | | |

26. There are two classes of flowering plants, monocots and dicots. Complete the table below, which summarizes information about the two groups of flowering plants.

| Class | Number of Cotyledons | Number of Floral Parts | Leaf Venation | Pollen Grains | Vascular Bundles |
|---|---|---|---|---|---|
| a. Monocots | | | | | |
| b. Dicots | | | | | |

## 22 - III. SHOOT PRIMARY STRUCTURE (pp. 382 - 383)

*Selected Words:* "internode," "compound" leaves, *cortex, pith*

### Boldfaced, Page-Referenced Terms

(382) leaf primordia _____

_____

(382) node _____

_____

(382) bud _____

_____

(382) leaf _____

_____

(382) deciduous plants _____

_____

(382) evergreen plants _____

(383) vascular bundles _____

## Labeling

Name the structures numbered in the illustrations of leaf development and leaf forms below.

1. _____
2. _____  _____
3. _____
4. _____
5. _____
6. _____
7. _____
8. _____
9. _____
10. _____
11. _____

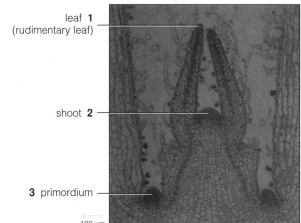

leaf **1**
(rudimentary leaf)

shoot **2**

**3** primordium

100 µm

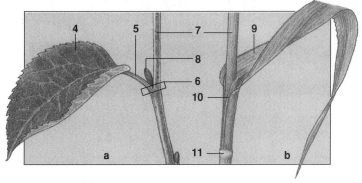

## Dichotomous Choice

Circle one of two possible answers given between parentheses in each statement.

12. The leaf illustrated on the left above is a (monocot/dicot).
13  The leaf illustrated on the right above is a (monocot/dicot).

## Labeling

Name the structures numbered in the illustrations of the monocot stem below. Choose from air space, epidermis, sclerenchyma cells, vessel, vascular bundle, sieve tube, ground tissue, and companion cell.

14. _____

15. _____ _____

16. _____ _____

17. _____ _____

18. _____ _____

19. _____

20. _____ _____

21. _____ _____

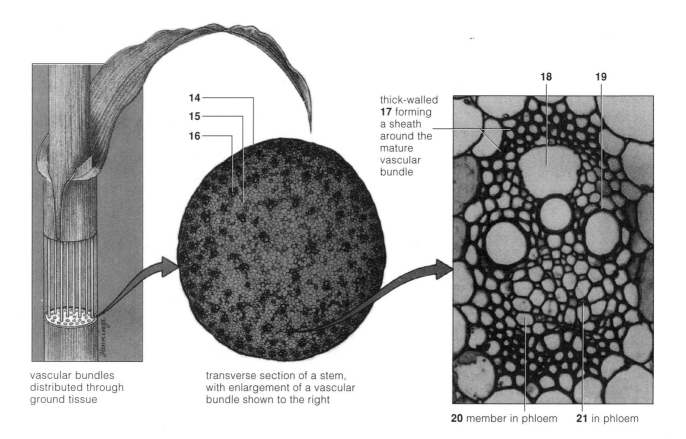

vascular bundles distributed through ground tissue

transverse section of a stem, with enlargement of a vascular bundle shown to the right

thick-walled **17** forming a sheath around the mature vascular bundle

**18**   **19**

**20** member in phloem   **21** in phloem

## Labeling

Name the structures numbered in illustrations of the herbaceous dicot stem below. Choose from sieve tube and companion cells in phloem, vascular bundle, xylem vessel, cortex, vascular cambium, pith, phloem fibers, and epidermis.

22. _____    26. _____

23. _____    27. _____ _____

24. _____ _____    28. _____ _____

25. _____    29. _____

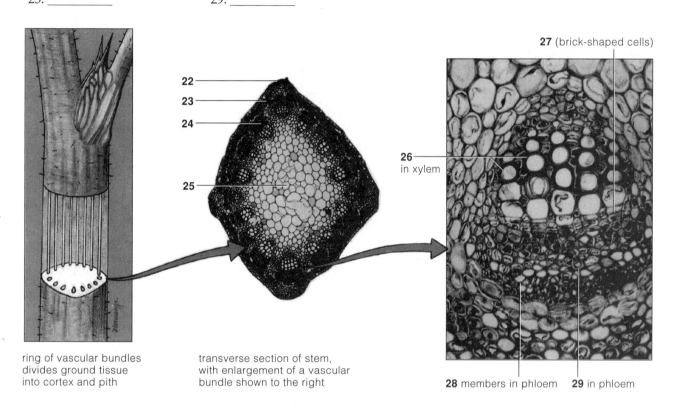

ring of vascular bundles divides ground tissue into cortex and pith

transverse section of stem, with enlargement of a vascular bundle shown to the right

**27** (brick-shaped cells)

**26** in xylem

**28** members in phloem    **29** in phloem

## 22 - IV. A CLOSER LOOK AT LEAVES (pp. 384 - 385)
### Commentary: USES AND ABUSES OF LEAVES (p. 385)

***Selected Words:*** <u>Atropa</u> <u>belladonna</u>, <u>Digitalis</u> <u>purpurea</u>, <u>Aloe</u> <u>vera</u>, <u>Agave</u>, <u>Nicotiana</u> <u>rusticum</u>, <u>Cannabis</u> <u>sativa</u>, <u>Hyosyamus</u> <u>niger</u>

### Boldfaced, Page-Referenced Terms

(385) mesophyll _____

_____

(385) veins _____

_____

## Label and Match

Identify each numbered part of the accompanying illustration (Figure 22.13, text). Complete the exercise by matching and entering the letter of the proper function description in the parentheses following each label.

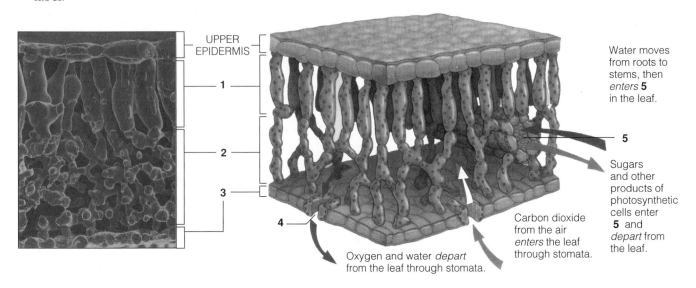

UPPER EPIDERMIS

1

2

3

4

5

Water moves from roots to stems, then *enters* **5** in the leaf.

Sugars and other products of photosynthetic cells enter **5** and *depart* from the leaf.

Carbon dioxide from the air *enters* the leaf through stomata.

Oxygen and water *depart* from the leaf through stomata.

1. _____ _____ ( )

2. _____ _____ ( )

3. _____ _____ ( )

4. _____ ( )

5. _____ ( )

A. Lowermost cuticle-covered cell layer

B. Loosely packed photosynthetic parenchyma cells just above the lower epidermal layer

C. Allows movement of oxygen and water vapor out of leaves and allows carbon dioxide to enter

D. Photosynthetic parenchyma cells just beneath the upper epidermis

E. Move water and solutes to photosynthetic cells and carry products away from them

## Matching

Choose the most appropriate answer for each term.

6. ___ nightshade leaves

7. ___ digitalis from foxglove leaves

8. ___ *Aloe vera* leaves

9. ___ rosy periwinkle leaves

10. ___ century plant

11. ___ Manila hemp leaf fibers

12. ___ Panamanian palm fronds

13. ___ Mexican cockroach plants

14. ___ neem tree leaves

15. ___ tobacco plants

16. ___ marijuana

17. ___ coca leaves

18. ___ henbane

A. Smoked by Mayan priests to carry their priestly thoughts to the gods

B. Kills a variety of insects, mites, and nematodes without killing natural predators

C. Soothe sun-damaged skin

D. Used to produce hats

E. Source of twine and rope

F. Produces mind-altering products; linked with low sperm counts

G. Extracts used to treat Parkinson's disease

H. Caused Hamlet's death

I. Cords and textiles

J. Kills fleas, lice, flies, and cockroaches

K. Helps stabilize heartbeat and circulation

L. Used to treat Hodgkin's disease

M. Source of cocaine

## 22 - V. ROOT PRIMARY STRUCTURE (pp. 386 - 387)

*Selected Word:* adventitious

### Boldfaced, Page-Referenced Terms

(386) primary root _____

_____

(386) lateral roots _____

_____

(386) taproot system _____

_____

(386) fibrous root system _____

_____

(387) root hairs _____

_____

(387) vascular cylinder _____

_____

## Label and Match

Identify each numbered part of the accompanying illustration. Complete the exercise by matching the letter of the proper description in the parentheses following each label. Some choices are used more than once.

1. _____ _____ ( )
2. _____ ( )
3. _____ ( )
4. _____ ( )
5. _____ ( )
6. _____ _____ _____ ( )
7. _____ _____ ( )
8. _____ ( )
9. _____ ( )
10. _____ _____ ( )

A. Dome-shaped cell mass produced by the apical meristem
B. Part of the vascular cylinder; gives rise to lateral roots
C. Part of the vascular cylinder; transports photosynthetic products
D. Ground tissue region surrounding the vascular cylinder
E. The absorptive interface with the root's environment
F. The region of dividing cells
G. Innermost part of the root cortex; helps control water and mineral movement into the vascular column
H. Epidermal cell extensions; greatly increases the surface available for taking up water and solutes

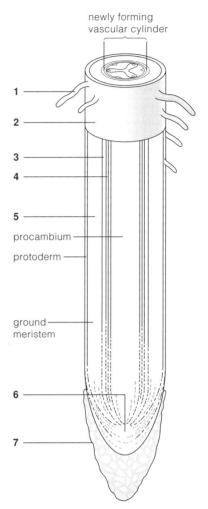

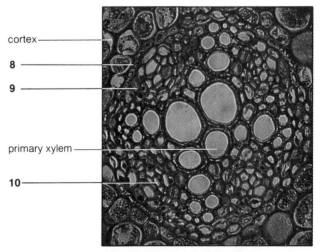

Vascular Cylinder

## Fill-in-the-Blanks

Oak trees, carrots, and dandelions are examples of plants whose primary root and its lateral branchings represent a(n) (11) _____ system.  In monocots such as grasses, the primary root is short-lived;  in its place, numerous (12) _____ roots arise from the stem of the young plant.  Such roots and their branches are somewhat alike in length and diameter and form a (13) _____ root system.  Some root epidermal cells send out absorptive extensions called root (14) _____.  Vascular tissues form a (15) _____ cylinder, a central column inside the root.  Ground tissues surrounding the cylinder are called the root (16) _____.  Dissolved (17) _____ easily diffuse through numerous spaces around cortex cells.  Water entering the root moves from cell to cell until it reaches the (18) _____, the innermost part of the root cortex.  This is a sheetlike layer, one cell thick, around the vascular cylinder.  Abutting walls of its cells are waterproof, so they force incoming water to pass through the cytoplasm of (19) _____ cells.  This arrangement helps (20) _____ the movement of water and dissolved substances into the vascular cylinder.  Just inside the endodermis is the (21) _____.  This part of the vascular cylinder gives rise to (22) _____ roots.

## 22 - VI. WOODY PLANTS (pp. 388 - 389)

*Selected Words:* nonwoody, woody, heartwood, sapwood, bark, girdling, fusiform initials, ray initials

## Boldfaced, Page-Referenced Terms

(388) annuals _____

_____

(388) biennials _____

_____

(388) perennials _____

_____

(388) lateral meristems _____

_____

(389) annual growth layers _____

_____

## Matching

Choose the most appropriate answer for each term.

1. ___ tree rings
2. ___ heartwood
3. ___ sapwood
4. ___ lateral meristem
5. ___ early wood
6. ___ nonwoody plants
7. ___ biennial
8. ___ woody plants
9. ___ girdling
10. ___ bark
11. ___ cork cambium
12. ___ fusiform initials
13. ___ late wood
14. ___ annual
15. ___ ray initials
16. ___ perennial

A. These cells produce periderm—a corky replacement for lost epidermis.
B. Plant life cycle completed in one growing season
C. The core of a mature tree, no living cells
D. Includes all tissues outside of the vascular cambium
E. Show secondary growth during two or more growing seasons
F. As the growing season progresses, the xylem cell diameters become smaller and the walls become thinner.
G. Vegetative growth and seed formation continue year after year; some have secondary tissues, others do not.
H. Vascular cambium cells that produce secondary xylem and phloem
I. Stripping off a band of phloem all the way around a tree's circumference; interrupts phloem transport
J. Vascular cambium cells that produce living parenchyma and other cells for lateral conduction
K. Cylindrical zone of xylem between heartwood and vascular cambium; contains some living parenchyma cells among nonliving conducting xylem cells
L. A plant life cycle completed in two growing seasons
M. Alternating bands representing annual growth layers
N. A cylinder of cells in older stems and roots that give rise to secondary xylem and phloem; xylem forms on the inner face, phloem on the outer
O. The first xylem cells produced at the start of the growing season; tend to have large diameters and thin walls
P. Herbaceous plants that show little or no secondary growth

## Label and Match

Identify each numbered part of the illustration of a tree trunk cross section. Complete the exercise by matching and entering the letter of the proper description in the parentheses following each label.

17. _____ _____ ( )      20. _____ ( )

18. _____ ( )                     21. _____ ( )

19. _____ ( )

A. Corky replacement for epidermis
B. Vascular tissue that conducts water and dissolved minerals absorbed from soil; gives mechanical support to the plant
C. Has meristematic cells that give rise to secondary xylem and phloem tissues
D. The vascular tissue that transports sugars and other solutes through the plant body
E. All living and nonliving tissues between the vascular cambium and the stem or root surface

---

# Self-Quiz

___ 1. _____ develops into the plant's surface layers.
   a. Ground tissue
   b. Dermal tissue
   c. Vascular tissue
   d. Pericycle

___ 2. Which of the following is *not* considered a ground cell type?
   a. Epidermis
   b. Parenchyma
   c. Collenchyma
   d. Sclerenchyma

___ 3. The _____ produces secondary xylem growth.
   a. apical meristem
   b. lateral meristem
   c. cork cambium
   d. endodermis

___ 4. The _____ is a leaflike structure that is part of the embryo; monocot embryos have one, dicot embryos have two.
   a. shoot tip
   b. root tip
   c. cotyledon
   d. apical meristem

___ 5. Leaves are differentiated and buds develop at specific points along the stem called _____.
   a. nodes
   b. internodes
   c. vascular bundles
   d. cotyledons

___ 6. Which of the following structures is *not* considered to be meristematic?
   a. vascular cambium
   b. lateral meristem
   c. cork cambium
   d. endodermis

___ 7. Which of the following statements about monocots is *false*?
   a. They are usually herbaceous.
   b. They develop one cotyledon in their seeds.
   c. Their vascular bundles are scattered throughout the ground tissue of their stems.
   d. They have a single central vascular cylinder in their stems.

___ 8. New plants grow and older plant parts lengthen through cell divisions at _____ meristems present at root and shoot tips; older roots and stems of woody plants increase in diameter through cell divisions at _____ meristems.
   a. lateral; lateral
   b. lateral; apical
   c. apical; apical
   d. apical; lateral

___ 9. Vascular bundles called _____ form a network through a leaf blade.
   a. xylem
   b. phloem
   c. veins
   d. stomata

___ 10. A primary root and its lateral branchings represent a _____ system.
   a. lateral root
   b. adventitious root
   c. taproot
   d. branch root

___ 11. Plants whose vegetative growth and seed formation continue year after year are _____ plants.
   a. annual
   b. perennial
   c. biennial
   d. herbaceous

___ 12. The _____ layer of a root divides to produce lateral roots.
   a. endodermis
   b. pericycle
   c. xylem
   d. cortex

# Chapter Objectives/Review Questions

This section lists general and detailed chapter objectives that can be used as review questions. You can make maximum use of these items by writing answers on a separate sheet of paper. Fill in answers where blanks are provided. To check for accuracy, compare your answers with information given in the chapter or glossary.

*Page*      *Objectives/Questions*

(378)      1. The aboveground parts of flowering plants are called _____; the plants' descending parts are called _____.

(378)      2. Distinguish between the ground tissue system, the vascular tissue system, and the dermal tissue system.

| (379) | 3. | Plants grow at localized regions of self-perpetuating embryonic cells called _____. |
|---|---|---|
| (379) | 4. | Lengthening of stems and roots originates at _____ meristems and all dividing tissues are derived from them; this is called _____ growth. |
| (379) | 5. | Increases in the diameter of a plant originate at _____ meristems. |
| (379) | 6. | Describe the role of vascular cambium and cork cambium in producing secondary tissues of the plant body. |
| (380) | 7. | Name and generally describe the simple tissues called parenchyma, collenchyma, and sclerenchyma. |
| (380) | 8. | Name the cell wall compound that was necessary for the evolution of rigid and erect land plants. |
| (380) | 9. | _____ tissue conducts soil water and dissolved minerals, and it mechanically supports the plant. |
| (381) | 10. | _____ tissue transports sugars and other solutes. |
| (381) | 11. | Name and describe the functions of the conducting cells in xylem and phloem. |
| (381) | 12. | All surfaces of primary plant parts are covered and protected by a dermal tissue system called _____. |
| (381) | 13. | What is the function of guard cells and stomata found within the epidermis of young stems and leaves? |
| (381) | 14. | The cork cells of _____ replace the epidermis of stems and roots showing secondary growth. |
| (381) | 15. | Distinguish between monocots and dicots by listing their characteristics and citing examples of each group. |
| (382) | 16. | Leaves develop from _____ located on the flanks of the apical meristem. |
| (382) | 17. | How does the simplest type of leaf differ from "compound" leaves? |
| (382) | 18. | What feature distinguishes deciduous plants from evergreen plants? |
| (383) | 19. | The primary xylem and phloem develop as vascular _____. |
| (384) | 20. | Describe the structure and major functions of leaf epidermis, mesophyll, and vein tissue. |
| (386) | 21. | The _____ root is the first to poke through the coat of a germinating seed; later, _____ roots erupt through the epidermis. |
| (386) | 22. | How does a taproot system differ from a fibrous root system? |
| (386 - 387) | 23. | What is the origin and function of root hairs? |
| (387) | 24. | Vascular tissues form a vascular _____, a central column inside the root. |
| (387) | 25. | Describe the passage of soil water through root epidermis to the xylem of the vascular cylinder; include the role of the endodermis. |
| (388) | 26. | Define these three categories of flowering plants: annuals, biennials, and perennials. |
| (388) | 27. | Each growing season, new tissues that increase the girth of woody plants originate at their _____ meristems. |
| (388) | 28. | Distinguish early wood from late wood; heartwood from sapwood. |
| (389) | 29. | State the functions of fusiform and ray initials of the vascular cambium. |
| (389) | 30. | Explain the origin of the annual growth layers (tree rings) seen in a cross section of a tree trunk. |

## Integrating and Applying Key Concepts

Try to imagine the specific behavioral restrictions that might be imposed if the human body resembled the plant body in having (1) open growth with apical meristematic regions, (2) stomata in the epidermis, (3) cells with chloroplasts, (4) excess carbohydrates stored primarily as starch rather than as fat, and (5) dependence on the soil as a source of water and inorganic compounds.

# 23

# PLANT NUTRITION AND TRANSPORT

UPTAKE OF WATER AND NUTRIENTS AT THE
   ROOTS
     Absorption Routes
     Specialized Absorptive Structures
*Focus on the Environment:* PUTTING DOWN ROOTS
CONSERVATION OF WATER IN STEMS AND
   LEAVES
     The Water-Conserving Cuticle
     Controlled Water Loss at Stomata

A THEORY OF WATER TRANSPORT
     Transpiration Defined
     Cohesion-Tension Theory of Water Transport
DISTRIBUTION OF ORGANIC COMPOUNDS
   THROUGH THE PLANT
     Translocation
     Pressure Flow Theory

## Interactive Exercises

### 23 - I. UPTAKE OF WATER AND NUTRIENTS AT THE ROOTS (pp. 392 - 395)

*Boldfaced, Page-Referenced Terms*

(392) plant physiology _____

_____

(392) macronutrients _____

_____

(392) micronutrients _____

_____

(394) root hairs _____

_____

(394) vascular cylinder _____

_____

(394) endodermis _____

_____

(394) Casparian strip _____

_____

(395) exodermis _____

_____

(395) mutualism _____

_____

(395) nitrogen fixation _____

_____

(395) root nodules _____

_____

(395) mycorrhizae _____

_____

## Fill-in-the-Blanks

The three essential elements that plants use as their main metabolic building blocks are oxygen, carbon, and (1) _____. Plant survival depends on the uptake of at least (number) (2) _____ other elements. These are typically available to plants as dissolved (3) _____ _____. Of these, six are present in easily detectable concentrations in plant tissues and are known as (4) _____. The remainder occur in very small amounts in plant tissues and are known as (5) _____.

## Complete the Table

6. Thirteen essential elements are available to plants as mineral ions. Complete the table below, which summarizes information about these important plant nutrients. Refer to Table 23.1 (p. 392) in the text.

| Mineral Element | Macronutrient or Micronutrient | Known Functions |
| --- | --- | --- |
| a. | | Roles in chlorophyll synthesis, electron transport |
| b. | | Activation of enzymes, role in maintaining water-solute balance |
| c. | | Role in chlorophyll synthesis; coenzyme activity |
| d. | | Role in root, shoot growth; role in photolysis |
| e. | | Role in chlorophyll synthesis; coenzyme activity |
| f. | | Component of enzyme used in nitrogen metabolism |
| g. | | Component of proteins, nucleic acids, coenzymes, chlorophyll |
| h. | | Component of most proteins, two vitamins |
| i. | | Roles in flowering, germination, fruiting, cell division, nitrogen metabolism |
| j. | | Role in formation of auxin, chloroplasts, and starch; enzyme component |
| k. | | Roles in cementing cell walls, regulation of many cell functions |
| l. | | Component of several enzymes |
| m. | | Component of nucleic acids, phospholipids, ATP |

## Sequence

Arrange in correct chronological sequence the path that nutrients take from the soil to cells in living plant tissues. Write the letter of the first step next to 7, the letter of the second step next to 8, and so on.

7. ___
8. ___
9. ___
10. ___
11. ___
12. ___

A. Cortex cells lacking Casparian strips
B. Endodermis cells with Casparian strips
C. Root hairs
D. ATP energy (from photosynthesis, respiration or both) actively transporting nutrients at membrane proteins
E. Exodermis cells with Casparian strips
F. Vascular cylinder

## Label and Match

Identify each numbered part of the illustrations below, which deal with the control of nutrient uptake by plant roots. Choose from the following: water movement, cytoplasm, vascular cylinder, endodermal cell wall, exodermis, endodermis, and Casparian strip. Complete the exercise by matching from the list below and entering the correct letter in the parentheses following each label.

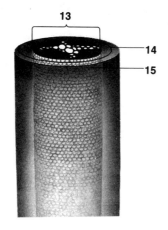
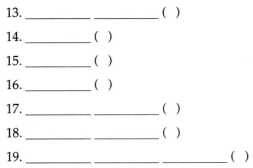
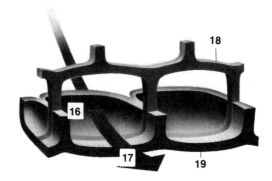

13. _____ _____ (  )

14. _____ (  )

15. _____ (  )

16. _____ (  )

17. _____ _____ (  )

18. _____ _____ (  )

19. _____ _____ _____ (  )

A. Cellular area through which water and dissolved nutrients must move due to Casparian strips
B. A layer of cortex cells just inside the epidermis; also equipped with Casparian strips
C. Waxy band acting as an impermeable barrier between the walls of abutting endodermal cells; forces water and dissolved nutrients through the cytoplasm of endodermal cells
D. Specific location of the waxy strips known as Casparian strips
E. Substance whose diffusion occurs through the cytoplasm of endodermal cells due to Casparian strips
F. Sheetlike layer of single cortex cells wrapped around the vascular cylinder
G. Tissues include the xylem, phloem, and pericycle.

### Dichotomous Choice

Circle one of two possible answers given between parentheses in each statement.

20. A form of symbiosis, (mutualism/parasitism), refers to permanent and intimate interactions between species in which two-way benefits pass between species.

21. (Gaseous nitrogen/Nitrogen "fixed" by bacteria) represents the chemical form of nitrogen plants can use in their metabolism.

22. Nitrogen-fixing, mutualistic bacteria reside in localized swellings on legume plants known as (root hairs/root nodules).

23. Mycorrhizae represent symbiotic relationships in which fungi and the roots they cover both benefit; the roots receive (sugars and nitrogen-containing compounds/scarce minerals).

24. (Root nodules/Root hairs) greatly increase a plant's capacity for absorbing water and nutrients from the soil.

## 23 - II. *Focus on the Environment:* PUTTING DOWN ROOTS (p. 396) CONSERVATION OF WATER IN STEMS AND LEAVES (pp. 396 - 397)

*Selected Word: topsoil*

### Boldfaced, Page-Referenced Terms

(396) cuticle _____

_____

(396) cutin _____

_____

(396) stomata _____

_____

(396) guard cells _____

_____

(397) CAM plants _____

_____

### True/False

If the statement is true, write a T in the blank. If the statement is false, make it correct by changing the underlined word(s) and writing the correct word(s) in the blank.

_____ 1. Of the water moving into a leaf, <u>2 percent</u> or more is lost by evaporation into the surrounding air.

_____ 2. When evaporation exceeds water uptake by roots, plant tissues <u>wilt</u> and water-dependent activities are seriously disrupted.

_____ 3. The surfaces of all plant epidermal cell walls have an outer layer of waxes embedded in cutin, the <u>cuticle</u>, which restricts water loss, limits inward diffusion of carbon dioxide, and limits outward diffusion of oxygen.

_____ 4. Plant epidermal layers are peppered with tiny openings called <u>guard cells</u> through which water leaves the plant and carbon dioxide enters.

_____ 5. When a pair of guard cells swells with water, the opening between them <u>closes</u>.

_____ 6. In most plants, stomata remain <u>open</u> during the daylight photosynthetic period; water is lost but they gain carbon dioxide.

_____ 7. Stomata stay <u>closed</u> at night in most plants.

_____ 8. Photosynthesis starts when the sun comes up; as the morning progresses, carbon dioxide levels <u>increase</u> in cells, including guard cells.

_____ 9. A drop in carbon dioxide level within guard cells triggers an inward active transport of potassium ions; water follows by osmosis and the fluid pressure <u>closes</u> the stoma.

_____ 10. When the sun goes down and photosynthesis stops, carbon dioxide levels rise; stomata <u>close</u> when potassium, then water, moves out of the guard cells.

_____ 11. CAM plants such as cacti and other succulents open stomata during the <u>day</u> when they fix carbon dioxide by way of a special C4 metabolic pathway.

_____ 12. CAM plants use carbon dioxide the following <u>night</u> when stomata are closed.

## 23 - III. A THEORY OF WATER TRANSPORT (pp. 398 - 399)

**Selected Word:** *cohesion*

### Boldfaced, Page-Referenced Terms

(398) transpiration _____

_____

(398) xylem _____

_____

(398) tracheids _____

_____

(398) vessel members _____

_____

(398) cohesion-tension theory _____

_____

### Fill-in-the-Blanks

The cohesion-tension theory explains (1) _____ transport to the tops of plants. The water travels inside tubelike strands of (2) _____, which are formed by hollow, dead, cells called tracheids and vessel members. The process begins with the drying power of air, which causes (3) _____, the evaporation of water from plant parts exposed to air. (4) _____ of hydrogen-bonded water molecules in the xylem of roots, stems, and leaves provides a continuous column of water. This places the xylem water in a state of (5) _____ that extends from veins in leaves, down through the stems, to roots. As water continues to escape from plant surfaces, water enters the roots to replace that which is lost.

# 23 - IV. DISTRIBUTION OF ORGANIC COMPOUNDS THROUGH THE PLANT (pp. 400 - 401)

*Selected Words:* source, sink

## Boldfaced, Page-Referenced Terms

(400) phloem _____

_____

(400) sieve tubes _____

_____

(400) companion cells _____

_____

(400) translocation _____

_____

(400) pressure flow theory _____

_____

## Fill-in-the-Blanks

Sucrose and other organic compounds resulting from (1) _____ are used throughout the plant. Most plant cells store their carbohydrates as (2) _____. Quantities of (3) _____ become stored in some fruits; (4) _____ store proteins and fats. (5) _____ molecules are too large to cross cell membranes and too insoluble to be transported to other regions of the plant body. (6) _____ are largely insoluble in water and cannot be transported from storage sites. (7) _____ proteins do not lend themselves to transport. Specific chemical reactions within plant cells, such as (8) _____, convert storage forms of organic compounds to their subunits, which are transportable forms. For example, the hydrolysis of starch liberates glucose units, which combine with fructose to form (9) _____, a transport form for sugars. Experiments with phloem-embedded aphid mouthparts revealed that (10) _____ was the most abundant carbohydrate being forced out of those conducting tubes.

## Matching

Choose the most appropriate answer for each.

11. ___ translocation

12. ___ sieve-tube members

13. ___ companion cells

14. ___ aphids

15. ___ source

16. ___ sink

17. ___ pressure flow theory

A. Any region where organic compounds are being loaded into the sieve-tube system

B. Nonconducting cells adjacent to sieve-tube members that supply energy to load sucrose at the source

C. Any region of the plant where organic compounds are being unloaded from the sieve-tube system and used or stored

D. Process occurring in phloem that distributes sucrose and other organic compounds through the plant (apparently under pressure)

E. States that pressure builds up at the source end of a sieve-tube system and pushes solutes toward a sink, where they are removed

F. Passive conduits for translocation within vascular bundles; water and organic compounds flow rapidly through large pores on their end walls

G. Insects used to verify that in most plant species sucrose is the main carbohydrate translocated under pressure

## Sequence

Consider the information in the short paragraph below (also see Figure 23.15, p. 402). Then arrange the events below in a logical, chronological sequence. Write the letter of the first step next to number 18, the letter of the second step next to 19, and so on.

Once nutrients enter a vascular cylinder, they are distributed to all living tissues in coordinated, inter-related ways that affect plant growth. Thus living cells expend ATP energy and actively transport certain nutrients at membrane proteins. In photosynthetic cells, the required ATP is produced during photosynthesis and aerobic respiration. In other cells, ATP forms mostly by aerobic respiration.

18. ___

19. ___

20. ___

21. ___

22. ___

23. ___

A. Photosynthesis

B. Absorption of minerals and water by roots

C. Respiration of sucrose by roots

D. Transport of minerals and water to leaves

E. ATP formation by roots

F. Transport of sucrose to roots

# Self-Quiz

___ 1. The three elements that are present in carbohydrates, lipids, proteins, and nucleic acids are _____.
   a. oxygen, carbon, and nitrogen
   b. oxygen, hydrogen, and nitrogen
   c. oxygen, carbon, and hydrogen
   d. carbon, nitrogen, and hydrogen

___ 2. Macronutrients are the six dissolved mineral ions that _____.
   a. play vital roles in photosynthesis and other metabolic events
   b. occur in only small traces in plant tissues

   c. become detectable in plant tissues
   d. can function only without the presence of micronutrients
   e. both a and c

___ 3. Gaseous nitrogen is converted to a plant-usable form by _____.
   a. root nodules
   b. mycorrhizae
   c. nitrogen-fixing bacteria
   d. Venus flytraps

___ 4. _____ prevent(s) inward-moving water from moving past the abutting walls of the root endodermal cells.
a. Cytoplasm
b. Plasma membranes
c. Osmosis
d. Casparian strips

___ 5. Most of the water moving into a leaf is lost through _____.
a. osmotic gradients being established
b. evaporation of water from plant parts exposed to air
c. pressure flow forces
d. translocation

___ 6. Stomata remain _____ during daylight, when photosynthesis occurs, but remain _____ during the night when carbon dioxide accumulates through aerobic respiration.
a. open; open
b. closed; open
c. closed; closed
d. open; closed

___ 7. By control of _____ levels inside the guard cells of stomata, the activity of stomata is controlled when leaves are losing more water than roots can absorb.
a. oxygen
b. potassium

c carbon dioxide
d. ATP

___ 8. Without _____, plants would rapidly wilt and die during hot, dry spells.
a. a cuticle
b. mycorrhizae
c. phloem
d. cotyledons

___ 9. The _____ theory of water transport states that hydrogen bonding allows water molecules to maintain a continuous fluid column as water is pulled from roots to leaves.
a. pressure flow
b. cohesion-tension
c. evaporation
d. abscission

___ 10. Leaves represent _____ regions; growing leaves, stems, fruits, seeds, and roots represent _____ regions.
a. source; source
b. sink; source
c. source; sink
d. sink; sink

---

## Chapter Objectives/Review Questions

| Page | | Objectives/Questions |
|---|---|---|
| (392) | 1. | Plants generally require _____ (number) essential elements. |
| (392) | 2. | Distinguish between macronutrients and micronutrients in relation to their role in plant nutrition. |
| (394 - 395) | 3. | Differentiate between the endodermis and the exodermis of the root cortex. |
| (394) | 4. | Due to the presence of the _____ strip in the walls of endodermal cells, water can move into the vascular cylinder only by crossing the plasma membrane and diffusing through the cytoplasm. |
| (395) | 5. | Describe the roles of root nodules and mycorrhizae in plant nutrition. |
| (396) | 6. | Even mildly stressed plants would rapidly wilt and die if it were not for the _____ covering their parts; describe additional functions of this structure. |
| (396) | 7. | Describe the chemical compounds found in cutin. |
| (396) | 8. | Evaporation from plant parts occurs mostly at _____, tiny epidermal passageways of leaves and stems. |
| (396) | 9. | Explain the mechanism by which stomata open during daylight and close during the night. |
| (397) | 10. | Describe the mechanisms by which CAM plants conserve water. |

(398)     11.   The evaporation of water from leaves as well as from stems and other plant parts is known as _____.

(398)     12.   In a plant's vascular tissues, water moves through pipelines collectively called _____ tissue.

(398)     13.   Henry Dixon's _____-_____ theory explains how water moves upward in an unbroken column through xylem to the tops of tall trees.

(398 - 399) 14.   Be able to give the key points of Dixon's explanation of upward water transport in plants.

(400)     15.   _____ is the main form in which sugars are transported through most plants.

(400)     16.   Describe the role of phloem, sieve-tube members, and companion cells in translocation.

(400)     17.   Define translocation.

(400)     18.   A _____ is where solutes are loaded into seive tubes; a _____ is any region of the plant where organic compounds are unloaded from the sieve-tube system and used or stored.

(400 - 401) 19.   According to the _____ _____ theory, differences in solute concentrations and pressure build up at the source end of a sieve-tube system and push solutes toward a sink, where they are removed.

(401)     20.   Companion cells supply the _____ that loads sucrose at the source.

(401)     21.   Name the gradients responsible for continuous flow of organic compounds through phloem.

---

## Integrating and Applying Key Concepts

How do you think maple syrup is made from maple trees? Which specific systems of the plant are involved, and why are maple trees tapped only at certain times of the year?

# 24

# PLANT REPRODUCTION
# AND DEVELOPMENT

## Interactive Exercises

### 24 - I. REPRODUCTIVE STRUCTURES OF FLOWERING PLANTS (pp. 404 - 407)
#### *Focus on the Environment:* POLLEN SETS ME SNEEZING (p. 407)

*Selected Words:* <u>Angraecum</u> <u>sesquipedale</u>, sexual reproduction, asexual reproduction, sepals, petals, stamens, carpels, "calyx," "corolla," *angiosperm, perfect* flowers, *imperfect* flowers

*Boldfaced, Page-Referenced Terms*

(404) coevolution _____

_____

(404) pollinator _____

_____

(406) sporophyte _____

_____

(406) flowers _____

_____

(406) gametophytes _____

_____

(407) stamens _____

_____

(407) pollen grains _____

_____

(407) carpels _____

_____

(407) ovary _____

_____

## Label and Match

Study the generalized life cycle for flowering plants as shown in Figure 24.3*a* in the text. Label each numbered part of the accompanying illustration. Complete the exercise by writing the letter of the proper description in the parentheses following each label.

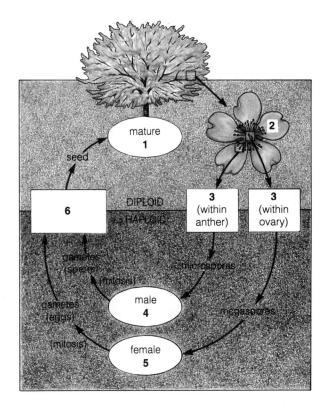

1. _____ (  )

2. _____ (  )

3. _____ (  )

4. _____ (  )

5. _____ (  )

6. _____ (  )

A. An event that produces a young sporophyte
B. A reproductive shoot produced by the sporophyte
C. The "plant"; a vegetative body that develops from a zygote
D. Cellular division event occurring within flowers to produce spores
E. Produces haploid eggs by mitosis
F. Produces haploid sperm by mitosis

## Labeling

Identify each numbered part of the illustration.

7. _____

8. _____

9. _____

10. _____

11. _____

12. _____

13. _____

14. _____

**12**
(male reproductive part)

**13**
(female reproductive part)

9

filament

**7**
(all combined are
the flowers corolla)

**8**
(all combined are
the flower's calyx)

10

style

**11**

**14**
(forms
within
ovary)

receptacle

## Matching

Choose the most appropriate answer for each term.

15. ___ sepals

16. ___ petals

17. ___ stamens

18. ___ ovule

19. ___ pollen sacs

20. ___ carpels

21. ___ ovaries

22. ___ perfect flowers

23. ___ imperfect flowers

A. Have both male and female parts
B. Four chambers within an anther where pollen grains develop
C. Collectively, the flower's "corolla"
D. Have male or female parts, but not both
E. Female reproductive parts; includes stigma, style, and ovary
F. Structure that matures to become a seed
G. Found just inside the flower's corolla, the male reproductive parts
H. Lower portion of the carpel where egg formation, fertilization, and seed development occur
I. Outermost leaflike whorl of floral organs; collectively, the calyx

# 24 - II. A NEW GENERATION BEGINS (pp. 408 - 409)

## Boldfaced, Page-Referenced Terms

(408) microspores _____

_____

(408) ovule _____

_____

(408) megaspores _____

_____

(408) endosperm _____

_____

(408) pollination _____

_____

(408) germination _____

_____

(409) double fertilization _____

_____

## Fill-in-the-Blanks

The numbered items on the illustration below represent missing information; complete the blanks in the following narrative to supply that information.

Within each (1) _____, mitotic divisions produce four masses of spore-forming cells, each mass forming within a (2) _____ _____. Each one of these diploid cells is known as a (3) _____ _____ cell and undergoes (4) _____ to produce four haploid (5) _____. Mitosis within each haploid microspore results in a two-celled haploid body, the immature male gametophyte. One of these cells will give rise to a (6) _____ _____; the other cell will develop into a (7) _____ - _____ cell. Mature microspores are eventually released from the pollen sacs of the anther as (8) _____. Pollination occurs, and after the pollen lands on a carpel's (9) _____, the pollen tube develops from one of the cells in the pollen grain. The other cell within the pollen grain divides to form two sperm cells. As the pollen tube grows through the carpel tissues, it contains the two sperm cells and a tube nucleus. The pollen tube with its included two sperm cells and the tube nucleus is known as the mature (10) _____ _____.

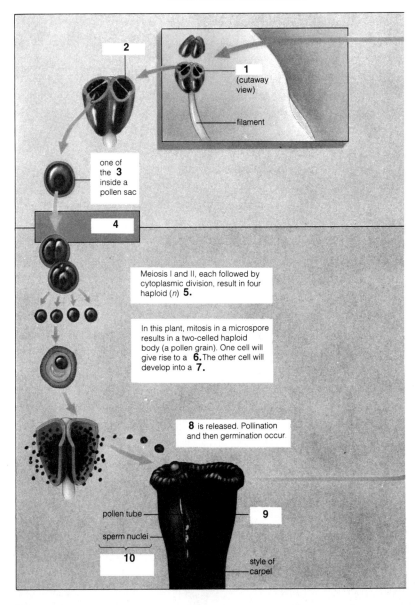

## Fill-in-the-Blanks

The numbered items on the illustration below represent missing information; complete the blanks in the following narrative to supply that information.

In the carpel of the flower, one or more dome-shaped, diploid tissue masses develop on the inner wall of the ovary. Each mass is the beginning of an (11) _____. A tissue forms inside a domed mass as it grows, and one or two protective layers called (12) _____ form around it. Inside each mass, a diploid cell (the megaspore mother cell) divides by (13) _____ to form four haploid spores, the (14) _____. Commonly, all but one (15) _____ disintegrates. The remaining (16) _____ undergoes (17) _____ three times without cytoplasmic division. At first, this structure is a cell with (18) _____ haploid nuclei. Cytoplasmic division results in a seven-cell (19) _____ _____, which represents the mature (20) _____ _____. One of these cells, the (21) _____

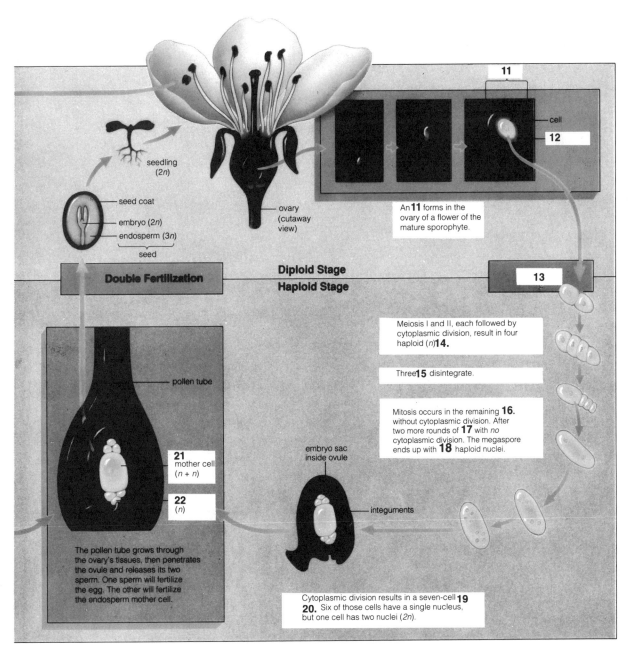

seedling
(2n)

seed coat

embryo (2n)

endosperm (3n)

seed

**Double Fertilization**

**Diploid Stage**

**Haploid Stage**

ovary
(cutaway
view)

An **11** forms in the ovary of a flower of the mature sporophyte.

cell

**11**

**12**

**13**

Meiosis I and II, each followed by cytoplasmic division, result in four haploid (n)**14.**

Three**15** disintegrate.

Mitosis occurs in the remaining **16.** without cytoplasmic division. After two more rounds of **17** with *no* cytoplasmic division. The megaspore ends up with **18** haploid nuclei.

pollen tube

**21**
mother cell
(n + n)

**22**
(n)

embryo sac
inside ovule

integuments

The pollen tube grows through the ovary's tissues, then penetrates the ovule and releases its two sperm. One sperm will fertilize the egg. The other will fertilize the endosperm mother cell.

Cytoplasmic division results in a seven-cell **19**
**20.** Six of those cells have a single nucleus, but one cell has two nuclei (2n).

mother cell, has two nuclei; following fertilization with one sperm, this cell will help form the 3n endosperm, a nutritive tissue for the forthcoming embryo. Another haploid cell within the embryo sac is the (22) _____, which will combine with the other sperm to form the diploid embryo.

## Short Answer

23. What guides the growth of the pollen tube down through the female floral tissues toward the chamber holding the egg? _____

_____

24. Describe the site of double fertilization known only in flowering plants. _____

_____

## Complete the Table

25. Complete the table below to summarize the unique double fertilization occurring only in flowering plant life cycles. Refer to Figure 24.6 (p. 408) in the text.

| Double Fertilization Products | Origin | Produce? | Function |
|---|---|---|---|
| a. Zygote (2n) nucleus | | | |
| b. Endosperm (3n) nucleus | | | |

# 24 - III. FROM ZYGOTE TO SEED (pp. 410 - 411)

## Boldfaced, Page-Referenced Terms

(410) cotyledons _____

_____

(410) seed _____

_____

(410) fruit _____

_____

## Complete the Table

1. Complete the following table, which summarizes concepts associated with seeds and fruits.

| Structure | Origin | Function(s) |
|---|---|---|
| a. Cotyledons | | |
| b. Seeds | | |
| c. Seed coat | | |
| d. Fruit | | |

## Labeling

Identify each numbered part of the illustration below.

2. _____    9. _____

3. _____    10. _____

4. _____    11. _____

5. _____    12. _____ _____

6. _____    13. _____

7. _____ _____    14. _____

8. _____ _____

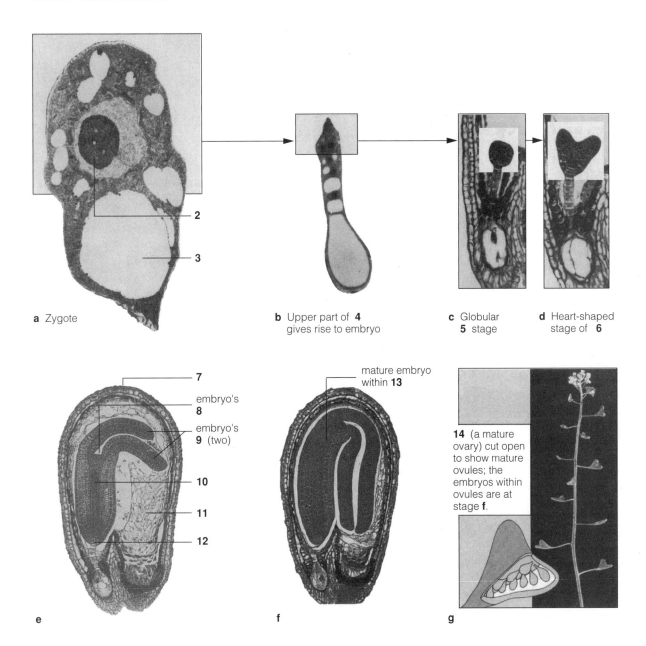

**a** Zygote

**b** Upper part of **4** gives rise to embryo

**c** Globular **5** stage

**d** Heart-shaped stage of **6**

7

embryo's **8**

embryo's **9** (two)

10

11

12

mature embryo within **13**

**14** (a mature ovary) cut open to show mature ovules; the embryos within ovules are at stage **f**.

e

f

g

## Sequence

Arrange the following developmental events in correct chronological sequence from first to last.

15. \_\_\_    A. Early on, cell divisions produce a single row of cells.

16. \_\_\_    B. The ovule's integuments thicken and harden into a seed coat.

17. \_\_\_    C. The lower cells (of the single row) develop into a stalklike structure that anchors the embryo and absorbs endosperm nutrients for it; the cells above the stalklike structure develop into the embryo proper.

18. \_\_\_    D. The embryo and food reserves are now a self-contained package. The ovule has become a seed.

19. \_\_\_    E. Following fertilization, the newly formed zygote embarks on a course of cell divisions leading to a mature embryo sporophyte.

20. \_\_\_    F. Primary meristems begin to form and two cotyledons (monocot embryos have one) have begun to develop from two tissue lobes.

21. \_\_\_    G. Food reserves accumulate in the expanding endosperm or cotyledons.

## Choice

For questions 22 - 28, choose from the following:

a. simple, wall dry; splits at maturity    b. simple, wall dry; intact at maturity

c. simple, fleshy; often leathery    d. aggregate    e. multiple

f. accessory, simple    g. accessory, aggregate

22. \_\_\_ apple, pear

23. \_\_\_ pea, magnolia, mustard

24. \_\_\_ strawberry

25. \_\_\_ sunflower, wheat, rice, maple

26. \_\_\_ pineapple, fig, mulberry

27. \_\_\_ grape, banana, lemon, cherry

28. \_\_\_ blackberry, raspberry

## Fill-in-the-Blanks

Long before humans began eating them, (29) _____ were undergoing natural selection as seed-dispersing mechanisms. They have coevolved with air currents, water currents, and with particular (30) _____. The wings of a dropping maple fruit spins sideways with air currents to arrive far enough that its enclosed embryo will not compete with the (31) _____ plant for soil water, minerals, and sunlight. Many (32) _____ have hooks, spines, hairs, and sticky surfaces that allow animals to move them to new locations. Fleshy fruits such as blueberries and cherries have their seeds dispersed as they are eaten and expelled by (33) _____. Digestive enzymes in the animal guts remove enough of the hard seed coats to increase the chance of successful (34) _____ after expulsion from the body.

# 24 - IV. ASEXUAL REPRODUCTION OF FLOWERING PLANTS (pp. 412 - 413)

*Selected Words:* Populus tremuloides, Larrea divaricata, Daucus carota

## Boldfaced, Page-Referenced Terms

(412) vegetative growth _____

_____

(412) parthenogenesis _____

_____

(412) tissue culture propagation _____

_____

## Matching

Match the following asexual reproductive modes of flowering plants.

1. ___ corm
2. ___ bulb
3. ___ parthenogenesis
4. ___ runner
5. ___ vegetative propagation on modified stems
6. ___ rhizome
7. ___ tuber
8. ___ tissue culture propagation (induced propagation)
9. ___ vegetative growth

A. In a general sense, new plants develop from tissues or organs that drop or separate from parent plants.
B. New shoots arise from axillary buds (enlarged tips of slender underground rhizomes).
C. New plants arise from cells in parent plant that were not irreversibly differentiated; it is a laboratory technique.
D. New plants arise at nodes of underground horizontal stem.
E. A new plant arises from an axillary bud on a short, thick, vertical underground stem.
F. New plants arise at nodes on an aboveground horizontal stem.
G. A new structure arises from an axillary bud on a short underground stem.
H. It involves asexual reproduction utilizing runners, rhizomes, corms, tubers, and bulbs.
I. The embryo develops without a nuclear or cellular fusion.

## Matching

Choose the most appropriate example for each modified stem.

10. ___ bulb
11. ___ rhizome
12. ___ tuber
13. ___ runner
14. ___ corm

A. Potato
B. Strawberry
C. Gladiolus
D. Onion, lily
E. Bermuda grass

## 24 - V. PATTERNS OF GROWTH AND DEVELOPMENT (pp. 414 - 415)

*Selected Words:* Zea mays, Phaseolus vulgaris

### Boldfaced, Page-Referenced Terms

(414) imbibition _____

_____

(414) primary root _____

_____

### Fill-in-the-Blanks

Before or after seed dispersal, the growth of the (1) _____ sporophyte idles. For seeds, (2) _____ is the resumption of growth after a time of arrested embryonic development. The development process is governed by (3) _____ and the environment. Environmental factors influencing this process include (4) _____ temperature, moisture and oxygen levels, and the number of daylight hours. Usually germination coincides with the return of spring (5) _____. In a process called (6) _____, water molecules move into the seed. As more and more water molecules move inside it, the seed swells and its (7) _____ ruptures. Once the seed coat splits, more oxygen reaches the embryo, and (8) _____ respiration moves into high gear. Cells then rapidly divide to provide growth of the embryo. Cells of the embryonic (9) _____ activate first, divide, elongate, and give rise to a (10) _____ root. Germination is concluded when the primary root breaks through the seed coat. Growth involves cell (11) _____ and cell (12) _____. There are two basic patterns of growth and development among flowering plants, that of a (13) _____ and a (14) _____. Basic growth and development patterns are (15) _____ and dictated by the plant's genes. Plant development requires cell (16) _____, as brought about by selective gene expression. The growth patterns can be adjusted in response to (17) _____ pressures. Enzymes, hormones, and other gene products within the cells carry out the (18) _____ response.

## 24 - VI. PLANT HORMONES (pp. 416 - 417)
### *Focus on Science:* FOOLISH SEEDLINGS, GORGEOUS GRAPES (p. 417)

*Selected Words:* Agent Orange, Gibberella fujikuroi

### Boldfaced, Page-Referenced Terms

(416) hormones _____

_____

(416) auxins _____

_____

(416) gibberellins _____

_____

(416) cytokinins _____

_____

(416) ethylene _____

_____

(416) abscisic acid _____

_____

(416) coleoptile _____

_____

(416) herbicide _____

_____

## Choice

For questions 1 - 14, choose from the following:

> a. auxins     b. gibberellins     c. cytokinins     d. abscisic acid     e. ethylene

1. ___ The ancient Chinese burned incense to hurry fruit ripening.

2. ___ Natural and synthetic versions are used to prolong the shelf life of cut flowers, lettuces, mushrooms, and other vegetables.

3. ___ Orchardists spray trees with IAA to thin out overcrowded seedlings in the spring.

4. ___ Inhibits cell growth, promotes bud dormancy, and prevents seeds from germinating prematurely; causes stomata to close and help conserve plant water

5. ___ IAA, the most important naturally occurring compound of its type

6. ___ Exposed to oranges and other citrus fruits to brighten the color of their rind before being displayed in the market

7. ___ Stimulates stem lengthening; influence plant responses to gravity and light and promote coleoptile elongation

8. ___ Causes stems to lengthen; help buds and seeds break dormancy and resume growth in the spring

9. ___ Used to prevent premature fruit drop—all fruit can be picked at the same time

10. ___ Stimulate cell division

11. ___ Used by food distributors to ripen green fruit after shipment

12. ___ Most abundant in root and shoot meristems and in the tissues of maturing fruits

13. ___ Under its influence, flowers, fruits, and leaves drop away from plants at prescribed times of the year

14. ___ Promote leaf expansion and retard leaf aging

## Matching

Choose the most appropriate answer for each.

15. ___ 2,4-D

16. ___ hormone

17. ___ apical dominance

18. ___ herbicide

19. ___ "foolish seedling" effect on rice plants

20. ___ IAA

21. ___ 2,4,5-T

22. ___ target cell

A. Mixed with 2,4-D to produce Agent Orange for defoliation in Vietnam

B. Gibberellin from fungal extracts

C. A cell with receptors to bind a given signaling molecule

D. Hormonal effect that inhibits lateral bud growth, which promotes stem elongation

E. Synthetic auxin used as an herbicide to selectively kill broadleaf weeds that compete with valuable crop plants

F. A signaling molecule released from one cell that changes the activity of target cells

G. The most important naturally occurring auxin

H. Any synthetic auxin compound used to selectively kill plants

# 24 - VII. ADJUSTMENTS IN THE RATE AND DIRECTION OF GROWTH (pp. 418 - 419)

## Boldfaced, Page-Referenced Terms

(418) plant tropism _____

_____

(418) gravitropism _____

_____

(418) statoliths _____

_____

(418) phototropism _____

_____

(419) flavoprotein _____

_____

(419) thigmotropism _____

_____

(419) vines _____

_____

(419) tendrils _____

_____

*Choice*

For questions 1 - 5, choose from the following:

    a. phototropism  b. gravitropism  c. thigmotropism  d. mechanical stress

1. ___ More intense sunlight on one side of a plant—stems curve toward the light

2. ___ Vines climbing around a fencepost as they grow upward

3. ___ Plants grown outdoors have shorter stems than plants grown in a greenhouse.

4. ___ A root turned on its side will curve downward.

5. ___ Briefly shaking a plant daily inhibits the growth of the entire plant.

6. ___ A potted seedling turned on its side—the growing stem curves upward

7. ___ Leaves turn until their flat surfaces face light.

## 24 - VIII. BIOLOGICAL CLOCKS AND THEIR EFFECTS (pp. 420 - 421)

*Selected Word: circadian*

*Boldfaced, Page-Referenced Terms*

(420) biological clocks _____

_____

(420) phytochrome _____

_____

(420) circadian rhythm _____

_____

(420) photoperiodism _____

_____

*Matching*

Choose the most appropriate answer for each term.

1. ___ photoperiodism

2. ___ Pr

3. ___ a plant response to mechanical stress

4. ___ phytochrome activation

5. ___ "long-day" plants

6. ___ circadian rhythms

7. ___ Pfr

8. ___ biological clocks

9. ___ rhythmic leaf movements

10. ___ phytochrome

A. Biological activities that recur in cycles of 24 hours or so
B. Flower in spring when daylength exceeds a critical value
C. Internal time-measuring mechanisms with roles in adjusting daily activities
D. Any biological response to a change in the relative length of daylight and darkness in the 24-hour cycle; active Pfr may be an alarm button for this process
E. A blue-green pigment that absorbs red or far-red wavelengths, with different results
F. An example of a circadian rhythm
G. Active form of phytochrome
H. Inhibition of plant growth due to strong winds, grazing animals, farm machinery, and shaking plants daily
I. Inactive form of phytochrome
J. May induce plant cells to take up free calcium ions or induce certain plant cell organelles to release them

## Labeling

Identify each numbered part of the illustration.

11. _____

12. _____

13. _____

14. _____

15. _____

```
              red
             light
  ┌─────┐  ═══════►  ┌─────┐        ┌─────────┐
  │ 11  │            │ 12  │  ───►  │   15    │
  └─────┘  ◄═══════  └─────┘        └─────────┘
(inactive)  far-red  (active)       (growth of
             light                   plant part is
                                     promoted
                                     or inhibited)

         13 reverts to 14
            in the dark
```

## Fill-in-the-Blanks

Photoperiodism is especially apparent in the (16) _____ process, which is often keyed to daylength changes throughout the year. Long-day plants flower in spring when daylength becomes (17) (choose one) ❑ shorter, ❑ longer than some critical value. Short-day plants flower in late summer or early autumn when daylength becomes (18) (choose one) ❑ shorter, ❑ longer than some critical value. Day-neutral plants flower whenever they become (19) _____ enough to do so without regard to daylength. Spinach is a (20) _____ - _____ plant because it will not flower and produce seeds unless it is exposed to fourteen hours of light every day for two weeks. Cocklebur is termed a (21) _____ - _____ plant because it flowers after a single night that is longer than 8-½ hours.

# 24 - IX. LIFE CYCLES END, AND TURN AGAIN (p. 422)

## Boldfaced, Page-Referenced Terms

(422) abscission _____

_____

(422) senescence _____

_____

(422) dormancy _____

_____

*Choice*

For questions 1 - 12. choose from the following:

        a. senescence      b. abscission      c. entering dormancy      d. breaking dormancy

1. ___ Dropping of leaves or other parts from a plant

2. ___ A process at work between fall and spring; temperatures become milder, and rain and nutrients become available again

3. ___ Strong cues are short days, long, cold nights, and dry, nitrogen-deficient soil

4. ___ The process proceeds at tissues in the base of leaves, flowers, fruits, or other plant parts; a special zone is involved.

5. ___ The sum total of processes leading to the death of plant parts or the whole plant

6. ___ A cue is the funneling of nutrients into reproductive parts

7. ___ When a plant stops growing under conditions that appear quite suitable for growth

8. ___ The forming of ethylene in cells near the abscission zones may trigger the process.

9. ___ Stopping nutrient drain to reproductive parts blocks this process.

10. ___ Abscisic acid may cause cells near break points to produce the required ethylene.

11. ___ Postponement can be caused by gardeners when they remove flower buds from plants to maintain vegetative growth.

12. ___ As an outcome of the redistribution of nutrients to newly forming flowers, fruits, and seeds, leaves usually wither and die.

---

# Self-Quiz

___ 1. A stamen is _____.
   a. composed of a stigma
   b. the mature male gametophyte
   c. the site where microspores are produced
   d. part of the vegetative phase of an angiosperm

___ 2. A gametophyte is _____.
   a. a gamete-producing plant
   b. haploid
   c. both a and b
   d. the plant produced by the fusion of gametes

___ 3. An immature fruit is a(n) _____ and an immature seed is a(n) _____.
   a. ovary; megaspore
   b. ovary; ovule
   c. megaspore; ovule
   d. ovule; ovary

___ 4. In flowering plants, one sperm nucleus fuses with that of an egg, and a zygote forms that develops into an embryo. Another sperm fuses with _____.

a. primary endosperm cell to produce three cells, each with one nucleus
b. a primary endosperm cell to produce one cell with one triploid nucleus
c. both nuclei of the endosperm mother cell, forming a primary endosperm cell with a single triploid nucleus
d. one of the smaller megaspores to produce what will eventually become the seed coat

___ 5. "Simple, aggregate, multiple, and accessory" refer to types of _____.
a. carpels
b. seeds
c. fruits
d. ovaries

___ 6. "When a leaf falls or is torn away from a jade plant, a new plant can develop from the leaf, from meristematic tissue." This statement refers to _____.
a. parthenogenesis
b. runners
c. tissue culture propagation
d. vegetative propagation

___ 7. Promoting fruit ripening and abscission of leaves, flowers, and fruits is a function ascribed to _____.
a. gibberellins
b. ethylene
c. abscisic acid
d. auxins

___ 8. _____ is demonstrated by a germinating seed whose first root always curves down while the stem always curves up.
a. Phototropism
b. Photoperiodism
c. Gravitropism
d. Thigmotropism

___ 9. Plants whose leaves are open during the day but folded at night are exhibiting a _____.
a. growth movement
b. circadian rhythm
c. biological clock
d. both b and c are correct

___ 10. 2,4-D, a potent dicot weed killer, is a synthetic _____.
a. auxin
b. gibberellin
c. cytokinin
d. phytochrome

___ 11. All the processes that lead to the death of a plant or any of its organs are called _____.
a dormancy
b. vernalization
c. abscission
d. senescence

___ 12. Phytochrome is converted to an active form, _____, at sunrise and reverts to an inactive form, _____, at sunset, night or in the shade.
a Pr; Pfr
b. Pfr; Pfr
c. Pr; Pr
d. Pfr; Pr

## Chapter Objectives/Review Questions

| Page | | Objectives/Questions |
|---|---|---|
| (404) | 1. | _____ refers to two or more species jointly evolving as an outcome of close ecological interactions. |
| (404) | 2. | Describe the role of a pollinator. |
| (406) | 3. | Be able to distinguish between sporophytes and gametophytes. |
| (406) | 4. | _____ are shoots on sporophytes that are specialized for reproduction. |
| (406) | 5. | Be able to identify the various parts of a typical flower and state their functions. |
| (407) | 6. | _____ reproduction requires formation of gametes, followed by fertilization. |
| (407) | 7. | Walled microspores form in pollen sacs and develop into sperm-producing bodies called _____ _____. |
| (407) | 8. | Distinguish between a flower that is *perfect* and one that is *imperfect*. |
| (407) | 9. | Describe the condition called allergic rhinitis. |
| (408) | 10. | Relate the sequence of events and structures involved that give rise to microspores and megaspores. |
| (408) | 11. | What structures represent the male gametophyte and female gametophyte in flowering plants? List the contents of each. |
| (408) | 12. | The endosperm mother cell in the embryo sac is composed of two _____. |
| (408) | 13. | _____ is the transfer of pollen grains to a receptive stigma. |
| (409) | 14. | Describe the *double fertilization* that occurs uniquely in the flowering plant life cycle. |
| (410) | 15. | Describe the formation of the embryo sporophyte; give the origin and formation of seeds and fruits. |
| (411) | 16. | Review the general types of fruits produced by flowering plants (Table 24.1, p. 411, in the text). |

(412)    17. Distinguish between parthenogenesis, vegetative propagation, and tissue culture propagation.

(412)    18. List representative plant examples of a runner, a rhizome, a corm, a tuber, and a bulb.

(414)    19. _____ is a resumption of growth after a time of arrested embryonic development.

(414)    20. Define *imbibition,* and describe its role in germination.

(414)    21. The primary _____ breaks through the seed coat first to signal the end of germination.

(414)    22. The basic patterns of growth and development are heritable, dictated by the plant's

_____.

(416)    23. Describe the general role of plant hormones and target cells.

(416 - 417) 24. Be able to give the general functions of auxins, gibberellins, cytokinins, abscisic acid, and ethylene.

(416)    25. An _____ is a compound that, at suitable concentration, kills some plants but not others.

(418 - 419) 26. Define *phototropism, gravitropism,* and *thigmotropism,* and cite examples of each.

(418)    27. State the function of statoliths.

(419)    28. Plants make the strongest phototropic response to light of _____ wavelengths; _____ is a yellow pigment molecule that absorbs blue wavelengths.

(420)    29. Plants have internal time-measuring mechanisms called biological _____.

(420)    30. What are circadian rhythms? Give an example.

(420)    31. _____ is a biological response to a change in the relative length of daylight and darkness in a 24-hour cycle.

(420)    32. Phytochrome is converted to an active form—_____—at sunrise, when red wavelengths dominate the sky. It reverts to an inactive form— _____ —at sunset, at night, or even in shade, where far-red wavelengths predominate.

(420)    33. Describe the photoperiodic responses of "long-day," "short-day," and "day-neutral" plants.

(422)    34. Define *abscission, senescence,* and *dormancy.*

## Integrating and Applying Key Concepts

An oak tree has grown up in the middle of a forest. A lumber company has just cut down all of the surrounding trees except for a narrow strip of wood that includes the oak. How will the oak be likely to respond as it adjusts to its changed environment? To what new stresses will it be exposed? Which hormones will most probably be involved in the adjustment?

# 25

# TISSUES, ORGAN SYSTEMS, AND HOMEOSTASIS

## Interactive Exercises

### 25 - I. EPITHELIAL TISSUE (pp. 426 - 429)

*Selected Words:* "internal environment," *anatomy, physiology, simple* epithelium, *stratified* epithelium

In addition to the boldfaced terms, the text features other important terms essential to understanding the assigned material. "Selected Words" is a list of these terms, which appear in the text in italics, in quotation marks, and occasionally in roman type. Latin binomials found in this section are underlined and in roman type to distinguish them from other italicized words.

### Boldfaced, Page-Referenced Terms

The page-referenced terms are important; they were in boldface type in the chapter. Write a definition for each term in your own words without looking at the text. Next, compare your definition with that given in the chapter or in the text glossary. If your definition seems inaccurate, allow some time to pass and repeat this procedure until you can define each term rather quickly (how fast you can answer is a gauge of your learning effectiveness).

(426) homeostasis _____

_____

(426) tissue _____

_____

(426) organ _____

_____

(427) organ system _____

(427) division of labor _____

(428) epithelium (pl., epithelia) _____

(428) tight junctions _____

(428) adhering junctions _____

(428) gap junctions _____

(429) exocrine glands _____

(429) endocrine glands _____

## True/False

If the statement is true, write a T in the blank. If the statement is false, correct it by changing the under-lined word(s) and writing the correct word(s) in the blank.

_____ 1.   <u>Physiology</u> is the study of the way body parts are arranged in an organism.

_____ 2.   Groups of <u>like</u> <u>cells</u> that work together to perform a task are known as an organ.

_____ 3.   Most animals are constructed of only four types of <u>tissue</u>; epithelial, connective, nervous, and muscle tissues.

_____ 4.   Mammalian skin contains <u>squamous</u> <u>epithelium</u> and other tissues.

_____ 5.   The more tight junctions there are in a tissue, the more <u>permeable</u> the tissue will be.

_____ 6   <u>Endocrine</u> glands secrete their products through ducts that empty onto an epithelial surface.

_____ 7.   <u>Endocrine-cell</u> products include digestive enzymes, saliva, and mucus.

## Fill-in-the-Blanks

Groups of like cells that work together to perform a task are known as a(n) (8) _____. Groups of different types of tissues that interact to carry out a task are known as a(n) (9) _____. Each cell engages in basic (10) _____ activities that assure its own survival.  The combined contributions of cells, tissues, organs, and organ systems help maintain a stable (11) _____ _____ that is required for individual cell survival.

   While a specific kind of tissue (for example, simple squamous epithelium) is composed of cells that look very similar and do similar jobs, a specific (12) _____ (for example, a kidney) is composed of different tissues that cooperate to do a specific job (in this case, to remove waste products from blood, produce urine, and maintain the composition of body fluids). Kidneys, a urinary bladder, and various tubes are grouped together into a(n) (13) _____ _____, which adds the functions of urine storage and elimination to the jobs that the kidneys do. A multicellular (14) _____ is most often composed of organ systems that cooperate to keep activities running smoothly in a coordinated fashion; this maintenance of stable operating conditions in the internal environment is known as (15) _____.

All the diverse body parts found in different animals can be assembled from a few (16) _____ types through variations in the way they are combined and arranged. Somatic cells compose the physical structure of the animal body; they become differentiated into four main types of specialized tissue: (17) _____, (18) _____, muscle, and nervous tissues.

## 25 - II. CONNECTIVE TISSUE (pp. 430 - 431)
### MUSCLE TISSUE (p. 432)
### NERVOUS TISSUE (p. 433)
### *Focus on Science:* FRONTIERS IN TISSUE RESEARCH (p. 433)

*Selected Words:* "ground substance," "involuntary," "message," *laboratory-grown epidermis, designer organs, diabetes mellitus*

### *Boldfaced, Page-Referenced Terms*

(430) loose connective tissue _____

_____

(430) dense, irregular connective tissue _____

_____

(430) dense, regular connective tissue _____

_____

(430) cartilage _____

_____

(431) bone _____

_____

(431) adipose tissue _____

_____

(431) blood _____

_____

(432) skeletal muscle tissue _____

_____

(432) smooth muscle tissue _____

_____

(432) cardiac muscle tissue _____

_____

(433) nervous tissue _____

_____

(433) neuron _____

_____

## True/False

If the statement is true, write a T in the blank. If the statement is false, correct it by changing the underlined word(s) and writing the correct word(s) in the blank.

_____ 1. Muscle bundles are identical to <u>individual</u> skeletal muscle cells.

_____ 2. Both skeletal and cardiac muscle tissues are <u>striated</u>.

_____ 3. Cardiac muscle cells are fused, end-to-end, but each cell contracts <u>independently</u> of other cardiac muscle cells.

_____ 4. Smooth muscle tissue is involuntary and not <u>striated</u>.

_____ 5. Smooth muscle tissue is located in the walls of the <u>intestine</u>.

_____ 6. Neurons conduct messages to other <u>neurons</u> or to muscles or glands.

## Fill-in-the-Blanks

Cells of connective tissues are scattered throughout an extensive extracellular (7) _____

_____. (8) _____ connective tissue contains a weblike scattering of strong, flexible protein fibers and a few highly elastic protein fibers and serves as a packing material that holds in place blood vessels, nerves, and internal organs. (9) _____ and (10) _____ are examples of dense, regular connective tissue that help connect elements of the skeletal and muscular systems.

## Label and Match

Label each of the illustrations below with one of the following: connective, epithelial, muscle, nervous, or gametes. Complete the exercise by writing *all* appropriate letters and numbers from each group below in the parentheses following each label.

11. _____ (   )

12. _____ (   )

13. _____ (   )

14. _____ (   )

15. _____ (   )

16. _____ (   )

17. _____ (   )

18. _____ (   )

19. _____ (   )

20. _____ (   )

21. _____ (   )

22. _____ (   )

23. _____ (   )

A. Adipose
B. Bone
C. Cardiac
D. Dense, regular
E. Loose
F. Simple columnar
G. Simple cuboidal
H. Simple squamous
I. Smooth
J. Skeletal

1. Absorption
2. Maintain diploid number of chromosomes in sexually reproducing populations
3. Communication by means of electrochemical signals
4. Energy reserve
5. Contraction for voluntary movements
6. Diffusion
7. Padding
8. Contract to propel substances along internal passageways; not striated
9. Attaches muscle to bone and bone to bone
10. In vertebrates, provides the strongest internal framework of the organism
11. Elasticity
12. Secretion
13. Pumps circulatory fluid; striated
14. Insulation
15. Transport of nutrients and waste products to and from body cells

11.

12.

13.

14.

15.

16.

17.

18.

19.

20.

21.

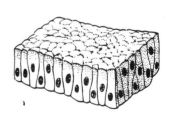

22.

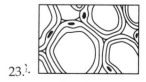

23.

## 25 - III. ORGAN SYSTEMS (pp. 434 - 435)

*Selected Words:* midsagittal, dorsal, ventral, "somatic"

### Boldfaced, Page-Referenced Terms

(434) integumentary system _____

_____

(434) muscular system _____

_____

(434) skeletal system _____

_____

(434) nervous system _____

_____

(434) endocrine system _____

_____

(434) circulatory system _____

_____

(435) lymphatic system _____

_____

(435) respiratory system _____

_____

(435) digestive system _____

_____

(435) urinary system _____

_____

(435) reproductive system _____

_____

(435) ectoderm _____

_____

(435) mesoderm _____

_____

(435) endoderm _____

_____

## Fill-in-the-Blanks

In (1) _____ _____ (immature reproductive cells that later develop into (2) _____ ), a special form of cell division known as (3) _____ occurs. In all other body tissues that consist of (4) _____ cells, the usual form of cell division, (5) _____ , occurs. The life of almost any animal begins with two gametes merging to form a fertilized egg, which undergoes reorganization and then divides by mitosis to form first undifferentiated cells, then three types of (6) _____ (groups of similar cells that perform similar activities) in the early embryo. Eventually, the outer layer of skin and the tissues of the nervous system are formed from the relatively unspecialized embryonic tissue known as (7) _____ located on the embryo's surface. The inner lining of the gut and the major organs formed from the embryonic gut develop from the internal embryonic tissue known as (8) _____ . Most of the internal skeleton, muscle, the circulatory, reproductive, and urinary systems, and the connective tissue layers of the gut and body covering are formed from (9) _____ , the embryonic tissue composed of cells that can move about like amoebae.

## Complete the Table

Supply the name of the "primary" tissue of the embryo that does the job indicated by becoming specialized in particular ways.

| Primary Tissue | Functions |
|---|---|
| 10. | Forms internal skeleton and muscle, circulatory, reproductive, and urinary systems |
| 11. | Forms inner lining of gut and linings of major organs formed from the embryonic gut |
| 12. | Forms outer layer of skin and the tissues of the nervous system |

## Labeling

Label each organ system described.

13. _____ Picks up nutrients absorbed from gut and transports them to cells throughout body

14. _____ Helps cells use nutrients by supplying them with oxygen and relieving them of $CO_2$ wastes

15. _____ Helps maintain the volume and composition of body fluids that bathe the body's cells

16. _____ Provides basic framework for the animal and supports other organs of the body

17. _____ Uses chemical messengers to control and guide body functions

18. _____ Protects the body from viruses, bacteria, and other foreign agents

19. _____ Produces younger, temporarily smaller versions of the animal

20. _____ Breaks down larger food molecules into smaller nutrient molecules that can be absorbed by body fluids and transported to body cells

21. _____ Consists of contractile parts that move the body through the environment and propel substances about in the animal

22. _____ Serves as an electrochemical communications system in the animal's body

23. _____ In the meerkat, served as a heat catcher in the morning and protective insulation at night

## Matching

Match the most appropriate function with each system shown on the next page.

24. ___ Male: production and transfer of sperm to the female. Female: production of eggs; provision of a protected nutritive environment for developing embryo and fetus. Both systems have hormonal influences on other organ systems.

25. ___ Ingestion of food, water; preparation of food molecules for absorption; elimination of food residues from the body

26. ___ Movement of internal body parts; movement of whole body; maintenance of posture; heat production

27. ___ Detection of external and internal stimuli; control and coordination of responses to stimuli; integration of activities of all organ systems

28. ___ Protection from injury and dehydration; body temperature control; excretion of some wastes; reception of external stimuli; defense against microbes

29. ___ Provisioning of cells with oxygen; removal of carbon dioxide wastes produced by cells; pH regulation

30. ___ Support, protection of body parts; sites for muscle attachment, blood cell production, and calcium and phosphate storage

31. ___ Hormonal control of body functioning; works with nervous system in integrative tasks

32. ___ Maintenance of the volume and composition of extracellular fluid

33. ___ Rapid internal transport of many materials to and from cells; helps stabilize internal temperature and pH

34. ___ Return of some extracellular fluid to blood; roles in immunity (defense against specific invaders of the body)

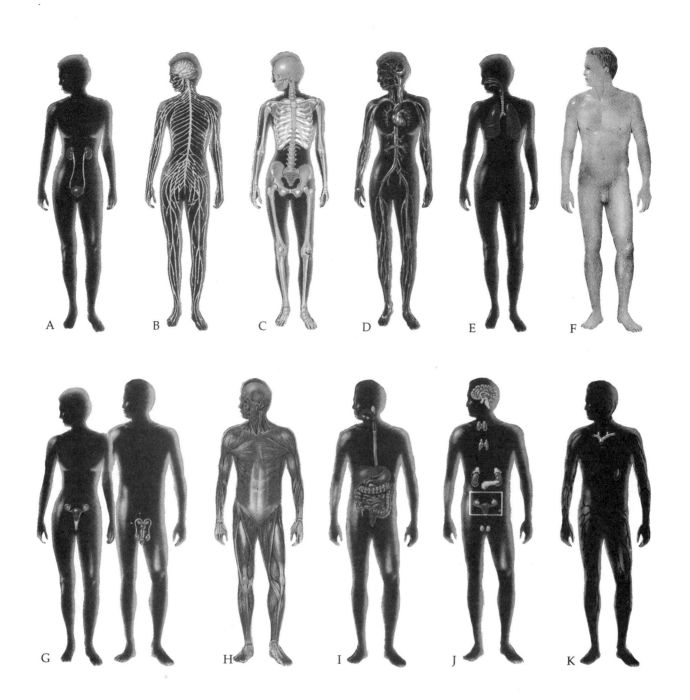

## Fill-in-the-Blanks

There are four major body cavities in humans. Lungs are located in the (35) _____ cavity, the brain is in the (36) _____ cavity, and, in the female, ovaries and the urinary bladder are in the (37) _____ cavity. In humans, the (38) _____ plane divides the body into right and left halves. The (39) _____ plane divides the body into (40) _____ (front) and posterior (back) parts. In humans, the (41) _____ plane divides it into dorsal (upper) and (42) _____ (lower) parts. The urinary system of an animal is responsible for the disposal of (43) _____ wastes; fecal material is not considered in the category. The endocrine system is generally responsible for internal (44) _____ control; together with the (45) _____ system, it integrates physiological processes.

## 25 - IV. HOMEOSTASIS AND SYSTEMS CONTROL (pp. 436 - 437)

*Selected Words:* interstitial, plasma, "set points"

## Boldfaced, Page-Referenced Terms

(436) extracellular fluid _____

_____

(436) internal environment _____

_____

(436) sensory receptors _____

_____

(436) stimulus _____

_____

(436) integrator _____

_____

(436) effectors _____

_____

(436) negative feedback mechanism _____

_____

(437) positive feedback mechanism _____

_____

## True/False

If the statement is true, write a T in the blank. If the statement is false, correct it by changing the under-lined word(s) and writing the correct word(s) in the blank.

_____ 1. The process of childbirth is an example of a <u>negative</u> feedback mechanism.

_____ 2. The human body's <u>effectors</u> are, for the most part, muscles and glands.

_____ 3. An integrator is constructed in such a way that it is informed of specific energy changes in the environment and relays messages about them to a <u>receptor.</u>

_____ 4. <u>Interstitial fluid</u> is a synonym for plasma.

### *Fill-in-the-Blanks*

In a (5) _____ feedback mechanism, a chain of events is set in motion that intensifies the original condition before returning to a set point; sexual arousal and childbirth are two examples. Generally, physiological controls work by means of (6) _____ feedback, in which an activity changes some condition in the internal environment, and the change causes the condition to be reversed; the maintenance of body temperature close to a "set point" is an example. Your brain is a(n) (7) _____, a control point where different bits of information are pulled together in the selection of a response. Muscles and (8) _____ are examples of effectors. Internal temperature control of the human body is achieved by (9) _____ in the skin and elsewhere sensing a temperature change at the body's surface and relaying the neural information to an integrator. In this example, the (10) _____ in the brain (see Figure 25.14) compares neural input against a set point. This part of the brain then sends output signals to (11) _____. Examples of these are the (12) _____ glands, which excrete water from the skin.

## Self-Quiz

___1. Which of the following is not included in connective tissues?
a. Bone
b. Blood
c. Cartilage
d. Skeletal muscle

___2. A surrounding material within which something originates, develops, or is contained is known as a _____.
a. lamella
b. ground substance
c. plasma
d. lymph

___3. Blood is considered to be a(n) _____ tissue.
a. epithelial
b. muscular
c. connective
d. none of these

___4. _____ are abundant in tissues of the heart and liver where they promote diffusion of ions and small molecules from cell to cell.
a. Adhesion junctions
b. Filter junctions
c. Gap junctions
d. Tight junctions

___5. Muscle that is not striped and is involuntary is _____.
a. cardiac
b. skeletal
c. striated
d. smooth

___6. Chemical and structural bridges link groups or layers of like cells, uniting them in structure and function as a cohesive

_____.
a. organ
b. organ system
c. tissue
d. cuticle

_____ 7. A fish embryo was accidentally stabbed by a graduate student in developmental biology. Later, the embryo developed into a creature that could not move and had no supportive or circulatory systems. Which embryonic tissue had suffered the damage?
   a. ectoderm
   b. endoderm
   c. mesoderm
   d. protoderm

_____ 8. A tissue whose cells are striated and fused at the ends by cell junctions so that the cells contract as a unit is called _____ tissue.
   a. smooth muscle
   b. dense fibrous connective
   c. supportive connective
   d. cardiac muscle

_____ 9. The secretion of tears, milk, sweat, and oil are functions of _____ tissues.
   a. epithelial
   b. loose connective
   c. lymphoid
   d. nervous

_____ 10. Memory, decision making, and issuing commands to effectors are functions of _____ tissue.
   a. connective
   b. epithelial
   c. muscle
   d. nervous

_____ 11. An animal that feels heated from the sun moves to an environment that tends to cool its body. This is an example of _____.
   a. intensifying an original condition
   b. positive feedback mechanism
   c. positive phototropic response
   d. negative feedback mechanism

_____ 12. Which group is arranged correctly from smallest structure to largest?
   a. muscle cells, muscle bundle, muscle
   b. muscle cells, muscle, muscle bundle
   c. muscle bundle, muscle cells, muscle
   d. none of the above

## Matching

Choose the most appropriate answer for each term.

13. ____ circulatory system

14. ____ digestive system

15. ____ endocrine system

16. ____ immune system

17. ____ integumentary system

18. ____ muscular system

19. ____ nervous system

20. ____ reproductive system

21. ____ respiratory system

22. ____ skeletal system

23. ____ urinary system

A. Picks up nutrients absorbed from gut and transports them to cells throughout body

B. Helps cells use nutrients by supplying them with oxygen and relieving them of $CO_2$ wastes

C. Helps maintain the volume and composition of body fluids that bathe the body's cells

D. Provides basic framework for the animal and supports other organs of the body

E. Uses chemical messengers to control and guide body functions

F. Protects the body from viruses, bacteria, and other foreign agents

G. Produces younger, temporarily smaller versions of the animal

H. Breaks down larger food molecules into smaller nutrient molecules that can be absorbed by body fluids and transported to body cells

I. Consists of contractile parts that move the body through the environment and propel substances about in the animal

J. Serves as an electrochemical communications system in the animal's body

K. In the meerkat, served as a heat catcher in the morning and protective insulation at night

# Chapter Objectives/Review Questions

This section lists general and detailed chapter objectives that can be used as review questions. You can make maximum use of these items by writing answers on a separate sheet of paper. Fill in answers where blanks are provided. To check for accuracy, compare your answers with information given in the chapter or glossary.

*Page*          *Objectives/Questions*

(426 - 427)  1.  Explain how the meerkat maintains a rather constant internal environment in spite of changing external conditions.

(426)        2.  Cells are the basic units of life; in a multicellular animal, like cells are grouped into a(n)
                 _____.

(427, 434)   3.  Explain how, if each cell can perform all its basic activities, organ systems contribute to cell survival.

(428 - 429)  4.  _____ tissues cover the body surface of all animals and line internal organs from gut cavities to vertebrate lungs; this tissue always has one _____ surface; the opposite surface adheres to a(n) _____ _____.

(428 - 433)  5.  Know the characteristics of the various types of tissues. Know the types of cells that compose each tissue type, and be able to cite some examples of organs that contain significant amounts of each tissue type.

(428)        6.  List the functions carried out by epithelial tissue, and state the general location of each type.

(428 - 429)  7.  Explain the nature of three different cell-to-cell junctions, and state the types of tissues in which these junctions occur.

(429)        8.  Explain the meaning of the term *gland,* cite three examples of glands, and state the extracellular products secreted by each.

(430)        9.  Connective tissue cells and fibers are surrounded by a(n) _____ _____.

(430 - 431) 10.  Describe the basic features of connective tissue, and explain how they enable connective tissue to carry out its various tasks.

(431)       11.  List three functions of blood.

(432)       12.  Distinguish among skeletal, cardiac, and smooth muscle tissues in terms of location, structure, and function.

(432)       13.  Muscle tissues contain specialized cells that can _____.

(433)       14.  Neurons are organized as lines of _____.

(434 - 435) 15.  List each of the eleven principal organ systems in humans, and match each to its main task.

(436)       16.  Describe the ways by which extracellular fluid helps cells survive.

(436 - 437) 17.  Draw a diagram that illustrates the mechanism of homeostatic control.

(436 - 437) 18.  Describe the relationships among receptors, integrators, and effectors in a negative feedback system.

# Interpreting and Applying Key Concepts

Explain why, of all places in the body, marrow is located on the interior of long bones. Explain why your bones are remodeled after you reach maturity. Why does your body not keep the same mature skeleton throughout life?

# 26

# PROTECTION, SUPPORT, AND MOVEMENT

**INTEGUMENTARY SYSTEMS**
  Functions of Skin
  Structure of Skin

*Focus on Health:* **SUNLIGHT AND SKIN**

**SKELETAL SYSTEMS**
  Skeletal Function
  Types of Skeletal Systems

**A CLOSER LOOK AT BONES AND THE JOINTS BETWEEN THEM**
  Structure and Functions of Bone
  Skeletal Joints

**SKELETAL-MUSCULAR SYSTEMS**
  How Muscles and Bones Interact
  Human Skeletal-Muscular System

**MUSCLE STRUCTURE AND FUNCTION**
  Functional Organization of a Skeletal Muscle
  Sliding-Filament Model of Contraction

**CONTROL OF MUSCLE CONTRACTION**

**PROPERTIES OF WHOLE MUSCLES**
  Muscle Tension and Muscle Fatigue
  Effects of Exercise and Aging

**ATP FORMATION AND LEVELS OF EXERCISE**

## Interactive Exercises

### 26 - I. INTEGUMENTARY SYSTEMS (pp. 440 - 443)
#### *Focus on Health:* SUNLIGHT AND SKIN (p. 443)

*Selected Words:* *body-builder's psychosis, albinism, cold sweats, acne, hirsutism, cold sores,* <u>Herpes</u> <u>simplex</u>, *epidermal skin cancers*

### *Boldfaced, Page-Referenced Terms*

(440) anabolic steroids _____

_____

(442) integument _____

(442) cuticle _____

(442) skin _____

(442) epidermis _____

_____

(442) dermis _____

_____

(443) hair _____

_____

## Fill-in-the-Blanks

Human skin is an organ system that consists of two layers: the outermost (1) _____, which contains mostly dead cells, and the (2) _____, which contains hair follicles, nerves, tiny muscles associated with the hairs, and various types of glands. The (3) _____ layer, with its loose connective tissue and store of fat in (4) _____ tissue, lies beneath the skin.

## Labeling

Label the numbered parts of the illustration.

5. _____
6. _____ _____
7. _____ _____
8. _____ _____
9. _____ _____
10. _____ _____
11. _____ _____
12. _____
13. _____
14. _____

## Choice

Match the following proteins (or protein derivatives) with the particular ability that each substance lends to the skin. For questions 15 - 21, choose from these letters:

a. collagen    b. elastin    c. keratin    d. melanin    e. hemoglobin

15. ____ Fibers that run through the dermis and lend a flexible, but substantive structure to it

16. ____ Protects against loss of moisture

17. ____ Protects against ultraviolet radiation and sunburn; contributes to skin color

18. ____ Helps skin to be stretchable, yet return to its previous shape

19. ____ Beaks, hooves, hair, fingernails, and claws contain a lot of this

20. ____ Helps ward off bacterial attack by making the skin surface rather impermeable

21. ____ Located in red blood cells; binds with $O_2$; contributes to skin color

## Fill-in-the-Blanks

(22) _____ _____ are synthetic hormones that mimic the effects of testosterone in building

greater (23) _____ mass in both men and women.  It is illegal for competitive athletes to use them

because of unfair advantage and because of the side effects.  In men, (24) _____, baldness,

shrinking (25) _____, and infertility are the first signs of damage.  Aside from these physical side

effects, some men experience uncontrollable (26) _____, delusions, and wildly manic behavior.  In

women, (27) _____ hair becomes more noticeable, (28) _____ _____ become irregular,

breasts may shrink, and the (29) _____ may become grossly enlarged.

## 26 - II. SKELETAL SYSTEMS (pp. 444 - 445)
## A CLOSER LOOK AT BONES AND THE JOINTS BETWEEN THEM (pp. 446 - 447)

*Selected Words: axial* [skeleton], *appendicular* [skeleton], *herniated disks, compact* bone, "Haversian canals,"
*spongy* bone, osteoblast, osteocyte, *osteoporosis, fibrous* joints, *cartilaginous* joints, *synovial* joints, *strain, sprain,*
*osteoarthritis, rheumatoid arthritis*

## Boldfaced, Page-Referenced Terms

(444) hydrostatic skeleton _____

_____

(444) exoskeleton _____

_____

(444) endoskeleton _____

_____

(444) intervertebral disks _____

_____

(446) bones _____

_____

(446) red marrow _____

_____

(446) yellow marrow _____

_____

(447) bone tissue turnover _____

_____

(447) joints _____

_____

(447) ligaments _____

_____

## Fill-in-the-Blanks

All motor systems are based on muscle cells that are able to (1) _____ and (2) _____, and on the presence of a medium against which the (3) _____ force can be applied. Longitudinal and (4) _____ muscle layers work as an antagonistic muscle system, in which the action of one motor element opposes the action of another. A membrane filled with fluid resists compression and can act as a(n) (5) _____ skeleton. Arthropods have opposing muscles attached to an (6) (choose one) ❏ endoskeleton ❏ exoskeleton. (7) _____ are hard compartments that enclose and protect the brain, spinal cord, heart, lungs, and other vital organs of vertebrates. Bones support and anchor (8) _____ and soft organs, such as eyes. (9) _____ systems are found in the long bones of mammals and contain living bone cells that receive their nutrients from the blood. (10) _____ _____ is a major site of blood cell formation. Bone tissue serves as a "bank" for (11) _____, (12) _____, and other mineral ions; depending on metabolic needs, the body deposits ions into and withdraws ions from this "bank."

Bones develop from (13) _____ secreting material inside the shaft and on the surface of the cartilage model. Bone can also give ions back to interstitial fluid as bone cells dissolve out component minerals and remodel bone in response to (14) _____ hormone deficiencies, lack of exercise, and lack of calcium in the diet. Extreme decreases in bone density result in (15) _____, particularly among older women.

The (16) _____ skeleton includes the skull, vertebral column, ribs, and breastbone; the (17) _____ skeleton includes the pectoral and pelvic girdles and the forelimbs and hindlimbs, when they exist in vertebrates.

(18) _____ joints are freely movable and are lubricated by a fluid secreted into the capsule of dense connective tissue that surrounds the bones of the joint. Bones are often tipped with (19) _____; as a person ages, the cartilage at (20) _____ joints may simply wear away, a condition called (21) _____. By contrast, in (22) _____ _____, the synovial membrane becomes inflamed, cartilage degenerates, and bone becomes deposited in the joint.

## Labeling

Identify each numbered part of the illustrations below and on the next page.

23. _____ _____
24. _____ _____
25. _____ _____
26. _____ _____
27. _____ _____ _____
28. _____ _____
29. _____ _____
30. _____ _____
31. _____

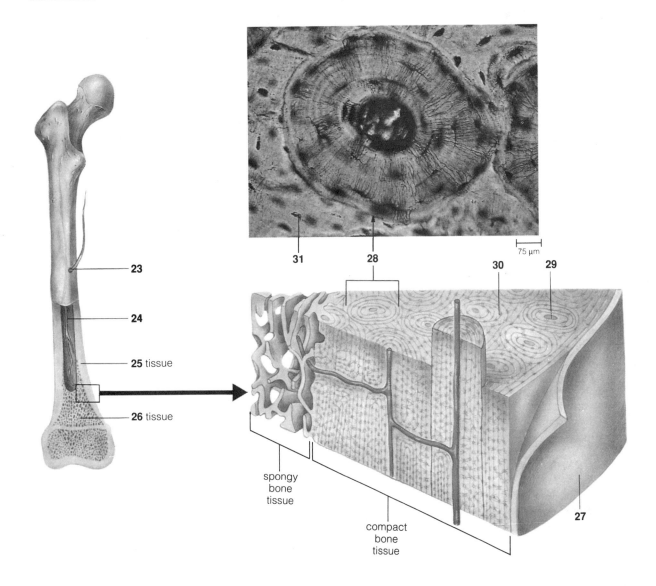

25 tissue

26 tissue

31

28

75 µm

30

29

spongy
bone
tissue

compact
bone
tissue

27

23

24

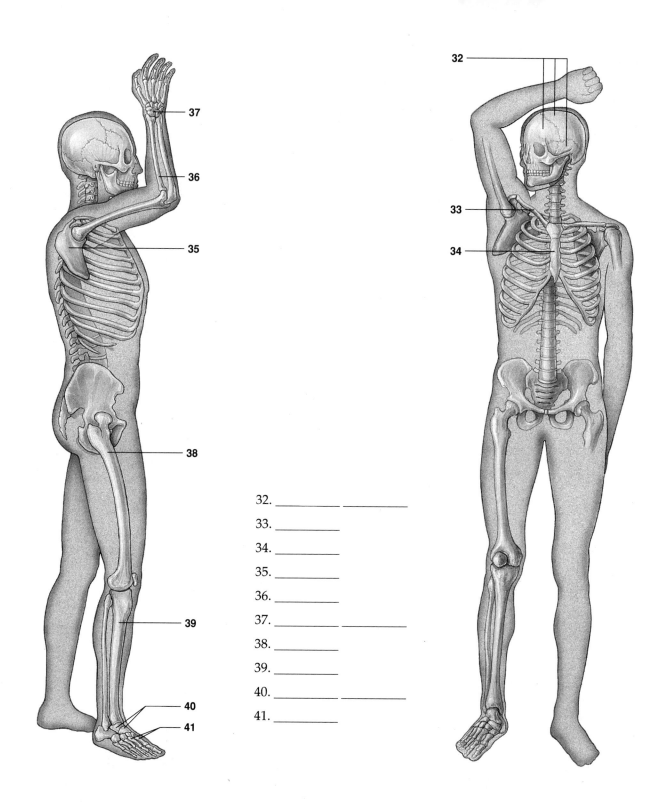

32. _____ _____

33. _____

34. _____

35. _____

36. _____

37. _____ _____

38. _____

39. _____

40. _____ _____

41. _____

# 26 - III. SKELETAL-MUSCULAR SYSTEMS (pp. 448 - 449)
## MUSCLE STRUCTURE AND FUNCTION (pp. 450 - 451)

*Selected Words:* *lever system,* myofibril, Z lines

## Boldfaced, Page-Referenced Terms

(448) skeletal muscles _____

_____

(448) tendons _____

_____

(450) sarcomeres _____

_____

(450) actin _____

_____

(450) myosin _____

_____

(451) sliding-filament model _____

_____

(451) cross-bridge formation _____

_____

## Fill-in-the-Blanks

(1) _____ muscle is striated, largely voluntary, and generally attached to (2) _____ or cartilage by means of (3) _____. (4) _____ muscle interacts with the skeleton to bring positional changes of body parts and to move the animal through its environment.

Together, the skeleton and its attached muscles are like a system of levers in which rigid rods, (5) _____, move about at fixed points, called (6) _____. Most attachments are close to joints, so a muscle has to shorten only a small distance to produce a large movement of some body part. When the (7) _____ _____ contracts, the elbow joint bends (flexes). As it relaxes and as its partner, the (8) _____ _____, contracts, the forelimb extends and straightens. As skeletal muscles contract, they transmit force to (9) _____ and make them move. (10) _____ attach skeletal muscles to bones.

## Sequence

Arrange in order of decreasing size.

11. ___     A. Muscle fiber (muscle cell)
12. ___     B. Myosin filament
13. ___     C. Muscle bundle
14. ___     D. Muscle
15. ___     E. Myofibril
16. ___     F. Actin filament

## Labeling

Identify each numbered part of the accompanying illustration.

17. _____ _____

18. _____ _____

19. _____ _____

20. _____ _____

21. _____ _____

22. _____ _____

23. _____ _____

24. (two of the _____
    _____)

25. _____

26. _____ _____

27. _____ _____

28. _____ _____

29. _____ _____

30. _____ _____

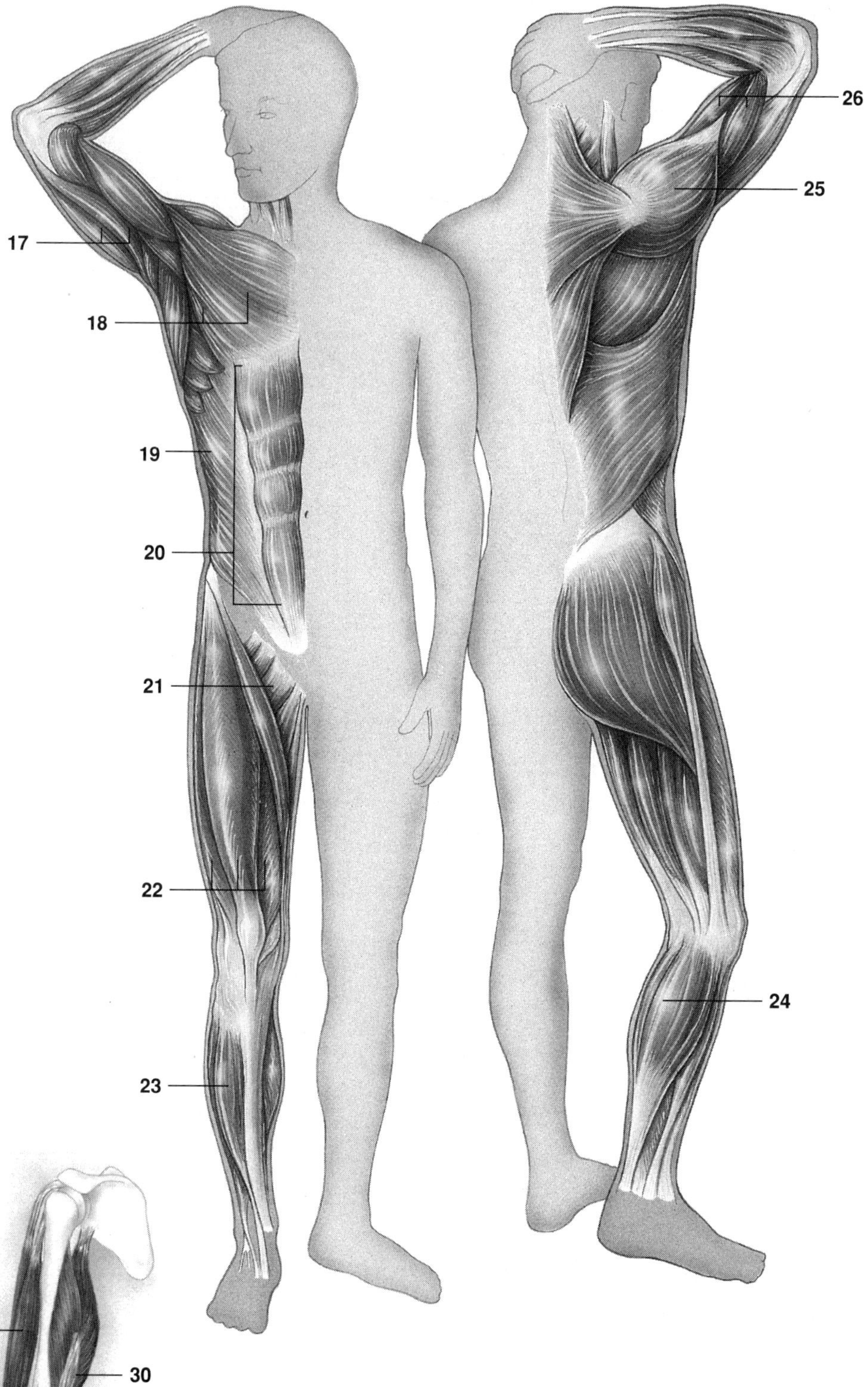

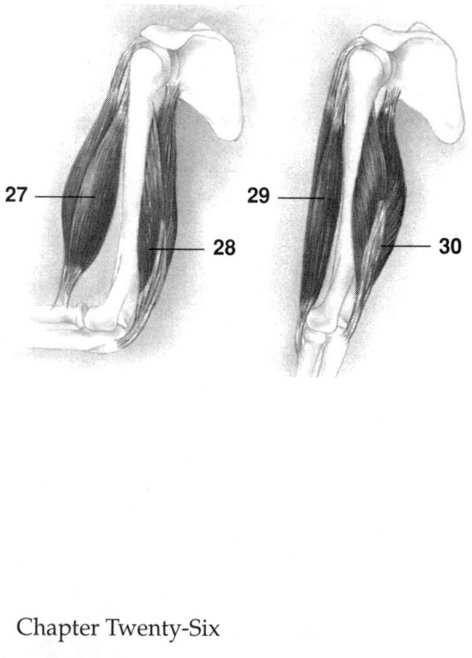

## Labeling/Fill-in-the-Blanks

Label the numbered parts of the accompanying illustrations.

31. _____ bundle

32. _____ _____

33. _____

34. _____

35. _____ _____

36. _____ _____

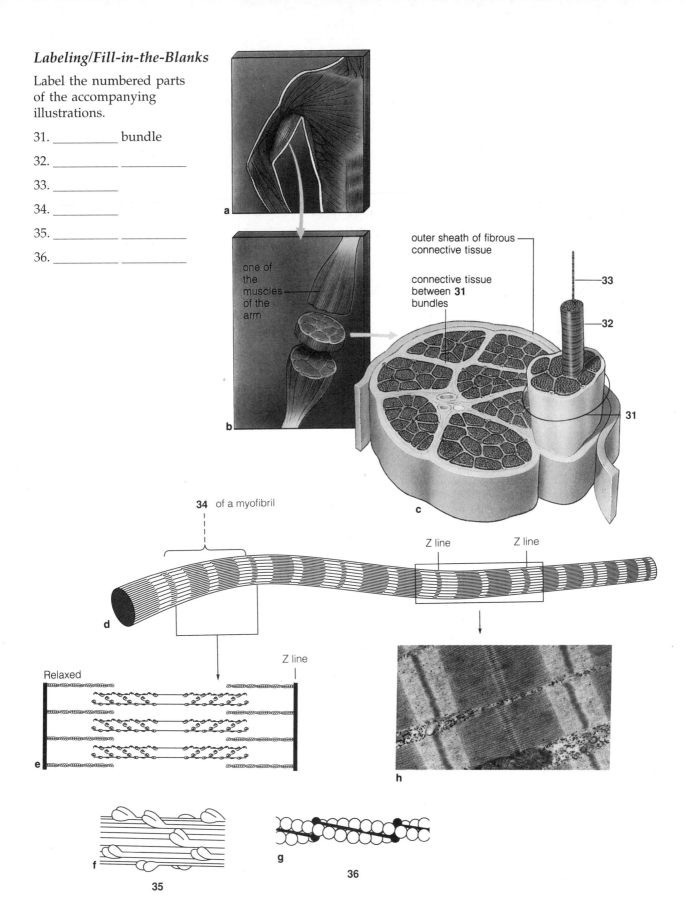

a

b

one of the muscles of the arm

outer sheath of fibrous connective tissue

connective tissue between **31** bundles

**33**

**32**

**31**

c

**34** of a myofibril

Z line    Z line

d

Relaxed

Z line

e

h

f

**35**

g

**36**

Each muscle cell contains (37) _____: threadlike structures packed together in parallel array. Every (37) is functionally divided into (38) _____, which appear to be striped and are arranged one after another along its length. Each myofibril contains (39) _____ filaments, which have cross-bridges and (40) _____ filaments, which are thin and lack cross-bridges.

According to the (41) _____ - _____ model, (42) _____ filaments physically slide along actin filaments and pull them toward the center of a (43) _____ during a contraction.

## 26 - IV. CONTROL OF MUSCLE CONTRACTION (p. 452)
### PROPERTIES OF WHOLE MUSCLES (p. 453)
### ATP FORMATION AND LEVELS OF EXERCISE (p. 454)

**Selected Words:** *excitable, isometric* contraction, *isotonic* contraction, *lengthening* contraction, "tetanus," *aerobic exercise, strength training*

### Boldfaced, Page-Referenced Terms

(452) action potential _____

_____

(452) sarcoplasmic reticulum _____

_____

(453) muscle tension _____

_____

(453) motor unit _____

_____

(453) muscle twitch _____

_____

(453) exercise _____

_____

(454) creatine phosphate _____

_____

(454) oxygen debt _____

_____

### Fill-in-the-Blanks

Muscle cells have three properties in common: contractility, elasticity, and (1) _____. Like a(n) (2) _____, a muscle cell is "excited" when a wave of electrical disturbance, a(n) (3) _____ _____, travels along its cell membrane. Neurons that send signals to muscles are (4) _____ neurons.

The energy that drives the forming and breaking of the cross-bridges comes immediately from (5) _____, which obtained its phosphate group from (6) _____ _____, which is stored in muscle cells.

(7) _____ of an entire muscle is brought about by the combined decreases in length of the individual sarcomeres that make up the myofibrils of the muscle cells. The parallel orientation of the muscle's component parts directs the force of contraction toward a (8) _____ that must be pulled in some direction.

When excited by incoming signals from motor neurons, muscle cell membranes cause (9) _____ ions to be released from the (10) _____ _____. These ions remove all obstacles that might interfere with myosin heads binding to the sites along (11) _____ filaments. When muscle cells relax/rest, calcium ions are sent back to the sarcoplasmic reticulum by (12) _____ _____. By controlling the (13) _____ _____ that reach the (14) _____ _____ in the first place, the nervous system controls muscle contraction by controlling calcium ion levels in muscle tissue.

The larger the (15) _____ of a muscle, the greater its strength. A (16) _____ neuron and the muscle cells under its control are called a (17) _____ _____. A(n) (18) _____ _____ is a response in which a muscle contracts briefly when reacting to a single, brief stimulus and then relaxes. The strength of the muscular contraction depends on how far the (19) _____ response has proceeded by the time another signal arrives. (20) _____ is the state of contraction in which a motor unit that is being stimulated repeatedly is maintained. In a (21) _____ contraction, the nervous system activates only a small number of motor units; in a stronger contraction, a larger number are activated at a high (22) _____ of stimulation.

## Labeling
Identify the numbered parts of the accompanying illustrations.

23. _____

24. _____ _____

25. _____
    _____

26. _____ _____
    _____

27. _____

28. _____ _____

29. _____ _____

30. _____ (_____)

31. _____

spinal cord
(section)

## Complete the Table

For each item in the table below, state its specific role in muscle contraction.

| | |
|---|---|
| Aerobic respiration | 32. |
| ATP | 33. |
| Calcium ions | 34. |
| Creatine phosphate | 35. |
| Glycogen | 36. |
| Lactate fermentation | 37. |
| Motor neuron | 38. |
| Myosin heads | 39. |
| Sarcomere | 40. |
| Sarcoplasmic reticulum | 41. |

## Fill-in-the-Blanks

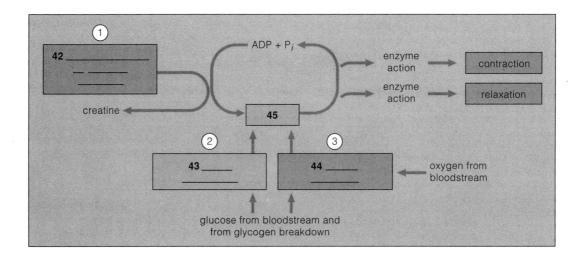

# Self-Quiz

For questions 1 - 7, choose from the following answers:

      a. bone
      b. cartilage
      c. epidermis
      d. dermis
      e. hypodermis

___1.   Fat cells in adipose tissue are most likely to be located in this.

___2.   Keratinized squamous cells are most likely to be located in this.

___3.   Melanin in melanocytes is most likely to be here.

___4.   Smooth muscles attached to hairs are probably here.

___5.   This makes up the original "model" of the skeletal framework.

___6.   The receiving ends of sensory receptors are most likely here.

___7.   This serves as a "bank" for withdrawing and depositing calcium and phosphate ions.

For questions 8 - 11, choose from the following answers:

      a. Ligaments
      b. Osteoblasts
      c. Osteocytes
      d. Red marrow
      e. Tendons

___8.   _____ secrete bone-dissolving enzymes.

___9.   Major site of blood cell formation.

___10.   _____ remove minerals (such as $Ca^{++}$ and $PO_4^{\equiv}$ ions) from blood and build bone.

___11.   _____ attach muscles to bone.

For questions 12 - 16, choose from the following answers:

      a. An action potential
      b. Cross-bridge formation
      c. The sliding-filament model
      d. Tension
      e. Tetanus

___12.   _____ is a mechanical force that causes muscle cells to shorten if is not exceeded by opposing forces.

___13.   A wave of electrical disturbance that moves along a neuron or muscle cell in response to a threshold stimulus.

___14.   _____ is a large contraction caused by repeated stimulation of motor units that are not allowed to relax.

___15.   _____ is assisted by calcium ions and ATP.

___16.   _____ explains how myosin filaments move to the centers of sarcomeres and back.

For questions 17 - 20, choose from the following answers:

      a. actin
      b. myofibril
      c. myosin
      d. sarcomere
      e. sarcoplasmic reticulum

___17.   A(n) _____ contains many repetitive units of muscle contraction.

___18.   The repetitive unit of muscle contraction.

___19.   Thin filaments that depend upon calcium ions to clear their binding sites so that they can attach to parts of thick filaments.

___20.   _____ stores calcium ions and releases them in response to an action potential.

# Chapter Objectives/Review Questions

| Page | | Objectives/Questions |
|---|---|---|
| (442) | 1. | Name four functions of human skin. |
| (442) | 2. | Describe the two-layered structure of human skin, and identify the items located in each layer. |
| (444 - 445) | 3. | Compare invertebrate and vertebrate motor systems in terms of skeletal and muscular components and their interactions. |
| (444 - 445) | 4. | Identify human bones by name and location. |

(447)    5.   Explain the various roles of osteoblasts, osteocytes, cartilage models, and long bones in the development of human bones.

(449)    6.   Refer to Figure 26.16 of your main text and indicate (a) a muscle used in sit-ups, (b) another used in dorsally flexing and inverting the foot, and (c) another used in flexing the elbow joint.

(450 - 451) 7.   Describe the fine structure of a muscle fiber; use terms such as *myofibril, sarcomere, Z lines, actin,* and *myosin.*

(450 - 452) 8.   Explain in detail the structure of muscles, from the molecular level to the organ systems level. Then explain how biochemical events occur in muscle contractions and how antagonistic muscle action refines movements.

(450 - 452) 9.   List, in sequence, the biochemical and fine structural events that occur during the contraction of a skeletal muscle fiber, and explain how the fiber relaxes.

(453)    10.  Distinguish twitch contractions from tetanic contractions.

---

## Integrating and Applying Key Concepts

If humans had an exoskeleton rather than an endoskeleton, would they move differently from the way they do now? Name any advantages or disadvantages that having an exoskeleton instead of an endoskeleton would present in human locomotion.

# 27

# CIRCULATION

## Interactive Exercises

### 27 - I. CIRCULATORY SYSTEMS—AN OVERVIEW (pp. 456 - 459)

*Selected Words: closed* circulatory system, *open* circulatory system

### Boldfaced, Page-Referenced Terms

(456) electrocardiogram (ECG) _____

_____

(458) circulatory system _____

_____

(458) interstitial fluid _____

_____

(458) blood _____

_____

(458) heart _____

_____

(458) capillary beds _____

_____

(458) capillaries _____

_____

(459) pulmonary circuit _____

_____

(459) systemic circuit _____

_____

(459) lymphatic system _____

_____

## Fill-in-the-Blanks

Cells survive by taking in from their surroundings what they need, (1) _____, and giving back to their surroundings materials what they don't need: (2) _____. In most animals, substances move rapidly to and from living cells by way of a (3) _____ circulatory system. (4) _____, a fluid connective tissue within the (5) _____ and blood vessels, is the transport medium.

Most of the cells of animals are bathed in a(n) (6) _____ _____; blood is constantly delivering nutrients and removing wastes from that fluid. The (7) _____ generates the pressure that keeps blood flowing. Blood flows (8) (choose one) ❏ rapidly ❏ slowly through large diameter vessels to and from the heart, but where the exchange of nutrients and wastes occurs, in the (9) _____ beds, the blood is divided up into vast numbers of small diameter vessels with tremendous surface area that enables the exchange to occur by diffusion. An elaborate network of drainage vessels attracts excess interstitial fluid and reclaimable (10) _____ and returns them to the circulatory system. This network is part of the (11) _____ system, which also helps clean the blood of disease agents.

Nutrients are absorbed into the blood from the (12) _____ and (13) _____ systems. Carbon dioxide is given to the (14) _____ system for elimination, and excess water, solutes and wastes are eliminated by the (15) _____ system.

The heart is a pumping station for two major blood transport routes: the (16) _____ circulation to and from the lungs, and the (17) _____ circulation to and from the rest of the body.

In the pulmonary circuit, the heart pumps (18) _____-poor blood to the lungs; then the (19) _____-enriched blood flows back to the (20) _____. During a cardiac cycle, contraction of the (21) _____ is the driving force for blood circulation; (22) _____ contraction helps fill the ventricles. In fishes, blood flows in (23) _____ circuit(s) away from and back to the heart.

## Labeling

Label the numbered parts in the illustrations shown at the right.

24. _____

25. _____ _____

26. _____

Describe the kind of circulatory system in:

27. Creature A. _____

_____

28. Creature B. _____

_____

## 27 - II. CHARACTERISTICS OF BLOOD (pp. 460 - 461)
### *Focus on Health:* BLOOD DISORDERS (p. 462)
### BLOOD TYPING AND BLOOD TRANSFUSIONS (pp. 462 - 463)

*Selected Words:* *hemorrhagic* anemia, *chronic* anemia, *hemolytic* anemia, $B_{12}$ *deficiency* anemia, *sickle-cell* anemia, *thalassemias, polycythemias,* blood doping, *infectious mononucleosis, leukemias, erythroblastosis fetalis*

### Boldfaced, Page-Referenced Terms

(460) plasma _____

_____

(460) red blood cells _____

_____

(460) white blood cells _____

_____

(460) platelets _____

_____

(461) stem cells _____

_____

(461) cell count _____

_____

(462) blood transfusions _____

_____

(462) agglutination _____

(462) ABO blood typing _____

(462) Rh Blood typing _____

## Fill-in-the-Blanks

Blood is a highly specialized fluid (1) _____ tissue that helps stabilize internal (2) _____ and equalize internal temperature throughout an animal's body. Blood volume for average-size adult humans amounts to approximately (4) (choose one) ❐ 2 - 3  ❐ 4 - 5  ❐ 6 - 7  ❐ 8 - 10  ❐ 11 - 15 quarts. Oxygen binds with the (5) _____ atom in a hemoglobin molecule. The red blood (6) _____ _____ in males is about 5.4 million cells per microliter of blood; in females, it is 4.8 million per microliter. The plasma portion constitutes approximately (7) _____ to _____ percent of the total blood volume. Erythrocytes are produced in the (8) _____ _____. (9) _____ _____ are immature cells not yet fully differentiated. (10) _____ and monocytes are highly mobile and phagocytic; they chemically detect, ingest, and destroy bacteria, foreign matter, and dead cells. (11) _____ (thrombocytes) are cell fragments that aid in forming blood clots.

In humans, red blood cells lack their (12) _____, but they contain enough materials to sustain them for about (13) _____ months. Platelets also have no (14) _____, but they last a maximum of (15) _____ days in the human bloodstream.

If you are blood type (16) _____, you have no effective antibodies against A or B markers in your plasma. If you are type (17) _____, you have antibodies against A and B markers in your plasma. (18) (choose one) ❐ Women  ❐ Men who are (19) (choose one) ❐ Rh$^+$  ❐ Rh$^-$ have to be careful so that they don't produce fetuses that develop erythroblastosis fetalis.

## Complete the Table

20. Complete the following table, which describes the components of blood.

| Components | Relative Amounts | Functions |
| --- | --- | --- |
| Plasma Portion (50%–60% of total volume): | | |
| Water | 91%–92% of plasma volume | Solvent |
| a. (albumin, globulins, fibrinogen, etc.) | 7%–8% | Defense, clotting, lipid transport, roles in extracellular fluid volume, etc. |
| Ions, sugars, lipids, amino acids, hormones, vitamins, dissolved gases | 1%–2% | Roles in extracellular fluid volume, pH, etc. |
| Cellular Portion (40%–50% of total volume): | | |
| b. | 4,800,000–5,400,000 per microliter | $O_2$, $CO_2$ transport |
| White blood cells: | | |
| c. | 3,000–6,750 | Phagocytosis |
| d. | 1,000–2,700 | Immunity |
| Monocytes (macrophages) | 150–720 | Phagocytosis |
| Eosinophils | 100–360 | Roles in inflammatory response, immunity |
| Basophils | 25–90 | Roles in inflammatory response, anticlotting |
| e. | 250,000–300,000 | Roles in clotting |

## Label and Match

Identify the numbered cell types in the illustration below. Complete the exercise by matching and entering the letter of the appropriate function in the parentheses following the given cell types. A letter may be used more than once. Cell types 21 and 27 have no matching letters.

21. _____

22. _____ _____ ( )

23. _____ ( )

24. _____ ( )

25. _____ ( )

26. _____ ( )

27. _____

28. _____ ( )

A. Phagocytosis
B. Plays a role in the inflammatory response
C. Plays a role in clotting
D. Immunity
E. $O_2$, $CO_2$ transport

22 _____ cells

eosinophils          24 _____          basophils

(mature in bone marrow)          (mature in thymus)

25 _____ lymphocytes          26 _____ lymphocytes

21 _____ cells

23 _____

27 _____

mature 28 _____

## 27 - III. HUMAN CARDIOVASCULAR SYSTEM (pp. 464 - 465)
## THE HEART IS A LONELY PUMPER (pp. 466 - 467)

*Selected Words:* *pulmonary* circuit, *systemic* circuit, capillaries

### Boldfaced, Page-Referenced Terms

(464) arteries _____

_____

(464) arterioles _____

_____

(464) venules _____

_____

(464) veins _____

_____

(464) aorta _____

_____

(466) atrium _____

_____

(466) ventricle _____

_____

(466) cardiac cycle _____

_____

(467) cadiac conduction system _____

_____

(467) cardiac pacemaker _____

_____

## Fill-in-the-Blanks

The heart is a pumping station for two major blood transport routes: the (1) _____ circulation to and from the lungs, and the (2) _____ circulation to and from the rest of the body.

    Look at Figure 27.11 in your text and use a red pen to redden all tubes in this illustration (except the pulmonary artery) that are indicated by "aorta" or "artery." Memorize the names, then fill in all of the blanks. Also, redden the parts of the heart that contain oxygen-rich blood.

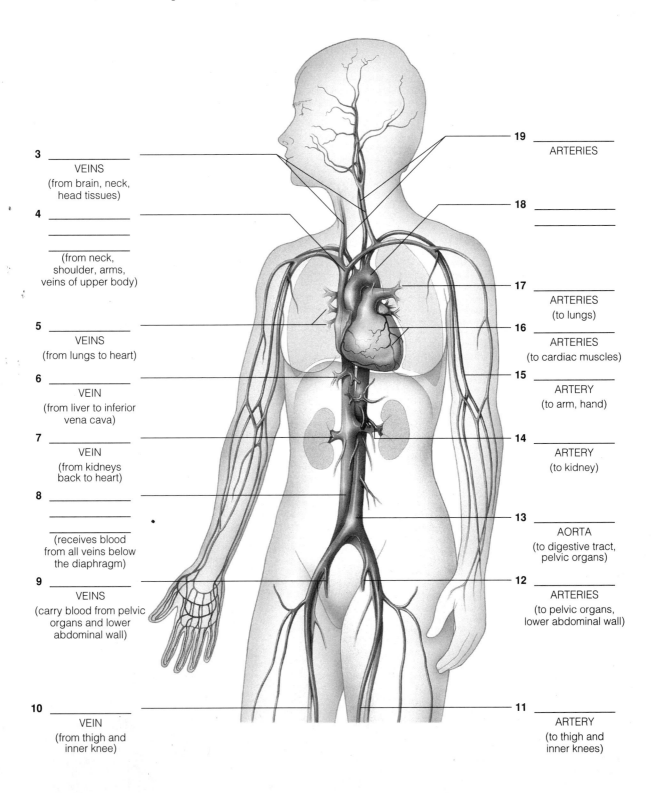

3 _____
VEINS
(from brain, neck,
head tissues)

4 _____
_____
(from neck,
shoulder, arms,
veins of upper body)

5 _____
VEINS
(from lungs to heart)

6 _____
VEIN
(from liver to inferior
vena cava)

7 _____
VEIN
(from kidneys
back to heart)

8 _____
_____
(receives blood
from all veins below
the diaphragm)

9 _____
VEINS
(carry blood from pelvic
organs and lower
abdominal wall)

10 _____
VEIN
(from thigh and
inner knee)

19 _____
ARTERIES

18 _____
_____

17 _____
ARTERIES
(to lungs)

16 _____
ARTERIES
(to cardiac muscles)

15 _____
ARTERY
(to arm, hand)

14 _____
ARTERY
(to kidney)

13 _____
AORTA
(to digestive tract,
pelvic organs)

12 _____
ARTERIES
(to pelvic organs,
lower abdominal wall)

11 _____
ARTERY
(to thigh and
inner knees)

## Labeling

Identify each numbered part of the accompanying illustration.

20. _____
21. _____ _____
    _____
22. _____ _____
23. _____ _____
24. _____ _____
    _____
25. _____ _____
26. _____ _____
    _____
27. _____ _____
    _____

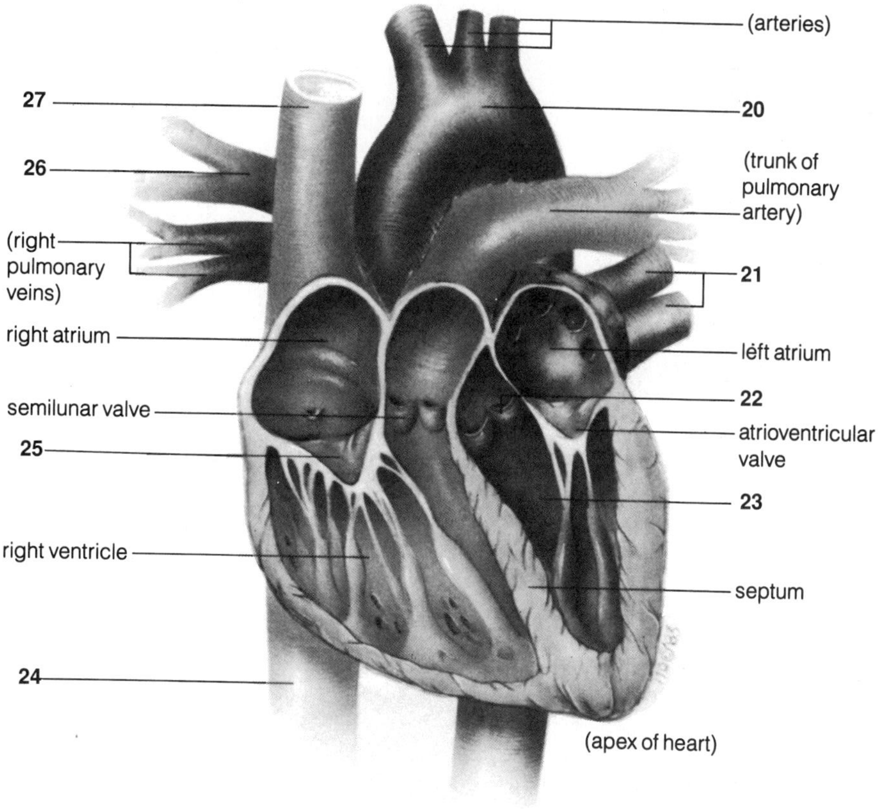

(Illustration of heart with labels:)

- (arteries)
- 27
- 26
- 20
- (trunk of pulmonary artery)
- (right pulmonary veins)
- 21
- right atrium
- léft atrium
- 22
- semilunar valve
- atrioventricular valve
- 25
- 23
- right ventricle
- septum
- 24
- (apex of heart)

## 27 - IV. BLOOD PRESSURE IN THE CARDIOVASCULAR SYSTEM (pp. 468 - 469)
### Focus on Health: CARDIOVASCULAR DISORDERS (pp. 470 - 471)
### HEMOSTASIS (p. 472)

**Selected Words:** *pulse, diffusion zones, edema, elephantiasis, hypertension, atherosclerosis, heart attacks, strokes, low-density lipoproteins (LDLs), high-density lipoproteins (HDLs), infiltrate, atherosclerotic plaque, thrombus, embolus, angina pectoris, angiography, coronary bypass surgery, laser angioplasty, balloon angioplasty, arrhythmias, bradycardia, tachycardia, atrial fibrillation, ventricular fibrillation*

## Boldfaced, Page-Referenced Terms

(468) blood pressure _____

_____

(468) vasodilation _____

_____

(468) vasoconstriction _____

_____

(472) hemostasis _____

_____

## Labeling

Identify each numbered part of the accompanying illustrations.

1. _____     4. _____

2. _____     5. _____ _____, _____ _____

3. _____     6. _____

a. 1

outer coat | 5 | basement membrane | endothelium | 6

c. 3

outer coat | smooth muscle rings over elastic layer | basement membrane | endothelium

b. 2

outer coat | smooth muscle between elastic layers | basement membrane | endothelium

d. 4

basement membrane | endothelium

## Fill-in-the-Blanks

A(n) (7) _____ carries blood away from the heart. A(n) (8) _____ is a blood vessel with such a small diameter that red blood cells must flow through it single-file; its wall consists of no more than a single layer of (9) _____ cells resting on a basement membrane. In each (10) _____ _____, small molecules move between the bloodstream and the (11) _____ fluid.

(12) _____ are in the walls of veins and prevent backflow. Both (13) _____ and (14) _____ serve as temporary reservoirs for blood volume.

A receiving zone of a vertebrate heart is called a(n) (15) _____; a departure zone is called a(n) (16)_____. Each contraction period is called (17) _____; each relaxation period is (18) _____.

(19) _____ are pressure reservoirs that keep blood flowing smoothly away from the heart while the (20) _____ are relaxing. (21) _____ are control points where adjustments can be made in

the volume of blood flow to be delivered to different capillary beds. They offer great resistance to flow, so there is a major drop in (22) _____ in these tubes.

One cause of (23) _____ is the rupture of one or more blood vessels in the brain. A (24) _____ _____ blocks a coronary artery. (25)_____ is a term for a formation that can include cholesterol, calcium salts, and fibrous tissue. It is not healthful to have a high concentration of (26) _____-density lipoproteins in the bloodstream. A clot that stays in place is a (27) _____, but a clot that travels in the bloodstream is an embolus. Bleeding is stopped by several mechanisms that are referred to as (28) _____; the mechanisms include blood vessel spasm, (29) _____ _____ _____, and blood (30) _____. Once the platelets reach a damaged vessel, through chemical recognition they adhere to exposed (31) _____ fibers in damaged vessel walls. Reactions cause rod-shaped proteins to assemble into long (32) _____ fibers. These trap blood cells and components of plasma. Under normal conditions, a clot eventually forms at the damaged site.

### True/False

If the statement is true, write a T in the blank. If the statement is false, correct it by changing the under-lined word(s) and writing the correct word(s) in the blank.

_____ 33. The pulse <u>rate</u> is the difference between the systolic and the diastolic pressure readings.

_____ 34. Because the total volume of blood remains constant in the human body, blood pressure must <u>also</u> <u>remain</u> <u>constant</u> throughout the circuit.

### More Fill-in-the-Blanks

Blood pressure is normally high in the (35) _____ immediately after leaving the heart, but then it drops as the fluid passes along the circuit through different kinds of blood vessels. Energy in the form of (36) _____ is lost as it overcomes (37) _____ to the flow of blood. Arterial walls are thick, muscular, and (38) _____, and have large diameters. Arteries present (choose one) (39) ❐ much ❐ little resistance to blood flow, so pressure (choose one) (40) ❐ drops a lot  ❐ does not drop much in the arterial portion of the systemic and pulmonary circuits.

The greatest drop in pressure occurs at (choose one) (41) ❐ capillaries  ❐ arterioles  ❐ veins. With this slowdown, blood flow can now be allotted in different amounts to different regions of the body in response to signals from the (42) _____ system and endocrine system or even changes in local chemical conditions.

When a person is resting, blood pressure is influenced most by reflex centers in the (43) _____ _____. When the resting level of blood pressure increases, reflex centers command the heart to (choose one) (44) ❐ beat more slowly  ❐ beat faster, and command smooth muscle cells in arteriole walls to (choose one) (45) ❐ contract  ❐ relax, which results in (choose one) (46) ❐ vasodilation ❐ vasoconstriction.

## Boldfaced, Page-Referenced Terms

(472) lymph _____

_____

(472) lymph vascular system _____

_____

(472) lymph capillaries _____

_____

(472) lymph vessels _____

_____

(473) lymphoid organs and tissues _____

_____

(473) lymph nodes _____

_____

(473) spleen _____

_____

(473) thymus _____

_____

## Labeling

Identify each numbered part of the accompanying illustrations.

1. _____

2. _____  _____  _____

3. _____  _____

4. _____  _____

5. _____

6. _____  _____

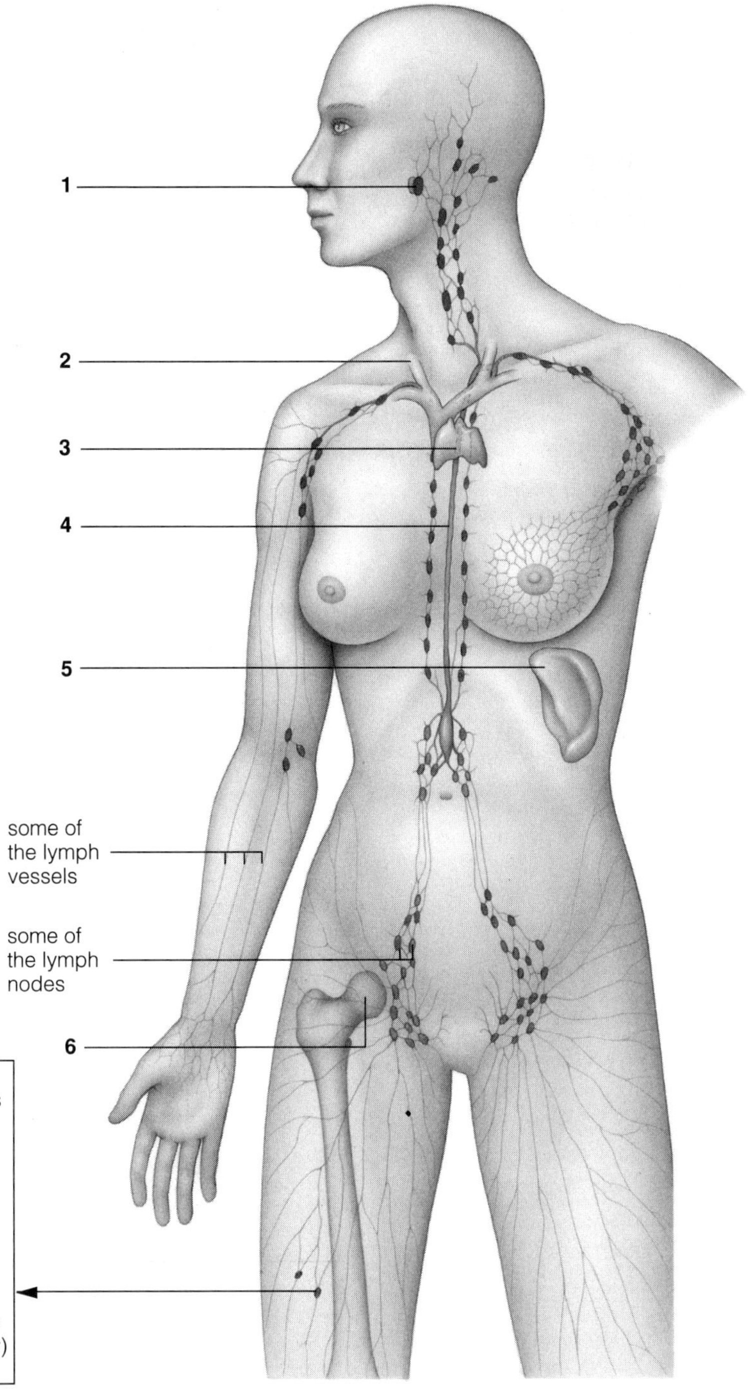

1

2

3

4

5

some of the lymph vessels

some of the lymph nodes

6

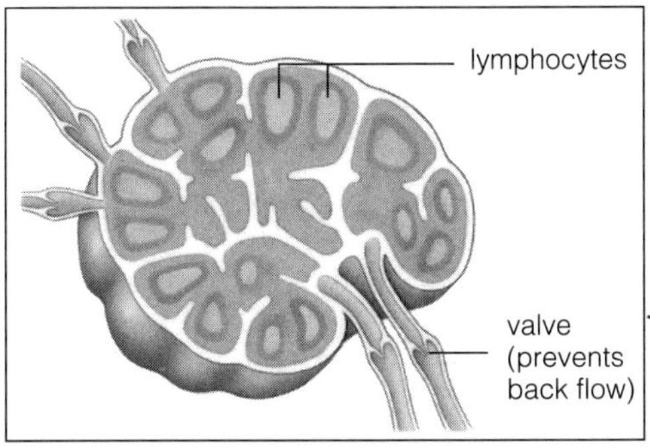

lymphocytes

valve (prevents back flow)

## Fill-in-the-Blanks

(7) _____ vessels reclaim fluid lost from the bloodstream, purify the blood of microorganisms, and transport (8) _____ from the (9) _____  _____ to the bloodstream.

# Self-Quiz

___1. Most of the oxygen in human blood is transported by _____.
   a. plasma
   b. serum
   c. platelets
   d. hemoglobin
   e. leukocytes

___2. Of all the different kinds of white blood cells, two classes of _____ are the ones that respond to *specific* invaders and confer *immunity* to a variety of disorders.
   a. basophils
   b. eosinophils
   c. monocytes
   d. neutrophils
   e. lymphocytes

___3. Open circulatory systems generally lack _____.
   a. a heart
   b. arterioles
   c. capillaries
   d. veins
   e. arteries

___4. Red blood cells originate in the _____.
   a. liver
   b. spleen
   c. yellow bone marrow
   d. thymus gland
   e. red bone marrow

___5. Hemoglobin contains _____.
   a. copper
   b. magnesium
   c. sodium
   d. calcium
   e. iron

___6. The pacemaker of the human heart is the _____.
   a. sinoatrial node
   b. semilunar valve

   c. inferior vena cava
   d. superior vena cava
   e. atrioventricular node

___7. During systole, _____.
   a. oxygen-rich blood is pumped to the lungs
   b. the heart muscle tissues contract
   c. the atrioventricular valves suddenly open
   d. oxygen-poor blood from all parts of the human body, except the lungs, flows toward the right atrium
   e. none of the above

___8. _____ are reservoirs of blood pressure in which resistance to flow is low.
   a. Arteries
   b. Arterioles
   c. Capillaries
   d. Venules
   e. Veins

___9. Begin with a red blood cell located in the superior vena cava and travel with it in proper sequence as it goes through the following structures. Which will be *last* in sequence?
   a. aorta
   b. left atrium
   c. pulmonary artery
   d. right atrium
   e. right ventricle

___10. The lymphatic system is the principal avenue in the human body for transporting _____.
   a. fats
   b. wastes
   c. carbon dioxide
   d. amino acids
   e. interstitial fluids

# Chapter Objectives/Review Questions

| Page | | Objectives/Questions |
|---|---|---|
| (457) | 1. | Describe how the circulatory, respiratory, digestive, and urinary systems cooperate to help a multicellular animal survive. |
| (458) | 2. | Distinguish between open and closed circulatory systems. Provide examples of animals with one or the other. |
| (459) | 3. | Describe how vertebrate circulatory systems have evolved from the fish model to the mammalian model. |
| (460) | 4. | Describe the composition of human blood, using percentages of total volume. |
| (461) | 5. | State where erythrocytes, leukocytes, and platelets are produced. |
| (462 - 463) | 6. | Describe how blood is typed for the ABO blood group and for the Rh factor. |
| (464 - 467) | 7. | List the factors that cause blood to leave the heart and the factors that cooperate to return blood to the heart. |
| (467) | 8. | Explain what causes a heart to beat. Then describe how the rate of heartbeat can be slowed down or speeded up. |
| (468 - 469) | 9. | Explain what causes high pressure and low pressure in the human circulatory system. Then show where major drops in blood pressure occur in humans. |
| (468 - 469) | 10. | Describe how the structures of arteries, capillaries, and veins differ. |
| (469) | 11. | Explain how veins and venules can act as reservoirs of blood volume. |
| (470 - 471) | 12. | Distinguish a stroke from a coronary artery blockage, or occlusion. |
| (470) | 13. | Describe how hypertension develops, how it is detected, and whether it can be corrected. |
| (471) | 14. | State the significance of high- and low-density lipoproteins to cardiovascular disorders. |
| (472 - 473) | 15. | Describe the composition and function of the lymphatic system. |

## Integrating and Applying Key Concepts

You observe that some people appear as though fluid had accumulated in their lower legs and feet. Their lower extremities resemble those of elephants. You inquire about what is wrong and are told that the condition is caused by the bite of a mosquito that is active at night. Construct a testable hypothesis that would explain (1) why the fluid was not being returned to the torso, as normal, and (2) what the mosquito did to its victims.

# 28

# IMMUNITY

## Interactive Exercises

### 28 - I. THREE LINES OF DEFENSE (pp. 476 - 478)
###       COMPLEMENT PROTEINS (p. 478)
###       INFLAMMATION (pp. 480 - 481)

*Selected Words:* smallpox," vaccination," pasteurization, *athlete's foot, nonspecific* target, *specific* target, *nonspecific* response, *specific pathogen* [response], "set point," *fever*

### *Boldfaced, Page-Referenced Terms*

(478) pathogens _____

_____

(478) lysozyme _____

_____

(479) complement system _____

_____

(479) lysis _____

_____

(480) neutrophils _____

_____

(480) eosinophils _____

_____

(480) basophils _____

_____

(480) macrophages _____

_____

(480) acute inflammation _____

_____

(481) histamine _____

_____

(481) interleukins _____

_____

## Fill-in-the-Blanks

Several barriers prevent pathogens from crossing the boundaries of your body. Intact skin and
(1) _____ membranes are effective barriers. (2) _____ is an enzyme that destroys the cell wall
of many bacteria. (3) _____ fluid destroys many food-borne pathogens in the gut. Normal
(4) _____ residents of the skin, gut, and vagina outcompete pathogens for resources and help keep
their numbers under control.

When a sharp object cuts through the skin and foreign microbes enter, some plasma proteins come
to the rescue and seal the wound with a (5) _____ mechanism. (6) _____ white blood cells
engulf bacteria soon thereafter. Also there are about twenty plasma proteins (collectively referred to as
the (7) _____ _____) that are activated one after another in a "cascade" of reactions to help
destroy invading microorganisms. (8) _____ are Y-shaped proteins that lock onto specific foreign
targets and thereby tag them for destruction by phagocytes or by activating the complement system.
The inflammatory response engages in battle both specific and nonspecific invaders. When the
complement system is activated, circulating basophils and mast cells in tissues release (9) _____,
which dilates (10) _____ and makes them "leaky," so fluid seeps out and causes the inflamed area
to become swollen and warm.

## Complete the Table

Complete the table by providing the specific functions carried out by each of the four different kinds of white blood cells listed below.

| Basophils | 11. |
|---|---|
| Eosinophils | 12. |
| Neutrophils | 13. |
| Macrophages | 14. |

## 28 - II. THE IMMUNE SYSTEM (pp. 482 - 483)
### LYMPHOCYTE BATTLEGROUNDS (p. 484)
### CELL-MEDIATED RESPONSES (pp. 484 - 485)

*Selected Words:* *specificity, memory, self* markers, *nonself* markers, *effector* cells, *memory* cells, *primary* [immune] response, *secondary* immune response, *antibody-mediated* [immune] response, cell-mediated [immune] responses, virgin T cells

### Boldfaced, Page-Referenced Terms

(482) B lymphocytes (B cells) _____

_____

(482) T lymphocytes (T cells) _____

_____

(482) immune system _____

_____

(482) antigen _____

_____

(482) MHC markers _____

_____

(482) antigen-MHC complexes _____

_____

(482) antigen-presenting cell _____

_____

(482) helper T cells _____

_____

(482) cytotoxic T cells _____

_____

(483) B cells _____

_____

(483) antibodies _____

_____

(484) TCRs (T-Cell Receptors) _____

_____

(484) perforins _____

_____

(485) natural killer cells (NK cells) _____

_____

## Fill-in-the-Blanks

If the (1) _____ defenses (fast-acting white blood cells such as neutrophils, eosinophils, and basophils, slower-acting macrophages, and the plasma proteins involved in clotting and complement) fail to repel the microbial invaders, then the body calls on its (2) _____ _____, which identifies *specific* targets to kill and *remembers* the identities of its targets. Your own unique (3) _____ _____ patterns identify your cells as "self" cells. Any other surface pattern is, by definition, (4) _____, and doesn't belong in your body.

The principal actors of the immune system are (5) _____ descended from stem cells (consult Figure 27.7) in the bone marrow, which have two different strategies of action to deal with their different kinds of enemies. (6) _____ _____ clones secrete antibodies and act principally against the extracellular (ones that stay outside the body cells) enemies that are pathogens in blood or on the cell surfaces of body tissues. (7) _____ _____ clones descend from lymphocytes that matured in the (8) _____ where they acquired specific markers on their cell surfaces; they defend principally against intracellular pathogens such as (9) _____, and against any (10) _____ cells and grafts of foreign tissue that are perceived as abnormal or foreign.

## Labeling

Identify each numbered part in the accompanying illustration.

11. _____ - _____ _____     15. _____ _____ _____

12. _____ _____ _____ _____     16. _____ _____ _____

13. _____ _____ _____     17. _____

14. _____     18. _____

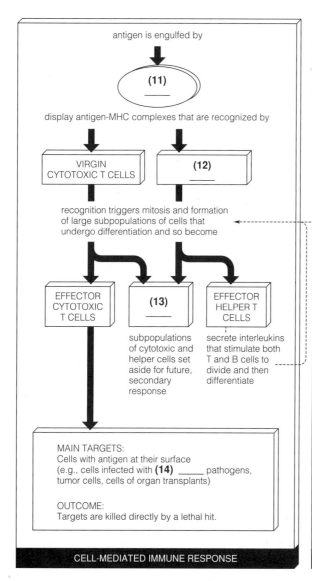

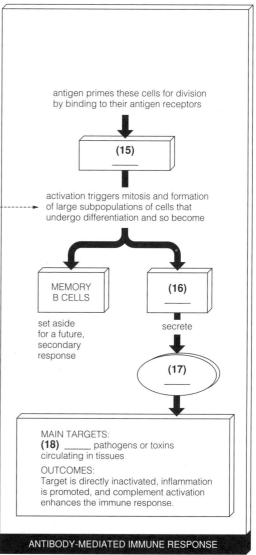

## Choice

For functions 19 - 23, choose the appropriate white blood cell type.

a. effector cytotoxic T cells    b. effector helper T cells    c. macrophages

d. memory cells    e. effector B cells

19. ___ Lymphocytes that directly destroy body cells already infected by certain viruses or by parasitic fungi

20. ___ A portion of B and T cell populations that were set aside as a result of a first encounter, now circulate freely and respond rapidly to any later attacks by the same type of invader

21. ___ Lymphocytes and their progeny that produce antibodies

22. ___ Lymphocytes that serve as master switches of the immune system; stimulate the rapid division of B cells and cytotoxic T cells

23. ___ Nonlymphocytic white blood cells that develop from monocytes, engulf anything perceived as foreign, and alert helper T cells to the presence of specific foreign agents

## Labeling

Identify each numbered part in the accompanying ilustration and its caption.

24. _____ _____

25. _____ _____

26. _____ - _____

27. _____

This illustration depicts activated

(24) _____ _____ cells

carrying out a(n) (28) _____ -

mediated immune response.

*Note:* If the same number is used more than once, that is because the label is the same in all such situations.

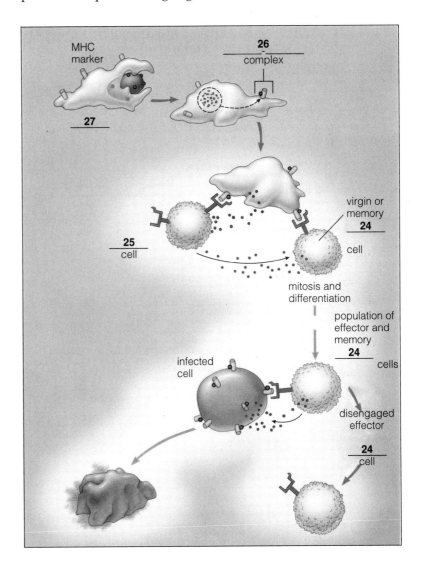

## Fill-in-the-Blanks

The (29) _____ _____ _____ is the route taken during a first-time contact with an

antigen; the secondary immune response is more rapid because patrolling battalions of (30) _____

_____ are in the bloodstream on the lookout for enemies they have conquered before. When these

meet up with recognizable (31) _____, they divide at once, producing large clones of B or T cells

within 2 - 3 days.

In addition to effector B cells and cytotoxic T cells (executioner lymphocytes that mature in the

thymus), (32) _____ _____ cells mature in other lymphoid tissues and search out any cell that

is either coated with complement proteins or antibodies, *or* bears any foreign molecular pattern. When

cytotoxic T cells find foreign cells, they secrete (33) _____ and other toxic substances to poison and

lyse the offenders.

## 28 - III. ANTIBODY-MEDIATED RESPONSES (pp. 486 - 487)
### *Focus on Health:* CANCER AND IMMUNOTHERAPY (p. 487)
### IMMUNE SPECIFICITY AND MEMORY (pp. 488 - 489)

*Selected Words:* virgin B cell, *IgM, Igd, IgG, IgA, IgE, carcinoma, sarcoma, leukemia, immunotherapy, antibody
"factories," monoclonal antibodies, therapeutic vaccines, clonal selection,* "immunological memory"

### *Boldfaced, Page-Referenced Terms*

(487) immunoglobulins (Igs) _____

_____

### *Fill-in-the-Blanks*

Antibodies are plasma proteins that are part of the (1) _____ group of proteins. Some of these

circulate in blood; others are present in other body fluids or bound to B cells. Although all antibodies

are proteins, each kind has (2) _____ _____ that only match up with a particular shape of

(3) _____. When freely-circulating antibody molecules bind antigen, they tag an invader for

destruction by (4) _____ and complement activation. Antibody-mediated responses generally target

(5) _____ _____ and toxins, which are freely circulating in tissues or in body fluids.

(6) _____ can't bind to pathogens or toxins hidden in a host cell.

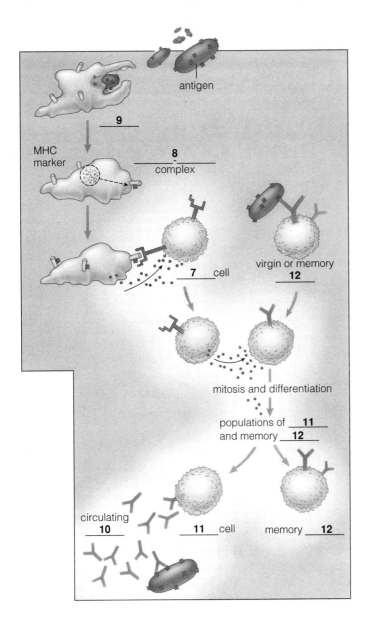

antigen

9 _____

MHC
marker

8 _____
-complex

virgin or memory
12 _____

7 _____ cell

mitosis and differentiation

populations of __11__
and memory __12__

circulating
10 _____

11 _____ cell

memory __12__

*Labeling*

Identify each numbered part in the accompanying ilustration and its caption.

7. _____ _____

8. _____ - _____

9. _____

10. _____

11. _____ _____

12. _____ _____

*Choice*

For questions 13 - 17, choose from the following:

　　　　a. IgA　　b. IgE　　c. IgG　　d. IgM　　e. Monoclonal antibodies

13. ___ Antibodies that activate complement proteins and neutralize many toxins; long-lasting; the only antibodies to cross the placenta during pregnancy and protect a fetus with the mother's acquired immunities; also in early mother's milk

14. ___ Y-shaped molecules produced by clones of a hybrid cell formed by fusing an effector B cell with a cell from a B cell tumor

15. ___ Y-shaped molecules that enter mucus-coated surfaces and neutralize infectious agents, also in mother's milk

16. ___ Antibodies that start the inflammatory response if parasitic worms attack the body; attach to basophils and mast cells that can generate an allergic response

17. ___ The first antibodies to be secreted during immune responses; trigger the complement cascade after binding antigen; also can bind targets together in clumps for removal by phagocytes

## Labeling

Identify each numbered part in the accompanying illustration and its caption.

18. _____

19. _____

20. _____

21. _____ _____

22. _____ _____

This illustration depicts selection of a (20) _____ cell, the descendants of which produced the specific (18) _____ that can combine with a specific antigen.

First exposure to an (19) _____ provokes a primary immune response:

(20) cell or T cell

(22) _____    (21) _____

Subsequent exposure to the same (19) _____ provokes a secondary immune response:

(22) _____ cells    (21) _____ cells

## Fill-in-the-Blanks

The (23) _____ _____ _____ explains how an individual has immunological memory, which is the basis of a secondary immune response; the theory also explains in part how *self* cells are distinguished from (24) _____ cells in the vertebrate immune response.

(25) _____ of gene segments drawn at random from receptor-encoding regions of DNA helps give each T or B cell a different gene sequence that will dictate one protein shape out of a (26) _____ possible antigen receptor shapes.
　　(number)

(27) _____ refers to cells that have lost control over cell division. Milstein and Kohler developed a means of producing large amounts of (28) _____ _____ .

# 28 - IV. IMMUNITY ENHANCED, MISDIRECTED, OR COMPROMISED (pp. 490 - 491)
## Focus on Health: AIDS—THE IMMUNE SYSTEM COMPROMISED (pp. 492 - 493)

*Selected Words:* *active* immunization, *passive* immunization, *asthma, hay fever, anaphylactic shock, Grave's disorder, myasthenia gravis, rheumatoid arthritis, severe combined immune deficiencies* (SCIDs), *AIDS*, HIV, Pneumocystis carinii

## Boldfaced, Page-Referenced Terms

(490) immunization _____

_____

(490) vaccine _____

_____

(490) allergens _____

_____

(490) allergy _____

_____

(490) autoimmune response _____

_____

## Fill-in-the-Blanks

Deliberately provoking the production of memory lymphocytes is known as (1) _____. In a(n)

(2) _____ immunization, a vaccine containing antigens is injected into the body or taken orally.

The first one elicits a (3) _____ _____ _____, and the second one (a booster shot) elicits

a secondary immune response, which causes the body to produce more effector cells and (4) _____

_____ to provide long-lasting protection.

(5) _____ is an altered secondary response to a normally harmless substance that may actually

cause injury to tissues. In a(n) (6) _____ _____, the body mobilizes its forces against certain

tissues of its own. (7) _____ _____ is an example of this kind of disorder in which antibodies

tag acetylcholine receptors on skeletal muscle cells and cause progressive weakness. (8) _____

_____ is a similar kind of disorder in which skeletal joints are chronically inflamed. AIDS is a

constellation of disorders that follow infection by the (9) _____ _____ _____. In the

United States, transmission has occurred most often among intravenous drug abusers who share needles

and among (10) _____ _____. HIV is one of the (11) _____ _____. Its genetic

material is RNA rather than DNA, and it has several copies of an enzyme (12) _____ _____,

which uses the viral RNA as a template for making DNA, which is then inserted into a host

chromosome. Currently, an estimated (13) _____ million plus people in the United States are HIV-

infected. Worldwide, an estimated (14) _____ are HIV-infected and more than (15) _____

million people are already dead from HIV infection.

## Complete the Table

Indicate with a checkmark ( √ ) the age(s) of vaccination.

Age Vaccination is Administered

| Disease | (a) 0 - 4 months after birth | (b) 6 - 18 months after birth | (c) 4 - 6 years after birth | (d) 11 - 12 years after birth |
|---|---|---|---|---|
| 16. Diphtheria | | | | |
| 17. *Hemophilus influenzae* | | | | |
| 18. Hepatitis B | | | | |
| 19. Measles | | | | |
| 20. Mumps | | | | |
| 21. Polio | | | | |
| 22. Rubella | | | | |
| 23. Tetanus | | | | |
| 24. Whooping cough | | | | |

## Complete the Graph

25. Given the graph on page 363, extend the curved lines (extrapolate) and estimate, for the year 2000, the number of people in the United States (a) who will have been diagnosed with AIDS and who will have died of AIDS. Also estimate (b) those numbers for the current year. Then subtract the current year's values from the year 2000 values, and state how many more Americans in that short time interval (c) will be diagnosed with AIDS and will die of AIDS.

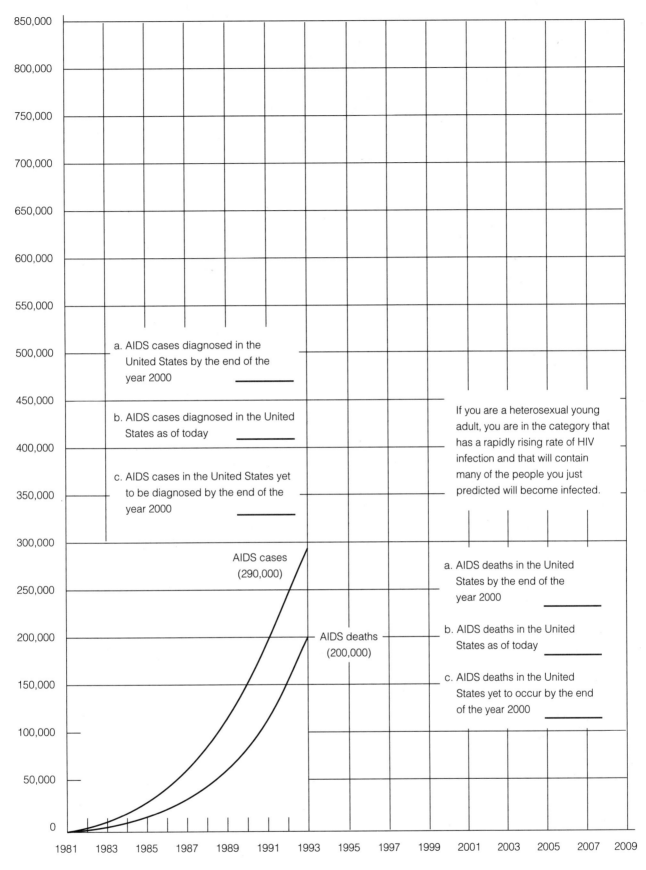

a. AIDS cases diagnosed in the
   United States by the end of the
   year 2000 _____

b. AIDS cases diagnosed in the United
   States as of today _____

c. AIDS cases in the United States yet
   to be diagnosed by the end of the
   year 2000 _____

If you are a heterosexual young
adult, you are in the category that
has a rapidly rising rate of HIV
infection and that will contain
many of the people you just
predicted will become infected.

AIDS cases
(290,000)

AIDS deaths
(200,000)

a. AIDS deaths in the United
   States by the end of the
   year 2000 _____

b. AIDS deaths in the United
   States as of today _____

c. AIDS deaths in the United
   States yet to occur by the end
   of the year 2000 _____

Cumulative number of AIDS cases and deaths in the United States from 1981 to the end of 1993. Even with massive public
education programs, the curve of new infections has continued to rise, although not as steeply.

# Self-Quiz

___1. All the body's phagocytes are derived from stem cells in the _____.
- a. spleen
- b. liver
- c. thymus
- d. bone marrow
- e. thyroid

___2. The plasma proteins that are activated when they contact a bacterial cell are collectively known as the _____ system.
- a. shield
- b. complement
- c. Ig G
- d. MHC
- e. HIV

___3. _____ are divided into two groups: T cells and B cells.
- a. Macrophages
- b. Lymphocytes
- c. Platelets
- d. Complement cells
- e. Cancer cells

___4. _____ produce and secrete antibodies that set up bacterial invaders for subsequent destruction by macrophages.
- a. B cells
- b. Phagocytes
- c. T cells
- d. Bacteriophages
- e. Thymus cells

___5. Antibodies are shaped like the letter _____.
- a. Y
- b. W
- c. Z
- d. H
- e. E

___6. The markers for every cell in the human body are referred to by the letters _____.
- a. HIV
- b. MBC
- c. RNA
- d. DNA
- e. MHC

___7. Effector B cells _____.
- a. fight against extracellular bacteria, viruses, some fungal parasites, and some protozoans
- b. develop from B cells
- c. manufacture and secrete antibodies
- d. do not divide and form clones
- e. all of the above

___8. Clones of B or T cells are _____.
- a. being produced continually
- b. sometimes known as memory cells if they keep circulating in the bloodstream
- c. only produced when their surface proteins recognize other specific proteins previously encountered
- d. produced and mature in the bone marrow
- e. both (b) and (c)

___9. Whenever the body is reexposed to a specific sensitizing agent, IgE antibodies cause _____.
- a. prostaglandins and histamine to be produced
- b. clonal cells to be produced
- c. histamine to be released
- d. the immune response to be suppressed
- e. none of the above

___10. The clonal selection theory explains _____.
- a. how self cells are distinguished from nonself
- b. how B cells differ from T cells
- c. how so many different kinds of antigen-specific receptors can be produced by lymphocytes
- d. how memory cells are set aside from effector cells
- e. how antigens differ from antibodies

## Matching

Choose the most appropriate answer for each term.

11. ___ allergy
12. ___ antibody
13. ___ antigen
14. ___ macrophage
15. ___ clone
16. ___ complement
17. ___ histamine
18. ___ MHC marker
19. ___ effector B cell
20. ___ T cell

A. Begins its development in bone marrow, but matures in the thymus gland
B. Cells that have directly or indirectly descended from the same parent cell
C. A potent chemical that causes blood vessels to dilate and let protein pass through the vessel walls
D. Y-shaped immunoglobulin
E. A nonself marker
F. The progeny of turned-on B cells
G. A group of about fifteen proteins that participate in the inflammatory response
H. An altered secondary immune response to a substance that is normally harmless to other people
I. The basis for self-recognition at the cell surface
J. Principal perpetrator of phagocytosis

---

# Chapter Objectives/Review Questions

| Page | | Objectives/Questions |
|---|---|---|
| (478) | 1. | Describe typical external barriers that organisms present to invading organisms. |
| (478) | 2. | List and discuss four nonspecific defense responses that serve to exclude microbes from the body. |
| (479 - 480) | 3. | Explain how the complement system is related to an inflammatory response. |
| (484 - 487) | 4. | Distinguish between the antibody-mediated response pattern and the cell-mediated response pattern. |
| (483) | 5. | Explain what is meant by primary immune response as contrasted with secondary immune response. |
| (487) | 6. | Explain what monoclonal antibodies are, and tell how they are currently being used in passive immunization and cancer treatment. |
| (489) | 7. | Describe the clonal selection theory, and tell what it helps to explain. |
| (490) | 8. | Describe two ways that people can be immunized against specific diseases. |
| (490 - 491) | 9. | Distinguish allergy from autoimmune disorder. |
| (492 - 493) | 10. | Describe how AIDS specifically interferes with the human immune system. |

---

# Integrating and Applying Key Concepts

Suppose you wanted to get rid of forty-seven warts that you have on your hands by treating them with monoclonal antibodies. Outline the steps you would have to take.

# 29

# RESPIRATION

## Interactive Exercises

### 29 - I. THE NATURE OF RESPIRATION (pp. 496 - 498)
###     INVERTEBRATE RESPIRATION (p. 499)
###     VERTEBRATE RESPIRATION (pp. 500 - 501)

*Selected Words:* hypoxia, *pressure* gradients, "partial pressure," *external* gills, *internal* gills

### Boldfaced, Page-Referenced Terms

(497) respiration _____

_____

(497) respiratory system _____

_____

(498) respiratory surface_____

_____

(498) ventilation _____

_____

(498) hemoglobin _____

_____

(499) integumentary exchange _____

_____

(499) gills _____

_____

(499) tracheal respiration _____

_____

(500) countercurrent flow _____

_____

(500) lungs _____

_____

(501) vocal cords _____

_____

(501) glottis _____

_____

## *Fill-in-the-Blanks*

A(n) (1) _____ is an outfolded, thin, moist membrane endowed with blood vessels; it may (as in fish) be protected by bony covering or (as in aquatic insects) it may be naked. Gas transfer is enhanced by (2) _____ _____, in which water flows past the bloodstream in the opposite direction. Insects have (3) _____ (chitin-lined air tubes leading from the body surface to the interior). At sea level, atmospheric pressure is approximately 760 mm Hg, and oxygen represents about (4) _____ percent of the total volume.

The energy to drive animal activities comes mainly from (5) _____ _____, which uses (6) _____ and produces (7) _____ _____ wastes. In a process called (8) _____, animals move (6) into their internal environment and give up (7) to the external environment.

All respiratory systems make use of the tendency of any gas to diffuse down its (9) _____ _____. Such a (9) exists between (6) in the atmosphere where pressure is (10) ❐ high  ❐ low, and the metabolically active cells in body tissues. Pressure is (11) ❐ highest  ❐ lowest where (6) is used up rapidly. Another (9) exists between (7) in body tissues where (12) ❐ high  ❐ low pressure exists and the atmosphere, with its (13) ❐ higher  ❐ lower amount of (7).

The more extensive the (14) _____ _____ of a respiratory surface membrane and the larger the differences in (15) _____ _____ across it, the faster a gas diffuses across the membrane. (16) _____ is an important transport pigment, each molecule of which can bind loosely with as many as four $O_2$ molecules in the lungs.

A(n) (17) _____ is an internal respiratory surface in the shape of a cavity or sac. In all lungs, (18) _____ carry gas to and from one side of the respiratory surface, and (19) _____ in blood vessels carries gas to and from the other side.

(20) _____ is the medical name for oxygen deficiency; it is characterized by faster breathing, faster heart rate, and anxiety at altitudes of 8,000 feet above sea level.

## Labeling

Identify the numbered parts of the accompanying illustration, which shows the respiratory system of many fishes.

21. _____
    _____

22. _____-
    _____
    _____

23. _____-
    _____
    _____

24. _____

25. _____

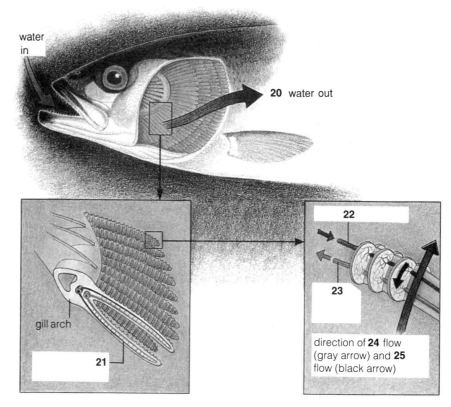

water in

**20** water out

gill arch

**22**

**23**

direction of **24** flow (gray arrow) and **25** flow (black arrow)

**21**

## 29 - II. HUMAN RESPIRATORY SYSTEM (pp. 502 - 503)

**Selected Words:** *laryngitis, lungs, pleurisy, respiratory* bronchioles, *inhale, inhalation, exhale exhalation*

### Boldfaced, Page-Referenced Terms

(502) pharynx _____

_____

(502) larynx _____

_____

(502) epiglottis _____

_____

(502) trachea _____

_____

(502) bronchus (pl., bronchi) _____

_____

(503) diaphragm _____

(503) bronchioles _____

(503) alveolus (pl., alveoli) _____

_____

## Fill-in-the-Blanks

During inhalation, the (1) _____ moves downward and flattens, and the (2) _____

_____ moves outward and upward. When these things happen, the chest cavity volume

(3) (choose one) ❏ increases, ❏ decreases, and the internal pressure (4) (choose one) ❏ rises, ❏ drops,

❏ stays the same. Every time you take a breath, you are (5) _____ the respiratory surfaces of your

lungs. The (6) _____ _____ surrounds each lung. In succession, air passes through the nasal

cavities, pharynx, and (7) _____, past the epiglottis into the (8) _____ (the space between the

true vocal cords), into the trachea, and then to the (9) _____, (10) _____, and alveolar ducts.

Exchange of gases occurs across the epithelium of the (11) _____.

## Labeling

Identify each numbered part of the accompanying illustration.

12. _____ _____

13. _____

14. _____

15. _____

16. _____ _____

17. _____

18. _____

19. _____

20. _____ _____

21. _____ _____

22. _____ _____

23. _____

24. _____ (singular), _____ (plural)

25. _____

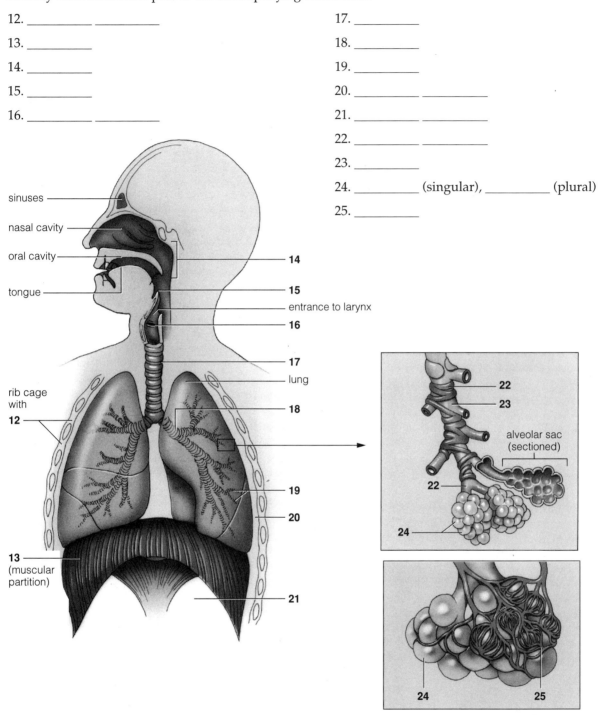

sinuses

nasal cavity

oral cavity

tongue

rib cage with

12

13 (muscular partition)

14

15

entrance to larynx

16

17

lung

18

19

20

21

22

23

alveolar sac (sectioned)

22

24

24

25

## Complete the Table

26. Complete the following table with the structures that carry out the functions listed.

| Structure | Function |
|---|---|
| a. | Thin-walled sacs where $O_2$ diffuses into body fluids and $CO_2$ diffuses out |
| b. | Increasingly branched airways that connect the trachea and alveoli |
| c. | Muscle sheet that contracts during inhalation |
| d. | Airway where breathing is blocked while swallowing and where sound is produced |
| e. | Airway that enhances speech sounds; connects nasal cavity with larynx |

## 29 - III. GAS EXCHANGE AND TRANSPORT (pp. 504 - 505)
### *Focus on Health:* WHEN THE LUNGS BREAK DOWN (pp. 506 - 507)

*Selected Words: carbon monoxide poisoning, "the bends," decompression sickness, bronchitis, emphysema, secondhand smoke, "smoker's cough," pot*

### Boldfaced, Page-Referenced Terms

(504) oxyhemoglobin ($HbO_2$) _____

_____

(504) carbonic anhydrase _____

_____

### Fill-in-the-Blanks

Oxygen is said to exert a(n) (1) _____ _____ of 760/21 or 160 mm Hg. (2) _____ alone moves oxygen from the alveoli into the bloodstream, and it is enough to move (3) _____

_____ in the reverse direction. (4) _____ boosts oxygen transport from the lungs by 70 times and boosts carbon dioxide transport away from tissues by 17 times. About 70 percent of the carbon dioxide in the blood is transported as (5) _____. Without (6) _____, the plasma would be able to carry only about 2 percent of the oxygen that whole blood carries. When oxygen-rich blood reaches a(n) (7) _____ tissue capillary bed, oxygen diffuses outward, and carbon dioxide moves from tissues into the capillaries. When the (8) _____ _____ of carbon dioxide is lower in the alveoli than in the neighboring blood capillaries, carbonic acid dissociates to form water and carbon dioxide. The rate of breathing is governed by clusters of cells that make up a respiratory center in the (9) _____, which monitors and coordinates signals coming in from arterial walls, from blood vessels, and other brain regions. The respiratory center regulates contractions of the diaphragm and intercostal muscles associated with inhalation and exhalation.

(10) _____ is the distension of lungs and the loss of gas exchange efficiency such that running, walking, and even exhaling are painful experiences. At least 90 percent of all (11) _____ _____ deaths are the result of cigarette smoking; only about 10 percent of afflicted individuals will survive.

When a diver ascends, (12) _____ tends to move out of the tissues and into the bloodstream. If the ascent is too rapid, many bubbles of nitrogen gas collect at the (13) _____, hence the common name, "the bends," for what is otherwise known as (14) _____ sickness.

## Self-Quiz

___1. Most forms of life depend on _____ to obtain oxygen and eliminate carbon dioxide.
a. active transport
b. bulk flow
c. diffusion
d. osmosis
e. muscular contractions

___2. _____ is the most abundant gas in Earth's atmosphere.
a. Water vapor
b. Oxygen
c. Carbon dioxide
d. Hydrogen
e. Nitrogen

___3. With respect to respiratory systems, countercurrent flow is a mechanism that explains how _____.
a. oxygen uptake by blood capillaries in the lamellae of fish gills occurs
b. ventilation occurs
c. intrapleural pressure is established
d. sounds originating in the vocal cords of the larynx are formed
e. all of the above

___4. _____ have the most efficient respiratory system.
a. Amphibians
b. Reptiles
c. Birds
d. Mammals
e. Humans

___5. Immediately before reaching the alveoli, air passes through the _____.
a. bronchioles
b. glottis
c. larynx
d. pharynx
e. trachea

___6. During inhalation, _____.
a. the pressure in the chest cavity is less than the pressure within the lungs
b. the pressure in the chest cavity is greater than the pressure within the lungs
c. the diaphragm moves upward and becomes more curved
d. the thoracic cavity volume decreases
e. all of the above

___7. Hemoglobin _____.
a. releases oxygen more readily in tissues with high rates of cellular respiration
b. tends to release oxygen in places where the temperature is lower
c. tends to hold on to oxygen when the pH of the blood drops
d. tends to give up oxygen in regions where partial pressure of oxygen exceeds that in the lungs
e. all of the above

___8. Oxygen moves from alveoli to the bloodstream _____.
a. whenever the concentration of oxygen is greater in alveoli than in the blood
b. by means of active transport
c. by using the assistance of carbaminohemoglobin
d. principally due to the activity of carbonic anhydrase in the red blood cells
e. by all of the above

___9. Oxyhemoglobin releases $O_2$ when
_____.
    a. carbon dioxide concentrations are high
    b. body temperature is lowered
    c. pH values are high
    d. $CO_2$ concentrations are low
    e. all of the above occur

___10. Nonsmokers live an average of _____ longer than people in their mid-twenties who smoke two packs of cigarettes each day.
    a. 6 months
    b. 1 - 2 years
    c. 3 - 5 years
    d. 7 - 9 years
    e. over 12 years

## Matching

11. ___ bronchioles

12. ___ bronchitis

13. ___ carbonic anhydrase

14. ___ emphysema

15. ___ glottis

16. ___ hypoxia

17. ___ intercostal muscles

18. ___ larynx

19. ___ oxyhemoglobin

20. ___ pharynx

21. ___ pleurisy

22. ___ tracheal respiration

23. ___ ventilation

24. ___ integumentary exchange

A. membrane that encloses human lung becomes inflamed and swollen; painful breathing generally results
B. $HbO_2$
C. occurs in terrestrial insects
D. throat passageway that connects to both the respiratory tract below and the digestive tract
E. inflammation of the two principal passageways that lead air into the human lungs
F. contract when air is leaving the lungs; relax when lungs are filling with air
G. the opening into the "voicebox"
H. finer and finer branchings that lead to alveoli
I. occurs in flatworms, earthworms, and many other invertebrates; gases diffuse directly across the body surface covering
J. an enzyme that increases the rate of production of $H_2CO_3$ from $CO_2$ and $H_2O$
K. lungs have become distended and inelastic so that walking, running and even exhaling are difficult
L. where sound is produced by vocal cords
M. movements that keep air or water moving across a respiratory surface
N. too little oxygen is being distributed in the body's tissues

# Chapter Objectives/Review Questions

| Page | | Objectives/Questions |
|---|---|---|
| (496; 499 - 500) | 1. | List some of the ways that respiratory systems are adapted to unusual environments. |
| (497) | 2. | Understand how the human respiratory system is related to the circulatory system and to aerobic respiration. |
| (498 - 501) | 3. | Understand the behavior of gases and the types of respiratory surfaces that participate in gas exchange. |
| (499, 501) | 4. | Describe how incoming oxygen is distributed to the tissues of insects, and contrast this process with the process that occurs in mammals. |
| (500 - 501) | 5. | Define *countercurrent flow*, and explain how it works. State where such a mechanism is found. |
| (502 - 503) | 6. | List all the principal parts of the human respiratory system, and explain how each structure contributes to transporting oxygen from the external world to the bloodstream. |

(503)      7.  Describe the relationship of the human lung to the pleural sac and to the chest (thoracic) cavity.

(504)      8.  Explain why oxygen diffuses from the bloodstream into the tissues far from the lungs. Then explain why carbon dioxide diffuses into the bloodstream from the same tissues.

(504 - 505)  9.  List the structures involved in detecting carbon dioxide levels in the blood and in regulating the rate of breathing. Name the location of each structure.

(504)     10.  Explain why oxygen diffuses from alveolar air spaces, through interstitial fluid, and across capillary epithelium. Then explain why carbon dioxide diffuses in the reverse direction.

(505)     11.  Describe what happens to carbon dioxide when it dissolves in water under conditions normally present in the human body.

(506 - 507) 12.  Distinguish bronchitis from emphysema. Then explain how lung cancer differs from emphysema.

## Integrating and Applying Key Concepts

Consider the amphibians—animals that generally have aquatic larval forms (tadpoles) and terrestrial adults. Outline the respiratory changes that you think might occur as an aquatic tadpole metamorphoses into a land-going juvenile.

# 30

# DIGESTION AND HUMAN NUTRITION

## Interactive Exercises

### 30 - I. THE NATURE OF DIGESTIVE SYSTEMS (pp. 510 - 513)
###        OVERVIEW OF THE HUMAN DIGESTIVE SYSTEM (p. 514)

*Selected Words:* anorexia nervosa, bulimia, crop, gizzard, "chewing cud"

*Boldfaced, Page-Referenced Terms*

(511) nutrition _____

_____

(511) digestive system _____

_____

(512) incomplete digestive system _____

_____

(512) complete digestive system _____

_____

(512) motility _____

_____

(512) secretion _____

_____

(512) absorption _____

_____

(512) elimination _____

_____

(513) ruminants _____

_____

(513) gut _____

_____


## Fill-in-the-Blanks

(1) _____ is a large concept that encompasses processes by which food is ingested, digested, absorbed, and later converted to the body's own (2) _____, lipids, proteins, and nucleic acids. A digestive system is some form of body cavity or tube in which food is reduced first to (3) _____ and then to small (4) _____. Digested nutrients are then (5) _____ into the internal environment. A(n) (6) _____ digestive system has only one opening, two-way traffic, and a highly branched gut cavity that serves both digestive and (7) _____ functions. A(n) (8) _____ digestive system has a tube or cavity with regional specializations and a(n) (9) _____ at each end. (10) _____ involves the muscular movement of the gut wall, but (11) _____ is the release into the lumen of enzyme fluids and other substances required to carry out digestive functions.

The human digestive system is a tube, 21 - 30 feet long in an adult, that has regions specialized for different aspects of digestion and absorption; they are, in order, the mouth, pharynx, esophagus, (12) _____, (13) _____ _____, large intestine, rectum, and (14) _____. Various (15) _____ structures secrete enzymes and other substances that are also essential to the breakdown and absorption of nutrients; these include the salivary glands, liver, gallbladder, and (16) _____. The (17) _____ system distributes nutrients to cells throughout the body. The (18) _____ system supplies oxygen to the cells so that they can oxidize the carbon atoms of food molecules, thereby changing them to the waste product, (19) _____ _____, which is eliminated by the same system. And if excess water, salts, and wastes accumulate in the blood, the (20) _____ system and skin will maintain the volume and composition of blood and other body fluids.

## Labeling

Identify each numbered structure in the accompanying illustration.

21. _____ _____

22. _____

23. _____

24. _____

25. _____

26. _____ _____

27. _____ _____

28. _____

29. _____

30. _____

31. _____

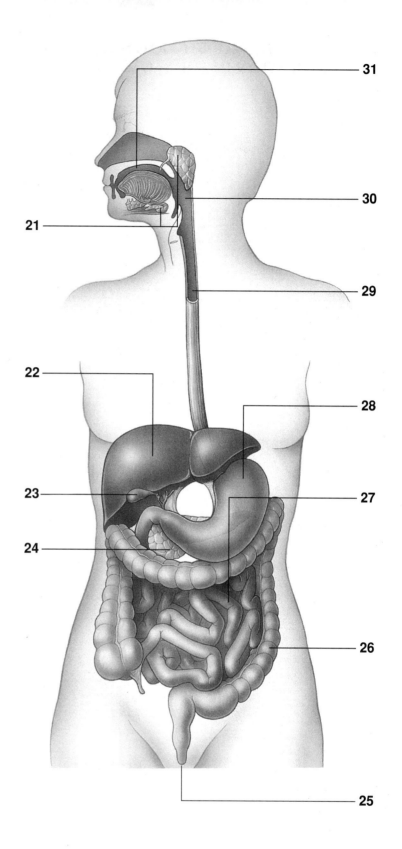

## 30 - II. INTO THE MOUTH, DOWN THE TUBE (p. 515)
## DIGESTION IN THE STOMACH AND SMALL INTESTINE (pp. 516 - 517)

*Selected Words:* caries, gingivitis, periodontal disease, heartburn, peptic ulcer, Helicobacter pylori, "emulsion"

### Boldfaced, Page-Referenced Terms

(515) mouth (oral cavity) _____

_____

(515) tooth _____

_____

(515) tongue _____

_____

(515) saliva _____

_____

(515) pharynx _____

_____

(515) esophagus _____

_____

(516) stomach _____

_____

(516) gastric fluid _____

_____

(516) chyme _____

(516) pancreas _____

(516) liver _____

(516) gallbladder _____

(516) bile _____

_____

(516) emulsification _____

_____

## Complete the Table

1. Complete the following table by naming the organs described.

| Organ | Main Functions |
| --- | --- |
| a. | Mechanically breaks down food, mixes it with saliva |
| b. | Moistens food; starts polysaccharide breakdown; buffers acidic foods in mouth |
| c. | Stores, mixes, dissolves food; kills many microorganisms; starts protein breakdown; empties in a controlled way |
| d. | Digests and absorbs most nutrients |
| e. | Produces enzymes that break down all major food molecules; produces buffers against hydrochloric acid from stomach |
| f. | Secretes bile for fat emulsification; secretes bicarbonate, which buffers hydrochloric acid from stomach |
| g. | Stores, concentrates bile from liver |
| h. | Stores, concentrates undigested matter by absorbing water and salts |
| i. | Controls elimination of undigested and unabsorbed residues |

## Fill-in-the-Blanks

Saliva contains an enzyme (2) _____ _____ that breaks down starch. Contractions force the larynx against a cartilaginous flap called the (3) _____, which closes off the trachea. The (4) _____ is a muscular tube that propels food to the stomach. Any alternating progression of contracting and relaxing muscle movements along the length of a tube is known as (5) _____. (6) _____ is an enzyme that works in the stomach to begin to break down proteins.

Carbohydrates include sugars and (7) _____, the name commonly given to polysaccharides. Rice, cereal, pasta, bread, and white potatoes are composed of many polysaccharide molecules that are too large to be absorbed into the internal environment. If these foods are chewed thoroughly, (8) _____ _____ in the mouth digests them to the (9) _____ (double sugar) level. Since no carbohydrate digestion occurs in the (10) _____, if you gulped down your food, starch digestion would again begin in the (11) _____ _____ where (12) _____ produced by the pancreas would do what should have been done in the mouth. Digestion of disaccharides to monosaccharides (simple sugars) also occurs in the (13) _____ _____. The enzymes responsible are (14) _____ with names such as sucrase, lactase, and maltase.

Proteins are digested to protein fragments, beginning in the (15) _____ by (16) _____ secreted by the lining of the stomach. Protein fragments are subsequently digested to smaller protein fragments in the (17) _____ _____ by enzymes known as trypsin and chymotrypsin produced by the (18) _____ . Eventually the smaller protein fragments are digested to (19) _____ _____ by means of carboxypeptidase produced by the pancreas and by aminopeptidase produced by glands in the intestinal lining.

Another name for <u>fat</u> is triglycerides. (20) _____ , produced by the pancreas but acting in the (21) _____ _____ , breaks down one triglyceride molecule into three (22) _____ _____ molecules and one glycerol molecule. (23) _____ , which is made by the liver, stored in the (24) _____ , and does not contain digestive enzymes emulsifies the fat droplet (converts it into small droplets coated with bile salts) thereby increasing the surface area of the substrate upon which (25) _____ can act.

Nutrients are also mostly digested and absorbed in the (26) _____ _____ . (23) is made by the liver, is stored in the gallbladder, and works in the (27) _____ _____ . (28) _____ _____ is an example of an enzyme that is made by the pancreas but works in the small intestine to convert protein fragments to amino acids. (29) _____ _____ are made in the pancreas, but convert DNA and RNA into nucleotides in the small intestine.

### True/False

If the statement is true, write a T in the blank. If the statement is false, correct it by changing the underlined word and writing the correct word in the answer blank.

_____ 30. Amylase digests starch, lipase digests lipids, and <u>proteases</u> break peptide bonds.

_____ 31. ATP is the end product of <u>digestion</u>.

## 30 - III. ABSORPTION IN THE SMALL INTESTINE (pp. 518 - 519)
### DISPOSITION OF ABSORBED ORGANIC COMPOUNDS (p. 520)
### THE LARGE INTESTINE (p. 521)
### *Focus on Health:* CANCER IN THE SYSTEM (p. 521)

*Selected Words:* "brush border" cell, *segmentation, appendicitis, colon cancer, stomach cancer, pancreatic cancer, esophageal cancer*

### Boldfaced, Page-Referenced Terms

(518) villi (sing., villus) _____

_____

(518) microvilli, (sing., microvillus) _____

_____

(518) micelle formation _____

_____

(521) colon _____

_____

(521) appendix _____

_____

(521) sphincter _____

_____

(521) bulk _____

_____

## Short Answer

1.  What is the pool of amino acids used for in the human body? (See Figure 30.11) _____

_____

2.  Which breakdown products result from carbohydrate and fat digestion? (See Table 30.1) _____

_____

3.  Monosaccharides, free fatty acids, and monoglycerides all have three uses; identify them. (See text, p. 520) _____

_____

## Fill-in-the-Blanks

Immediately after a meal, the body's cells take up the (4) _____ being absorbed from the small intestine and use it as an energy source. Excess amounts of (4) and other organic compounds are converted mainly to (5) _____, which get stored in adipose tissue. Some of the excess is converted to (6) _____, which is stored mainly in the (7) _____ and in muscle tissue. Absorbed amino acids are synthesized into structural (8) _____ and enzymes, as well as into some nitrogen-containing (9) _____, which act as chemical messengers, and (10) _____, which are assembled into DNA, RNA, and ATP. Between meals, the body uses the (11) _____ reservoirs as the main energy source; energy from them is transferred to ATP during cellular respiration. Amino acid conversions in the liver form (12) _____, which is potentially toxic to cells; the liver immediately converts this substance to (13) _____, a much less toxic waste product that is expelled by the urinary system from the body.

## True/False

If the statement is true, write a T in the blank. If the statement is false, correct it by changing the under-lined word(s) and writing the correct word(s) in the blank.

_____ 14. The appendix has no known <u>digestive</u> functions.

_____ 15. Water and <u>sodium</u> ions are absorbed into the bloodstream from the lumen of the large intestine.

_____ 16. Fatty acids and monoglycerides recombine into fats inside epithelial cells lining the <u>colon</u>.

## 30 - IV. HUMAN NUTRITIONAL REQUIREMENTS (pp. 522 - 523)
### VITAMINS AND MINERALS (pp. 524 - 525)
### *Focus on Science:* TANTALIZING ANSWERS TO WEIGHTY QUESTIONS (pp. 526 - 527)

*Selected Words:* "food pyramids," *sucrose polyester,* "complete" protein, "incomplete" protein, "set point," *ob gene*

### *Boldfaced, Page-Referenced Terms*

(522)  kilocalories _____

_____

(523)  essential fatty acids _____

_____

(523)  essential amino acids _____

_____

(523)  net protein utilization (NPU) _____

_____

(524)  vitamins _____

_____

(524)  minerals _____

_____

(526)  obesity _____

_____

(527)  leptin _____

_____

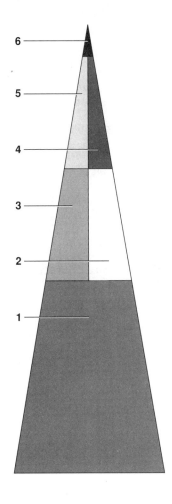

*Fill-in-the-Blanks*

Use the (1) _____ _____ diagram at the left, as revised in 1992, to devise a well-balanced diet for yourself. Group 1, the trapezoidal base, represents the group of complex (2) _____, which includes rice, pasta, cereal and (3) _____. (4) From this group, (choose 1) ❑ 0, ❑ 2 - 3, ❑ 2 - 4, ❑ 3 - 5, ❑ 6 - 11 servings are needed every day to supply energy and fiber. Group 2 represents the (5) _____ group. Use the choices in (4) to indicate the number of servings (6) _____ that are needed from this group each day. Group 3 is the (7) _____ group, from which (8) _____ servings are needed each day. Choices include mango, oranges, and (9) _____, cantaloupe, pineapple, or 1 cup of fresh (10) _____. Group 4 includes foods that are a source of nitrogen: nuts, poultry, fish, legumes, and (11) _____. From this group, the body's (12) _____ and nucleic acids are constructed. (13) _____ servings are required every day because the human body cannot synthesize eight of the 20 essential (14) _____ _____ that are used to construct proteins, and must get them in their food supplies. The foods in Group 5, the (15) _____, yogurt and cheese group, supply calcium, vitamins A, D, $B_2$, and $B_{12}$. You need (16) _____ servings every day. The foods in Group 6 provide extra calories but few vitamins and minerals; (17) _____ servings are needed every day.

*Complete the Table*

18. Complete the following table by determining how many kilocalories the people described should take in daily, given the stated exercise level, in order to maintain their weight. Consult page 526 of the text.

| Height | Age | Sex | Level of Physical Activity | Present Weight (lb.) | Number of Kilocalories/Day |
|--------|-----|-----|----------------------------|----------------------|----------------------------|
| 5'6" | 25 | Female | Moderately active | 138 | a. |
| 5'10" | 18 | Male | Very active | 145 | b. |
| 5'8" | 53 | Female | Not very active | 143 | c. |

*Fill-in-the-Blanks*

Researchers at Harvard University found in 1995 a strong link between the risk of heart attack and gaining even 11 to 18 pounds in adult life; (19) _____ who gained weight after age 18 were by far the group at greatest risk. (20) _____ _____ are the body's main sources of energy; they should make up (21) _____ to _____ percent of the human daily caloric intake. (22) _____ and cholesterol are components of animal cell membranes.

Fat deposits are used primarily as (23) _____ _____, but they also cushion many organs and provide insulation. Lipids should constitute less than (24) _____ percent of the human diet. One tablespoon a day of polyunsaturated oil supplies all (25) _____ _____ _____ that the body cannot synthesize. (26) _____ are digested to twenty common amino acids, of which eight are (27) _____, cannot be synthesized, and must be supplied by the diet. Animal proteins such as (28) _____ and (29) _____ contain high amounts of essential amino acids (that is, they are complete). (30) _____ are organic substances needed in small amounts in order to build enzymes or help them catalyze metabolic reactions. (31) _____ are inorganic substances needed for a variety of uses.

## Related Problems

You are a 19-year-old male, very sedentary (TV, sleep, and computers), with a 6'1", medium frame and you weigh 195 lbs.

32. Use Fig. 30.15 , page 526, to calculate the number of calories required to sustain your desired weight. Are you underweight, overweight or just right? _____
33. How many kilocalories are you allowed to ingest every day to maintain your weight? _____

## Complete the Table

Use the new, improved food pyramid (p. 522) to construct a one-day diet that would eventually allow you to reach that weight if you ate a similar diet every day. Place your choices in the table below as a diet for the person described above.

| How many servings from each group below are you allowed to have daily? | What, specifically, could you choose to eat? |
| --- | --- |
| 34a. complex carbohydrates | 34b. |
| 35a. fruits | 35b. |
| 36a. vegetables | 36b. |
| 37a. dairy group | 37b. |
| 38a. assorted proteins | 38b. |
| 39a. The "sin" group at the top | 39b. |

# Self-Quiz

___1. The process that moves nutrients into the blood or lymph is _____.
   a. ingestion
   b. absorption
   c. assimilation
   d. digestion
   e. none of the above

___2. The enzymatic digestion of proteins begins in the _____.
   a. mouth
   b. stomach
   c. liver
   d. pancreas
   e. small intestine

___3. The enzymatic digestion of starches begins in the _____.
   a. mouth
   b. stomach
   c. liver
   d. pancreas
   e. small intestine

___4. The greatest amount of absorption of digested nutrients occurs in the _____.
   a. stomach
   b. pancreas
   c. liver
   d. colon
   e. small intestine

___5. Glucose moves through the membranes of the small intestine mainly by _____.
   a. peristalsis
   b. osmosis
   c. diffusion
   d. active transport
   e. bulk flow

___6. Which of the following is not found in bile?
   a. lecithin
   b. salts
   c. digestive enzymes
   d. cholesterol
   e. pigments

___7. The average American consumes approximately _____ pounds of sugar per year.
   a. 25
   b. 50
   c. 75
   d. 100
   e. 125

___8. Of the following, _____ has (have) the highest net protein utilization.
   a. cashew nuts
   b. eggs
   c. fish
   d. beans
   e. bread

___9. Which of the following acts to slow emptying of the stomach so that food is not moved faster than it can be processed?
   a. fear, depression, and other emotional upsets
   b. gastrin secreted from the lining of the stomach triggers the secretion of stomach acid
   c. glucose and fat in the small intestine call for insulin secretion, which helps cells absorb glucose
   d. secretion prods the pancreas to secrete bicarbonate
   e. cholecystokinin (CCK) enhances secretin's action and causes the gallbladder to contract

___10. Which of the following structures and/or activities are NOT associated with the small intestine?
   a. micelle formation
   b. maximum compaction of bulk
   c. bile salts assist the absorption of fatty acids and monoglycerides
   d. triglycerides and proteins combine, forming chylomicrons
   e. substances cross the free surface of each villus by active transport, osmosis, and diffusion

___11. The element needed by humans for blood clotting, nerve impulse transmission, and bone and tooth formation is _____.
   a. magnesium
   b. iron
   c. calcium
   d. iodine
   e. zinc

*Matching*

Choose the most appropriate answer for each term.

12. ___ anorexia nervosa

13. ___ antioxidants

14. ___ bulimia

15. ___ complex carbohydrates

16. ___ essential amino acids

17. ___ essential fatty acids

18. ___ mineral

19. ___ rickets

20. ___ scurvy

21. ___ vitamin

A. Linoleic acid is one example
B. Phenylalanine, lysine, and methionine are 3 of 8
C. Combine with free radicals; counteract their destructive effects on DNA and cell membranes
D. Obsessive dieting + skewed perception of body weight
E. Vitamin C deficiency
F. Vitamin D deficiency in young children
G. Organic substances that help enzymes to do their jobs; required in small amounts for good health
H. Feasting followed by vomiting or taking laxatives
I. Inorganic substances required for good health
J. Long chains of simple sugars; in pasta and white potatoes

---

# Chapter Objectives/Review Questions

*Page*        *Objectives/Questions*

(512)        1.  Distinguish between incomplete and complete digestive systems, and tell which is characterized by (a) specialized regions and (b) two-way traffic.

(512)        2.  Define and distinguish among motility, secretion, digestion, and absorption.

(514 - 517)  3.  List all parts (in order) of the human digestive system through which food actually passes. Then list the auxiliary organs that contribute one or more substances to the digestive process.

(514 - 516)  4.  Explain how, during digestion, food is mechanically broken down. Then explain how it is chemically broken down.

(516 - 519)  5.  Describe how the digestion and absorption of fats differ from the digestion and absorption of carbohydrates and proteins.

(516, 519)   6.  List the items that leave the digestive system and enter the circulatory system during the process of absorption.

(514 - 517)  7.  Tell which foods undergo digestion in each of the following parts of the human digestive system, and state what the food is broken into: oral cavity, stomach, small intestine, large intestine.

(515 - 517)  8.  List the enzyme(s) that act in (a) the oral cavity, (b) the stomach, and (c) the small intestine. Then tell where each enzyme was originally made.

(517 - 519)  9.  Describe the cross-sectional structure of the small intestine, and explain how its structure is related to its function.

(521)       10.  State which processes occur in the colon (large intestine).

(522 - 525) 11.  Compare the contributions of carbohydrates, proteins, and fats to human nutrition with the contributions of vitamins and minerals.

(522)       12.  Reproduce from memory the Food Pyramid Diagram as revised in 1992. Identify each of the six components, list the numerical range of servings permitted from each group, and also list some of the choices available.

(522 - 523) 13.  Summarize current ideas for promoting health by eating properly.

(522 - 526) 14. Summarize the daily nutritional requirements of a 25-year-old man who is 5' 11" and works at a desk job and exercises very little. State what he needs in energy, carbohydrates, proteins, and lipids, and name at least six vitamins and six minerals that he needs to include in his diet every day.

(526) 15. Construct an ideal diet for yourself for one 24-hour period. Calculate the number of calories necessary to maintain your weight (see p. 526), and then use the Food Pyramid (p. 522) to choose exactly what to eat and how much.

(523 - 525) 16. State what is meant by net protein utilization, and distinguish vitamins from minerals.

(525) 17. Name five minerals that are important in human nutrition, and state the specific role of each.

(527) 18. Explain how the human body manages to meet the energy and nutritional needs of the various body parts even though the person may be feasting sometimes and fasting at other times.

## Integrating and Applying Key Concepts

Suppose you could not eat solid food for two weeks and you had only water to drink. List in correct sequential order the measures your body would take to try to preserve your life. Mention the command signals that are given as one after another critical point is reached, and tell which parts of the body are the first and the last to make up for the deficit.

# 31

# THE INTERNAL ENVIRONMENT

## Interactive Exercises

### 31 - I. URINARY SYSTEM OF MAMMALS (pp. 530 - 533)

*Selected Words: internal* environment, "metabolic water," ammonia, reflex, *glomerular* capillaries, *peritubular capillaries*

### Boldfaced, Page-Referenced Terms

(531) interstitial fluid _____

_____

(531) blood _____

(531) extracellular fluid _____

(532) urinary excretion _____

_____

(532) urea _____

_____

(533) urinary system _____

(533) kidneys _____

(533) urine _____

(533) ureter _____

(533) urinary bladder _____

(533) urethra _____

(533) nephrons _____

_____

(533) Bowman's capsule _____

_____

(533) glomerulus _____

(533) proximal tubule _____

(533) loop of Henle _____

(533) distal tubule _____

(533) collecting duct _____

_____

## Fill-in-the-Blanks

The body gains water by absorbing water from the slurry in the lumen of the small intestine and from

(1) _____ during condensation reactions. The mammalian body loses water mostly by excretion of

(2) _____, evaporation through the skin and (3) _____, elimination of feces from the gut, and

(4) _____ as the body is cooled. (5) _____ behavior, in which the brain compels the

individual to seek liquids, influences the gain of water.

The body gains solutes by absorption of substances from the gut, by the secretion of hormones and

other substances, and by (6) _____, which produces $CO_2$ and other waste products of degradative

reactions. Besides $CO_2$, there are several major metabolic wastes that must be eliminated:

(7) _____, formed when amino groups are detached from amino acids; (8) _____, which is

produced in the liver during reactions that link two ammonia molecules to $CO_2$ and release a molecule

of water, and (9) _____ _____, which is formed in reactions that break down nucleic acids.

## Labeling

Identify each numbered part of the accompanying illustrations.

10. _____      14. _____

11. _____      15. _____

12. _____ _____      16. _____

13. _____

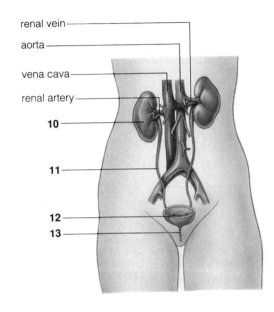

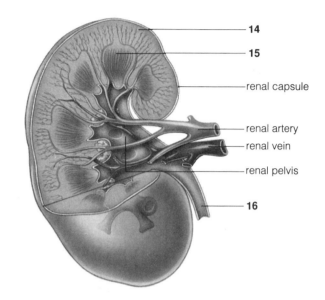

renal vein

aorta

vena cava

renal artery

**10**

**11**

**12**

**13**

**14**

**15**

renal capsule

renal artery

renal vein

renal pelvis

**16**

## Labeling

Identify each numbered part of the accompanying illustration.

17. _____ _____

18. _____ _____

19. _____ _____

20. _____ _____

21. _____ _____

22. _____ _____

_____

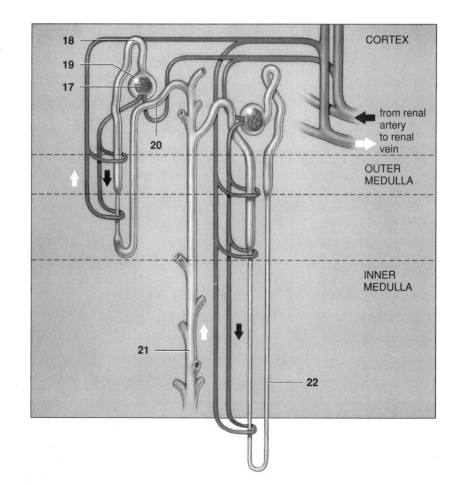

CORTEX

**18**

**19**

**17**

**20**

from renal artery

to renal vein

OUTER MEDULLA

INNER MEDULLA

**21**

**22**

## 31-II. URINE FORMATION (pp. 534 - 535)
### *Focus on Health:* WHEN THE KIDNEYS BREAK DOWN (p. 536)

*Selected Words:* adrenal cortex, *hypertension, kidney stones,* "dialysis," *kidney dialysis machine, hemodialysis, peritoneal dialysis*

### Boldfaced, Page-Referenced Terms

(534) filtration _____

_____

(534) tubular reabsorption _____

_____

(534) tubular secretion _____

_____

(535) ADH (antidiuretic hormone) _____

_____

(535) aldosterone _____

_____

(535) thirst behavior _____

_____

### True/False

If the statement is true, write a T in the blank. If the statement is false, correct it by changing the underlined word(s) and writing the correct word(s) in the blank.

_____ 1.  When the body rids itself of excess water, urine becomes more <u>dilute</u>.

_____ 2.  Water reabsorption into capillaries is achieved by <u>diffusion</u> <u>and</u> <u>active</u> <u>transport</u>.

### Fill-in-the-Blanks

In mammals, urine formation occurs in a pair of (3) _____. Each contains about a million tubelike blood-filtering units called (4) _____. The function of (3) depends on intimate links between the (4) and the (5) _____. In every (4) blood flows from a(n) (6) _____ into a set of capillaries inside the (7) _____ _____, then into a second set of capillaries that thread around the tubular parts of the nephron, then back to the bloodstream, leaving the kidney. Urine composition and volume depend on three processes: *filtration* of blood at the (8) _____ of a nephron, with (9) _____ _____ providing the force for filtration; *reabsorption,* in which water and (10) _____ move out of tubular parts of the nephron and back into adjacent (11) _____ _____; and (12) _____, in which excess ions and a few foreign substances move out of those capillaries and back into the nephron so that they are disposed of in the urine. (13) _____ carry urine away from the kidney to the (14) _____ _____, where it is stored until it is released via a tube called the (15) _____, which carries urine to the outside.

Two hormones, ADH and (16) _____, adjust the reabsorption of water and (17) _____ along the distal tubules and collecting ducts. An increase in the secretion of aldosterone causes (18) (choose 1) ❐ more, ❐ less sodium to be excreted in the urine. When the body cannot rid itself of excess sodium, it inevitably retains excess water, and this leads to a rise in (19) _____ _____. Abnormally high blood pressure is called (20) _____; it can damage the kidneys, the vascular system, and brain. One way to control hypertension is to restrict the intake of (21) _____ _____. Increased secretion of (22) _____ enhances water reabsorption at distal tubules and collecting ducts when the body must conserve water. When excess water must be excreted, ADH secretion is (23) (choose one) ❐ stimulated, ❐ inhibited.

By adjusting the blood's (24) _____ and composition, kidneys help maintain conditions in the extracellular fluid.

## 31-III. THE BODY'S ACID-BASE BALANCE (p. 536)
### ON FISH, FROGS, AND KANGAROO RATS (p. 537)
### MAINTAINING THE BODY'S CORE TEMPERATURE (pp. 538 - 539)

*Selected Words:* acids, bases, *bicarbonate–carbon dioxide* buffer system, hypothermia, frostbite, *hyperthermia, fever,* "thermostat"

## Boldfaced, Page-Referenced Terms

(536) acid-base balance _____

_____

(536) buffer system _____

_____

(538) core temperature _____

_____

(538) radiation _____

_____

(538) conduction _____

_____

(538) convection _____

_____

(538) evaporation _____

_____

(538) ectotherms _____

_____

(538) behavioral temperature regulation _____

_____

(538) endotherms _____

_____

(538) heterotherms _____

_____

(538) peripheral vasoconstriction _____

_____

(539) pilomotor response _____

_____

(539) shivering _____

(539) peripheral vasodilation _____

_____

(539) evaporative heat loss _____

_____

(539) sweat glands _____

_____

## Fill-in-the-Blanks

The (1) _____ control the acid-base balance of body fluids by controlling the levels of dissolved

ions, especially (2) _____ ions. The extracellular pH of humans must be maintained between 7.37

and (3) _____. (4) (choose 1) ❏ Acids  ❏ Bases lower the pH and (5) (choose 1) ❏ acids  ❏ bases

raise it. If you were to drink a gallon of orange juice, the pH would be (6) (choose 1) ❏ raised

❏ lowered, but the effect is minimized when excess (7) _____ ions are neutralized by

(8) _____ ions in the bicarbonate–carbon dioxide buffer system. Only the (9) _____ system

eliminates excess H+ and restores buffers. Desert-dwelling kangaroo rats have very long

(10) _____ _____ _____ so that nearly all (11) _____ that reaches their very long

collecting ducts is reabsorbed. In freshwater, bony fishes and amphibians tend to gain (12) _____

and lose (13) _____; they produce (14) ❏ very dilute  ❏ very concentrated urine.

In the brain of mammals, the (15) _____ is the seat of temperature control. Thermoreceptors

located deep in the body are called (16) _____ thermoreceptors. The (17) _____ _____

contains smooth muscles that erect hairs or feathers and create an insulative layer of still air that helps

prevent heat loss. (18) _____ _____ is a response to cold stress in which the bloodstream's

convective delivery of heat to the body's surface is reduced. A drop in body temperature below

tolerance levels is referred to as (19) _____.

## True/False

If the statement is true, write a T in the blank. If the statement is false, correct it by changing the underlined word(s) and writing the correct word(s) in the blank.

_____ 20. Polar bears are <u>endotherms</u>.

_____ 21. When the core temperature of the human body drops <u>a few</u> <u>degrees</u>, brain function is affected and thinking patterns become confused.

_____ 22. When the core temperature of the human body causes <u>hyperthermia</u> to occur, the hypothalamus may reset the "thermostat" that dictates what the core temperature is supposed to be.

## Matching

Choose the most appropriate answer for each term.

23. ___ conduction
24. ___ convection
25. ___ ectotherm
26. ___ endotherm
27. ___ evaporation
28. ___ heterotherm
29. ___ radiation

A. Body temperature determined more by heat exchange with the environment than by metabolic heat
B. Heat transfer by air or water heat-bearing currents away from or toward a body
C. Body temperature determined largely by metabolic activity and by precise controls over heat produced and heat lost
D. Direct transfer of heat energy between two objects in direct contact with each other
E. The emission of energy in the form of infrared or other wavelengths that are converted to heat by the absorbing body
F. Body temperature fluctuating at some times and heat balance controlled at other times
G. In changing from the liquid state to the gaseous state, the energy required is supplied by the heat content of the liquid.

# Self-Quiz

___1. _____ forms as amino groups are stripped from amino acids.
   a. water
   b. uric acid
   c. urea
   d. ammonia
   e. carbon dioxide

___2. An entire subunit of a kidney that purifies blood and restores solute and water balance is called a _____.
   a. glomerulus
   b. loop of Henle
   c. nephron
   d. ureter
   e. none of the above

___3. In humans, the thirst center is located in the _____.
   a. adrenal cortex
   b. thymus
   c. heart

   d. adrenal medulla
   e. hypothalamus

___4. The longer the _____, the greater an animal's capacity to conserve water and to concentrate solutes to be excreted in the urine.
   a. loop of Henle
   b. proximal tubule
   c. ureter
   d. Bowman's capsule
   e. collecting tubule

___5. During reabsorption, sodium ions cross the proximal tubule walls into the interstitial fluid principally by means of _____.
   a. phagocytosis
   b. countercurrent multiplication
   c. bulk flow
   d. active transport
   e. all of the above

___6. Filtration of the blood in the kidney takes place in the _____.
a. loop of Henle
b. proximal tubule
c. distal tubule
d. Bowman's capsule
e. all of the above

___7. _____ primarily controls the concentration of solutes in urine.
a. Insulin
b. Glucagon
c. Antidiuretic hormone
d. Aldosterone
e. Epinephrine

___8. Hormonal control over excretion primarily affects _____.
a. Bowman's capsules
b. distal tubules
c. proximal tubules
d. the urinary bladder
e. loops of Henle

___9. The last portion of the excretory system passed by urine before it is eliminated from the body is the _____.

a. renal pelvis
b. bladder
c. ureter
d. collecting ducts
e. urethra

___10. Desert animals excrete _____ as their principal nitrogenous waste in a highly concentrated urine.
a. urea
b. uric acid
c. ammonia
d. amino acids
e. ADH

___11. Normally, the extracellular pH of the human body must be maintained between _____ and _____; only the urinary system eliminates excess _____ and restores _____.
a. 6.45 - 7.30; $NH_4^+$; urea
b. 7.37 - 7.43; $H^+$; buffers
c. 7.50 - 7.85; $H^+$; glucose
d. 7.90 - 8.30; $NH_4^+$; urea
e. 8.15 - 8.35; $OH^-$; glucose

---

# Chapter Objectives/Review Questions

| Page | | Objectives/Questions |
|---|---|---|
| (530 - 533, 535) | 1. | List some of the factors that can change the composition and volume of body fluids. |
| (532) | 2. | List three soluble by-products of animal metabolism that are eliminated from urinary systems. |
| (533) | 3. | List successively the parts of the human urinary system that constitute the path of urine formation and excretion. |
| (534 - 535) | 4. | Locate the processes of filtration, reabsorption, and tubular secretion along a nephron, and tell what makes each process happen. |
| (534 - 535) | 5. | State explicitly how the hypothalamus, the adrenal cortex, and distal tubules of the nephrons are interrelated in regulating water and solute levels in body fluids. |
| (536) | 6. | Describe the role of the kidney in maintaining the pH of the extracellular fluids between 7.37 and 7.43. |
| (536) | 7. | List two kidney disorders, and explain what can be done if kidneys become too diseased to work properly. |

---

# Integrating and Applying Key Concepts

The hemodialysis machine used in hospitals is expensive and time-consuming. So far, artificial kidneys capable of allowing people who have nonfunctional kidneys to purify their blood by themselves, without having to go to a hospital or clinic, have not been developed. Which aspects of the hemodialysis procedure do you think have presented the most problems in development of a method of home self-care? If you had an unlimited budget and were appointed head of a team to develop such a procedure and its instrumentation, what strategy would you pursue?

# 32

# NEURAL CONTROL
# AND THE SENSES

---

## Interactive Exercises

### 32 - I. INVERTEBRATE BEGINNINGS (pp. 542 - 545)
###     NEURONS—THE COMMUNICATION SPECIALISTS (pp. 546 - 547)
###     ACTION POTENTIALS (pp. 548 - 549)

*Selected Words:* crack, *input* zones, *conducting* zone, *trigger* zone, *output* zone, *unipolar, bipolar, multipolar, graded* signal, *local* signal, *threshold* level, *all-or-nothing* event

## Boldfaced, Page-Referenced Terms

(543) nervous system _____

_____

(543) sensory neurons _____

_____

(543) interneurons _____

_____

(543) motor neurons _____

_____

(543) neuroglia _____

_____

(544) nerve _____

_____

(544) nerve net _____

_____

(544) reflex pathways _____

_____

(544) ganglia (sing., ganglion) _____

_____

(544) brain _____

_____

(546) dendrite _____

(546) axon _____

(546) resting membrane potential _____

_____

(546) action potential _____

_____

(547) sodium-potassium pumps _____

_____

## Fill-in-the-Blanks

Nerve cells that conduct messages are called (1) _____. (2) _____ cells, which support and

nurture the activities of neurons, make up less than half the volume of the nervous system.

(3) _____ neurons respond to specific kinds of environmental stimuli, (4) _____ connect

different neurons in the spinal cord and brain, and (5) _____ neurons are linked with muscles or

glands.  All neurons have a (6) _____ _____ that contains the nucleus and the metabolic means to carry out protein synthesis. (7) _____ are short, slender extensions of (6), and together these two neuronal parts are the neurons' "input zone" for receiving (8) _____. The (9) _____ is a single long cylindrical extension away from the (10) _____ _____; in motor neurons, the (11) _____ has finely branched (12) _____ that terminate on muscle or gland cells and are "output zones," where messages are sent on to other cells.

A neuron at rest establishes unequal electric charges across its plasma membrane, and a (13) _____ _____ is maintained. Another name for (13) is the (14) _____ _____ _____; it represents an ability for the membrane to be disturbed. Weak disturbances of the neuronal membrane might set off only slight changes across a small patch, but strong disturbances can cause a(n) (15)_____ _____, which is an abrupt, short-lived reversal in the polarity of charge across the plasma membrane of the neuron. For a fraction of a second, the cytoplasmic side of a bit of membrane becomes positive with respect to the outside. The (16) _____ that travels along the neural membrane is nothing more than short-lived changes in the membrane potential. When action potentials reach the end of a motor neuron, they cause (17) _____ to be released that serve as chemical signals to adjacent muscle cells. Muscles (18) _____ in response to the signals.

How is the resting membrane potential established, and what restores it between action potentials? The concentrations of (19) _____ ions ($K^+$), sodium ions (20) (___$^+$), and other charged substances are not the same on the inside and outside of the neuronal membrane. (21) _____ proteins that span the membrane affect the diffusion of specific types of ions across it. (22) _____ proteins that span the membrane pump sodium and potassium ions against their concentration gradients across it by using energy stored in ATP. A neuronal membrane has many more positively charged (23) _____ ions inside than out and many more positively charged (24) _____ ions outside than inside. There are about (25) _____ times more potassium ions on the cytoplasmic side as outside, and there are about (26) _____ times more sodium ions outside as inside. Some channel proteins leak ions through them all the time; others have (27) _____ that open only when stimulated. Transport proteins called (28) _____ - _____ _____ counter the leakage of ions across the neuronal membrane and maintain the resting membrane potential. In all neurons, stimulation at an input zone produces (29) _____ signals that do not spread very far (half a millimeter or less). (30) _____ means that signals can vary in magnitude—small or large—depending on the intensity and (31) _____ of the stimulus. When stimulation is intense or prolonged, graded signals can spread into an adjacent (32) _____ _____ of the membrane—the site where action potentials can be initiated.

A(n) (33) _____ _____ is an all-or-nothing, brief reversal in membrane potential; it is also known as a(n) (34) _____ _____. Once an action potential has been achieved, it is an (35) _____ - _____ - _____ event; its amplitude will not change even if the strength of the stimulus changes. The minimum change in membrane potential needed to achieve an action potential is the (36) _____ value.

## Labeling

Identify the parts of the neuron illustrated below.

37. _____ _____

38. _____ _____

39. _____

40. _____-_____ _____

41. _____ _____

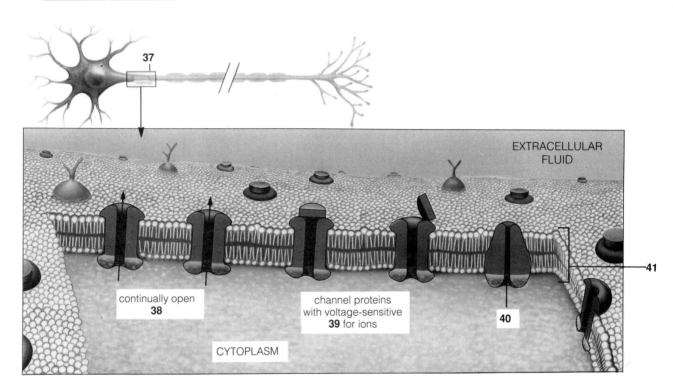

Identify the numbered parts of the illustration.

42. _____ _____

43. _____

44. _____ _____ _____

45. _____

46. _____

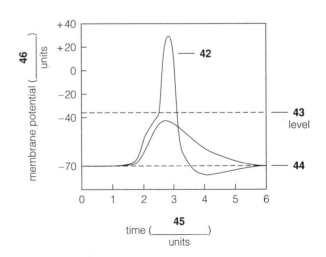

## 32 - II. CHEMICAL SYNAPSES (pp. 550 - 551)
### PATHS OF INFORMATION FLOW (pp. 552 - 553)
#### Focus on Health: CASES OF SKEWED INFORMATION FLOW (p. 554)

**Selected Words:** *pre*synaptic cell, *excitatory, inhibitory,* neuromuscular junction, *post*synaptic *cell,* neuromodulators, *depolarizing, hyperpolarizing,* divergent circuits, convergent circuits, reverberating circuits, *multiple sclerosis,* Schwann cells, *botulism, tetanus*

### Boldfaced, Page-Referenced Terms

(550) neurotransmitters _____

_____

(550) chemical synapses _____

_____

(551) synaptic integration _____

_____

(552) myelin sheath _____

_____

(553) stimulus _____

(553) response _____

_____

### Fill-in-the-Blanks

The junction specialized for transmission between a neuron and another cell is called a (1) _____

_____. Usually, the signal being sent to the receiving cell is carried by chemical messengers called

(2) _____ . (3) _____ is an example of this type of chemical messenger that diffuses across

the synaptic cleft, combines with protein receptor molecules on the muscle cell membrane, and soon

thereafter is rapidly broken down by enzymes. At an (4) _____ synapse, the membrane potential

is driven toward the threshold value and increases the likelihood that an action potential will occur. At

an (5) _____ synapse, the membrane potential is driven away from the threshold value, and the

receiving neuron is less likely to achieve an action potential. A specific transmitter substance can have

either excitatory or inhibitory effects depending on which type of protein channel it opens up in the

(6) _____ membrane.

(7) _____ are neuromodulators that inhibit perceptions of pain and may have roles in memory

and learning, emotional states, temperature regulation, and sexual behavior. (8) _____ _____

at the cellular level is the moment-by-moment tallying of all excitatory and inhibitory signals acting on a

neuron. Incoming information is (9) _____ by cell bodies, and the charge differences across the

membranes are either enhanced or inhibited.

Some narrow-diameter neurons are wrapped in lipid-rich (10) _____ produced by specialized neuroglial cells called Schwann cells; each of these is separated from the next by a(n) (11) _____ _____— a small gap where the axon is exposed to extracellular fluid.

A (12) _____ is an involuntary sequence of events elicited by a stimulus. During a (13) _____ _____, a muscle contracts involuntarily whenever conditions cause a stretch in length; many of these help you maintain an upright posture despite small shifts in balance.

Imbalances can occur at chemical synapses; a neurotoxin produced by *Clostridium tetani* interferes with the effect of (14)_____ on motor neurons, which may cause tetanus—a prolonged, spastic paralysis that can lead to death.

## Labeling

Label the parts of the nerve illustrated at the right and below.

15. _____

16. _____ _____

17. _____ _____

18. _____

19. _____ _____

20. _____ _____

outer wrapping of the nerve

17

a nerve fascicle (many **18** bundled in connective tissue)

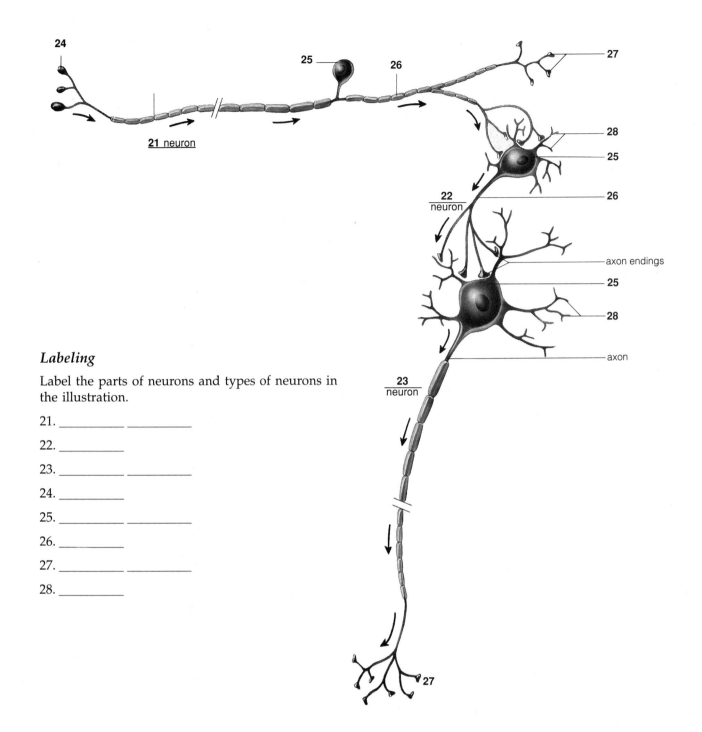

**21** neuron

**22** neuron

**23** neuron

axon endings

axon

## Labeling

Label the parts of neurons and types of neurons in the illustration.

21. _____ _____

22. _____

23. _____ _____

24. _____

25. _____ _____

26. _____

27. _____ _____

28. _____

## Matching

Match the choices below with the correct number in the diagram. (Two of the numbers match with two lettered choices.)

29. _____
30. _____
31. _____
32. _____
33. _____
34. _____
35. _____
36. _____
37. _____

A. Response
B. Action potentials generated in motor neuron and propagated along its axon toward muscle
C. Motor neuron synapses with muscle cells
D. Muscle cells contract
E. Local signals in receptor endings of sensory neuron
F. Muscle spindle stretches
G. Action potentials generated in all muscle cells innervated by motor neuron
H. Stimulus
I. Axon endings synapse with motor neuron
J. Spinal cord
K. Action potential propagated along sensory neuron toward spinal cord

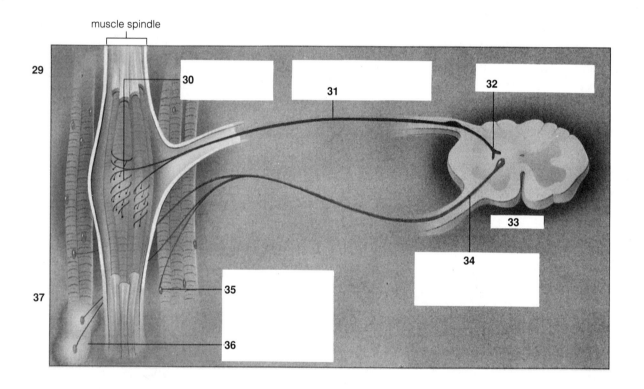

muscle spindle

## 32 - III. FUNCTIONAL DIVISIONS OF VERTEBRATE NERVOUS SYSTEMS (pp. 554 - 555)
### THE MAJOR EXPRESSWAYS (pp. 556 - 557)

*Selected Words:* "brain stem," *afferent* nerves, *efferent* nerves, *spinal* nerves, *cranial* nerves, "housekeeping" tasks, *fight-flight response,* "rebound effect," *meningitis,* "white matter," "gray matter"

### *Boldfaced, Page-Referenced Terms*

(554) medulla oblongata_____

(554) cerebellum _____

_____

(554) pons _____

_____

(555) cerebrum _____

_____

(555) thalamus _____

_____

(555) hypothalamus _____

_____

(555) central nervous system _____

_____

(555) peripheral nervous system _____

_____

(556) somatic nerves _____

_____

(556) autonomic nerves _____

_____

(556) parasympathetic nerves _____

_____

(556) sympathetic nerves _____

_____

(557) spinal cord _____

_____

## Fill-in-the-Blanks

All motor-nerves-to-skeletal-muscle pathways and all sensory pathways make up the (1) _____ nervous system. The remaining nerve tissue, which generally is not under conscious control, is collectively known as the (2) _____ nervous system. It is subdivided into two parts: (3) _____ nerves, which respond to emergency situations, and (4) _____ nerves, which oversee the restoration of normal body functioning. The (5) _____ _____ _____ consists of the brain and spinal cord. A skull encloses the brain, and the (6) _____ _____ encloses and protects the spinal cord in vertebrates. In humans, thirty-one pairs of spinal nerves connect with the spinal cord and are grouped by anatomical region; twelve pairs of (7) _____ nerves connect parts of the head and neck with brain centers.

## Matching

Choose the most appropriate answer for each term.

8. ___ cervical        A. Chest
                       B. Neck
9. ___ coccygeal       C. Pelvic
10. ___ lumbar         D. Tail
                       E. Waist
11. ___ sacral

12. ___ thoracic

## Labeling

Identify the divisions of the nervous system in the posterior view at the right.

13. _____ _____

14. _____ _____

15. _____ _____

16. _____ _____

17. _____ _____

18. _____ _____

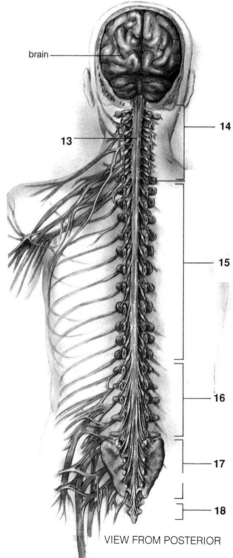

brain

13

14

15

16

17

18

VIEW FROM POSTERIOR

## Labeling

Label each numbered part of the accompanying illustration.

19. _____        22. _____

20. _____        23. _____

21. _____

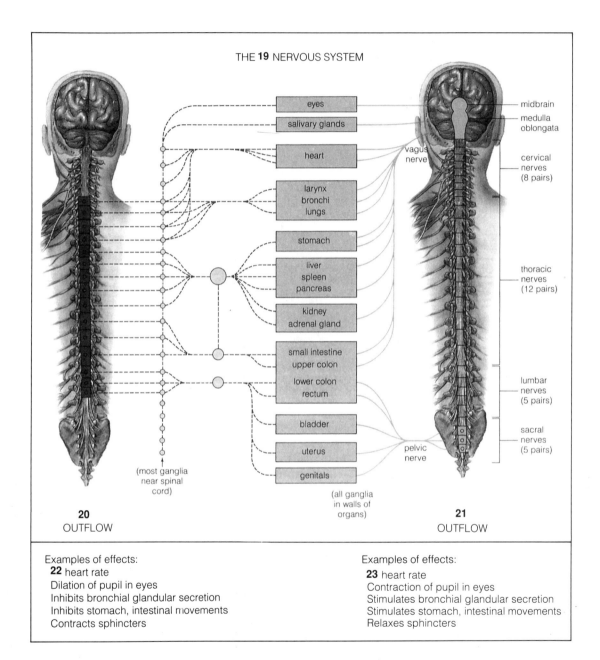

THE **19** NERVOUS SYSTEM

eyes
salivary glands
heart
larynx
bronchi
lungs
stomach
liver
spleen
pancreas
kidney
adrenal gland
small intestine
upper colon
lower colon
rectum
bladder
uterus
genitals

midbrain
medulla oblongata
vagus nerve
cervical nerves (8 pairs)
thoracic nerves (12 pairs)
lumbar nerves (5 pairs)
sacral nerves (5 pairs)
pelvic nerve

(most ganglia near spinal cord)

(all ganglia in walls of organs)

**20**
OUTFLOW

**21**
OUTFLOW

Examples of effects:
**22** heart rate
Dilation of pupil in eyes
Inhibits bronchial glandular secretion
Inhibits stomach, intestinal movements
Contracts sphincters

Examples of effects:
**23** heart rate
Contraction of pupil in eyes
Stimulates bronchial glandular secretion
Stimulates stomach, intestinal movements
Relaxes sphincters

## Fill-in-the-Blanks

The (24) _____ _____ is a region of local integration and reflex connections with nerve pathways leading to and from the brain. Its (25) _____ _____, which contains myelinated sensory and motor axons, is the expressway that connects the brain with the peripheral nervous system. The (26) _____ _____ includes neuronal cell bodies, dendrites, nonmyelinated axon terminals, and neuroglial cells; this is the integrative zone that deals mainly with (27) _____ for limb movements (such as walking) and organ activity (such as bladder emptying).

## Labeling

Identify the numbered parts of the accompanying illustrations.

28. _____ _____
29. _____
30. _____
31. _____ _____
32. _____
33. _____ _____
34. _____ _____

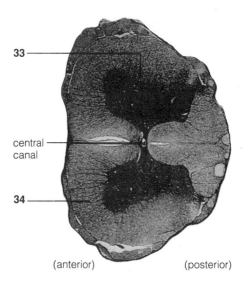

central canal

(anterior)          (posterior)

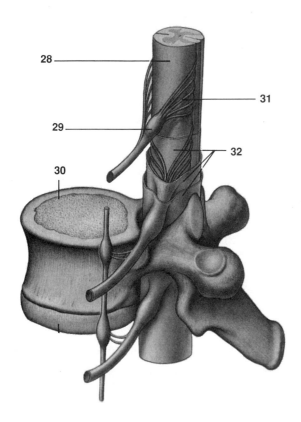

# 32 - IV. THE HUMAN BRAIN (pp. 558 - 559)
## MEMORY (p. 560)
### *Focus on Health:* DRUGS, THE BRAIN, AND BEHAVIOR (pp. 560 - 561)

*Selected Words:* "<u>cerebral</u> <u>hemispheres</u>," forebrain, midbrain, hindbrain, occipital lobe, temporal lobe, parietal lobe, frontal lobe, "gut reactions," short-term storage, long-term storage, *retrograde amnesia, Alzheimer's disease, drug addiction,* "speed"

## Boldfaced, Page-Referenced Terms

(558) cerebral cortex _____

_____

(559) limbic system _____

_____

(559) reticular formation _____

_____

(560) memory _____

_____

(560) stimulants _____

_____

(560) depressants _____

_____

(560) hypnotics _____

_____

(560) analgesics _____

_____

(560) psychedelics _____

_____

(560) hallucinogens _____

_____

## Fill-in-the-Blanks

The hindbrain is an extension and enlargement of the upper spinal cord; it consists of the (1) _____ _____, which contains the control centers for the heartbeat rate, blood pressure, and breathing reflexes.  The hindbrain also includes the (2) _____, which helps coordinate motor responses associated with refined limb movements, maintenance of posture, and spatial orientation.  The (3) _____ is a major routing station for nerve tracts passing between the cerebellum and cerebral cortex.

The (4) _____ evolved as a center that coordinates reflex responses to images and sounds.  The midbrain, pons, and medulla oblongata make up the brain stem; within its core, the (5) _____

_____ is a major network of interneurons that extends beyond its entire length to the cerebral cortex and helps govern an organism's level of nervous system functioning.

The (6) _____ contains two cerebral hemispheres, the surfaces of which are composed of gray matter. Underlying the cerebral hemispheres is the (7) _____, a region that monitors internal organs and influences hunger, thirst, and (8) _____ behaviors, and the (9) _____, the coordinating center where some motor pathways converge and relay sensory signals to the cerebrum. Both the spinal cord and the brain are bathed in (10) _____ _____.

The storage of individual bits of information somewhere in the brain is called (11) _____. Experiments suggest that at least two stages are involved in its formation. One is a (12) _____ - _____ _____ period, lasting only a few hours; information then becomes spatially and temporally organized in neural pathways. The other is a (13) _____ - _____ _____; information then is put in a different neural representation and is permanently filed in the brain.

(14) _____ are analgesics produced by the brain that inhibit regions concerned with our emotions and perception of (15) _____. Imbalances in (16) _____ can produce emotional disturbances.

## Matching

Choose the lettered statement that describes each part's function.

17. ___ cerebellum

18. ___ corpus callosum

19. ___ hypothalamus

20. ___ limbic system

21. ___ medulla oblongata

22. ___ motor cortex

23. ___ olfactory lobes

24. ___ prefrontal cortex

25. ___ primary auditory cortex

26. ___ primary visual cortex

27. ___ primary somatic sensory cortex

28. ___ thalamus

A. Monitors visceral activities; influences behaviors related to thirst, hunger, reproductive cycles, and temperature control
B. Inhibits unsuitable behaviors, helps to plan movements
C. Receives inputs from cochleas of inner ears
D. Issues commands to muscles
E. Relays and coordinates sensory signals to the cerebrum
F. Receives inputs from receptors for smell and taste
G. Receives and processes input from body-feeling areas
H. Broad channel of white matter that keeps the two cerebral hemispheres communicating with each other
I. Coordinates nerve signals for maintaining balance, posture, and refined limb movements
J. Receives inputs from retinas of eyeballs for seeing
K. Connects pons and spinal cord; contains reflex centers involved in respiration, stomach secretion, and cardiovascular function
L. Contains brain centers that coordinate activities underlying emotional expression

## Labeling

Identify each numbered part of the accompanying illustration.

29. _____

30. _____  _____

31. _____

32. _____

33. _____

34. _____  _____

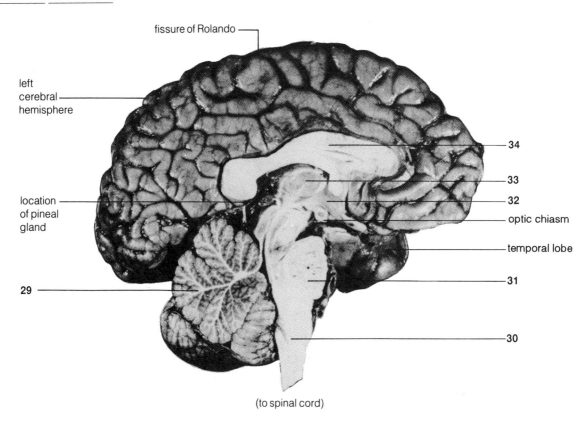

fissure of Rolando ——

left
cerebral —
hemisphere

location —
of pineal
gland

29 ——

— 34

— 33

— 32

— optic chiasm

— temporal lobe

— 31

— 30

(to spinal cord)

## Matching

Choose the most appropriate category for each drug.

35. ___ amphetamines

36. ___ barbituates

37. ___ caffeine

38. ___ cocaine

39. ___ ethyl alcohol

40. ___ heroin

41. ___ LSD

42. ___ marijuana

43. ___ nicotine

A. Depressant or hypnotic drug
B. Narcotic analgesic drug
C. Psychedelic or hallucinogenic drug
D. Stimulant

# 32 - V. SENSORY SYSTEMS (p. 562)
## SOMATIC SENSATIONS (p. 563)
## HEARING AND BALANCE (pp. 564 - 565)

*Selected Words:* *special senses,* amplitude, *frequency, somatic* sensations, *pain* "perception," acoustical organs, acoustical receptors, "ultrasounds," *equilibrium, outer* ear, *middle* ear, *inner* ear, organ of Corti, cochlea, *motion sickness*

## Boldfaced, Page-Referenced Terms

(562) sensory systems _____

_____

(562) stimulus (pl., stimuli) _____

(562) sensation _____

_____

(562) chemoreceptor _____

_____

(562) mechanoreceptor _____

_____

(562) photoreceptor _____

_____

(562) thermoreceptor _____

_____

(562) nociceptor _____

_____

(564) ears _____

_____

(564) echolocation _____

_____

## Fill-in-the-Blanks

Finely branched peripheral endings of sensory neurons that detect specific kinds of stimuli are

(1) _____. A (2) _____ is any form of energy change in the environment that the body

actually detects. (3) _____ detect substances dissolved in water or air; (4) _____ detect

pressure, stretching, and vibrational changes; (5) _____ detect the energy of visible and ultraviolet

light; (6) _____ detect radiant energy associated with temperature changes. A (7) _____ is

conscious awareness of change in internal or external conditions; this is not to be confused with

(8) _____, which is an understanding of what sensation means. A sensory system consists of

sensory receptors for specific stimuli, (9) _____ _____ that conduct information from those

receptors to the brain, and (10) _____ _____ where information is evaluated.

## Matching

Choose the most appropriate term to complete each statement.

11. ___ Vision is associated with _____.

12. ___ Pain is associated with _____.

13. ___ Odors are detected by _____.

14. ___ Hearing is detected by _____.

15. ___ $CO_2$ concentration in the blood is detected by _____.

16. ___ Environmental temperature is detected by _____.

17. ___ Internal body temperature is detected by _____.

18. ___ Touch is detected by _____.

19. ___ Rods and cones are _____.

20. ___ Hair cells in the ear's organ of Corti are _____.

21. ___ Pacinian corpuscles in the skin are _____.

22. ___ Olfactory receptors are _____.

23. ___ Any stimuli that cause tissue damage are _____.

24. ___ The movement of fluid in the inner ear is associated with _____.

A. chemoreceptors
B. mechanoreceptors
C. nociceptors
D. photoreceptors
E. thermoreceptors

## Fill-in-the-Blanks

The somatic sensations (awareness of (25) _____, pressure, heat, (26) _____, and pain) start with receptor endings that are embedded in (27) _____ and other tissues at the body's surfaces, in (28)_____ muscles, and in the walls of internal organs. All skin (29) _____ are easily deformed by pressure on the skin's surface; these make you aware of touch, vibrations, and pressure. (30) _____ nerve endings serve as "heat" receptors, and their firing of action potentials increases with increases in temperature. (31) _____ is the perception of injury to some body region. The brain sometimes gets confused and may associate perceived pain with a tissue some distance from the damaged area; this phenomenon is called (32) _____ _____. Usually the nerve pathways to both the injured and the mistaken area pass through the same segment of spinal cord. (33) _____ in skeletal muscle, joints, tendons, ligaments, and (34) _____ are responsible for awareness of the body's position in space and of limb movements.

## Label and Match

Identify each numbered part in the illustrations to the right. Complete the exercise by writing the appropriate letter in the parentheses that follow the labels.

35. _____ _____ _____ ( )

36. _____ _____ ( )

37. _____ _____ ( )

38. _____

39. _____

40. _____ _____ ( )

A. React continually to ongoing stimuli
B. Contribute to sensations of vibrations
C. Involved in sensing heat, light pressure, and pain
D. Stimulated at the beginning and end of sustained pressure

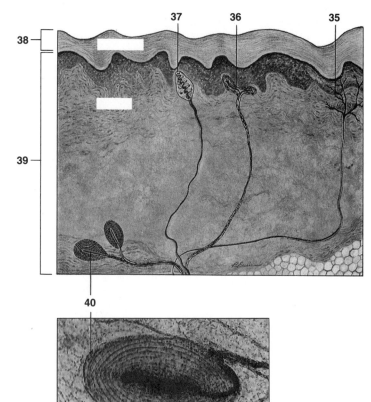

## Fill-in-the-Blanks

The (41) _____ senses a stimulus based on which nerve pathways carry the signals, the (42) _____ of signals from each axon of that pathway, and the (43) _____ of axons carrying the signals to (41). The special senses of (44) _____ and balance involve variously-shaped tubes that contain fluid and hair cells. The (45) _____ (perceived loudness) of sound depends on the height of the sound wave. The (46) _____ (perceived pitch) of sound depends on how fast the wave changes occur. The faster the vibrations, the (47) (choose one) ❏ higher  ❏ lower the sound. Hair cells are (48) (choose one) ❏ nociceptors  ❏ mechanoreceptors  ❏ thermoreceptors that detect vibrations. The hammer, anvil, and stirrup are located in the (49) (choose one) ❏ inner  ❏ middle ear. The (50) _____ is a coiled tube that resembles a snail shell and contains the (51) _____ _____ _____—the organ that changes vibrations into electrochemical impulses. Overstimulated structures that detect rotational acceleration in humans may cause (52) _____ _____.

## Labeling

Identify each numbered part in the accompanying illustrations.

53. _____ _____

54. _____

55. _____ _____

56. _____ _____

57. _____ _____

58. _____ _____

59. _____ _____

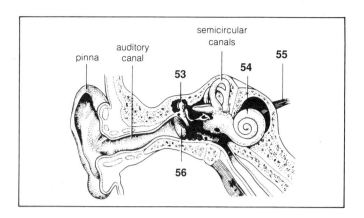

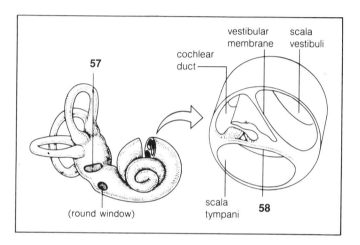

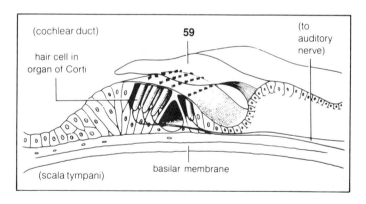

## 32 - VI. VISION (pp. 566 - 567)
### CASE STUDY — FROM SIGNALING TO VISUAL PERCEPTION (pp. 568 - 569)

**Selected Words:** *vision*, eyespot, photoreceptor, camera eyes, *astigmatism*, visual accommodation, *nearsightedness*, *farsightedness*, *histoplasmosis*, <u>Herpes</u>, *cataracts*, *glaucoma*, *red-green color blindness*, receptive field, visual field

### Boldfaced, Page-Referenced Terms

(566) eyes _____

_____

(566) retina _____

_____

(566) compound eyes _____

_____

(568) rod cells _____

_____

(568) cone cells _____

_____

## Fill-in-the-Blanks

(1) _____ is a stream of photons—discrete energy packets.  A (2) _____ is a cell in which photons are absorbed by pigment molecules and photon energy is transformed into the electrochemical energy of a nerve signal.  (3) _____ requires precise light focusing onto a layer of photoreceptive cells that are dense enough to sample details of the light stimulus, followed by image formation in the brain.  (4) _____ are simple clusters of photosensitive cells, usually arranged in a cuplike depression in the epidermis.  (5) _____ are well-developed photoreceptor organs that allow at least some degree of image formation.  The (6) _____ is a transparent cover of the lens area, and the (7) _____ consists of tissue containing densely packed photoreceptors.  In the vertebrate eye, lens adjustments assure that the (8) _____ _____ for a specific group of light rays lands on the retina.  (9) _____ _____ refers to the lens adjustments that bring about precise focusing onto the retina.  (10) _____ people focus light from nearby objects posterior to the retina.

(11) _____ cells are concerned with daytime vision and, usually, color perception.  The (12) _____ is a funnel-shaped pit on the retina that provides the greatest visual acuity.

## Labeling

Identify each numbered part of the accompanying illustration.

13. _____ _____
14. _____
15. _____
16. _____
17. _____ _____
18. _____ _____
19. _____
20. _____
21. _____ _____
22. _____ _____
23. _____

# Self-Quiz

___1. Which of the following is *not* true of an action potential?
   a. It is a short-range message that can vary in size.
   b. It is an all-or-none brief reversal in membrane potential.
   c. It doesn't decay with distance.
   d. It is self-propagating.

___2. The principal place in the human ear where sound waves are amplified is _____.
   a. the pinna
   b. the ear canal
   c. the middle ear
   d. the organ of Corti
   e. none of the above

___3. The place where vibrations are translated into patterns of nerve impulses is _____.
   a. the pinna
   b. the ear canal
   c. the middle ear
   d. the organ of Corti
   e. none of the above

___4. The resting membrane potential _____.
   a. exists as long as a charge difference sufficient to do work exists across a membrane
   b. occurs because there are more potassium ions outside the neuronal membrane than there are inside
   c. occurs because of the unique distribution of receptor proteins located on the dendrite exterior
   d. is brought about by a local change in membrane permeability caused by a greater-than-threshold stimulus

___5. An action potential is brought about by _____.
   a. a sudden membrane impermeability
   b. the movement of negatively charged proteins through the neuronal membrane
   c. the movement of lipoproteins to the outer membrane
   d. a local change in membrane permeability caused by a greater-than-threshold stimulus

___6. The _____ are the protective coverings of the brain.
   a. ventricles
   b. meninges
   c. tectums
   d. olfactory bulbs
   e. pineal glands

___7. The left hemisphere of the brain is responsible for _____.
   a. music
   b. mathematics
   c. language skills
   d. abstract abilities
   e. artistic ability and spatial relationships

___8. To produce a split-brain individual, an operation would need to sever the _____.
   a. pons
   b. fissure of Rolando
   c. hypothalamus
   d. reticular formation
   e. corpus callosum

___9. Accommodation involves the ability to _____.
   a. change the sensitivity of the rods and cones by means of transmitters
   b. change the width of the lens by relaxing or contracting certain muscles
   c. change the curvature of the cornea
   d. adapt to large changes in light intensity
   e. all of the above

___10. Nearsightedness is caused by _____.
   a. eye structure that focuses an image in front of the retina
   b. uneven curvature of the lens
   c. eye structure that focuses an image posterior to the retina
   d. uneven curvature of the cornea
   e. none of the above

*Matching*

Choose the most appropriate statement for each term.

11. ___ amphetamine

12. ___ axon

13. ___ cell body

14. ___ cerebellum

15. ___ cerebrum

16. ___ cornea

17. ___ fovea

18. ___ hypothalamus

19. ___ interneuron

20. ___ iris

21. ___ medulla oblongata

22. ___ parasympathetic

23. ___ retina

24. ___ sclera

25. ___ sympathetic

A. The part of the brain that controls the basic responses necessary to maintain life processes (breathing, heartbeat) is _____.

B. The adjustable ring of contractile and connective tissues that controls the amount of light entering the eye is the _____.

C. Rods and cones are located in the _____.

D. The integrative zone of a neuron is the _____.

E. The white protective fibrous tissue of the eye is the _____.

F. The center for balance and coordination in the human brain is the _____.

G. The conducting zone of a neuron is the _____.

H. The outer transparent protective covering of part of the eyeball is the _____.

I. The highest concentration of cones is in the _____.

J. _____ are responsible for integration in the nervous system.

K. The center of consciousness and intelligence is the _____.

L. _____ nerves generally dominate internal events when environmental conditions permit normal body functioning.

M. _____ is a stimulant.

N. Monitor's internal organs, acts as gatekeeper to the limbic system, helps the reasoning centers of the brain to dampen rage and hatred, and governs hunger, thirst, and sex derives.

O. _____ nerves generally dominate internal events when fight-or-flight situations occur.

# Chapter Objectives/Review Questions

*Page*          *Objectives/Questions*

(543, 546 - 547)   1. Outline some of the ways by which information flow is regulated and integrated in the human body.

(544 - 545)   2. Summarize current best guesses about how the evolution of animal nervous systems may have occurred.

(546)   3. Draw a neuron and label it according to its three general zones, its specific structures, and the specific function(s) of each structure.

(546)   4. Define *resting membrane potential*; explain what establishes it and how it is used by the cell neuron.

(546 - 547)   5. Describe the distribution of the invisible array of large proteins, ions, and other molecules in a neuron, both at rest and as a neuron experiences a change in potential.

(546)   6. Define *action potential* by stating its three main characteristics.

(547)   7. Define *sodium-potassium pump*, and state how it helps maintain the resting membrane potential.

(548 - 549)   8. Explain the chemical basis of the action potential. Look at Figure 32.9 in your text and determine which part of the curve represents the following:

a. the point at which the stimulus was applied;

b. the events prior to achievement of the threshold value;

c. the opening of the ion gates and the diffusing of the ions;

d. the change from net negative charge inside the neuron to net positive charge and back again to net negative charge; and

e. the active transport of sodium ions out of and potassium ions into the neuron.

| (548) | 9. | Explain how graded signals differ from action potentials. |
| (550 - 551) | 10. | Understand how a nerve impulse is received by a neuron, conducted along a neuron, and transmitted across a synapse to a neighboring neuron, muscle, or gland. |
| (553) | 11. | Explain what a reflex is by drawing and labeling a diagram and telling how it functions. |
| (555) | 12. | Contrast the central and peripheral nervous systems. |
| (556 - 557) | 13. | Explain how parasympathetic nerve activity balances sympathetic nerve activity. |
| (557) | 14. | Describe the basic structural and functional organization of the spinal cord. In your answer, distinguish spinal cord from vertebral column. |
| (558) | 15. | For each part of the brain state how the behavior of a normal person would change if he or she suffered a stroke in that part of the brain. |
| (560 - 561) | 16. | List the major classes of psychoactive drugs, and provide an example of each class. |
| (558 - 559) | 17. | Describe how the cerebral hemispheres are related to the other parts of the forebrain. |
| (563 - 564) | 18. | Distinguish the types of stimuli detected by tactile and stretch receptors from those detected by hearing and equilibrium receptors. |
| (564 - 565) | 19. | Follow a sound wave from pinna to organ of Corti; mention the name of each structure it passes and state where the sound wave is amplified and where the pattern of pressure waves is translated into electrochemical impulses. |
| (566) | 20. | Explain what a visual system is and list three of the four aspects of a visual stimulus that are detected by different components of a visual system. |
| (544, 566 - 567) | 21. | Contrast the structure of compound eyes with the structures of invertebrate eyespots and of the human eye. |
| (567) | 22. | Define *nearsightedness* and *farsightedness*, and relate each to eyeball structure. |
| (568 - 569) | 23. | Explain the general principles that affect how light is detected by photoreceptors and changed into electrochemical messages. |

## Integrating and Applying Key Concepts

Suppose that anger is eventually determined to be caused by excessive amounts of specific transmitter substances in the brains of angry people. Can violent murderers now argue that they have been wrongfully punished because they were victimized by their brain's transmitter substances and could not have acted in any other way? Suppose an antidote is prescribed to curb violent tempers in easily angered people. Suppose also that such a person forgets to take the antidote pill and subsequently murders a family member. Can the murderer still claim to be victimized by transmitter substances?

# 33

# ENDOCRINE CONTROL

## Interactive Exercises

### 33 - I. THE ENDOCRINE SYSTEM (pp. 574 - 575)

*Selected Words:* neurotransmitters, local signaling molecules, pheromones, endon, krinein

### Boldfaced, Page-Referenced Terms

(574) hormones _____

_____

(575) endocrine system _____

_____

## Matching

Choose the most appropriate answer for each term.

1. ___ hormones
2. ___ neurotransmitters
3. ___ target cells
4. ___ local signaling molecules
5. ___ pheromones

A. Signaling molecules released from axon endings of neurons that act swiftly on target cells
B. Released by many types of body cells and alter conditions within localized regions of tissues
C. Exocrine gland secretions; signaling molecules that act on cells of other animals of the same species and help integrate social behavior
D. Secretions from endocrine glands, endocrine cells, and some neurons that the bloodstream distributes to nonadjacent target cells
E. Cells that have receptors for any given type of signaling molecule

## Complete the Table

6. Complete the table below by identifying the numbered components of the endocrine system shown in the illustration on the facing page as well as the hormones produced by each gland (see text Figure 33.2).

| Gland Name | Number | Hormones Produced |
|---|---|---|
| a. Hypothalamus | | |
| b. Pituitary, anterior lobe | | |
| c. Pituitary, posterior lobe | | |
| d. Adrenal glands (cortex) | | |
| e. Adrenal glands (medulla) | | |
| f. Ovaries (two) | | |
| g. Testes (two) | | |
| h. Pineal | | |
| i. Thyroid | | |
| j. Parathyroids (four) | | |
| k. Thymus | | |
| l. Pancreatic islets | | |

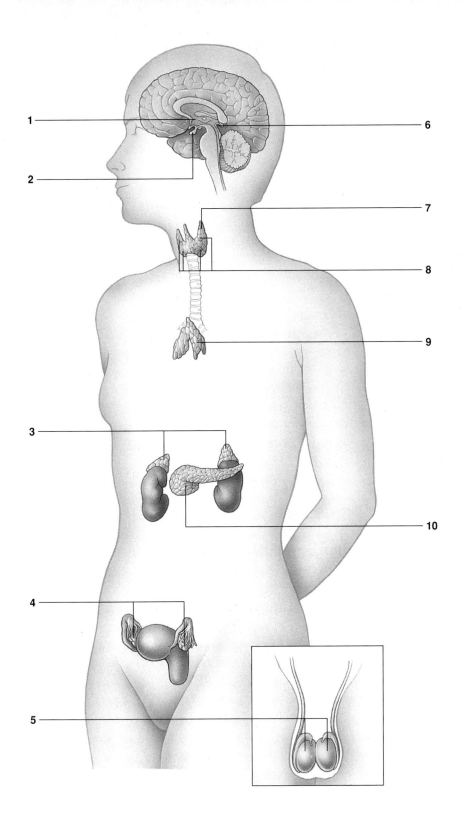

# 33 - II. SIGNALING MECHANISMS (pp. 576 - 577)

*Selected Words:* testicular feminization syndrome

### Boldfaced, Page-Referenced Terms

(576) steroid hormones _____

_____

(577) peptide hormones _____

_____

(577) second messengers _____

_____

### Choice

For questions 1 - 8, choose from the following:

a. steroid hormones     b. peptide hormones

1. ___ Lipid-soluble molecules synthesized from cholesterol; can diffuse directly across the lipid bilayer of a target cell's plasma membrane

2. ___ Include amines, peptides, proteins, and glycoproteins

3. ___ One example involves testosterone, defective receptors, and a condition called testicular feminization syndrome.

4. ___ Hormones that often activate second messengers

5. ___ Protein hormones that bind to receptors at the plasma membrane with the complex moving into the cell by endocytosis to cause further action

6. ___ Lipid-soluble hormones that move through the target cell's plasma membrane to the nucleus where it binds to some type of protein receptor. The hormone-receptor complex moves into the nucleus and interacts with specific DNA regions to stimulate or inhibit transcription of mRNA.

7. ___ A hormone-receptor complex that activates transport proteins or triggers the opening of gated channel proteins spanning the membrane

8. ___ Involves molecules (such as cyclic AMP) that activate many enzymes in cytoplasm, causing alteration in some cell activity

# 33 - III. THE HYPOTHALAMUS AND PITUITARY GLAND (pp. 578 - 579)

*Selected Words:* *posterior* lobe, *anterior* lobe, *second* capillary bed, *releasers*, *inhibitors*

### Boldfaced, Page-Referenced Terms

(578) hypothalamus _____

_____

(578) pituitary gland _____

_____

## Choice and Matching

Label each hormone given below with an A if it is secreted by the anterior lobe of the pituitary, a P if it is released from the posterior pituitary, or an I if it is released from intermediate tissue. Complete the exercise by entering the letter of the hormone's corresponding action in the parentheses following each label.

1. ___ ( ) ACTH

2. ___ ( ) ADH

3. ___ ( ) FSH

4. ___ ( ) STH (GH)

5. ___ ( ) LH

6. ___ ( ) MSH

7. ___ ( ) oxytocin

8. ___ ( ) PRL

9. ___ ( ) TSH

A. Stimulates egg and sperm formation in ovaries and testes
B. Targets are pigmented cells in skin and other surface coverings; induces color changes in response to external stimuli and affects behavior
C. Stimulates and sustains milk production in mammary glands
D. Stimulates ovulation and corpus luteum formation in females; promotes testosterone secretion and sperm release in males
E. Induces uterine contractions and milk movement into secretory ducts of the mammary glands
F. Stimulates release of thyroid hormones from the thyroid gland
G. Acts on the kidneys to conserve water required in control of extracellular fluid volume
H. Stimulates release of adrenal steroid hormones from the adrenal cortex
I. Promotes growth in young; induces protein synthesis and cell division; roles in adult glucose and protein metabolism

## Dichotomous Choice

Circle one of two possible answers given between parentheses in each statement.

10. The (hypothalamus/pituitary gland) region of the brain monitors internal organs and activities related to their functioning, such as eating and sexual behavior; it also secretes some hormones.
11. The (posterior/anterior) lobe of the pituitary stores and secretes two hormones that are produced by the hypothalamus.
12. The (posterior/anterior) lobe of the pituitary produces and secretes its own hormones that govern the release of hormones from other endocrine glands.
13. Humans lack the (posterior/intermediate) lobe of the pituitary gland, one that is possessed by many other vertebrates.
14. Most hypothalamic hormones acting in the anterior pituitary lobe are (releasers/inhibitors) and cause target cells there to secrete hormones of their own.
15. Some hypothalamic hormones slow down secretion from their targets in the anterior pituitary; these are classed as (releasers/inhibitors).

## 33 - IV. EXAMPLES OF ABNORMAL PITUITARY OUTPUT (p. 580)

*Selected Words:* gigantism, pituitary dwarfism, acromegaly, diabetes insipidus

## Complete the Table

1.  Complete the table below to summarize examples of abnormal pituitary output.

| Condition | Hormone/Abnormality | Characteristics |
|---|---|---|
| a. | Excessive somatotropin produced during childhood | Affected adults are proportionally similar to a normal person but larger |
| b. | Insufficient somatotropin produced during childhood | Affected adults are proportionally similar to a normal person but much smaller |
| c. | Diminished ADH secretion by a damaged posterior pituitary lobe | Large volumes of dilute urine are secreted, causing life-threatening dehydration |
| d. | Excessive somatotropin output during adulthood when long bones can no longer lengthen | Abnormal thickening of bone, cartilage, and other connective tissues in the hands, feet, and jaws |

## 33 - V. SOURCES AND EFFECTS OF OTHER HORMONES (p. 581)

### Complete the Table

Complete the table below by matching the gland/organ and the hormone(s) produced by it to the descriptions of hormone action. Refer to Table 33.3 and p. 581 in the text.

| Glands/Organs | Hormones |
|---|---|
| A.  adrenal cortex | a.  thyroxine and triiodothyronine |
| B.  adrenal medulla | b.  glucagon |
| C.  thyroid | c.  PTH |
| D.  parathyroids | d.  androgens (includes testosterone) |
| E.  testes | e.  somatostatin |
| F.  ovaries | f.  thymosins |
| G.  pancreas (alpha cells) | g.  glucocorticoids |
| H.  pancreas (beta cells) | h.  estrogens, general (includes progesterone) |
| I.  pancreas (delta cells) | i.  epinephrine |
| J.  thymus | j.  melatonin |
| K.  pineal | k.  insulin |
| | l.  progesterone |
| | m.  mineralocorticoids (including aldosterone) |
| | n.  calcitonin |
| | o.  norepinephrine |

| Gland/Organ | Hormone(s) | Hormone Action |
|---|---|---|
| 1. | | Elevates calcium levels in blood |
| 2. | | Influences carbohydrate metabolism of insulin-secreting cells |
| 3. | | Required in egg maturation and release; preparation of uterine lining for pregnancy and its maintenance in pregnancy; influences growth and development; genital development; maintains sexual traits |
| 4. | | Promote protein breakdown and conversion to glucose in most cells |
| 5. | | Lowers blood sugar level in muscle and adipose tissue |
| 6. | | In general, required in sperm formation, and genital development and maintenance of sexual traits; influences growth and development |
| 7. | | In most cells, regulates metabolism, plays roles in growth and development |
| 8. | | In the gonads, influences daily biorhythms and influences gonad development and reproductive cycles |
| 9. | | In liver, muscle, and adipose tissues; raises blood level of sugar and fatty acids; increases heart rate and force of contraction |
| 10. | | Targets lymphocytes, has roles in immunity |
| 11. | | Raises blood sugar level |
| 12. | | Prepares, maintains uterine lining for pregnancy; stimulates breast development |
| 13. | | In kidneys, promotes sodium reabsorption and control of salt-water balance |
| 14. | | In smooth muscle cells of blood vessels, promotes constriction or dilation of blood vessels |
| 15. | | In bone, lowers calcium levels in blood |

## 33 - VI. FEEDBACK CONTROL OF HORMONE SECRETIONS (pp. 582 - 583)

*Selected Words:* *hypoglycemia, stress response, fight-flight response,* thyroid-stimulating hormone (TSH), *goiter, hypothyroidism, hyperthyroidism*

### Boldfaced, Page-Referenced Terms

(582) negative feedback _____

_____

(582) positive feedback _____

_____

(582) adrenal cortex _____

_____

(582) adrenal medulla _____

_____

(582) thyroid gland _____

_____

(583) gonads _____

_____

## Matching

Choose the most appropriate answer for each term.

1. ___ adrenal medulla

2. ___ ACTH

3. ___ nervous system

4. ___ glucocorticoids

5. ___ CRH

6. ___ cortisol

7. ___ hypoglycemia

8. ___ cortisol-like drugs

9. ___ adrenal cortex

10. ___ homeostatic feedback loops

A. Occurs when the glucose blood level falls below a set point, and a negative feedback mechanism kicks in

B. Outer portion of each adrenal gland; some of its cells secrete glucocorticoids and other hormones

C. Govern the secretion of the hypothalamus and the pituitary

D. Initiates a stress response during severe stress, painful injury, or prolonged illness

E. Inner portion of the adrenal gland; neurons located here release epinephrine and norepinephrine

F. Stimulates the adrenal cortex to secrete cortisol; this helps raise the level of glucose by preventing muscle cells from taking up more blood glucose

G. Used to counter asthma and other chronic inflammatory disorders

H. Help maintain blood glucose level and help suppress inflammatory responses

I. Secreted by the hypothalamus in response to falling glucose blood level

J. Blocks the uptake and use of glucose by muscle cells; also stimulates liver cells to form glucose from amino acids

## Fill-in-the-Blanks

Thyroxine and triiodothyronine are the main hormones secreted by the human (11) _____ gland. They are critical for normal development of many tissues, and they control overall (12) _____ rates in humans and other warm-blooded animals. The synthesis of thyroid hormones requires (13) _____, which is obtained from the diet. In the absence of that element, blood levels of these hormones decrease. The anterior pituitary responds by secreting (14) _____. When thyroid hormones cannot be synthesized, the feedback signal continues—and so does TSH secretion. TSH secretion continues and overstimulates the thyroid gland and causes (15) _____, an enlargement of the thyroid gland. (16) _____ results from insufficient blood level concentrations of thyroid hormones. Such (17) _____ adults are often overweight, sluggish, dry-skinned, intolerant of cold, and sometimes confused and depressed. (18) _____ results from excess concentrations of thyroid hormones. Affected adults show an increased heart rate, elevated blood pressure, weight loss despite normal caloric intake, heat intolerance, and profuse sweating. They are also nervous, agitated, and have trouble sleeping. The primary reproductive organs are known as (19) _____. These organs produce and secrete (20) _____, essential to reproduction. These organs are known as the (21) _____ in human males and (22) _____ in females. Both types of organs produce (23) _____, or sex cells, and sex (24) _____. The latter affects development of (25) _____ sexual traits.

## 33 - VII. RESPONSES TO LOCAL CHEMICAL CHANGES (pp. 584 - 585)

***Selected Words:*** *rickets, exocrine* cells, *endocrine* cells, *alpha* cells, *glucagon, beta* cells, *insulin, delta* cells, ketones, *diabetes mellitus,* "type 1 diabetes," juvenile-onset diabetes, "type-2 diabetes"

### Boldfaced, Page-Referenced Terms

(584) parathyroid glands _____

_____

(584) local signaling molecules _____

_____

(584) pancreatic islet _____

_____

### Fill-in-the-Blanks

Four (1) _____ glands are positioned next to the posterior (or back) of the human thyroid; they secrete (2) _____ in response to a low (3) _____ level in blood. This hormone stimulates cells of the skeleton, kidneys, and absorptive lining of the small intestine to release (4) _____ and other minerals to interstitial fluid and then blood. PTH prods living bone cells to secrete (5) _____ that digest bone tissue to release calcium and other minerals. PTH induces some kidney cells to secrete enzymes by enhancing (6) _____ reabsorption from the filtrate flowing through kidney nephrons. Under the influence of PTH, some kidney cells secrete enzymes that help activate substances required for the formation of the hormone, vitamin (7) _____. This hormone stimulates (8) _____ cells to increase calcium absorption from the gut lumen. In a child with vitamin D deficiency, too little calcium and phosphorus are absorbed and so rapidly growing bones develop improperly. This ailment is called (9) _____ and is characterized by bowed legs, a malformed pelvis, and in many cases a malformed skull and rib cage.

## Complete the Table

10. Complete the table below to summarize function of the pancreatic islets.

| Pancreatic Islet Cells | Hormone Secreted | Hormone Action |
|---|---|---|
| a. Alpha cells | | |
| b. Beta cells | | |
| c. Delta cells | | |

## Dichotomous Choice

Circle one of two possible answers given between parentheses in each statement.

11. Insulin deficiency can lead to diabetes mellitus, a disorder in which the glucose level (increases/decreases) in the blood, then in the urine.
12. In a person with diabetes mellitus, urination becomes (reduced/excessive), so the body's water-solute balance becomes disrupted; people become abnormally dehydrated and thirsty.
13. Lacking a steady glucose supply, body cells of a person with diabetes mellitus begin breaking down fats and proteins for (energy/water).
14. Weight loss occurs and (amino acids/ketones) accumulate in blood and urine; this promotes excessive water loss with a life-threatening disruption of brain function.
15. After a meal, the blood glucose level increases. The pancreatic beta cells secrete (glucagon/insulin); targets use glucose or store it as glycogen.
16. Between meals, the blood glucose level decreases. The stimulated pancreas alpha cells secrete (glucagon/insulin); targets convert glycogen back to glucose, which then enters the blood.
17. In (type 1 diabetes/type 2 diabetes), the body mounts an immune response against its own insulin-secreting beta cells and destroys them.
18. Juvenile-onset diabetes is also known as (type 1 diabetes/type 2 diabetes).
19. In (type 1 diabetes/type 2 diabetes), insulin levels are close to or above normal, but target cells fail to respond to insulin.
20. (Type 1 diabetes/Type 2 diabetes) usually is manifested during middle age and is less dramatically dangerous than the other type; beta cells produce less insulin as a person ages.

# 33 - VIII. HORMONAL RESPONSES TO ENVIRONMENTAL CUES (pp. 586 - 587)

**Selected Words:** melatonin, "jet lag," *winter blues*, ecdysone

**Boldfaced, Page-Referenced Terms:**

(586) pineal gland _____

_____

(587) molting _____

_____

*Choice*

For questions 1 - 10, choose from the following:

a. melatonin     b. ecdysone

1. ___ The hormone that controls molting

2. ___ Hormone secreted by the pineal gland

3. ___ High blood levels of this hormone in winter (long nights) suppress sexual activity in hamsters

4. ___ Winter blues

5. ___ Chemical interactions that cause an old cuticle to detach from the epidermis and muscles

6. ___ Hormone release controlled by neurons in the brain

7. ___ Suppresses growth of a bird's gonads in fall and winter

8. ___ Jet lag

9. ___ Waking up at sunrise

10. ___ Stored by insects and crustaceans

---

# Self-Quiz

___ 1. The _____ governs the release of hormones from other endocrine glands; it is controlled by the _____.
   a. pituitary, hypothalamus
   b. pancreas, hypothalamus
   c. thyroid, parathyroid glands
   d. hypothalamus, pituitary
   e. pituitary, thalamus

___ 2. Neurons of the _____ produce ADH and oxytocin, which are stored within axon endings of the _____.
   a. anterior pituitary, posterior pituitary
   b. adrenal cortex, adrenal medulla
   c. posterior pituitary, hypothalamus
   d. posterior pituitary, thyroid
   e. hypothalamus, posterior pituitary

___ 3. If you were lost in the desert and had no fresh water to drink, the level of _____ in your blood would increase as a means to conserve water.
   a. insulin
   b. corticotropin
   c. oxytocin
   d. antidiuretic hormone
   e. salt

___ 4. If all sources of calcium were eliminated from your diet, your body would secrete more _____ in an effort to release calcium stored in your body and send it to the tissues that require it.
   a. parathyroid hormone
   b. aldosterone
   c. calcitonin
   d. mineralocorticoids
   e. none of the above

For questions 5 - 7, choose from the following answers:
   a. estrogen
   b. PTH
   c. FSH
   d. somatotropin
   e. prolactin

___ 5. _____ stimulates bone cells to release calcium and phosphate and the kidneys to conserve it.

___ 6. _____ stimulates and sustains milk production in mammary glands.

___ 7. Protein synthesis and cell division are activities stimulated by _____.

For questions 8 - 10, choose from the following answers:

    a. adrenal medulla
    b. adrenal cortex
    c. thyroid
    d. anterior pituitary
    e. posterior pituitary

___ 8. The _____ produces glucocorticoids, which help maintain the blood level of glucose and suppress inflammatory responses.

___ 9. The gland that is most closely associated with emergency situations is the _____.

___ 10. The _____ gland regulates the basic metabolic rate.

*Matching*

Choose the most appropriate answer for each term.

11. ___ ACTH

12. ___ ADH

13. ___ thymosins

14. ___ oxytocin

15. ___ cortisol

16. ___ epinephrine and norepinephrine

17. ___ estrogen

18. ___ glucagon

19. ___ insulin

20. ___ melatonin

21. ___ parathyroid hormone

22. ___ STH (GH)

23. ___ calcitonin

24. ___ testosterone

25. ___ thyroxine

26. ___ progesterone

27. ___ TSH

A. Raises the glucose level in the blood

B. Influences daily biorhythms, gonad development, and reproductive cycles

C. Affects development of male sexual traits

D. Increases heart rate and controls blood volume; the "emergency hormones"

E. Produced by gonad; essential for egg maturation and maintenance of secondary sex characteristics in the female

F. The water conservation hormone; released from posterior pituitary

G. Lowers blood sugar by encouraging cells to take in glucose; responsible for synthesis of proteins and fats

H. Stimulates adrenal cortex to secrete cortisol

I. Elevates calcium levels in blood by stimulating calcium reabsorption from bone and kidneys and calcium absorption from gut

J. Influences overall metabolic rate, growth, and development

K. Roles in immunity

L. Triggers uterine contractions during labor and causes milk release during nursing

M. Prepares and maintains uterine lining for pregnancy; stimulates breast development

N. Inhibits uptake of more blood glucose by muscle cells

O. Lowers calcium levels in blood; bone is the target

P. Secreted by anterior pituitary; stimulates release of thyroid hormones

Q. Secreted by anterior pituitary; enhances growth in young animals, especially of cartilage and bone

# Chapter Objectives/Review Questions

| Page | | Objectives/Questions |
|---|---|---|
| (574) | 1. | Hormones, neurotransmitters, local signaling molecules, and pheromones are all known as _____ molecules that carry out integration. |
| (574) | 2. | _____ cells are any cells that have receptors for a specific signaling molecule and that may alter their behavior in response to it. |
| (575) | 3. | Be able to locate and name the components of the human endocrine systems on a diagram such as text Figure 33.2. |
| (575) | 4. | Collectively, sources of hormones came to be viewed as the _____ system. |

(576 - 577) 5. Contrast the proposed mechanisms of hormonal action on target cell activities by (a) steroid hormones and (b) peptide hormones that are proteins or are derived from proteins.

(578 - 579) 6. Explain how, even though the anterior and posterior lobes of the pituitary are compounded as one gland, the tissues of each part differ in character.

(578) 7. The _____ and the pituitary gland interact as a major neural- endocrine control center.

(578) 8. Identify the hormones released from the posterior lobe of the pituitary, and state their target tissues.

(578) 9. Identify the hormones produced by the anterior lobe of the pituitary, and tell which target tissues or organs each acts on.

(579) 10. Most hypothalamic hormones acting in the anterior lobe are _____, causing target cells to secrete hormones of their own, but others are _____, slowing down secretion from their targets.

(580) 11. Pituitary dwarfism, gigantism, and acromegaly are all associated with abnormal secretion of _____ by the pituitary gland.

(581) 12. Be familiar with the major human hormone sources, their secretions, main targets, and primary actions as shown on text Table 33.3.

(582) 13. With _____ feedback, an increase or decrease in the concentration of a secreted hormone triggers events that inhibit further secretion.

(582) 14. With _____ feedback, an increase in the concentration of a secreted hormone triggers events that stimulate further secretion.

(582) 15. The adrenal _____ secretes glucocorticoids.

(582) 16. Define *hypoglycemia;* relate the role of the hypothalamus and the anterior pituitary in this condition.

(582) 17. The adrenal _____ contains neurons that secrete epinephrine and norepinephrine.

(582) 18. The _____ system initiates the stress response.

(582) 19. Describe the role of cortisol in a stress response.

(582) 20. List the features of the fight-flight response.

(583) 21. Excess _____ overstimulates the thyroid gland; this causes an enlargement known as a form of _____.

(583) 22. Describe the characteristics of hypothyroidism and hyperthyroidism.

(583) 23. The _____ are primary reproductive organs that produce and secrete hormones with essential roles in reproduction.

(584) 24. Name the glands that secrete PTH, and state the function of this hormone.

(584) 25. Describe an ailment called *rickets* and its cause.

(584) 26. Give two examples that illustrate the effects of *local signaling molecules.*

(584) 27. Be able to name the hormones secreted by alpha, beta, and delta pancreatic cells; list the effect of each.

(584 - 585) 28. Describe the symptoms of diabetes mellitus, and distinguish between type 1 and type 2 diabetes.

(586) 29. The pineal gland secretes the hormone _____; relate two examples of the action of this hormone.

(586) 30. Explain the cause of jet lag and winter blues.

(587) 31. The invertebrate hormone _____ is related to the control of a phenomenon known as _____, which occurs among crustaceans and insects.

---

# Integrating and Applying Key Concepts

Suppose you suddenly quadruple your already high daily consumption of calcium. State which body organs would be affected, and tell how they would be affected. Name two hormones whose levels would most probably be affected, and tell whether your body's production of them would increase or decrease. Suppose you continue this high rate of calcium consumption for ten years. Can you predict the organs that would be subject to the most stress as a result?

# 34

# REPRODUCTION AND DEVELOPMENT

## Interactive Exercises

### 34 - I. THE BEGINNING: REPRODUCTIVE MODES (pp. 590 - 592)
### STAGES OF DEVELOPMENT—AN OVERVIEW (p. 593)

*Selected Words:* *reproductive timing*

## Boldfaced, Page-Referenced Terms

(592) sexual reproduction _____

_____

(592) asexual reproduction _____

_____

(592) yolk _____

_____

(593) embryos _____

_____

(593) gamete formation _____

_____

(593) fertilization _____

_____

(593) cleavage _____

_____

(593) gastrulation _____

_____

(593) ectoderm _____

_____

(593) endoderm _____

_____

(593) mesoderm _____

_____

(593) organ formation _____

_____

(593) growth and tissue specialization _____

_____

## Fill-in-the-Blanks

New sponges budding from parent sponges and a flatworm dividing into two flatworms represent examples of (1) _____ reproduction. This type of reproduction is useful when gene-encoded traits are strongly adapted to a limited, more or less consistent set of (2) _____ conditions. Separation into male and female sexes often requires special reproductive structures such as a penis, control mechanisms for gamete maturation and release, energy and mechanisms for production of chemical and structural signals, and engaging in behaviors such as courtship. This biological cost to a particular species is offset by a selective advantage: (3) _____ in traits among the offspring.

(4) _____ _____ is considered the first stage of animal development as eggs or sperm develop in reproductive organs in a parent's body. (5) _____ begins when a sperm penetrates an egg. This process ends when the egg nucleus fuses with the sperm nucleus. This fusion marks the formation of a (6) _____, the first cell of a new animal. (7) _____ includes the repeated mitotic divisions of a zygote to convert it into the multicelled (8) _____. In many species, this process results in a (9) _____, a tiny ball of cells that is not wider in diameter than the zygote was. As cleavage draws to a close, the pace of cell division slows and the embryo enters (10) _____, a stage of major cellular reorganization.

Embryonic cells then become arranged into two or three primary tissues, or (11) "_____ layers," which will give rise to all of the tissues of the adult. (12) _____ is the outermost primary tissue layer and the one that forms first in the embryos of all animals. It will eventually give rise to the outer layer of the integument and the (13) _____ system. (14) _____ is the innermost primary tissue layer. It is the embryonic forerunner of the (15) _____ inner lining and of the organs derived from it. (16) _____ is the intermediate germ layer and the forerunner of muscle; circulatory, reproductive, and excretory organs; most of the skeleton; and the connective tissue layers of the gut and integument. The three primary tissue layers split into subpopulations of cells. This marks the beginning of (17) _____ formation. By now, different sets of cells have become unique in structure and function, and their descendants are giving rise to different kinds of tissues and organs. During the final stage, called growth and tissue (18) _____, organs increase in size and take on specialized properties. This stage continues into (19) _____. Each stage of embryonic development builds on structures that were formed during the (20) _____ preceding it. Development cannot proceed properly unless each stage is successfully completed before the next begins.

### Sequence

Arrange the following events in correct chronological sequence. Write the letter of the first step next to 21, the letter of the second step next to 22, and so on.

21. ___    A. Gastrulation
22. ___    B. Fertilization
23. ___    C. Cleavage
24. ___    D. Growth, tissue specialization
25. ___    E. Organ formation
26. ___    F. Gamete formation

## Complete the Table

27. Complete the table below by entering the correct germ layer (ectoderm, mesoderm, or endoderm) that forms the tissues and organs listed.

| Tissues/Organs | Germ Layer |
|---|---|
| a. Muscle, circulatory organs | |
| b. Nervous system tissues | |
| c. Inner lining of the gut | |
| d. Circulatory organs (blood vessels, heart) | |
| e. Outer layer of the integument | |
| f. Reproductive and excretory organs | |
| g. Organs derived from the gut | |
| h. Most of the skeleton | |
| i. Connective tissues of the gut and integument | |

## 34 - II. A VISUAL TOUR OF FROG AND CHICK DEVELOPMENT (pp. 594 - 595) EARLY MARCHING ORDERS (pp. 596 - 597)

*Selected Words: animal* pole, *vegetal* pole, gastrula

### Boldfaced, Page-Referenced Terms

(596) oocyte _____

_____

(596) sperm _____

_____

(596) gray crescent _____

_____

(597) cytoplasmic localization _____

_____

(597) blastula _____

_____

(597) blastocyst _____

_____

(597) gastrulation _____

_____

## Matching

Choose the most appropriate answer for each.

1. ___ oocyte
2. ___ sperm
3. ___ animal pole
4. ___ vegetal pole
5. ___ gray crescent
6. ___ cleavage
7. ___ cytoplasmic localization
8. ___ blastula
9. ___ blastocyst
10. ___ gastrulation

A. The blastula stage of mammals
B. The region closest to the nucleus of an animal egg
C. A developmental stage of cell rearrangements in an embryo; three germ layers are formed
D. An immature animal egg
E. By virtue of cell location in an embryo, cells forming during cleavage receive different maternal instructions.
F. A ball of cells with a fluid-filled cavity formed by successive cleavages
G. Where yolk and other substances accumulate in an animal egg
H. Consists of paternal DNA and a bit of equipment that helps this cell reach and penetrate an egg
I. Follows fertilization; mitotic cell divisions carve up the zygote with no change in embryo volume
J. Site where frog embryo gastrulation normally begins; an area of intermediate pigmentation holding vital chemical messages

## Fill-in-the-Blanks

The numbered items in the illustrations of developmental stages below represent missing information; complete the numbered blanks in the narrative to supply this information.

Within about an hour following fertilization of a frog egg, the (11) _____ _____ establishes the body axis for the embryo. This is always the site where gastrulation begins. Concerning the polarity of the young embryo, the (12) _____ pole is the one closest to the nucleus while the (13) _____ pole is where yolk and other substances accumulate. (14) _____ of cells leads to a blastula, a ball of cells in which a cavity, the (15) _____, has appeared. Cells move about and become rearranged during (16) _____ formation. This time also marks the appearance of the three primary germ layers—ectoderm, mesoderm, and endoderm. The primitive gut, or (17) _____, also forms during this developmental stage and a (18) _____ plug is also evident. (19) _____ canal developments now take place, and the fluid-filled body cavity in which vital organs will be suspended appears. In chick eggs, cleavage produces a layer of cells, the chick (20) _____, at the surface of the yolk.

## True/False

If the statement is true, write a T in the blank. If the statement is false, make it correct by changing the underlined word(s) and writing the correct word(s) in the blank.

_____ 21. The cleavage furrow defines the plane where the <u>nucleus</u> will pinch in two.

_____ 22. The small amount of yolk in a sea urchin egg develops into the <u>blastula</u>.

_____ 23. The blastula stage of mammals, called a blastocyst, has <u>three</u> distinct regions.

_____ 24. Mammal embryos develop from the <u>hollow</u> <u>sphere</u> of a mammal blastocyst.

_____ 25. In sea urchin embryos, surface cells migrate inward and form a <u>gastrula</u>.

## 34 - III. HOW SPECIALIZED TISSUES AND ORGANS FORM (pp. 598 - 599)

***Selected Words:*** <u>Xenopus</u> <u>laevis</u>, *programmed* cell death, *homeobox* genes, <u>Caenorhabditis</u> <u>elegans</u>, <u>Drosophila</u>, *antennapedia* gene

### Boldfaced, Page-Referenced Terms

(598) cell differentiation _____

_____

(598) morphogenesis _____

_____

(599) pattern formation _____

_____

(599) embryonic induction _____

_____

## Choice

For questions 1 - 15, choose from the following:

a. cell differentiation     b. morphogenesis     c. pattern formation

1. ____ Homeobox genes that govern responses of groups of cells along the embryo's anterior-posterior axis

2. ____ Controlled cell death that eliminates tissues and cells that are used for only short periods in the embryo

3. ____ As an embryo develops, tissues and organs change in size, shape, and proportion, and they become organized in patterns.

4. ____ When a cell selectively uses certain genes and synthesizes proteins that are not found in other cell types

5. ____ Accomplished by cell differentiation and morphogenesis

6. ____ Selective gene expression during development is known as embryonic induction.

7. ___ Mutated homeobox genes can transform one body segment into the likeness of another.

8. ___ Initially a human embryo's hands and feet look like paddles; then certain cells die on cue, leaving separate toes and fingers.

9. ___ A process involving cell divisions, tissue growth, cell migrations, and changes in cell size and shape

10. ___ A frog intestinal cell nucleus from a differentiated cell was transplanted into enucleated, unfertilized frog eggs and demonstrated that the nucleus still contained all the genes that could direct the development of a complete frog.

11. ___ Migrating Schwann cells stick to adhesion proteins that are on the surface of axons but not on blood vessels.

12. ___ In *Drosophila*, a mutated gene product activated the wrong set of homeobox genes in cells that should produce two antennae on the head; they gave rise to a pair of legs on the head instead.

13. ___ Morphogens are signaling molecules that diffuse through tissues and trigger gene activity, sculpting specialized tissues and body parts.

14. ___ "Fate maps" that show where each kind of differentiated cell in the forthcoming body segments of *Drosophila* will originate

15. ___ During human eye development, certain cells initiate the synthesis of long crystallin protein fibers found in the eye's lens; only those cells could activate the required genes to produce crystallin.

## 34 - IV. REPRODUCTIVE SYSTEM OF HUMAN MALES (pp. 600 - 601)
## MALE REPRODUCTIVE FUNCTION (pp. 602 - 603)

***Selected Words:*** *secondary* sexual trait, *prostate cancer, testicular cancer, Sertoli* cells, *Leydig* cells

### *Boldfaced, Page-Referenced Terms*

(600) testes _____

_____

(602) testosterone _____

_____

(602) LH (luteinizing hormone) _____

_____

(602) FSH (follicle-stimulating hormone) _____

_____

## Fill-in-the-Blanks

The numbered items in the illustration that follows represent some of the missing information in the narrative below; complete the numbered blanks (7 - 17) in the narrative to supply this information. One illustrated structure is numbered twice to aid identification.

The adult human male has a pair of gonads known as (1) _____. These gonads produce sperm and sex hormones necessary for the development of (2) _____ sexual traits. Prior to birth, the testes descend into the (3) _____. Gonads are fully formed at birth but reach full size and become functional (4) _____ to (5) _____ years later. For sperm cells to develop properly, the scrotum's interior temperature must be a few degrees (6) _____ than the rest of the body's normal core temperature. Within each testis and following repeated (7) _____ divisions of undifferentiated diploid cells just inside the (8) _____ tubule walls, (9) _____ occurs to form haploid, mature (10) _____. Within the seminiferous tubules, adjacent (11) _____ cells provide nourishment and chemical signals for the developing sperm cells.

Males produce sperm continuously from puberty onward. Sperm leaving a testis enter a long, coiled duct, the (12) _____; the sperm are stored in the last portion of this organ where secretions from glandular cells in the duct wall trigger actions that put the finishing touches on sperm. When a male is sexually aroused, muscle contractions quickly propel the sperm through a thick-walled tube, the

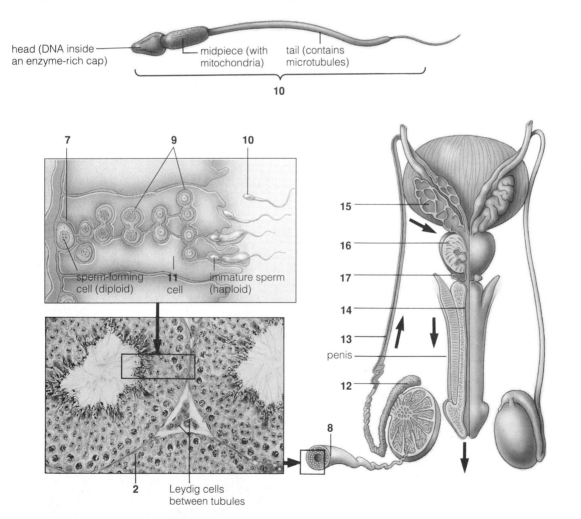

head (DNA inside an enzyme-rich cap) — midpiece (with mitochondria) — tail (contains microtubules)
**10**

7   9   10

sperm-forming cell (diploid)   **11** cell   immature sperm (haploid)

**2**   Leydig cells between tubules

15
16
17
14
13
penis
12
**8**

(13) _____ _____, then to ejaculatory ducts, and finally the (14) _____, which opens at the tip of the penis. During the trip to the urethra, glandular secretions become mixed with the sperm to form semen. A pair of (15) _____ _____ secrete fructose to nourish the sperm and prostaglandins to induce contractions in the female reproductive tract.

(16) _____ gland secretions help neutralize vaginal acids. (17) Two _____ glands secrete mucus to lubricate the penis, aid vaginal penetration, and improve sperm motility.

## Dichotomous Choice

Circle one of two possible answers given between parentheses in each statement.

18. Testosterone is secreted by (Leydig/hypothalamus) cells.

19. (Testosterone/FSH) governs the growth, form, and functions of the male reproductive tract.

20. Sexual behavior, aggressive behavior, and secondary sexual traits are associated with (LH/testosterone).

21. LH and FSH are secreted by the (anterior/posterior) lobe of the pituitary gland.

22. The (testes/hypothalamus) governs sperm production by controlling interactions among testosterone, LH, and FSH.

23. When blood levels of testosterone (increase/decrease), the hypothalamus stimulates the pituitary to release LH and FSH, which travel the bloodstream to the testes.

24. Within the testes, (LH/FSH) acts on Leydig cells; they secrete testosterone, which enters the sperm-forming tubes.

25. FSH enters the sperm-forming tubes and diffuses into (Sertoli/Leydig) cells to improve testosterone uptake.

26. When blood testosterone levels (increase/decrease) past a set point, negative feedback loops to the hypothalamus slow down testosterone secretion.

## Labeling

Identify each numbered part of the accompanying illustrations.

27. _____
28. _____
29. _____
30. _____
31. _____
32. _____
33. _____
34. _____
35. _____
36. _____
37. _____
38. _____
39. _____

head ( **27** inside an enzyme-rich cap)     midpiece (with **28** )     tail (contains **29** )

**a** Blood level of **30** falls. The hypothalamus secretes **39** , a releasing hormone.

(−)

(+)

**31**

(−)

(−)

**f** High sperm count makes **33** cells secrete inhibin, with inhibitory effect on GnRH, LH, and FSH secretions.

**g** High level of **38** inhibits GnRH secretion.

(+)

**b** **32** stimulates LH, FSH secretion from pituitary's anterior lobe.

**c** LH prompts **37** cells in testes to make, release **36**.

**d** **34** cells bind FSH, then make and release protein that enhances testosterone-binding capacity of germ cells.

**e** **35** stimulates sperm formation.

## 34 - V. REPRODUCTIVE SYSTEM OF HUMAN FEMALES (pp. 604 - 605)
### FEMALE REPRODUCTIVE FUNCTION (pp. 606 - 607)
### VISUAL SUMMARY OF THE MENSTRUAL CYCLE (pp. 608 - 609)

*Selected Words:* oocyte, *estrous* cycle, *follicular* phase, *ovulation*, *luteal* phase, follicle, *menopause*, *endometriosis*, *primary* oocyte, *secondary* oocyte

### Boldfaced, Page-Referenced Terms

(604) ovaries _____

_____

(604) endometrium _____

_____

(604) menstrual cycle _____

_____

(604) estrogens _____

_____

(604) progesterone _____

_____

(607) corpus luteum _____

_____

## Fill-in-the-Blanks

The numbered items in the illustration that follows represent missing information; complete the numbered blanks in the narrative to supply this information. Some illustrated structures are numbered more than once to aid identification.

An immature egg (oocyte) is released from one (1) _____ of a pair. From each ovary, an

(2) _____ forms a channel for transport of the immature egg to the (3) _____, a hollow, pear-

shaped organ where the embryo grows and develops. The lower narrowed part of the uterus is the

(4) _____. The uterus has a thick layer of smooth muscle, the (5) _____, lined inside with

connective tissue, glands, and blood vessels; this lining is called the (6) _____. The (7) _____,

a muscular tube, extends from the cervix to the body surface; this tube receives sperm and functions as

part of the birth canal. At the body surface are external genitals (vulva) that include organs for sexual

stimulation. Outermost is a pair of fat-padded skin folds, the (8) _____ _____. Those folds

enclose a smaller pair of skin folds, the (9) _____ _____. The smaller folds partly enclose the

(10) _____, an organ sensitive to stimulation. The location of the (11) _____ is about midway

between the clitoris and the vaginal opening.

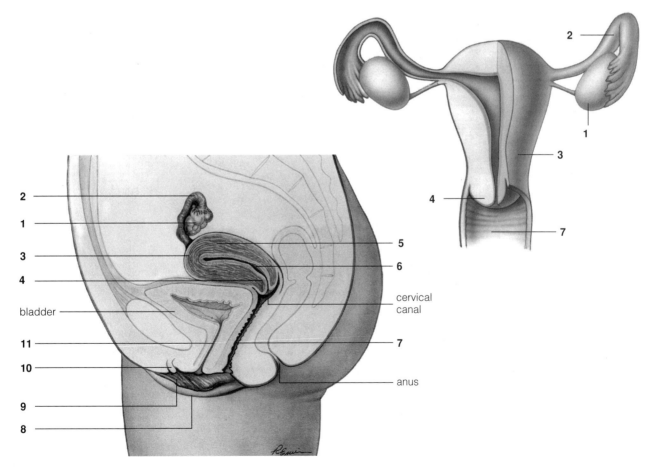

(12) _____ occurs in the ovaries so that a normal female infant has about 2 million primary oocytes with the division process halted in the (13) (choose one) I ❑ II ❑ stage. By age seven, only about (14) _____ remain. A primary oocyte surrounded by a nourishing layer of granulosa cells is called a(n) (15) _____.

When a female enters puberty, the (16) _____ secretes a hormone (GnRH) that makes the (17) _____ _____ secrete follicle-stimulating hormone (FSH) and luteinizing hormone (LH). These hormones are carried by the blood to all parts of the body, but each month, one (or more) follicles respond by growing and adding glycoprotein deposits and the noncellular coating, the (18) _____ _____. Cells outside the zona pellucida secrete the steroid hormones called (19) _____. The primary oocyte completes (20) _____ _____ 8 - 10 hours before (21) _____ (the time when the haploid secondary oocyte is released from the ovary). (22) _____ occurs in response to a surge of (23) _____ being produced by the (24) _____ gland on day 12 or 13 of a 28-day cycle. Estrogens released by cells outside the zona pellucida during the early phase of the menstrual cycle also stimulates growth of the (25) _____ and its glands. Just prior to the midcycle LH surge, cells of the (26) _____ _____ begin progesterone and estrogen secretion.

Following ovulation, granulosa cells of the follicle differentiate into a yellowish glandular structure, the (27) _____ _____, which lasts about twelve days. This structure secretes progesterone and some estrogen. (28) _____ prepares the reproductive tract for the arrival of a blastocyst and maintains the endometrium during a pregnancy. The hypothalamus signals for minimal FSH secretion to stop other follicles from developing. If pregnancy does not occur, the (29) _____ _____ self-destructs during the final days of the menstrual cycle by secreting prostaglandins.

Following this, progesterone and estrogen blood levels decline rapidly and the endometrium begins to break down. The first 3 to 6 days of the cycle are occupied by (30) _____, the deterioration and expulsion of (31) _____ tissues that line the uterus. During the next week, (32) _____ stimulate new (33) _____ tissues to be constructed. By (34) _____, the supply of oocytes is dwindling, hormone secretions slow down, and in time menstrual cycles and fertility are over.

## 34 - VI. PREGNANCY HAPPENS (p. 609)
### FORMATION OF THE EARLY EMBRYO (pp. 610 - 611)
### EMERGENCE OF THE VERTEBRATE BODY PLAN (p. 612)
### ON THE IMPORTANCE OF THE PLACENTA (p. 613)
### EMERGENCE OF DISTINCTLY HUMAN FEATURES (pp. 614 - 615)
### *Focus on Health:* MOTHER AS PROTECTOR, PROVIDER, POTENTIAL THREAT (pp. 616 - 617)

*Selected Words:* blastocyst, *amniotic* cavity, *pregnancy tests*, gastrulation, *teratogens, rubella, thalidomide, fetal alcohol syndrome, second-hand smoke*

## Boldfaced, Page-Referenced Terms

(609) ovum (pl., ova) _____

_____

(610) embryonic period _____

_____

(610) fetal period _____

_____

(610) implantation _____

_____

(611) amnion _____

_____

(611) yolk sac _____

_____

(611) chorion _____

_____

(611) allantois _____

_____

(611) HCG (human chorionic gonadotropin) _____

_____

(613) placenta _____

_____

## Fill-in-the-Blanks

Following sexual intercourse, (1) _____ are in the vagina. Fertilization most often takes place in the upper part of the (2) _____. When living sperm contact an oocyte they release acrosomal (3) _____ that clear a path through the zona pellucida. Usually only one sperm fuses with the secondary oocyte; only centrioles and nucleus from the sperm enter the oocyte's cytoplasm. This event stimulates the secondary oocyte and first polar body to complete meiosis II, which produces three polar bodies and a mature egg, or (4) _____. Sperm and egg nuclei fuse, their chromosomes restore the diploid number within a zygote. Five or six days after conception, (5) _____ begins as the blastocyst sinks into the endometrium. Extensions from the chorion fuse with the endometrium of the uterus to form a (6) _____, the organ of interchange between mother and fetus. By the beginning of the (7) _____ trimester, all major organs have formed; the offspring is now referred to as a(n) (8) _____.

## Matching

Choose the most appropriate answer for each.

9. ___ embryonic period

10. ___ fetal period

11. ___ human blastocyst

12. ___ implantation

13. ___ amnion

14. ___ yolk sac

15. ___ chorion

16. ___ allantois

17. ___ HCG

18. ___ at-home pregnancy tests

A. Secreted by the cells of the blastocyst; stimulates the corpus luteum to maintain secretion of estrogen and progesterone

B. Human extraembryonic membrane formed as the lining of a cavity around the amnion and yolk sac; villi forms on this membrane that becomes part of the placenta

C. Process by which the blastocyst adheres to the uterine lining and some of its cells send out projections that invade the mother's tissues

D. Period of developmental time in which organs grow and become specialized; lasts from the eighth week after fertilization until birth

E. Based on a "dip-stick" that changes color when HCG is present

F. Period of developmental time in which organs form; lasts from the third to the end of the eighth week

G. In humans, an extraembryonic membrane that functions in early formation of blood and the urinary bladder

H. Consists of a surface layer of cells and an interior cluster called the inner cell mass

I. Human extraembryonic membrane that encloses the embryo; fluid-filled to serve as a buoyant, protective cradle for the embryo

J. In humans, an extraembryonic membrane that becomes a site of blood cell formation and gives rise to germ cells

## Labeling

Identify each numbered part of the accompanying illustrations.

19. _____ _____

20. _____ _____

21. _____ _____

22. _____ _____

23. _____ _____

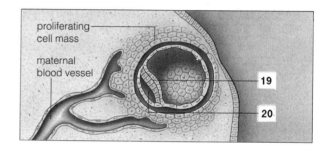

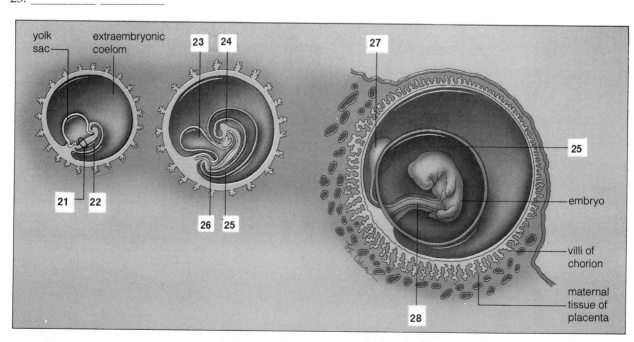

24. _____          29. _____

25. _____          30. _____ _____

26. _____          31. _____

27. _____ _____   32. _____

28. _____ _____   33. _____

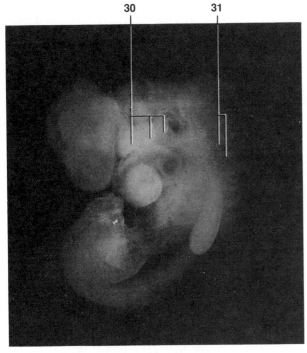

A human embryo at **29** weeks after conception.          A human embryo at **32** weeks after conception.

## Sequence

Arrange the following human developmental stages of early embryo formation in the proper chronological sequence (day 1 to day 14). Find the letter of the first stage and write it next to 34, and so on.

34. ___     A. A blastocyst develops from the morula.

35. ___     B. The four-cell stage is completed by the second cleavage division to produce four cells.

36. ___     C. Cleavage begins about 24 hours after fertilization.

37. ___     D. A connecting stalk has formed between the embryonic disk and chorion; chorionic villi
38. ___        begin to form.

39. ___     E. Blood-filled spaces begin to form in maternal tissue; the chorionic cavity starts to form.

40. ___     F. Cleavage divisions have produced a ball of six to twelve cells.

41. ___     G. The yolk sac, embryonic disk, and amniotic cavity have started to form.

42. ___     H. A morula, a ball of sixteen to thirty-two cells is produced; the morula gives rise to the
               embryo and attached membranes.

        I. The blastocyst begins to burrow into the endometrium.

## Sequence

Arrange the following human developmental stages demonstrating the emergence of the vertebrate body plan in the proper chronological sequence (day 15 to days 24 - 25). Find the letter of the first stage and write it next to 43, and so on.

43. ____
44. ____
45. ____

A. Morphogenesis occurs; the neural tube and somites form; mesoderm somites give rise to the axial skeleton, skeletal muscles, and much of the dermis.

B. Pharyngeal arches form to contribute to formation of the face, neck, mouth, nasal cavities, larynx, and pharynx.

C. The primitive streak forms to mark the onset of gastrulation in vertebrate embryos.

## Sequence

Arrange the following human developmental stages demonstrating the emergence of distinctly human features in the proper chronological sequence (four weeks to full term, or completion of the fetal period). Find the letter of the first stage and write it next to 46, and so on.

46. ____
47. ____
48. ____
49. ____
50. ____

A. Movements begin as nerves make functional connections with developing muscles; legs kick, arms wave, fingers grasp, and the mouth puckers.

B. The length of the fetus increases from 16 centimeters to 50 centimeters, and weight increases from about 7 ounces to 7.5 pounds.

C. The human embryo has a tail and pharyngeal arches, and limbs, fingers, and toes are sculpted from embryonic paddles; the circulatory system becomes more complex, and the head develops.

D. The human embryo is distinctly human as compared to other vertebrate embryos; upper and lower limbs are well-formed; fingers and then toes have separated; primordial tissues of all internal and external structures have developed; the tail has become stubby.

E. Human features are visible at the boundary of the embryonic and fetal periods; the embryo floats in amniotic fluid and the chorion covers the amnion.

## Labeling

Identify each numbered placental component in the illustration below.

51. _____     56. _____

52. _____  _____     57. _____  _____

53. _____     58. _____

54. _____     59. _____  _____

55. _____     60. _____

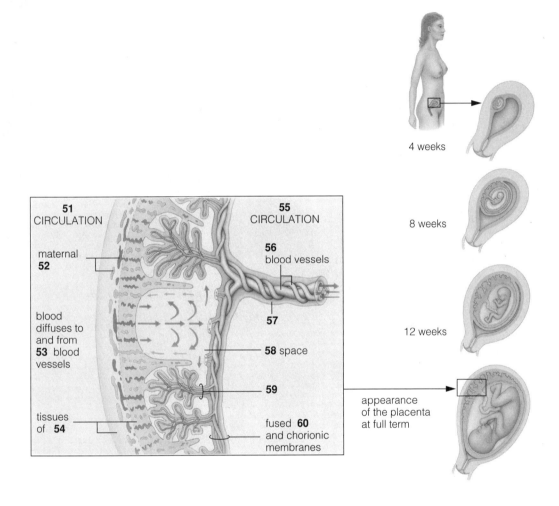

## Choice

For questions 61 - 71, choose from the following threats to human development:

a. nutrition    b. infections    c. prescription drugs    d. alcohol

e. cocaine    f. cigarette smoke    g. all of the preceding threats

61. ___ About 60 - 70 percent of newborns of these women have FAS.

62. ___ In the case of rubella, there is a 50 percent chance some organs won't form properly.

63. ___ In cases of exposure to the second-hand type, children were smaller, ended in more postdelivery deaths, and had twice as many heart abnormalities.

64. ___ Increasing B-complex intake of the mother before conception and during early pregnancy reduces the risk that an embryo will develop severe neural tube defects.

65. ___ The "crack" type disrupts nervous system development of a child.

66. ___ Teratogens

67. ___ Women who used the tranquilizer thalidomide during the first trimester gave birth to infants with missing or severely deformed arms and legs.

68. ___ A pregnant woman must eat enough so that her body weight increases by 20 - 25 pounds, on average.

69. ___ Facial deformities, poor coordination, and sometimes, heart defects

70. ___ As birth approaches, a fetus makes greater demands of this type from the mother that will influence the remaining developmental events.

71. ___ Vaccination before pregnancy can prevent it.

## 34 - VII. FROM BIRTH ONWARD (pp. 618 - 619)
## CONTROL OF HUMAN FERTILITY (pp. 620 - 621)

*Selected Words:* breast cancer, prenatal, postnatal, abstinence, rhythm method, withdrawal, douching, vasectomy, tubal ligation, spermicidal foam, spermicidal jelly, diaphragm, condoms, birth control pill, Depo-provera, Norplant, morning-after pill

### Boldfaced, Page-Referenced Terms

(619) aging _____

_____

### Dichotomous Choice

Circle one of two possible answers given between parentheses in each statement.

1. Pregnancy normally ends about (twenty-eight/thirty-eight) weeks after the time of fertilization.

2. The birth process begins with contractions of the (amnion/uterus).

3. After labor begins, contractions continue for the next (two to five/two to eighteen) hours.

4. The canal that the fetus will pass through when it is fully dilated is within the (vagina/cervix).

5. The (amnion/chorion) ruptures just before birth, and "water" gushes from the vagina.

6. In most cases, the fetus is expelled within (one hour/five hours) after full dilation.

7. Contractions also force fluid, blood, and the placenta from the body, typically within (sixty/fifteen) minutes following birth.

8. Following birth, the (allantois/umbilical cord) is severed from the fetus.

9. During pregnancy, the growth of mammary glands and ducts in the mother's breasts are stimulated by (prolactin/estrogen and progesterone).

10. The anterior pituitary secretes (prolactin/oxytoxin) to stimulate milk production.

11. As a newborn suckles, the pituitary releases (prolactin/oxytoxin) to trigger contractions forcing milk into breast tissue ducts and shrinking the uterus to normal size.

12. Each year in the United States, well over (1,000,000/100,000) women develop breast cancer.

13. Infant, child, and pubescence represent (prenatal/postnatal) stages of human development.

14. The embryo and the fetus represent (prenatal/postnatal) stages of human development.

15. A gradual deterioration of the human body is known as (aging/maturation).

## Matching

Choose the most appropriate answer for each.

16. ___ abstinence

17. ___ rhythm method

18. ___ withdrawal

19. ___ douching

20. ___ vasectomy

21. ___ tubal ligation

22. ___ spermicidal foam and jelly

23. ___ diaphragm

24. ___ condoms

25. ___ birth control pill

26. ___ Depo-Provera and Norplant

27. ___ RU-486

A. A flexible, dome-shaped device inserted into the vagina and positioned over the cervix before intercourse; 84 percent effective
B. Progestin injections or implants that inhibit ovulation; 95 - 96 percent effective
C. A woman's oviducts are cauterized or cut and tied off; extremely effective
D. Rinsing the vagina with a chemical right after intercourse; next to useless
E. Thin, tight-fitting sheaths worn over the penis during intercourse; 85 - 93 percent reliable
F. Avoiding intercourse during a woman's fertile period; 74 percent effective
G. The morning-after pill that interferes with hormonal signals; still controversial due to undesirable side effects
H. Cutting and tying off each vas deferens of a man; extremely effective
I. An oral contraceptive made of synthetic estrogens and progestins that suppress oocyte maturation and ovulation; 94 percent effective
J. No sexual intercourse; foolproof
K. Removing the penis from the vagina before ejaculation; an ineffective method
L. Chemicals toxic to sperm are transferred from an applicator into the vagina just before intercourse; 82 - 83 percent effective

## 34 - VIII. *Focus on Health:* SEXUALLY TRANSMITTED DISEASES (pp. 622 - 623)
### *Focus on Bioethics:* MEDICAL INTERVENTIONS THAT PROMOTE OR END PREGNANCY (p. 624)

*Selected Words:* Herpes, type II *Herpes, behavioral* controls, *genital warts,* Neisseria gonorrhoeae, *gonorrhea, syphilis,* Treponema pallidum, *pelvic inflammatory disease, genital herpes,* Chlamydia trachomatis

### Boldfaced, Page-Referenced Terms

(622) sexually transmitted diseases _____

_____

(624) in vitro fertilization _____

_____

## Matching

Match each of the following with all applicable diseases listed below.

1. _____ Can damage the brain and spinal cord in ways leading to various forms of insanity and paralysis

2. _____ Has no cure

3. _____ Can cause violent cramps, fever, vomiting, and sterility due to scarring and blocking of the oviducts

4. _____ Caused by a motile, corkscrew-shaped bacterium, *Treponema pallidum*

5. _____ Affects about 14 million women a year, causing scarred oviducts, abnormal pregnancies, and sterility

6. _____ Infected women typically have miscarriages, stillbirths, or sickly infants

7. _____ Caused by a bacterium with pili, *Neisseria gonorrhoeae*

8. _____ Caused by direct contact with the viral agent; about 5 million to 20 million people in the United States are infected by it

9. _____ Produces a chancre (localized ulcer) one to eight weeks following infection

10. _____ Chronic infections by this can lead to cervical cancer

11. _____ Caused by an intracellular parasite that infects 3 million to 10 million Americans per year, especially college students

12. _____ Can lead to lesions in the eyes that cause blindness in babies born to mothers with this

13. _____ Acyclovir decreases the healing time and may also decrease the pain and viral shedding from the blisters

14. _____ Following infection, the parasites migrate to lymph nodes, which become enlarged and tender; may lead to pronounced tissue swelling

15. _____ May be cured by antibiotics but can infect again

16. _____ Can be treated with tetracycline and sulfonamides

17. _____ Generally preventable by correct condom usage

A. AIDS
B. Chlamydial infection
C. Genital herpes
D. Gonorrhea
E. Pelvic inflammatory disease
F. Syphilis

# Self-Quiz

___ 1. The process of cleavage most commonly produces a(n) _____.
   a. zygote
   b. blastula
   c. gastrula
   d. third germ layer
   e. organ

___ 2. Cell differentiation refers to _____.
   a. Controlled cell death
   b. activating genes and synthesizing proteins not found in other cells
   c. activating morphogens
   d. activating homeobox genes
   e. none of the above

___ 3. The formation of three germ (embryonic) tissue layers occurs during _____.
   a. gastrulation
   b. cleavage
   c. pattern formation
   d. morphogenesis
   e. neural plate formation

___ 4. The differentiation of a body part in response to signals from an adjacent body part is _____.
   a. contact inhibition
   b. ooplasmic localization
   c. embryonic induction
   d. pattern formation
   e. none of the above

___ 5. A homeobox mutation _____.
   a. may cause a leg to develop on the head where an antenna should grow
   b. affects the expression of imaginal disks
   c. affects morphogenesis
   d. may alter the path of development
   e. all of the above

___ 6. Shortly after fertilization, the zygote is subdivided into a multicelled embryo during a process known as _____.
   a. meiosis
   b. parthenogenesis
   c. embryonic induction
   d. cleavage
   e. invagination

___ 7. Muscles differentiate from _____ tissue.
   a. ectoderm
   b. mesoderm
   c. endoderm
   d. parthenogenetic
   e. yolky

___ 8. The gray crescent is _____.
   a. formed where the sperm penetrates the egg
   b. next to the dorsal lip of the blastopore
   c. the yolky region of the egg
   d. where the first mitotic division begins
   e. formed opposite from where the sperm enters the egg

___ 9. The nervous system differentiates from _____ tissue.
   a. ectoderm
   b. mesoderm
   c. endoderm
   d. yolky
   e. homeobox

For questions 10 - 14, choose from the following answers:
   a. AIDS
   b. Chlamydial infection
   c. Genital herpes
   d. Gonorrhea
   e. Syphilis

___ 10. _____ is a disease caused by a spherical bacterium (*Neisseria*) with pili; it is curable by prompt diagnosis and treatment.

___ 11. _____ is a disease caused by a spiral bacterium (*Treponema*) that produces a localized ulcer (a chancre).

___ 12. _____ is an incurable disease caused by a retrovirus (an RNA-based virus).

___ 13. _____ is a disease caused by an obligate, intracellular parasite that migrates to regional lymph nodes, which swell and become tender.

___ 14. _____ is an extremely contagious viral infection (DNA-based) that causes sores on the facial area and reproductive tract; it is also incurable.

For question 15 - 17, choose from the following answers:
   a. blastocyst
   b. allantois
   c. yolk sac
   d. oviduct
   e. cervix

___ 15. The _____ lies between the uterus and the vagina.

___ 16. The _____ is a pathway from the ovary to the uterus.

___ 17. The _____ results from the process known as cleavage.

For questions 18 - 21, choose from the following answers:

    a. Leydig cells
    b. seminiferous tubules
    c. vas deferens
    d. epididymis
    e. prostate

___ 18. The _____ connects a structure on the surface of the testis with the ejaculatory duct.

___ 19. Testosterone is produced by the _____.

___ 20. Meiosis occurs in the _____.

___ 21. Sperm mature and become motile within the _____.

---

# Chapter Objectives/Review Questions

| Page | | Objectives/Questions |
|---|---|---|
| (592) | 1. | Understand how asexual reproduction differs from sexual reproduction. Be able to list and explain the advantages and problems associated with having separate sexes. |
| (593) | 2. | Be able to list and briefly describe the stages of development, beginning with gamete formation. |
| (593) | 3. | Relate the germ layers, ectoderm, endoderm, and mesoderm to the process of gastrulation; list the adult tissues and organs arising from each. |
| (593) | 4. | Describe early embryonic development and distinguish among the following: oocyte, sperm, fertilization, cleavage, blastula, blastocyst, gastrulation, organ formation, and growth and tissue specialization. |
| (594 - 595) | 5. | Compare the early stages of frog and chick development (see Figures 34.4 and 34.5) with respect to egg size and type of cleavage pattern. |
| (596) | 6. | The fertilized frog egg has an animal pole, a vegetal pole, and a gray _____ of intermediate pigmentation where _____ normally begins. |
| (596) | 7. | State the importance of the gray crescent area as it relates to the first mitotic division of a fertilized frog egg. |
| (597) | 8. | By virtue of their location, the cells that form during cleavage receive different _____ instructions. |
| (597) | 9. | Cytoplasmic _____ influences how each cell will interact with others during the _____ stage and subsequent developmental stages. |
| (597) | 10. | The blastula stage of mammals is called a _____. |
| (597) | 11. | Define *gastrulation,* and state what process begins at this stage that did not happen during cleavage. |
| (598) | 12. | Define *differentiation,* and give two examples of cells in a multicellular organism that have undergone differentiation. |
| (598) | 13. | _____ refers to programmed changes in an embryo's size, shape, and proportion and to the specialization and positioning of tissues in space. |
| (598 - 599) | 14. | _____ cell death eliminates tissues and cells that are used for only short periods in the embryo; _____ formation is the specialization of tissues and their orderly positioning in space. |
| (599) | 15. | Selective gene expression during development is called embryonic _____. |
| (599) | 16. | Describe the role of *homeobox* genes. |
| (599) | 17. | What does a "fate map" show? |
| (600) | 18. | Distinguish between primary and secondary sexual traits and between gonads and accessory reproductive organs. |
| (600 - 601) | 19. | Follow the path of a mature sperm from the seminiferous tubules to the urethral exit. List every structure encountered along the path, and state the contribution to the nurture of the sperm. |

(601)   20.   _____ cancer alone kills 40,000 older men annually in the U.S.; among young men, about 5,000 cases of _____ cancer are diagnosed each year.

(602)   21.   Compare the function and location of *Leydig* cells and *Sertoli* cells.

(602)   22.   Review the stages and chromosome terminology of spermatogenesis (sperm formation).

(602 - 603) 23.   Name the four hormones that directly or indirectly control male reproductive function. Diagram the negative feedback mechanisms that link the hypothalamus, anterior pituitary, and testes in controlling gonadal function.

(604)   24.   _____ are the primary reproductive organs in females; they produce oocytes and sex _____.

(604)   25.   _____ lines the uterus, the chamber in which embryos develop.

(604 - 605) 26.   Distinguish the follicular phase of the menstrual cycle from the luteal phase and explain how the two cycles are synchronized by hormones from the anterior pituitary, hypothalamus, and ovaries.

(604)   27.   A menstrual cycle begins with _____, breakdown and rebuilding of the endometrium, and maturation of an _____.

(604)   28.   Following release of a mature egg, the menstrual cycle ends when a _____ _____ forms and the _____ is primed for pregnancy.

(606 - 608) 29.   State which hormonal event brings about ovulation and which other hormonal events bring about the onset and finish of menstruation.

(606)   30.   List the hormones that stimulate follicle growth during a menstrual cycle.

(607)   31.   Describe signals for endometrial repair and growth.

(609)   32.   List the physiological factors that bring about erection of the penis during sexual stimulation and the factors that bring about ejaculation.

(609)   33.   List the similar events that occur in both male and female orgasm.

(609)   34.   Trace the path of a sperm from the urethral exit to the place where fertilization normally occurs. Mention in correct sequence all major structures of the female reproductive tract that are passed along the way, and state the principal function of each structure.

(610 - 611) 35.   Describe the events that occur during the first month of human development. State how much time cleavage and gastrulation require, when organ development begins, and what is involved in implantation and placenta formation.

(612 - 613   36.   Explain why the mother must be particularly careful of her diet, health habits, and life-style
616 - 617)         during the first trimester after fertilization (especially during the first six weeks).

(614 - 615) 37.   Distinguish an embryo from a fetus.

(618, 601) 38.   Describe how a woman examines herself for breast cancer and how a man examines himself for testicular cancer.

(620)   39.   Identify the factors that encourage and discourage methods of human birth control.

(620)   40.   Describe two different types of sterilization (controlling fertility by surgical intervention).

(621)   41.   Identify the three most effective birth control methods used in the United States and the four least effective birth control methods.

(622 - 623) 42.   For each STD described in the *Commentary*, know the causative organism and the symptoms of the disease.

(624)   43.   State the physiological circumstances that would prompt a couple to try in vitro fertilization.

---

# Integrating and Applying Key Concepts

1.   If embryonic induction did not occur in a human embryo, how would the eye region appear? What would happen to the forebrain and epidermis? If programmed cell death did not happen in a human embryo, how would its hands appear? Its face?

2.   What rewards do you think a society should give a woman who has at most two children during her lifetime? In the absence of rewards or punishments, how can a society encourage women not to have abortions and yet ensure that the human birth rate does not continue to increase?

# 35

# POPULATION ECOLOGY

## Interactive Exercises

### 35 - I. CHARACTERISTICS OF POPULATIONS (pp. 628 - 630)
### POPULATION SIZE AND EXPONENTIAL GROWTH (pp. 630 - 631)

*Selected Words:* pre-reproductive, reproductive, and post-reproductive ages, births, immigration, deaths, emigration, "exponential"

In addition to the boldfaced terms, the text features other important terms essential to understanding the assigned material. "Selected Words" is a list of these terms, which appear in the text in italics, in quotation marks, and occasionally in roman type. Latin binomials found in this section are underlined and in roman type to distinguish them from other italicized words.

### Boldfaced, Page-Referenced Terms

The page-referenced terms are important; they were in boldface type in the chapter. Write a definition for each term in your own words without looking at the text. Next, compare your definition with that given in the chapter or in the text glossary. If your definition seems inaccurate, allow some time to pass and repeat this procedure until you can define each term rather quickly (how fast you can answer is a gauge of your learning effectiveness).

(629) ecology _____

(630) population _____

_____

(630) habitat _____

_____

(630) population size _____

_____

(630) population density _____

_____

(630) population distribution _____

_____

(630) age structure _____

_____

(630) reproductive base _____

_____

(630) zero population growth _____

_____

_____

(630) net reproduction per individual per unit time ($r$) _____

_____

_____

(630) exponential growth _____

_____

(631) doubling time _____

_____

(631) biotic potential _____

_____

## Matching

Choose the most appropriate answer for each term or concept.

1. ___ population
2. ___ habitat
3. ___ population size
4. ___ population density
5. ___ population distribution
6. ___ age structure
7. ___ reproductive base
8. ___ births and immigration
9. ___ deaths and emigration
10. ___ zero population growth
11. ___ net reproduction per individual

   per unit time

12. ___ $G = rN$
13. ___ exponential growth
14. ___ doubling time
15. ___ biotic potential

A. Composed of pre-reproductive and reproductive age categories
B. The general pattern in which individuals are dispersed in a habitat
C. Decreases population size
D. Symbolized by $r$
E. The number of individuals in the gene pool of a species
F. Population growth over increments of time plots a J-shaped curve
G. A time interval in which the number of births and number of deaths are in balance
H. The type of place where a species normally lives
I. The maximum rate of increase per individual under ideal conditions
J. The number of individuals in each of several to many age categories
K. Increases population size
L. A group of individuals of the same species occupying a given area
M. A method of representing population growth
N. The number of individuals in a specified area or volume
O. The time it takes a population to double its size

## Problems

For exercises 16 - 20, consider the equation $G = rN$, where $G$ = the population growth rate, $r$ = the $N$ reproduction per individual per unit of time, and $N$ = the number of individuals in the population.

16. Assume that $r$ remains constant at 0.2.

   a. As the value of $G$ increases, what happens to the value of $N$? _____

   _____

   b. If the value of $G$ decreases, what happens to the value of $N$? _____

   _____

   c. If the net reproduction per individual stays the same and the population grows faster, then what must happen to the number of individuals in the population?_____

17. If a society decides it is necessary to lower its value of $N$ through reproductive means because supportive resources are dwindling, it must lower either its net reproduction per individual per unit or its

   _____

18. The equation $G = rN$ expresses a direct relationship between $G$ and $rxN$. If $G$ remains constant and $N$ increases, what must the value of $r$ do? (In this situation, $r$ varies inversely with $N$.) _____

   _____

19. Look at line (a) in the graph at the right. After seven hours have elapsed, approximately how many individuals are in the population? _____

_____

20. Look at line (b) in the same graph.

   a. After 24 hours have elapsed, approximately how many individuals are in the population? _____

   b. After 28 hours have elapsed, approximately how many individuals are in the population? _____

## 35 - II. LIMITS ON THE GROWTH OF POPULATIONS (pp. 632 - 633)
##       LIFE HISTORY PATTERNS (pp. 634 - 635)
### *Focus on Science:* THE GUPPIES OF TRINIDAD (p. 635)

*Selected Words:* *sustainable, density-dependent* population growth, *bubonic plague, pneumonic plague,* Yersinia pestis, *density-independent* factors, *Type I* curves, *Type II* curves, *Type III* curves

### *Boldfaced, Page-Referenced Terms*

(632) limiting factor _____

_____

(632) carrying capacity _____

_____

(632) logistic growth _____

_____

(634) cohort _____

_____

(634) life table _____

_____

(634) survivorship curves _____

_____

### *Fill-in-the-Blanks*

If (1) _____ factors (essential resources in short supply) act on a population, population growth tapers off. (2) _____ _____ refers to the maximum number of individuals of a population that can be sustained indefinitely by the environment. S-shaped growth curves are characteristic of (3) _____ population growth. The plot of (3) growth levels off once the (4) _____ _____

is reached. In the equation $G = r_{max}N[(K - N)/K]$, as the value of $N$ approaches the value of $K$, and $K$ and $r_{max}$ remain constant, the value of $G$ (5) (choose one) ❐ increases ❐ decreases ❐ cannot be determined by humans, even if they know algebra. As the value of $r_{max}$ increases and $G$ and $N$ remain constant, the value of $K$ (the carrying capacity) (6) (choose one) ❐ increases ❐ decreases ❐ cannot be determined by humans, even if they know algebra. In an overcrowded population, predators, parasites, and disease agents serve as (7) _____-_____ controls. When an event such as a freak summer snowstorm in the Colorado Rockies causes more deaths or fewer births in a butterfly population (with no regard to crowding or dispersion patterns),the controls are said to be (8) _____-_____.
(9) _____ _____ use information summarized from life tables, which show trends in mortality and life expectancy. Type (10) _____ populations have low survivorship early in life. Food availability is a density-(11) _____ factor that works to cut back population size when it approaches the environment's (12) _____ _____. Environmental disruptions such as forest fires and floods are density-(13) _____ factors that may push a population above or below its tolerance range for a given variable.

## Matching

Choose the most appropriate answer for each.

14. ___ cohort
15. ___ type III survivorship curves
16. ___ life tables
17. ___ type I survivorship curves
18. ___ type II survivorship curves

A. Survivorship curves that reflect a fairly constant death rate at all ages; typical of some song birds, lizards, and small mammals
B. Survivorship curves that reflect high survivorship until fairly late in life; produce a few large offspring provided with extended parental care; examples are elephants and humans
C. A group tracked by researchers from birth until the last survivor dies
D. Survivorship curves that reflect a high death rate early in life; typical of sea stars and other invertebrate animals, insects, many fishes, plants, fungi
E. Summaries of age-specific patterns of birth and death

## Matching

Choose the most appropriate answer for each.

19. ___ cohort example
20. ___ example(s) of density-independent controls
21. ___ example(s) of density-dependent controls
22. ___ selective effects of predation by killifish and pike cichlids

A. Drought, floods, earthquakes
B. Bubonic plague and pneumonic plague
C. Food availability
D. Adrenal enlargement in wild rabbits in response to crowding
E. Heavy applications of pesticides in your back-yard
F. All of the 1987 human babies of New York City
G. Natural selection in the life history patterns of Trinidadian guppies
H. Freak snowstorms in the Rocky Mountains

# 35 - III. HUMAN POPULATION GROWTH (pp. 636 - 637)
## CONTROL THROUGH FAMILY PLANNING (pp. 638 - 639)
## POPULATION GROWTH AND ECONOMIC DEVELOPMENT (pp. 640 - 641)
## SOCIAL IMPACT OF NO GROWTH (p. 641)

*Selected Words:* <u>Vibrio</u> <u>cholerae</u>, "replacement rate," *preindustrial* stage, *transitional* stage, *industrial* stage, *postindustrial* stage

## Boldfaced, Page-Referenced Terms

(638)  family planning programs _____

_____

(638)  total fertility rate _____

_____

(640)  demographic transition model _____

_____

_____

_____

## Graph Construction

1.   Plot the following data on the graph provided on page 462.

| Year | Estimated World Population |
|------|----------------------------|
| 1650 | 500,000,000 |
| 1850 | 1,000,000,000 |
| 1930 | 2,000,000,000 |
| 1975 | 4,000,000,000 |
| 1986 | 5,000,000,000 |
| 1993 | 5,500,000,000 |
| 1995 | 5,700,000,000 |

a. Estimate the year that Earth contained 3 billion humans. _____

b. Estimate the year that Earth will house 8 billion humans. _____

c. Do you expect Earth to house 8 billion humans within your lifetime? _____

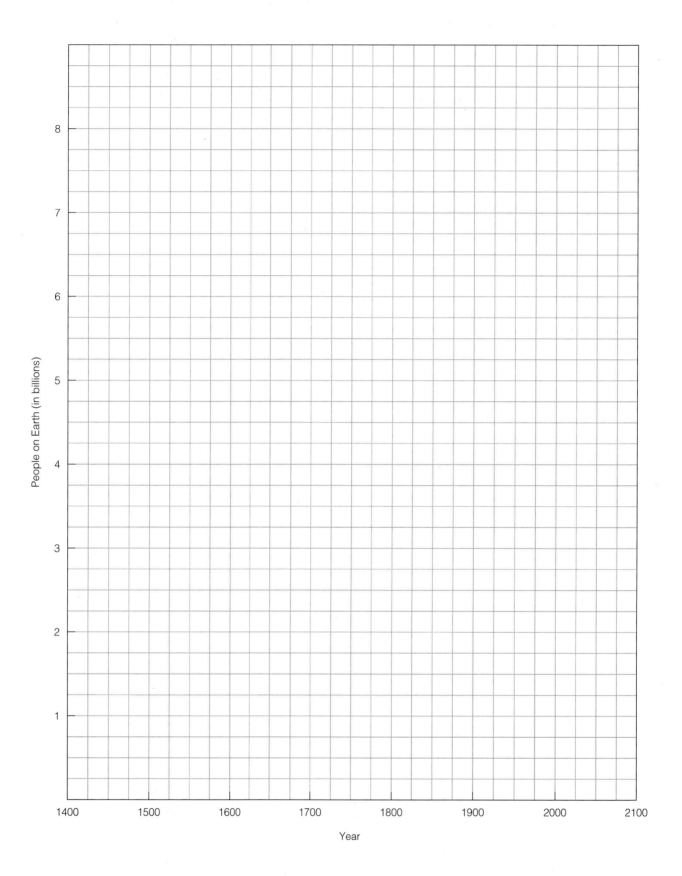

## True/False

If the statement is true, write a T in the blank. If the statement is false, make it correct by changing the underlined word(s) and writing the correct word(s) in the blank.

_____ 2. During 1995, the human population reached <u>8.2 billion</u>.

_____ 3. Even if we could double our present food supply, death from starvation could still reach <u>20 million to 40 million</u> people a year.

_____ 4. Compared with the geographic spread of other organisms, it has taken the human population an extremely <u>long</u> period of time to expand into new environments.

_____ 5. Managing food supplies through agriculture has had the effect of increasing the <u>carrying capacity</u> for human populations.

_____ 6. Humans have sidestepped the <u>biotic potential</u> of their environment by bringing many disease agents under control and by tapping into concentrated, existing stores of energy.

## Sequence

Arrange the following stages of the demographic transition model in correct chronological sequence. Write the letter of the first step next to 7, the letter of the second step next to 8, and so on.

7. ____    A. Industrial stage: population growth slows and industrialization is in full swing

8. ____    B. Preindustrial stage: harsh living conditions, high birth rates and low death rates, slow population growth

9. ____    C. Postindustrial stage: zero population growth is reached; birth rate falls below death rate, and population size slowly decreases

10. ____   D. Transitional stage: industrialization begins, food production rises, and health care improves; death rates drop, birth rates remain high to give rapid population growth

## Fill-in-the-Blanks

If the present rate of human population growth continues, (11) _____ (number) million individuals will be born next year, another (12) _____ (number) the year after that, and so on into the future. The resultant crowding will have adverse effects on resource supplies and will invite the return of severe density-(13) _____ controls. An example is the largest (14) _____ epidemic that is now sweeping through South Asia, which may claim up to 5 million lives. Many governments attempt to control population growth rates by restricting (15)_____, but others attempt to reduce population pressures by encouraging (16) _____. However, most countries focus on reducing (17)_____ rates through economic security and family planning. The key points of the (18)_____ _____ model are that changes in population growth can be correlated with changes that occur during four stages of economic development. The United States, Canada, Australia, Japan, the Soviet Union, and most countries of Western Europe are in the (19)_____ stage of the demographic transition model; Mexico and other less-developed countries are in the (20) _____ stage.

To achieve global zero population growth, the average replacement rate would have to be about (21) _____ (number) children per woman in less-developed countries, and (22) _____ (number) in more-developed countries. Age structure diagrams of actively growing human populations show that more than a third of the world population falls in the broad (23) _____ base category (high numbers of men and women of reproductive age). A simple method of slowing reproduction is to encourage (24) _____ reproduction. Either humans must make an effort to limit population growth to fit environmental (25) _____ _____, or the environment will eventually do it for us. Most governments are trying to lower birth rates with (26) _____ _____ programs. The average number of children born to women during their reproductive years estimated on the basis of current age-specific birth rates is known as the total (27) _____ rate. This is a useful measure of global trends. In 1995, this estimate was an average rate of (28) _____ children per woman in the United States.

## Matching

Choose the most appropriate examples to match with each population age structure (text, p. 639, and Figure 35.9).

29. ___ zero growth

30. ___ rapid growth (more than a third of the world population)

31. ___ negative growth

32. ___ slow growth

A. United States, Australia, and Canada
B. Germany, Bulgaria, and Hungary
C. Kenya, Nigeria, and Saudi Arabia
D. Denmark, Austria, and Italy

---

# Self-Quiz

___ 1. The total number of individuals in the same gene pool that occupy a given area at a given time is _____.
  a. the population density
  b. the population growth
  c. the population birth rate
  d. the population size

___ 2. The average number of individuals in the same species per unit area at a given time is _____.
  a. the population density
  b. the population growth
  c. the population birth rate
  d. the population size

___ 3. A population that is growing exponentially in the absence of limiting factors can be illustrated accurately by a(n) _____.
  a. S-shaped curve
  b. J-shaped curve
  c. curve that terminates in a plateau phase
  d. tolerance curve

___ 4. How are the individuals in a population most often dispersed?
  a clumped
  b. very uniform
  c. nearly uniform
  d. random

___ 5. A situation in which the birth rate plus immigration over the long term equal the death rate plus emigration is called
_____.
  a. an intrinsic limiting factor
  b. exponential growth
  c. saturation
  d. zero population growth

___ 6. The rate of increase for a population ($r$) refers to the _____ the birth rate and death rate plus immigration and minus emigration.
  a. sum of
  b. product of
  c. doubling time between
  d. difference between

___ 7. _____ is a way to express the growth rate of a given population.
   a. Doubling time
   b. Population density
   c. Population size
   d. Carrying capacity

___ 8. Any population that is not restricted in some way will grow exponentially _____.
   a. except in the case of bacteria
   b. irrespective of doubling time
   c. if the death rate is even slightly greater than the birth rate
   d. all of the above

___ 9. The maximum number of individuals of a population (or species) that a given environment can sustain indefinitely defines _____.
   a. the carrying capacity of the environment
   b. exponential growth
   c. the doubling time of a population
   d. density-independent factors

___ 10. In natural communities, some feedback mechanisms operate whenever populations change in size; they are _____.
   a. density-dependent factors
   b. density-independent factors
   c. always intrinsic to the individuals of the community
   d. always extrinsic to the individuals of the community

___ 11. Which of the following is *not* an intrinsic factor that can influence population size?
   a. behavior
   b. metabolism
   c. perdition
   d. fertility

___ 12. Which of the following is *not* characteristic of logistic growth?
   a. S-shaped curve
   b. leveling off of growth as carrying capacity is reached

c. unrestricted growth
d. slow growth of a low-density population followed by rapid growth

___ 13. The population growth rate (G) is equal to the _____ net population growth rate per individual (r) and number of individuals (N).
   a. sum of
   b. product of
   c. doubling of
   d. difference between

___ 14. $G = r_{max} N (K - N/K)$ represents _____.
   a. exponential growth
   b. population density
   c. population size
   d. logistic growth

___ 15. The beginning of industrialization, a rise in food production, improvement of health care, rising birth rates, and declining death rates describes the _____ stage of the demographic transition model.
   a. preindustrial
   b. transitional
   c. industrial
   d. postindustrial

___ 16. A group of individuals that is typically tracked from the time of birth until the last species dies is a _____.
   a. cohort
   b. species
   c. type I curve group
   d. density-independent group

___ 17. The survivorship curve typical of industrialized human populations is type _____.
   a. I
   b. II
   c. III
   d. none of the above types

## Dichotomous Choice

Circle one of two possible answers given between parentheses in each statement.

18. Population (size/density) is the number of individuals per unit of area or volume.
19. Population (size/density) refers to the number of members that make up the gene pool.
20. The general pattern of dispersal of members of a population through the habitat, for example, clumped or random, is its (density/distribution).
21. Dividing a population into pre-reproductive, reproductive, and post-reproductive categories characterizes its (age/distribution) structure.
22. The reproductive base of a population refers to the number of individuals in the (pre-reproductive/reproductive) age structure category.
23. When the number of population members stabilizes due to a balance in births/immigrations and deaths/emigrations, (biotic potential/zero population growth) is demonstrated.
24. Any population whose growth is not restricted in some way will show a pattern of (exponential/logistic) growth.
25. Any increase in population size enlarges the population (distribution/reproductive) base.
26. When the course of exponential growth is plotted on a graph, a (J-shaped/S-shaped) curve is obtained.
27. When limiting factors (essential resources in short supply) act on a population, population growth (increases/decreases).

---

## Chapter Objectives/Review Questions

This section lists general and detailed chapter objectives that can be used as review questions. You can make maximum use of these items by writing answers on a separate sheet of paper. Fill in answers where blanks are provided. To check for accuracy, compare your answers with information given in the chapter or glossary.

| Page | | Objectives/Questions |
|---|---|---|
| (630) | 1. | Be able to define *population, habitat, population size, population density*, and *population distribution*. |
| (630) | 2. | List the three ways that individuals can be distributed in space, and provide an example of each. |
| (630) | 3. | In a population, the pre-reproductive ages and the reproductive ages together are counted as the _____ _____. |
| (630) | 4. | Define *zero population growth*, and describe how achieving it would affect the human population of the United States. |
| (630) | 5. | Calculate a population growth rate ($G$); use values for birth, death, and number of individuals ($N$) that seem appropriate. |
| (630) | 6. | In the equation $G = rN$, as long as $r$ holds constant, any population will show _____ growth. |
| (631) | 7. | State how a population's doubling time is affected by an increasing death rate. |
| (630 - 633) | 8. | Contrast the conditions that promote J-shaped curves with those that promote S-shaped curves in populations. |
| (632) | 9. | Understand the meaning of the logistic growth equation, and know how to calculate values of $G$ by using the logistic growth equation. Understand the meaning of $r_{max}$ and $K$. |
| (632) | 10. | Define *limiting factors*, and tell how they influence population curves. |
| (633) | 11. | Define *density-dependent controls*, give two examples, and indicate how density-dependent factors act on populations. |
| (633) | 12. | Define *density-independent controls*, give two examples, and indicate how such controls affect populations. |
| (634 - 635) | 13. | Understand the significance and use of life tables; describe and be able to interpret the three survivorship curves. |

(634 - 635) 14. Explain how the construction of life tables and survivorship curves can be useful to humans in managing the distribution of scarce resources.

(636) 15. Be able to list three possible reasons that growth of the human population is out of control.

(638) 16. Most governments are trying to lower birth rates by using _____ _____ programs.

(640) 17. Describe the four stages of the demographic transition model.

(630, 639) 18. Define *age structure,* and explain why this is the principal reason it would be 70 to 100 years before the world population would stabilize even if the world average became 2.5 children per family.

(638) 19. Explain how the timing of reproduction can affect the degree of intraspecific competition for available resources.

## Integrating and Applying Key Concepts

Assume that the world has reached zero population growth. The year is 2110, and there are 10.5 billion individuals of *Homo pollutans* on Earth. You have seen stories on the community television screen about how people used to live 120 years ago. List the ways that life has changed, and comment on the events that no longer happen because of the enormous human population.

# 36

# COMMUNITY INTERACTIONS

FACTORS THAT SHAPE COMMUNITY STRUCTURE
    The Niche
    Categories of Species Interactions

MUTUALISM

COMPETITIVE INTERACTIONS
    Categories of Competition
    Competitive Exclusion
    Resource Partitioning

PREDATION AND PARASITISM
    Dynamics of Predator-Prey Interactions
    Parasite-Host Interactions
    Parasites as Biological Control Agents

*Focus on the Environment:* THE COEVOLUTIONARY
    ARMS RACE

FORCES CONTRIBUTING TO COMMUNITY
    STABILITY
    The Successional Model
    The Climax-Pattern Model
    Cyclic, Nondirectional Changes

COMMUNITY INSTABILITY
    How Keystone Species Tip the Balance
    How Species Introductions Tip the Balance

PATTERNS OF BIODIVERSITY
    Mainland and Marine Patterns
    Island Patterns

---

## Interactive Exercises

### 36 - I. FACTORS THAT SHAPE COMMUNITY STRUCTURE (pp. 644 - 646)
###       MUTUALISM (p. 647)

*Selected Words:* *potential* niche, *realized* niche, neutral relationship, *obligatory*

### Boldfaced, Page-Referenced Terms

(646) habitat _____

_____

(646) community _____

_____

(646) niche _____

_____

(646) commensalism _____

_____

(646) mutualism _____

_____

(646) interspecific competition _____

_____

(646) predation _____

_____

(646) parasitism _____

_____

(646) symbiosis _____

_____

## Fill-in-the-Blanks

A (1) _____ is characterized by the kinds and diversity of species, as well as by the numbers and dispersion of their individuals throughout the habitat. The (2) _____ of a population is the sort of place where it is typically located; in contrast, the (3) _____ of a population is defined by its role in a community, including all the ecological requirements and interactions that influence that population in its community. Tree-roosting birds are (4) _____ with trees; the trees get nothing but are not harmed. The flowering plants and their pollinators form (5) _____ relationships from which both participating populations benefit (two-way exploitation). (6) In _____ competition, disadvantages flow both ways between species. An extreme interaction that benefits one species, the predator, is known as (7) _____; another extreme interaction where one species, the parasite, directly benefits is called (8) _____. Commensalism, mutualism, and parasitism are all forms of (9) _____; it means "living together." When the individuals of one species cannot grow and reproduce unless they spend their entire life intimately with individuals of other species, it is known as (10) _____ mutualism. An example is a (11) _____—an intermeshing of two kinds of absorptive structures, fungal hyphae and young roots.

## Matching

Choose the most appropriate answer to match with each term.

12. ___ habitat
13. ___ community
14. ___ commensalism
15. ___ mutualism
16. ___ interspecific competition
17. ___ predation
18. ___ parasitism

A. An interaction that directly benefits one species but does not harm or help the other
B. Has adverse effects on both of the interacting species
C. The parasite benefits, the host is harmed
D. The type of place one finds a particular organism; characterized by physical and chemical features as well as other species
E. The predator benefits, the prey is harmed
F. An interaction from which both species benefit
G. The associations of all populations of species in any given habitat

*Complete the Table*

19. Complete the following table to describe how each of the organisms listed is intimately dependent on the other for survival and reproduction in an obligatory, mutualistic-symbiotic interaction.

| Organism | Dependency |
|---|---|
| a. Yucca moth | |
| b. Yucca plant | |

## 36 - II. COMPETITIVE INTERACTIONS (pp. 648 - 649)
## PREDATION AND PARASITISM (pp. 650 - 651)
### *Focus on the Environment:* THE COEVOLUTIONARY ARMS RACE (pp. 652 - 653)

**Selected Words:** *intraspecific, interspecific,* Paramecium, *carrying capacity, effective* defenses, host species, "speed" mimicry, Lithops, Dendrobates, Ranunculus

### Boldfaced, Page-Referenced Terms

(649) competitive exclusion _____

_____

(649) resource partitioning _____

_____

(650) predator _____

_____

(650) parasite _____

_____

(650) coevolution _____

_____

(652) camouflage _____

_____

(652) mimicry _____

_____

### Fill-in-the-Blanks

In (1) _____ _____, one population or individual exploits the same limited resources as another or intervenes with another sufficiently to keep it from gaining access to the resources. According to the concept of (2) _____ _____, when two species are competing for the same resource, one tends to exclude the other from the area of niche overlap. To a greater or lesser extent, one would have the advantage; the other would be forced to modify its (3) _____. When two or more populations share resources in different ways, in different areas, or at different times, coexistence is also possible through (4) _____ _____. Populations of the Canadian lynx and snowshoe hare undergo irregular fluctuations providing support for a popular model of (5) _____-_____ interactions. When predation prevents a prey population from overshooting its (6) _____ _____, the predator and prey populations tend to coexist at relatively stable levels.

(7) _____ often occurs through reciprocal selection pressures operating on two ecologically interacting populations. Prey populations attempt to avoid predation by adaptations for flight, hiding, fighting, and/or (8) _____. (9) _____ refers to adaptations in form, patterning, color, or behavior that enable an organism to blend with its background and escape detection. Sometimes, weaponless prey species strikingly resemble unpalatable or dangerous species; this type of (10) _____ helps protect prey.

## Matching

Choose the most appropriate answer for each of the following. The same letter may be used more than once. Use only one letter per blank.

11. ___ Polar bears against the snow

12. ___ Cornered earwigs, skunks, and stink beetles producing awful odors

13. ___ Different species of New Guinea pigeons coexisting

14. ___ Tapeworms and humans

15. ___ Yucca moth and yucca plant

16. ___ Canadian lynx and snowshoe hare

17. ___ Resemblance of *Lithops*, a desert plant, to a small rock

18. ___ Striped skunk, yellow-banded wasp, and bright-orange monarch butterfly

19. ___ Smartweed has a taproot system that branches in topsoil and in soil below the roots of other species

20. ___ A mycorrhiza relationship

21. ___ Two species of *Paramecium* requiring identical resources cannot coexist indefinitely

22. ___ Different species of Hairston's salamanders compete, yet they coexist at suppressed population sizes

23. ___ Baboon on the run turns to give canine tooth display to a pursuing leopard

24. ___ An aggressive yellow jacket is the probable model for similar-looking but edible flies

25. ___ Least bittern with coloration similar to surrounding withered reeds

A. parasitism
B. mutualism
C. predator-prey relationship
D. camouflage
E. warning coloration
F. mimicry
G. moment-of-truth defense
H. competitive exclusion
I. resource partitioning

## Dichotomous Choice

Circle one of two possible answers given between parentheses in each statement.

26. Intraspecific competition is (more/less) fierce than interspecific competition.
27. When all individuals have equal access to a required resource and some are better at exploiting it, the interaction tends to (reduce/increase) the common supply of the shared resource unless it is abundant.
28. Two species are (less/more) likely to coexist in the same habitat when they are very similar in their use of scarce resources.
29. Gause used two species of *Paramecium* competing for (the same/different) bacterial cells to illustrate competitive exclusion.

30. Gause also found that two species of *Paramecium* that did not overlap as much in their requirements were (more/less) likely to coexist.

# 36 - III. FORCES CONTRIBUTING TO COMMUNITY STABILITY (pp. 654 - 655)
## COMMUNITY INSTABILITY (pp. 656 - 657)

*Selected Words:* *primary* succession, *secondary* succession, colonizers *facilitate*, Pisaster, Mytilus, Littorina littorea, Enteromorpha, *jump* dispersal, Cryphonectria parasitica

## Boldfaced, Page-Referenced Terms

(654) ecological succession _____

_____

(654) pioneer species _____

_____

(654) climax community _____

_____

(654) climax-pattern model _____

_____

(656) keystone species _____

_____

(656) geographic dispersal _____

_____

## Fill-in-the-Blanks

When a community develops in a predictable sequence from pioneers to a stable end array of species over some region, this is known as (1) _____. (2) _____ species are opportunistic colonizers of vacant habitats that are noted for their high dispersal rates and rapid growth. As time passes, the (3) _____ are replaced by species that are more competitive. These species are in turn replaced in an orderly progression of species change until a stable, self-perpetuating (4) _____ community results. When pioneer species begin to colonize a barren habitat such as a volcanic island, it is known as (5) _____ succession. Once established, the pioneers improve conditions for other species and often set the stage for their own (6) _____. When an area within a community is disturbed and then recovers to move again toward the stable, self-perpetuating climax community, it is known as (7) _____ succession. Some scientists hypothesize that the colonizers (8) _____ their own replacement, while other scientists subscribe to the idea that the sequence of (9) _____ depends on what species arrive first to compete successfully with species that could replace them.

According to the (10) _____-_____ model, a community is adapted to a total pattern of environmental factors—including climate, soil, topography, wind, species interactions, and recurring disturbances such as fires and chance events. Small-scale changes contribute to the internal dynamics of the (11) _____ as a whole. For example, a tree falling in a tropical forest opens a gap in the forest canopy that allows more light to reach that gap. Here the growth of previously suppressed small trees, and germination of (12) _____ or shade-intolerant species is encouraged. Giant sequoia trees in climax communities of the California Sierra Nevada are best maintained by modest (13) _____ that eliminate trees and shrubs that compete with young sequoias but do not damage the older sequoias. Thus, recurring, small-scale changes are built into the overall workings of many communities.

## Choice

For questions 14 - 23, choose from the following:

a. primary succession        b. secondary succession

14. ___ Following a disturbance, a patch of habitat or a community moves once again toward the climax state.

15. ___ Successional changes begin when a pioneer population colonizes a barren habitat.

16. ___ Involves populations that are adapted to growing in habitats that cannot support most other populations

17. ___ Many plants in this succession arise from seeds or seedlings that are already present when the process begins.

18. ___ Early successional populations inhibit the growth of later ones, which become dominant only when some disturbance removes the established competitors.

19. ___ The first plants are typically small with short life cycles; each year they produce an abundance of quickly dispersed small seeds.

20. ___ A successional pattern that occurs in ponds, shallow lakes, abandoned fields, and in parts of established forests

21. ___ On land, early and late species often are able to grow together under prevailing conditions.

22. ___ Might occur on a new volcanic island or on land exposed by the retreat of a glacier

23. ___ Includes populations adapted to growing in areas exposed to intense sunlight, wide temperature swings, and nutrient-deficient soil

## 36 - IV. PATTERNS OF BIODIVERSITY (pp. 658 - 659)

*Selected Words:* resource availability, species diversity, rates of speciation, background extinction

## Boldfaced, Page-Referenced Terms

(659) distance effect _____

_____

(659) area effect _____

_____

## Short Answer

1. List three factors responsible for creating the higher species diversity values as related to the distance of land and sea from the equator.

   a. _____

   _____

   _____

   _____

   b. _____

   _____

   _____

   _____

   c. _____

   _____

   _____

   _____

## Fill-in-the-Blanks

The number of coexisting species is highest in the (2) _____, and it systematically declines toward the poles. In 1965, a volcanic eruption formed a new island, Surtsey, southwest of (3) _____. The new island served as a laboratory for studying (4) _____. Within six months, bacteria, fungi, seeds, flies, and some seabirds became established on the island. After two years one vascular plant appeared and two years after that, a moss plant was observed. These organisms were all colonists from (5) _____. Islands at great distances from potential colonists receive few colonists; the few that arrive are adapted for long distance (6) _____. This is called the (7) _____ effect. Larger islands tend to support more species than small islands at equivalent distances from source areas. This is the (8) _____ effect.

## Problem

9. After consideration of the distance effect, the area effect, and species diversity patterns as related to the equator, answer the following question. There are two islands (B and C) of the same size and topography that are equidistant from the African coast (A), as shown in the illustration at right. Which will have the higher species diversity values?

   _____

   _____

# Self-Quiz

___ 1. All the populations of different species that occupy and are adapted to a given habitat are referred to as a(n) _____.
a. biosphere
b. community
c. ecosystem
d. niche

___ 2. The range of all factors that influence whether a species can obtain resources essential for survival and reproduction is called the _____ of a species.
a. habitat
b. niche
c. carrying capacity
d. ecosystem

___ 3. A one-way relationship in which one species benefits and the other is directly harmed is called _____.
a. commensalism
b. competitive exclusion
c. parasitism
d. mutualism

___ 4. A lopsided interaction that directly benefits one species but does not harm or help the other much, if at all, is _____.
a. commensalism
b. competitive exclusion
c. predation
d. mutualism

___ 5. An interaction in which both species benefit is best described as _____.
a. commensalism
b. mutualism
c. predation
d. parasitism

___ 6. A striped skunk being pursued by a predator suddenly turns and releases its foul-smelling odor. This is an example of

_____.
a. warning coloration
b. mimicry
c. camouflage
d. moment-of-truth defense

___ 7. When an inexperienced predator attacks a yellow-banded wasp, the predator receives the pain of a stinger and will not attack again. This is an example of _____.
a. mimicry
b. camouflage
c. a prey defense
d. warning coloration
e. both c and d

___ 8. _____ is represented by foxtail grass, mallow plants, and smartweed because their root systems exploit different areas of the soil in a field.
a. Succession
b. Resource partitioning
c. A climax community
d. A disturbance

___ 9. During the process of community succession, _____.
a. pioneer populations adapt to growing in habitats that cannot support most species
b. pioneers set the stage for their own replacement
c. later successional populations crowd out the pioneers
d. species composition eventually is stable in the form of the climax community
e. all of the above

___ 10. Gause utilized two species of *Paramecium* in a study that described _____.
a. interspecific competition and competitive exclusion
b. resource partitioning
c. the establishment of territories
d. coevolved mutualism

___ 11. The most striking patterns of species diversity on land and in the seas relate to

_____.
a. distance effect
b. area effect
c. immigration rate for new species
d. distance from the equator

___ 12. In a community, friction between two competing populations might be minimized by _____.
a. partitioning the niches in time
b. partitioning the niches spatially
c. fights to the death
d. both a and b

___ 13. Robins probably have a(n) _____ relationship with humans.
a. parasitic
b. mutualistic
c. obligate
d. commensal

___ 14. The relationship between an insect and the plants it pollinates (for example, apple blossoms, dandelions, and honeysuckle) is best described as _____.
a. mutualism
b. competitive exclusion
c. parasitism
d. commensalism

___ 15. The relationship between the yucca plant and the yucca moth that pollinates it is best described as _____.
a. camouflage
b. commensalism
c. competitive exclusion
d. coevolution

---

# Chapter Objectives/Review Questions

| Page | | Objectives/Questions |
|---|---|---|
| (646) | 1. | The type of place where you would normally find a maple is its _____. |
| (646) | 2. | List five factors that shape the structure of a biological community. |
| (646) | 3. | The full range of environmental and biological conditions under which its members can live, grow, and reproduce is called the _____ of that species. |
| (646) | 4. | The interaction of a bird's nest and a tree is known as _____. |
| (646) | 5. | In forms of _____, each of the participating species reaps benefits from the interaction. |
| (646) | 6. | Define *symbiosis*. |
| (647) | 7. | Define and cite an example of *obligatory mutualism*. |
| (648 - 649) | 8. | Describe a study that demonstrates laboratory evidence in support of the competitive exclusion concept. |
| (649) | 9. | Cite one example of resource partitioning. |
| (650) | 10. | A predator gets food from other living organisms, its _____. |
| (650 - 651) | 11. | List three factors that influence the outcome of predator-prey interactions. |
| (650) | 12. | Suggest why coevolving might serve the interests of the populations concerned. |
| (650) | 13. | _____ tend to interact with their hosts in ways that produce less-than-fatal effects. |
| (652) | 14. | Be able to completely define and give examples of the following prey defenses: warning coloration and mimicry, moment-of-truth defenses, and camouflage. |
| (654) | 15. | A _____ community is a stable, self-perpetuating array of species in equilibrium with one another and their habitat. |
| (654) | 16. | Distinguish between primary and secondary succession. |
| (654) | 17. | By a _____-_____ model, a community is adapted to a total pattern of environmental factors. |
| (655) | 18. | Describe how fire disturbances positively affect a community of giant sequoias. |
| (656) | 19. | Explain how the presence of a keystone species can increase biodiversity in a community; relate an example. |
| (656 - 657) | 20. | Explain how the introduction of nonnative species can be disastrous. List five specific examples of species introductions into the United States that have had adverse results (see Table 36.2). |
| (658) | 21. | List three factors that underlie the existing patterns of biodiversity. |

(658 - 659) 22. Estimate qualitatively the differences in species diversity and abundance of organisms likely to exist on two islands with the following characteristics: Island A has an area of 6,000 square miles, and Island B has an area of 60 square miles; both islands lie at 10° north latitude and are equidistant from the same source area of colonizers.

## Integrating and Applying Key Concepts

1. If you were Ruler of All People on Earth, how would you organize industry and human populations in an effort to solve our most pressing pollution problems?

2. Is there a *fundamental niche* that is occupied by humans? If you think so, describe the minimal abiotic and biotic conditions required by populations of humans in order to live and reproduce. (Note that *"thrive* and *be happy"* are not criteria.) If you do not think so, state why.

   These minimal niche conditions can be viewed as resource categories that must be protected by populations if they are to survive.
   a. Do you believe that the cold war between the United States and the Soviet Union primarily involved protection of minimal niche conditions? If so, how do you think these *minimal* conditions might have been guaranteed for all humans willing and able to accept certain responsibilities as their contribution toward enabling the fulfillment of this guarantee?

   b. Do you believe that the cold war was based on other, more (or less) important factors? If so, identify what you think those factors are, and explain why you consider them more (or less) important than minimal niche conditions.

# 37

# ECOSYSTEMS

## Interactive Exercises

### 37 - I. THE NATURE OF ECOSYSTEMS (pp. 662 - 665)
### ENERGY FLOW THROUGH ECOSYSTEMS (pp. 666 - 667)
### *Focus on Science:* ENERGY FLOW AT SILVER SPRINGS, FLORIDA (p. 668)

*Selected Words:* herbivores, carnivores, omnivores, parasites, energy input, nutrient input

### *Boldfaced, Page-Referenced Terms*

(664) primary producers _____

_____

(664) consumers _____

_____

(664) decomposers _____

_____

(664) detritivores _____

_____

(664) ecosystem _____

_____

(664) trophic levels _____

_____

(664) food chain _____

_____

(665) food webs _____

_____

(666) primary productivity _____

_____

(666) grazing food webs _____

_____

(666) detrital food webs _____

_____

(666) ecological pyramid _____

_____

(667) energy pyramid _____

_____

## Fill-in-the-Blanks

In an ecosystem, autotrophs known as primary (1) _____ are able to secure energy from the
environment for the entire ecosystem of which it is a part. All other organisms in the system are not
self-feeders and, in general, are known as (2) _____. There are several types of these that extract
energy from compounds put together by the primary producers: (3) _____ eat plants;
(4) _____ eat animals; (5) _____ eat both plants and animals; (6) _____ extract energy
from living hosts that they live in or on. Still other heterotrophs, known as (7) _____, include
fungi and bacteria that obtain energy by breaking down the remains or products of organisms. Some
heterotrophs, the (8) _____, obtain nutrients from decomposing particles of organic matter.
Ecosystems are (9) _____ systems, so they cannot be self-sustaining. They require (10) _____
input (the sun) and (11) _____ inputs (minerals); they also have outputs of the same components.
However, (12) _____ cannot be recycled and much is lost to the environment; (13) _____ do
get recycled, but some are still lost.

   Within virtually any ecosystem, the members fit into a hierarchy of energy transfers called
(14) _____ levels. A straight-line sequence of "who eats whom" in an ecosystem (for example, cow
eats grass, man eats cow) is called a (15) _____ _____. In reality, the same food source is
most likely part of more than one chain, especially at low trophic levels. It is more accurate to think of
different food chains as cross-connecting with one another—that is, as a (16) _____ _____.

   The rate at which an ecosystem's primary producers capture and store a given amount of energy in
a specified time interval is the primary (17) _____. Energy from a primary source flows in one

direction through two kinds of food webs. In (18) _____ food webs, the energy flows from plants to herbivores, and then through an assortment of carnivores. In (19) _____ food webs, it flows mainly from plants through detritivores and decomposers. These two kinds of food webs often cross-connect.

Ecologists often represent the trophic structure of an ecosystem in the form of an ecological (20) _____. In such forms, the primary (21) _____ form a broad base for successive tiers of consumers above them. Some pyramids are based on (22) _____, or the weight of all the members at each trophic level. Sometimes, a pyramid of (23) _____ can be "upside down," with the (24) (choose one) ❐ smallest ❐ largest tier on the bottom. A more useful way to depict an ecosystem's trophic structure is with an (25) _____ pyramid. These pyramids show energy losses at each transfer to a different (26) _____ level in the ecosystem. They have a (27) (choose one) ❐ small ❐ large energy base at the bottom and are always "right-side up." The ecological study of energy flow at Silver Springs, Florida, found that about (28) (choose one) ❐ 1 ❐ 5 percent of all incoming solar energy was captured by producers prior to transfer to the next trophic level. It was also reported that only about (29) (choose one) ❐ 6 ❐ 8 to (30) (choose one) ❐ 10 ❐ 16 percent of the energy entering one trophic level becomes available for organisms at the next level. In general, the study showed that the efficiency of the energy transfers is so (31) (choose one) ❐ high ❐ low, ecosystems have usually no more than (32) (choose one) ❐ 4 ❐ 6 consumer trophic levels.

## Complete the Table

33. Complete the following table of productivity definitions.

| Productivity Type | Definition |
|---|---|
| a. Primary productivity | |
| b. Gross primary productivity | |
| c. Net primary productivity | |

## Choice

For questions 34 - 48, choose from the following:

a. primary producer     b. consumer     c. detritivore     d. decomposer

34. ___ Mule deer

35. ___ Earthworm

36. ___ Parasites

37. ___ The only category lacking heterotrophs

38. ___ Omnivores

39. ___ Tapeworm

40. ___ Herbivores

41. ___ Wolf

42. ___ Fungi and bacteria

43. ___ Carnivores

44. ___ Green plants

45. ___ Homo sapiens

46. ___ Crabs

47. ___ Autotrophs

48. ___ Grasshopper

## Dichotomous Choice

Circle one of two possible answers given between parentheses in each statement.

49. Ecosystems are (open/closed) systems, and so are not self-sustaining.
50. Minerals carried by erosion into a lake represent nutrient (input/output).
51. Energy (can/cannot) be recycled.
52. Nutrients typically (can/cannot) be cycled.
53. Ecosystems have energy and nutrient inputs as well as nutrient output, and they (have/lack) energy output.

## Matching

Match each organism in the Antarctic food chain given below with the principal trophic level it occupies. Some letters may not be needed. (See Figure 37.3 in the text.)

54. ___ emperor penguins
55. ___ krill
56. ___ blue whale
57. ___ diatoms
58. ___ leopard seal
59. ___ fishes, small squids
60. ___ killer whale

A. Chemosynthetic autotrophs
B. Herbivores
C. Photosynthetic autotrophs
D. Secondary consumers
E. Tertiary consumers

# 37 - II. BIOGEOCHEMICAL CYCLES—AN OVERVIEW (p. 669)
## HYDROLOGIC CYCLE (pp. 670 - 671)
## CARBON CYCLE (pp. 672 - 673)
### *Focus on the Environment:* FROM GREENHOUSE GASES TO A WARMER PLANET? (pp. 674 - 675)
## NITROGEN CYCLE (pp. 676 - 677)
## PHOSPHORUS CYCLE (p. 677)
## PREDICTING THE IMPACT OF CHANGE IN ECOSYSTEMS (p. 678)

*Selected Words:* *nutrients, hydrologic* cycle, *atmospheric* cycles, *sedimentary* cycles, "greenhouse gases," Anabaena, Nostoc, Rhizobium, Azotobacter, Plasmodium japonicum, Rickettsia rickettsii

## *Boldfaced, Page-Referenced Terms*

(669) biogeochemical cycle _____

_____

(670) hydrologic cycle _____

_____

(670) watershed _____

_____

(672) carbon cycle _____

_____

(674) greenhouse effect _____

_____

(674) global warming _____

_____

(676) nitrogen cycle _____

_____

(676) nitrogen fixation _____

_____

(676) decomposition _____

_____

(676) ammonification _____

_____

(676) nitrification _____

_____

(676) denitrification _____

_____

(677) phosphorus cycle _____

_____

(678) ecosystem modeling _____

(678) biological magnification _____

## Short Answer

1.  In what form(s) are elements used as nutrients usually available to producers? _____

    _____

    _____

2.  How is the ecosystem's reserve of nutrients maintained? _____

    _____

    _____

3.  How does the amount of a nutrient being cycled through most major ecosystems compare with the amount entering or leaving in a given year? _____

    _____

    _____

4.  What are the common input sources for an ecosystem's nutrient reserves? _____

    _____

    _____

5.  What are the output sources of nutrient loss for land ecosystems? _____

    _____

    _____

## Complete the Table

6.  Complete the following table to summarize the functions of the three types of biogeochemical cycles.

| Biogeochemical Cycle Type | General Function(s) |
| --- | --- |
| a. Hydrologic cycle | |
| b. Atmospheric cycles | |
| c. Sedimentary cycles | |

## Matching

Choose the most appropriate answer for questions 7 - 10; 11 - 14 may have more than one answer.

7. ___ solar energy

8. ___ source of most water

9. ___ watershed

10. ___ water and plants taking up water

11. ___ forms of precipitation falling to land

12. ___ deforestation

13. ___ have important roles in the global hydrologic cycle

14. ___ forms of atmospheric water

A. Mostly rain and snow
B. Mechanisms by which nutrients move into and out of ecosystems
C. Water vapor, clouds, and ice crystals
D. Where precipitation of a specified region becomes funneled into a single stream or river
E. May have long-term disruptive effects on nutrient availability for an entire ecosystem
F. Slowly drives water through the atmosphere, on or through land mass surface layers, to oceans, and back again
G. Ocean currents and wind patterns
H. Evaporation from the ocean

## Fill-in-the-Blanks

The (15) _____ _____ studies demonstrated that stripping the land of vegetation disrupts nutrient retention by an entire ecosystem for a long time; in the watershed, (16) _____ _____ efficiently mined the soil for calcium, storing it in a growing biomass of tree tissues. (17) _____ and weathering of rocks brought calcium replacements back into the watershed. In (18) _____ cycles, the elements lack a gaseous phase. These watershed studies showed that (19) _____ greatly influence the movement of nutrients through the ecosystem phase of biogeochemical cycles. Nutrient outputs changed when experimental watersheds were stripped of vegetation. Compared to undisturbed watersheds, (20) _____ times as much calcium was lost by way of stream outflow. Thus, (21) _____ stabilize the soil of land ecosystems by absorbing dissolved minerals to minimize loss of soil nutrients in runoff.

## Matching

Choose the most appropriate answer for each.

22. ___ greenhouse gases

23. ___ carbon dioxide fixation

24. ___ carbon cycle

25. ___ ways carbon enters the atmosphere

26. ___ greenhouse effect

27. ___ carbon dioxide ($CO_2$)

28. ___ oceans and plant biomass accumulation

A. Form of most of the atmospheric carbon
B. Aerobic respiration, fossil fuel burning, and volcanic eruptions
C. $CO_2$, CFCs, $CH_4$, and $N_2O$
D. Photosynthesizers incorporate carbon atoms into organic compounds
E. Annual "holding stations" for about half of all atmospheric carbon
F. Carbon reservoirs → atmosphere and oceans → through organisms → carbon reservoirs
G. Warming of Earth's lower atmosphere due to accumulation of certain gases

## Fill-in-the-Blanks

In the (29) _____ _____, Earth's surface is warmed by sunlight and radiates (30) _____ (infrared wavelengths) to the atmosphere and space. Greenhouse gases such as (31) _____, which are used as refrigerants, solvents, and plastic foams, and (32) _____ _____, which is released from burning fossil fuels, deforestation, car exhaust, and factory emissions, are together responsible for the alarming global (33) _____ trend.

(34) _____ can assimilate nitrogen from the air in the process known as (35) _____ _____. In (36) _____, either ammonia or ammonium ions are stripped of electrons, and (37) _____ ($NO_2^-$) is released as a product of the reactions. Under some conditions, nitrate is converted into nitorgen and some (38) _____ _____ by denitrifying bacteria.

## Choice

For questions 39 - 43, choose from the following:

        a. nitrogen cycle description     b. nitrogen fixation     c. ammonification

                d. nitrification     e. denitrification

39. ___ Ammonia or ammonium in soil are stripped of electrons, and nitrite ($NO_2^-$) is the result; other bacteria convert nitrite to nitrate ($NO_3^-$).

40. ___ Bacteria convert nitrate or nitrite to $N_2$ and a bit of nitrous oxide ($N_2O$).

41. ___ Bacteria and fungi break down nitrogen-containing wastes and plant and animal remains; released amino acids and proteins are used for growth with the excess given up as ammonia or ammonium ions that plants can use.

42. ___ Occurs in the atmosphere (largest reservoir); only certain bacteria, volcanic action, and lightning can convert $N_2$ into forms that can enter food webs.

43. ___ A few kinds of bacteria convert $N_2$ to ammonia ($NH_3$), which dissolves quickly in water to form ammonium ($NH_4^+$).

## Short Answer

44. List reasons that an insufficient soil nitrogen supply is a problem for land plants.

_____

_____

_____

_____

_____

## Fill-in-the-Blanks

The Earth's crust is the primary (45) _____ for phosphorus and other minerals that move through ecosystems as part of (46) _____ cycles. In rock formations on land, phosphorus is typically in the form of (47) _____. By weathering and erosion of rock formations, phosphates enter rivers and streams that transport them to the (48) _____. Phosphorus collects mainly on (49) _____ shelves. Following long time periods, crustal plates can uplift part of the seafloor and expose the phosphate. All (50) _____ require phosphorus for synthesizing phospholipids, NADPH, ATP, nucleic acids, and other compounds.

Plants take up dissolved, (51) _____ forms of phosphate very rapidly. Herbivores obtain phosphorus by eating (52) _____; carnivores get it by eating (53) _____. These organisms excrete phosphorus in urine and feces. (54) _____ also releases phosphorus to the soil and again, plants can take it in to recycle it within the (55) _____.

During World War II, DDT was sprayed in the tropical Pacific to control (56) _____ responsible for transmitting the organisms that cause a dangerous disease, (57) _____. Because of its stability, DDT is a prime candidate for (58) _____ _____—the increasing concentration of a nondegradable substance as it moves up through trophic levels. (59) _____ _____ is a method of identifying and combining crucial bits of information about an ecosystem through computer programs and models in order to predict the outcome of the next disturbance.

# Self-Quiz

___ 1. A network of interactions that involve the cycling of materials and the flow of energy between a community and its physical environment is a(n) _____.
   a. population
   b. community
   c. ecosystem
   d. biosphere

___ 2. _____ consume dead or decomposing particles of organic matter.
   a. Herbivores
   b. Parasites
   c. Detritivores
   d. Carnivores

___ 3. The members of feeding relationships are structured in a hierarchy, the steps of which are called _____.
   a. organism levels
   b. energy source levels
   c. eating levels
   d. trophic levels

___ 4. In the Antarctic, blue whales feed mainly on _____.
   a. petrels
   b. krill
   c. seals
   d. fish and small squids

___ 5. Which of the following is a primary consumer?
   a. cow
   b. dog
   c. hawk
   d. all of the above

___ 6. In a natural community, the primary consumers are _____.
   a. herbivores
   b. carnivores
   c. scavengers
   d. decomposers

___ 7. A straight-line sequence of who eats whom in an ecosystem is sometimes called a(n) _____.
   a. trophic level
   b. food chain
   c. ecological pyramid
   d. food web

___ 8. Of the 1,700,000 kilocalories of solar energy that entered an aquatic ecosystem in Silver Springs, Florida, investigators determined that about _____ percent of incoming solar energy was trapped by photosynthetic autotrophs.
   a. 1
   b. 10
   c. 25
   d. 74

___ 9. A biogeochemical cycle that deals with phosphorus and other nutrients that do not have gaseous forms is the _____ type.
   a. sedimentary
   b. hydrologic
   c. nutrient
   d. atmospheric

___ 10. _____ is a process in which nitrogenous waste products or organic remains of organisms are decomposed by soil bacteria and fungi that use the amino acids being released for their own growth and release the excess as ammonia or ammonium, which plants take up.
   a. Nitrification
   b. Ammonification
   c. Denitrification
   d. Nitrogen fixation

___ 11. In the carbon cycle, carbon enters the atmosphere through _____.
   a. carbon dioxide fixation
   b. respiration, burning, and volcanic eruptions
   c. oceans and accumulation of plant biomass
   d. release of greenhouse gases

___ 12. _____ refers to an increase in concentration of a nondegradable (or slowly degradable) substance in organisms as it is passed along food chains.
   a. Ecosystem modeling
   b. Nutrient input
   c. Biogeochemical cycle
   d. Biological magnification

## Chapter Objectives/Review Questions

| Page | | Objectives/Questions |
|---|---|---|
| (664) | 1. | Members of an ecosystem fit somewhere in a hierarchy of energy transfers (feeding relationships) called _____ levels. |
| (664 - 665) | 2. | List the principal trophic levels in an ecosystem of your choice; state the source of energy for each trophic level, and give one or two examples of organisms associated with each trophic level. |
| (664) | 3. | Explain why nutrients can be completely recycled but energy cannot. |
| (664) | 4. | An _____ is a complex of organisms and their physical environment, all interacting through a flow of energy and a cycling of materials. |
| (665) | 5. | Distinguish between food chains and food webs. |
| (664 - 665) | 6. | Understand how materials and energy enter, pass through, and exit an ecosystem. |
| (666) | 7. | Distinguish between net and gross primary productivity. |
| (666) | 8. | Compare grazing food webs with detrital food webs. Present an example of each. |
| (666 - 667) | 9. | Ecological pyramids that are based on _____ are determined by the weight of all the members of each trophic level; _____ pyramids reflect the energy losses at each transfer to a different trophic level. |
| (669) | 10. | In _____ cycles, the nutrient is transferred from the environment to organisms, then back to the environment—which serves as a large reservoir for it. |
| (670 - 671) | 11. | Be able to discuss water movements through the hydrologic cycle. |
| (671) | 12. | Explain what studies in the Hubbard Brook watershed have taught us about the movement of substances (water, for example) through a forest ecosystem. |

(672) 13. The carbon cycle traces carbon movement from reservoirs in the _____ and oceans, through organisms, then back to reservoirs.

(674) 14 Certain gases cause heat to build up in the lower atmosphere, a warming action known as the _____ effect.

(676) 15. A major element found in all proteins and nucleic acids moves in an atmospheric cycle called the _____ cycle.

(676) 16. Define the chemical events that occur during nitrogen fixation, nitrification, ammonification, and denitrification.

(677) 17. Describe the geochemical and ecosystem phase of the phosphorus cycle.

(677) 18. Explain why agricultural methods in the United States tend to put more energy into mechanized agriculture in the form of fertilizers, pesticides, food processing, storage, and transport than is obtained from the soil in the form of energy stored in foods.

(678) 19. Through _____ modeling, crucial bits of information about different ecosystem components are identified and used to build computer models for predicting outcomes of ecosystem disturbances.

(678) 20. Describe how DDT damages ecosystems; discuss biological magnification.

---

# Integrating and Applying Key Concepts

In 1971, Diet for a Small Planet was published. Frances Moore Lappé, the author, felt that people in the United States of America wasted protein and ate too much meat. She said, "We have created a national consumption pattern in which the majority, who can pay, overconsume the most inefficient livestock products [cattle] well beyond their biological needs (even to the point of jeopardizing their health), while the minority, who cannot pay, are inadequately fed, even to the point of malnutrition." Cases of marasmus (a nutritional disease caused by prolonged lack of food calories) and kwashiorkor (caused by severe, long-term protein deficiency) have been found in Nashville, Tennessee, and on an Indian reservation in Arizona, respectively. Lappé's partial solution to the problem was to encourage people to get as much of their protein as possible directly from plants and to supplement that with less meat from the more efficient converters of grain to protein (chickens, turkeys, and hogs) and with seafood and dairy products.

Most of us realize that feeding the hungry people of the world is not just a matter of distributing the abundance that exists—that it is being prevented in part by political, economic, and cultural factors. Devise two full days of breakfasts, lunches, and dinners that would enable you to exploit the lowest acceptable trophic levels to sustain yourself healthfully.

# 38

# THE BIOSPHERE

## Interactive Exercises

### 38 - I. AIR CIRCULATION PATTERNS AND REGIONAL CLIMATES (pp. 680 - 683) THE OCEAN, LAND FORMS, AND REGIONAL CLIMATES (pp. 684 - 685)

*Selected Words:* hydrosphere, lithosphere, atmosphere

*Boldfaced, Page-Referenced Terms*

(681) biosphere _____

_____

(681) climate _____

_____

(682) ozone layer _____

_____

(682) temperature zones _____

_____

(684) ocean _____

_____

(685) rain shadow _____

_____

(685) monsoons _____

_____

## Matching

Choose the one most appropriate answer for each.

1. ___ biosphere

2. ___ climate

3. ___ biome

4. ___ Earth's regional climates

5. ___ temperature zones

6. ___ Earth's seasonal climate variations

7. ___ global patterns of air circulation

8. ___ greenhouse effect

9. ___ "mini-monsoons"

10. ___ monsoon

11. ___ ozone layer

12. ___ rain shadow

13. ___ surface currents and drifts in the ocean

A. Due to tilt of Earth's axis, which yields annual incoming solar radiation variation; provides daylength and temperature differences

B. Caused by molecules of the lower atmosphere absorbing some heat and then reradiating it back to Earth; this heat drives Earth's weather systems

C. Due to the interaction of atmospheric circulation patterns, ocean currents, and landforms such as mountains, valleys, and other land formations

D. The entire Earth realm where organisms live

E. Defined by the differences in solar heating at different latitudes and modified air circulation patterns

F. Due to warm equatorial air rising, then spreading southward and northward; Earth's rotation creates worldwide belts of prevailing east and west winds

G. A large region of the Earth that is a subdivision of a geographic realm; characterized mainly by the climax vegetation of ecosystems within its boundaries

H. Reduction in rainfall on the leeward side of high mountains

I. Created by Earth's rotation, prevailing surface winds, and water temperature variations

J. Refers to prevailing weather conditions such as temperature, humidity, wind speed, cloud cover, and rainfall

K. Recurring sea breezes along coastlines; morning warm air above the land rises and cooler marine air moves in; after sunset, breezes flow from land to sea

L. Contains atmospheric molecules that absorb potentially lethal ultraviolet wavelength

M. Very wet summers alternate with very dry winters

## Fill-in-the-Blanks

The numbered items in the illustrations below represent missing information. Complete the matching numbered blanks in the narrative to supply this information.

The sun's rays are more direct in (14) _____ regions than in polar regions. The global pattern of air circulation begins as (15) _____ equatorial air (16) _____ and spreads northward and southward giving up much (17) _____ as precipitation (warm air can hold more moisture than cold air) that supports luxuriant tropical forests. The cooled, drier air then (18) _____ at latitudes of about 30°, becomes warmer and drier in areas where deserts form. Air even farther from the equator

picks up some (19) _____, and (20) _____ to higher altitudes, cools, and then gives up
(21) _____ at latitudes of about 60° to create another moist belt. The cooled, dry air then
(22) _____ at the polar regions, where the low temperatures and almost nonexistent precipitation
give rise to the cold, dry, polar deserts. The Earth's rotation then modifies the air circulation by
deflecting it into worldwide belts of prevailing (23) _____ and (24) _____ winds. The world's
temperature zones, beginning at the equator and moving toward the poles, are the (25) _____, the
(26) _____ temperate, the (27) _____ temperate, and the (28) _____. Finally, the amount
of the (29) _____ radiation reaching the surface varies annually, owing to the Earth's
(30) _____ around the sun. This leads to seasonal changes in daylength, prevailing wind
directions, and temperature. These factors influence the locations of different ecosystems.

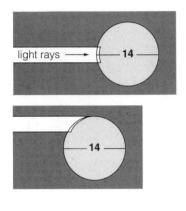

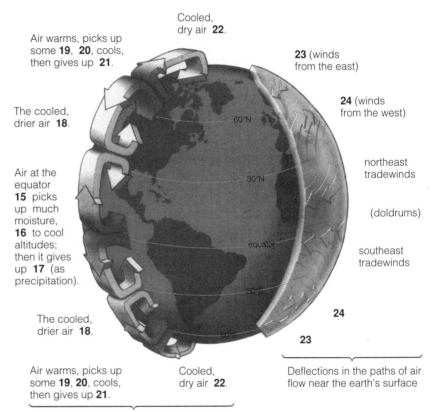

Cooled,
dry air 22.

Air warms, picks up
some 19, 20, cools,
then gives up 21.

23 (winds
from the east)

24 (winds
from the west)

The cooled,
drier air 18.

northeast
tradewinds

Air at the
equator
15 picks
up much
moisture,
16 to cool
altitudes;
then it gives
up 17 (as
precipitation).

(doldrums)

southeast
tradewinds

60°N

30°N

equator

The cooled,
drier air 18.

24

23

Air warms, picks up
some 19, 20, cools,
then gives up 21.

Cooled,
dry air 22.

Deflections in the paths of air
flow near the earth's surface

Initial pattern of air circulation

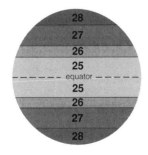

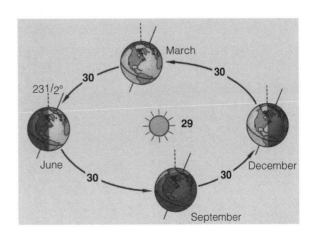

## Identification

Identify the numbered currents shown in the map by matching their numbers with the lettered choices below.

31. ___
32. ___
33. ___
34. ___
35. ___
36. ___
37. ___

A. Benguela current
B. California current
C. Canary current
D. Gulf Stream
E. Humboldt current
F. Japan current
G. Labrador current

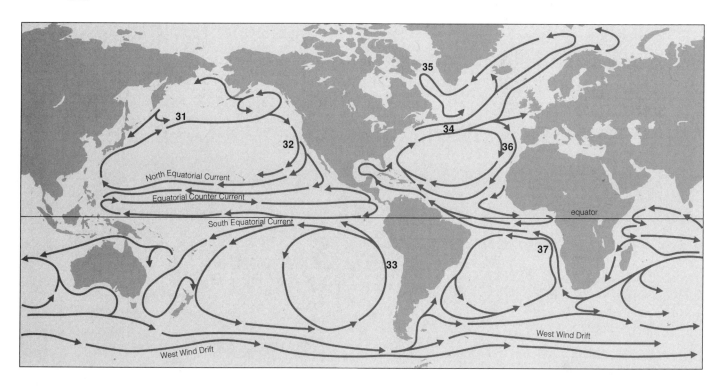

## 38 - II. THE WORLD'S BIOMES (pp. 686 - 687)
### SOILS OF MAJOR BIOMES (p. 688)
### DESERTS (p. 689)
### DRY SHRUBLANDS, DRY WOODLANDS, AND GRASSLANDS (pp. 690 - 691)

*Selected Words:* *shortgrass* prairie, *tallgrass* prairie, *monsoon* grasslands

### Boldfaced, Page-Referenced Terms

(686) biogeography _____

_____

(686) biogeographic realms _____

_____

(687) biome _____

_____

(688) soil _____

_____

(689) deserts _____

_____

(689) desertification _____

_____

(690) dry shrublands _____

_____

(690) dry woodlands _____

_____

(690) grasslands _____

_____

(690) savannas _____

_____

## Matching

Choose the most appropriate answer for each term.

1. ___ biogeography
2. ___ biogeographic realms
3. ___ biome
4. ___ soil
5. ___ deserts
6. ___ desertification
7. ___ dry shrublands
8. ___ dry woodlands
9. ___ grasslands
10. ___ savannas

A. Complex mixture of rock, mineral ions, and organic matter in some state of physical and chemical breakdown
B. Areas receiving less than 25 to 60 centimeters of rain per year
C. Wholesale conversion of grasslands and other productive biomes to dry wastelands
D. Sweep across much of the interior of continents, in the zones between deserts and temperate forests; warm temperatures prevail in summer, winters very cold
E. Areas that dominate when annual rainfall is about 40 to 100 centimeters
F. The study of the global distribution of species, each of which is adapted to regional conditions
G. Subdivision of biogeographic realms; a large region of land characterized by the climax vegetation of the ecosystems within its boundaries
H. Formed in land regions where the potential for evaporation exceeds rainfall
I. Six vast land areas on the Earth, each with distinguishing plants and animals
J. Broad belts of grasslands with a smattering of shrubs or trees; rainfall averages 90 to 150 centimeters a year with prolonged seasonal droughts common

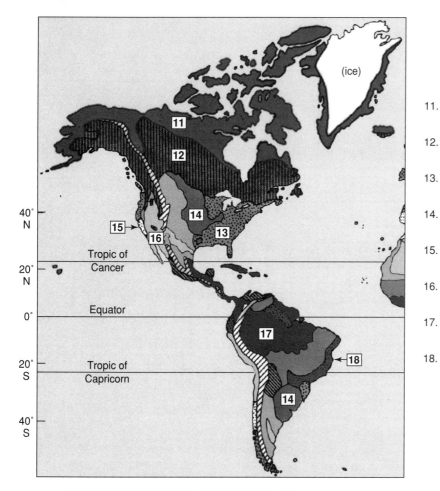

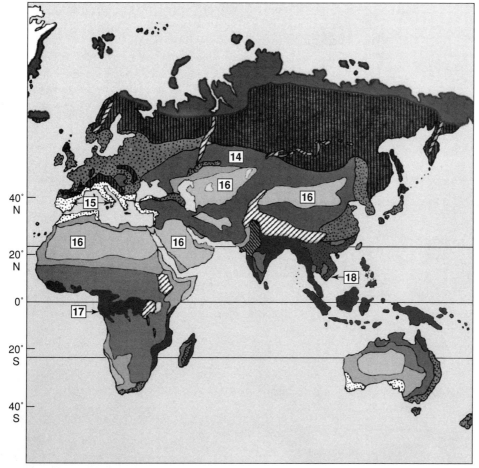

## Identification

Identify the numbered regions in the world map by matching their numbers with the lettered choices below.

A. Desert
B. Dry woodlands and shrublands (e.g., chaparral)
C. Evergreen broadleaf forest (e.g., tropical rain forest)
D. Evergreen coniferous forest (e.g., boreal forest, montane coniferous forest)
E. Temperate deciduous forest
F. Temperate grassland
G. Tropical deciduous forest
H. Tundra

11. ____

12. ____

13. ____

14. ____

15. ____

16. ____

17. ____

18. ____

## Choice

For questions 19 - 28, choose from the following:

      a. deserts    b. dry shrublands    c. dry woodlands    d. grasslands    e. savannas

19. ___ Monsoon type that forms dense stands of tall, coarse plants in parts of southern Asia where heavy rains alternate with a dry season

20. ___ Biome where the potential for evaporation greatly exceeds rainfall

21. ___ Steinbeck's *Grapes of Wrath* speaks eloquently of the disruption of this biome.

22. ___ Local names for this biome include fynbos and chaparral; plants are woody, multibranched, and only a few meters tall.

23. ___ A biome in which dominant trees can be tall but do not form a dense canopy; includes eucalyptus woodlands of southwestern Australia and oak woodlands of California and Oregon

24. ___ Home to deep-rooted evergreen shrubs, fleshy-stemmed, shallow-rooted cacti, saguaro, short prickly pear, and ocotillo

25. ___ Broad belts of grasslands with a smattering of shrubs and trees; prolonged seasonal droughts are common

26. ___ The dominant animals are grazing and burrowing types; grazing and periodic fires maintain the fringes of this biome.

27. ___ Within this biome, fast-growing grasses dominate where rainfall is low but acacia and other shrubs dominate where there is slightly more moisture.

28. ___ In summer, lightning-sparked, wind-driven firestorms can sweep quickly through these biomes and burn shrubs with highly flammable leaves.

## 38 - III. TROPICAL RAIN FORESTS AND OTHER BROADLEAF FORESTS (pp. 692 - 693)
### CONIFEROUS FORESTS (p. 694)
### TUNDRA (p. 695)

*Selected Words:* *tropical* deciduous forests, *monsoon* forests, *temperate* deciduous forests, *taiga*, *tuntura*, *arctic* tundra, *alpine* tundra

### Boldfaced, Page-Referenced Terms

(692) evergreen broadleaf forests _____

_____

(692) tropical rain forest _____

_____

_____

(693) deciduous broadleaf forests _____

_____

(694) coniferous forests _____

_____

(694) boreal forests _____

_____

(694) pine barrens _____

_____

(695) tundra _____

_____

(695) permafrost _____

_____

## Choice

For questions 1 - 10, choose from the following:

<blockquote>
a. deciduous broadleaf forests    b. evergreen coniferous forests

c. evergreen broadleaf forests    d. tundra
</blockquote>

1. ___ Nearly continuous sunlight in summer; short plants grow and flower profusely with rapidly ripening seeds

2. ___ Near the equator, highly productive, rainfall regular and heavy

3. ___ Highly productive forest; decomposition and mineral cycling are rapid in the hot, humid climate

4. ___ Within this biome, complex forests of ash, beech, chestnut, elm, and deciduous oaks once stretched across northeastern North America.

5. ___ Boreal forests or taiga; conifers are the primary producers in this biome

6. ___ Spruce and balsam fir dominate this North American biome

7. ___ Biome of the temperate zone; cold winter temperatures; many trees drop all their leaves in winter

8. ___ Lies between the polar ice cap and belts of boreal forests in North America, Europe, and Asia

9. ___ Soils are highly weathered, humus-poor, and not good nutrient reservoirs.

10. ___ Sandy, nutrient-poor soil of New Jersey's coastal plain supports scrub forests known as the pine barrens.

## Fill-in the-Blanks

Warm temperatures combined with uniform, abundant rainfall encourage the growth of the highly productive (11) _____ _____ forests. This type of biome exists where the annual mean temperature is about (12) _____ °C and a (13) _____ of 80 percent or more; it produces more litter than any other forest biome. However, (14) _____ and mineral cycling are rapid in the hot, humid climate, and its soils are not a good reservoir of nutrients. In (15) _____ _____ forests, most plants lose their leaves during a pronounced dry season. In (16) _____ _____ forests, decomposition is not as rapid as in the humid tropics, and many nutrients are conserved in accumulated litter on the forest floor. Conifers are the primary producers of (17) _____ forests. (18) _____ forests stretch across northern Europe, Asia, and North America. These "swamp forests" are also known by the name (19) _____ . In the Northern Hemisphere, (20) _____ coniferous forests extend southward through the great mountain ranges. (21) _____ and fir dominate in the north and at higher elevations. They give way to fir and (22) _____ in the south and at lower elevations. New Jersey's coastal plain is composed of sandy, nutrient-poor soil that supports (23) _____ _____, or scrub forests in which grasses and low shrubs grow beneath the open stands of pitch pine and oak trees. Just beneath the surface of the (24) _____ (a Finnish word for treeless plain) is the (25) _____, a permanently frozen layer. Here anaerobic conditions and low temperatures limit nutrient cycling. A similar biome at high elevations in mountains throughout the world is the (26) _____ tundra. However, here the permanently frozen layer, the (27) _____, is lacking. Dominant plants form low cushions or mats that withstand strong winds.

## 38 - IV. FRESHWATER PROVINCES (pp. 696 - 697)
##    THE OCEAN PROVINCES (pp. 698 - 699)

*Selected Words:* *phyto*plankton, *zoo*plankton, *oligotrophic* lakes, *eutrophic* lakes, *benthic* province, *pelagic* province, *submerged* mountains and valleys and plains

### Boldfaced, Page-Referenced Terms

(696) lake _____

_____

(696) plankton _____

_____

(696) spring overturn _____

_____

(696) fall overturn _____

_____

(696) eutrophication _____

_____

(696) streams _____

_____

(698) ultraplankton _____

_____

(699) hydrothermal vents _____

_____

### Fill-in-the-Blanks

A (1) _____ is a body of freshwater with three zones. The shallow, usually well-lit (2) _____ zone extends around the shore to the depth at which rooted aquatic plants stop growing. The diversity of organisms is greatest here. The (3) _____ zone is the open, sunlit water past the littoral and extends to a depth where photosynthesis is insignificant. Aquatic communities of (4) _____ abound here. The (5) _____ zone includes all open water below the depth at which wavelengths suitable for photosynthesis can penetrate. Bacterial decomposers in bottom sediments of this zone release nutrients into the water.

Water is densest at 4°C; at this temperature, it sinks to the bottom of its basin, displacing the nutrient-rich bottom water upward and giving rise to spring and fall (6) _____. In spring, ice melts, daylength increases, and the surface waters of a lake slowly warm to (7) _____ °C. Surface winds cause a (8) _____ overturn in which strong vertical water movements carry dissolved oxygen from a lake's surface layer to its depths, and nutrients released by decomposition are brought from the bottom sediments to the surface layer. The surface layer warms above 4°C by midsummer, becomes less dense, and the lake has developed a (9) _____ that prevents vertical mixing. When autumn comes, the upper layer (10) _____, becomes denser, then sinks, and the thermocline vanishes. This is termed the (11) _____ overturn. Water then mixes vertically, allowing dissolved oxygen to move (12) (choose one) ❑ up  ❑ down and nutrients to move (13) (choose one) ❑ up ❑ down.

Primary productivity of a lake corresponds with the seasons. Following a spring overturn, longer daylengths and cycled nutrients support (14) (choose one) ❑ lower  ❑ higher rates of photosynthesis. By late summer, nutrient shortages are limiting photosynthesis. After the fall overturn, nutrient cycling drives a (15) (choose one) ❑ short  ❑ long burst of primary activity. Lakes have a trophic nature. (16) _____ lakes are deep, nutrient-poor, and low in primary productivity. Lakes that are (17) _____ are often shallow, nutrient-enriched, and high in primary productivity. Human activities can determine the trophic condition of lakes. The term (18) _____ refers to nutrient enrichment of a lake resulting in reduced water transparency and a community rich in phytoplankton. (19) _____ are flowing-water ecosystems that begin as freshwater springs or seeps. Three habitat types are found between headwaters and the river's end: riffles, pools, and (20) _____.

## Label and Match

Label each numbered item in the illustration below. Complete the exercise by matching and entering the letter of the proper description in the parentheses following each label.

21. _____ province (  )

22. _____ province (  )

23. _____ zone (  )

24. _____ zone (  )

A. The entire volume of ocean water
B. All the water above the continental shelves
C. Includes all sediments and rocks of the ocean bottom; begins with continental shelves and extends to deep-sea trenches
D. Water of the ocean basin

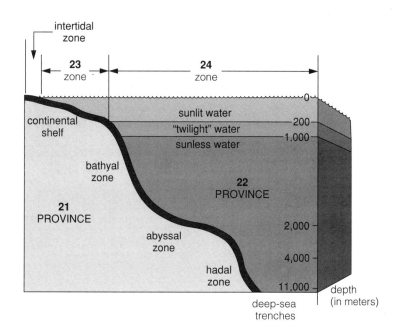

*Choice*

For questions 25 - 43, choose from the following:

> a. stream ecosystems    b. lake ecosystems    c. ocean provinces

25. ___ Riffles, pools, and runs

26. ___ Hydrothermal vent communities of chemosynthetic bacteria in the abyssal zone

27. ___ Seventy percent of the primary productivity is contributed by ultraplankton

28. ___ Can be oligotrophic or eutrophic

29. ___ Submerged mountains, valleys, and plains

30. ___ Begin as freshwater springs or seeps

31. ___ Sewage and logging on adjacent lands can contribute to eutrophication of this water.

32. ___ Average flow volume and temperature depend on rainfall, snowmelt, geography, altitude, and even shade cast by plants.

33. ___ Spring and fall overturns

34. ___ Has benthic and pelagic provinces

35. ___ Especially in forests, these waters import most of the organic matter that supports food webs.

36. ___ Created by geologic processes such as retreating glaciers

37. ___ They grow and merge as they flow downslope and then often combine.

38. ___ The final successional stage is a filled-in basin.

39. ___ Vast "pastures" of phytoplankton and zooplankton become the basis of detrital food webs.

40. ___ Since cities formed, these waters have been sewers for industrial and municipal wastes.

41. ___ Has a thermocline by midsummer

42. ___ Littoral, limnetic, and profundal zones

43. ___ Bathyal, abyssal, and hadal zones

## 38 - V. CORAL REEFS AND BANKS (pp. 700 - 701)
### LIFE ALONG THE COASTS (pp. 702 - 703)
### *Focus on the Environment:* EL NIÑO AND SEESAWS IN THE WORLD'S CLIMATES (pp. 704 - 705)

**Selected Words:** <u>Corallina</u>, <u>Lophelia</u>, *atolls, fringing reefs, barrier reefs, rocky* shores, *sandy* and *muddy* shores, *El Niño Southern Oscillation* (ENSO)

### Boldfaced, Page-Referenced Terms

(700) coral reef _____

_____

(702) estuary _____

_____

(702) intertidal zone _____

_____

(703) upwelling _____

_____

(703) El Niño _____

_____

## Complete the Table

1.  Complete the following table, which describes three types of coral reefs.

| Reef Type | Description |
|---|---|
| a. Atolls | |
| b. Fringing reefs | |
| c. Barrier reefs | |

## Choice

For questions 2 - 21, choose from the following:

    a. coral reefs and banks     b. estuary    c. intertidal zone    d. coastal upwelling    e. ENSO

2.  ___ The famous fogbanks along the California coast are one outcome.

3.  ___ *Lophelia* has constructed large ones in the cold waters of Norway's fjords.

4.  ___ Waves batter its resident organisms fiercely during storms.

5.  ___ Pulses of heat from the Earth may be the trigger for warming the southern surface waters; more warm air arises here than anyplace else.

6.  ___ Organisms living there must contend with the tides.

7.  ___ Constructed mainly from the remains of countless organisms

8.  ___ Primary producers here include phytoplankton, salt-tolerant plants, and some algae that grow in mud and on plant surfaces.

9.  ___ In general, an upward movement of cold, deep, often nutrient-rich ocean waters occurring in equatorial currents as well as along the coasts of both continents

10. ___ A partly enclosed coastal region where seawater swirls and mixes with nutrient-rich freshwater from rivers, streams, and land runoff

11. ___ The larger ones are found in clear, warm waters between latitudes 25° north and south.

12. ___ Salt marshes are common.

13. ___ El Niño

14. ___ Wind friction causes surface waters to begin moving and under the force of the Earth's rotation, moving water is deflected west, away from a coast.

15. ___ Many of these structures are being destroyed by human activities.

16. ___ Commercial fishing industries of Peru and Chile depend on it.

17. ___ A recurring seesaw in atmospheric pressure in the western equatorial Pacific

18. ___ Home to exotic and rare fish sold in pet stores and eaten in Japanese restaurants

19. ___ Sandy and muddy shores

20. ___ Reverses the usual westward flow of air and water, and displaces the cold, deep Humboldt current; this stops nutrient upwelling along the western coast of South America

21. ___ Found along rocky and sandy coastlines

---

## Self-Quiz

___ 1. The distribution of different types of ecosystems is influenced by _____.
   a. air currents
   b. variations in the amount of solar radiation reaching the Earth throughout the year
   c. ocean currents
   d. all of the above

___ 2. In a(n)_____, water draining from the land mixes with seawater carried in on tides.
   a. pelagic province
   b. rift zone
   c. upwelling
   d. estuary

___ 3. A biome with grasses as primary producers and scattered trees adapted to prolonged dry spells is known as a _____.
   a. warm desert
   b. savanna
   c. tundra
   d. taiga

___ 4. Located at latitudes of about 30° north and south, limited vegetation, and rapid surface cooling at night describes a _____ biome.
   a. shrubland
   b. savanna
   c. taiga
   d. desert

___ 5. In tropical rain forests, _____.
   a. competition for available sunlight is intense
   b. diversity is limited because the tall forest canopy shuts out most of the incoming light
   c. conditions are extremely favorable for growing luxuriant crops
   d. all of the above

___ 6. In a lake, the open sunlit water with its suspended phytoplankton is referred to as its _____ zone.
   a. epipelagic
   b. limnetic
   c. littoral
   d. profundal

___ 7. The lake's upper layer cools, the thermocline vanishes, lake water mixes vertically, and once again dissolved oxygen moves down and nutrients move up. This describes the _____.
   a. spring overturn
   b. summer overturn
   c. fall overturn
   d. winter overturn

___ 8. The _____ is a permanently frozen, water-impermeable layer just beneath the surface of the _____ biome.
   a. permafrost; alpine tundra
   b. hydrosphere; alpine tundra
   c. permafrost; arctic tundra
   d. taiga; arctic tundra

___ 9. _____ refers to a season of heavy rain that corresponds to a shift in prevailing winds over the Indian Ocean.
a. Geothermal ecosystem
b. Upwelling
c. Taiga
d. Monsoon

___ 10. All of the water above the continental shelves is in the _____.
a. neritic zone of the benthic province
b. oceanic zone of the pelagic province
c. neritic zone of the pelagic province
d. oceanic zone of the benthic province

___ 11. Complex forests of ash, beech, birch, chestnut, elm, and oaks are found in the _____.
a. tropical deciduous forest
b. monsoon forest
c. temperate deciduous forest
d. evergreen broadleaf forest

___ 12. Permafrost is characteristic of the _____ forest.
a. boreal
b. tundra
c. deciduous broadleaf
d. evergreen broadleaf

# Chapter Objectives/Review Questions

| Page | | Objectives/Questions |
|---|---|---|
| (681) | 1. | The _____ is the entire realm in which organisms live. |
| (681) | 2. | _____ refers to prevailing weather conditions, such as temperature, humidity, wind speed, cloud cover, and rainfall. |
| (682) | 3. | State the reason that most forms of life depend on the ozone layer. |
| (682) | 4. | _____ energy drives Earth's weather systems. |
| (683) | 5. | Be able to describe the causes of global air circulation patterns. |
| (683) | 6. | Describe how the tilt of the Earth's axis affects annual variation in the amount of incoming solar radiation. |
| (684) | 7. | Atmospheric and oceanic _____ patterns influence the distribution of different types of ecosystems. |
| (685) | 8. | Mountains, valleys, and other land formations influence _____ climates. |
| (685) | 9. | Describe the cause of the rain shadow effect. |
| (686) | 10. | Broadly, there are six distinct land realms, the _____ realms, as named by W. Sclater and Alfred Wallace. |
| (687) | 11. | Realms are divided into _____. |
| (688) | 12. | _____ is a mixture of rock, mineral ions, and organic matter in some state of physical and chemical breakdown. |
| (689 - 695) | 13. | Be able to list the major biomes, and briefly characterize them in terms of climate, topography, and organisms. |
| (689) | 14. | The wholesale conversion of grasslands and other productive biomes to dry wastelands is known as _____. |
| (696) | 15. | A _____ is a body of freshwater with littoral, limnetic, and profundal zones. |
| (696) | 16. | Define *plankton*, *phytoplankton*, and *zooplankton*. |
| (696) | 17. | Describe the spring and fall overturn in a lake in terms of causal conditions and physical outcomes. |
| (697) | 18. | _____ refers to nutrient enrichment of a lake or some other body of water. |
| (697) | 19. | _____ lakes are often deep, poor in nutrients, and low in primary productivity; _____ lakes are often shallow, rich in nutrients, and high in primary productivity. |
| (697) | 20. | Describe a stream ecosystem. |
| (698) | 21. | Be able to fully describe the benthic and pelagic provinces of the ocean. |
| (698) | 22. | Within the pelagic province, all the water above the continental shelves is the _____ zone; the _____ zone is the water of the ocean basin. |

(698)      23.  As much as 70 percent of the ocean's primary productivity may be the contribution of

           _____ .

(698 - 699) 24.  Describe the unusual hydrothermal vent ecosystems.

(700)      25.  Know the origin of coral reefs and banks.

(702)      26.  Be able to descriptively distinguish between an estuary and the intertidal zone.

(703)      27.  State the significance of ocean upwelling.

(704 - 705) 28.  Describe conditions of ENSO occurrence and its significance to the ocean.

---

## Integrating and Applying Key Concepts

One species, *Homo sapiens,* uses about 40 percent of Earth's entire productivity, and its members have invaded every biome, either by living there or by dumping waste products there. Many of Earth's residents are being denied the minimal resources they need to survive, while human populations continue to increase exponentially. Can you suggest a better way of keeping Earth's biomes healthy while providing at least the minimal needs of all Earth's residents (not just humans)? If so, outline the requirements of such a system and devise a way in which it could be established.

# 39

# HUMAN IMPACT ON THE BIOSPHERE

## Interactive Exercises

### 39 - I. AIR POLLUTION—PRIME EXAMPLES (pp. 708 - 711) OZONE THINNING—GLOBAL LEGACY OF AIR POLLUTION (p. 712)

*Selected Words:* ozone hole

*Boldfaced, Page-Referenced Terms*

(710) pollutants _____

(710) thermal inversion _____

(710) industrial smog _____

(710) photochemical smog _____

(710) PANs (peroxyacyl nitrates) _____

(710) dry acid deposition _____

(710) acid rain _____

(712) ozone thinning _____

(712) chlorofluorocarbons (CFCs) _____

## Fill-in-the-Blanks

(1) _____ are substances with which ecosystems have had no prior evolutionary experience. Adaptive mechanisms are not in place to deal with them. When a layer of dense, cool air gets trapped beneath a layer of warm air, the situation is known as a(n) (2) _____ _____; this has been a key factor in some of the worst air pollution disasters. Where (3) _____ are cold and wet, (4) _____ _____ develops as a gray haze over industrialized cities that burn coal and other fossil fuels for manufacturing, heating, and generating electric power. In warm climates, (5) _____ _____ develops as a brown haze over large cities located in natural bowls. The key culprit is nitric oxide. After release from vehicles, it reacts with (6) _____ in the air to form (7) _____ _____. When (7) is exposed to sunlight, it reacts with hydrocarbons, and (8) _____ oxidants result. Most hydrocarbons come from spilled or partially burned (9) _____. The main oxidants are ozone and (10) _____ (peroxyacyl nitrates). Even traces can sting eyes, irritate lungs, and damage crops. Oxides of (11) _____ and (12) _____ are among the worst pollutants. Coal-burning power plants, metal smelters, and factories emit most (13) _____ dioxides. Motor vehicles, gas- and oil-burning power plants, and (14) _____-rich fertilizers produce (15) _____ oxides.

During dry weather, fine particles of oxides may be briefly airborne and then fall to Earth as dry (16) _____ _____. When the oxides dissolve in atmospheric water, they form weak solutions of (17) _____ and (18) _____ acids. Strong winds may distribute them over great distances; when they fall to Earth in rain and snow, it is called wet acid deposition, or (19) _____ _____. (19) can be 10 to 100 times more acidic than normal rainwater that has a pH of about (20) _____. The deposited acids eat away at marble buildings, metals, rubber, plastic, even nylon stockings. They also have the potential to disrupt the physiology of organisms and the chemistry of ecosystems. (21) _____ (CFCs) are compounds of chlorine, fluorine, and carbon that are odorless and invisible. They are major factors in reduction of the ozone layer in the atmosphere.

## Choice

For questions 22 - 43, choose from the following aspects of atmospheric pollution:

        a. thermal inversion     b. industrial smog     c. photochemical smog

        d. acid deposition     e. chlorofluorocarbons     f. ozone layer

22. ___ Develops as a brown, smelly haze over large cities

23. ___ Includes the "dry" and "wet" types

24. ___ Contributes to ozone reduction more than any other factor

25. ___ Where winters are cold and wet, this develops as a gray haze over industrialized cities that burn coal and other fossil fuels.

26. ___ Weather conditions trap a layer of cool, dense air under a layer of warm air.

27. ___ The cause of London's 1952 air pollution disaster, in which 4,000 died

28. ___ Each year, from September through mid-October, it thins down by as much as half above Antarctica.

29. ___ By some estimates, nearly all of this released between 1955 and 1990 is still making its way up to the stratosphere.

30. ___ Intensifies a phenomenon called smog

31. ___ Today most of this forms in cities of China, India, and other developing countries, as well as in Hungary, Poland, and other countries of eastern Europe.

32. ___ Contains airborne pollutants, including dust, smoke, soot, ashes, asbestos, oil, bits of lead and other heavy metals, and sulfur oxides

33. ___ Depending on soils and vegetation, some regions are more sensitive than others to this.

34. ___ Have contributed to some of the worst local air pollution disasters

35. ___ Chemically attack marble buildings, metals, mortar, rubber, plastic, and even nylon stockings

36. ___ Its reduction allows more ultraviolet radiation to reach the Earth's surface.

37. ___ Reaches harmful levels where the surrounding land forms a natural basin, as it does around Los Angeles and Mexico City

38. ___ Tall smokestacks were added to power plants and smelters in an unsuccessful attempt to solve this problem.

39. ___ Those already in the air will be there for over a century, before they are neutralized by natural processes.

40. ___ A dramatic rise in skin cancers, eye cataracts, immune system weakening, and harm to photosynthesizers is related to its reduction.

41. ___ Widely used as propellants in aerosol spray cans, refrigerator coolants, air conditioners, industrial solvents, and plastic foams; enter the atmosphere slowly and resist breakdown

42. ___ The main culprit is nitric oxide, produced mainly by cars and other vehicles with internal combustion engines.

43. ___ Oxides of sulfur and nitrogen dissolve in water to form weak solutions of sulfuric acid and nitric acid that may fall with rain or snow.

## 39 - II. WHERE TO PUT SOLID WASTES, WHERE TO PRODUCE FOOD (p. 713)
### DEFORESTATION—CONCERTED ASSAULT ON FINITE RESOURCES (pp. 714 - 715)
### *Focus on the Environment:* YOU AND THE TROPICAL RAIN FOREST (p. 716)

*Selected Words:* *subsistence* agriculture, *animal-assisted* agriculture, *mechanized* agriculture, <u>Ficus</u> <u>benjamina</u>, genetic "resources"

### *Boldfaced, Page-Referenced Terms*

(713) green revolution _____

(714) deforestation _____

(714) shifting cultivation _____

## Matching

Choose the most appropriate answer for each.

1. ___ throwaway mentality
2. ___ green revolution
3. ___ shifting cultivation
4. ___ animal-assisted agriculture
5. ___ deforestation
6. ___ recycling
7. ___ forested watersheds
8. ___ subsistence agriculture
9. ___ new genetic resources
10. ___ mechanized agriculture

A. Runs on energy inputs from sunlight and human labor
B. An affordable, technologically feasible alternative to "throwaway technology"
C. Act like giant sponges that absorb, hold, and gradually release water
D. Potential benefits to be obtained by genetic engineering and tissue culture methods in the rain forests
E. Runs on energy inputs from oxen and other draft animals
F. An attitude prevailing in the United States and other developed countries that greatly adds to solid waste accumulation
G. Requires massive inputs of fertilizers, pesticides, fossil fuel energy, and ample irrigation to sustain high-yield crops
H. Research directed toward improving crop plants for higher yields and exporting modern agricultural practices and equipment to developing countries
I. Trees are cut and burned, then ashes tilled into the soil; crops are grown for one to several seasons on quickly leached soils.
J. Removal of all trees from large land tracts; leads to loss of fragile soils and disrupts watersheds

## 39 - III. TRADING GRASSLANDS FOR DESERTS (p. 717) A GLOBAL WATER CRISIS (pp. 718 - 719)

*Selected Words: primary treatment, secondary treatment, tertiary treatment*

### Boldfaced, Page-Referenced Terms

(717) desertification _____

(718) desalination _____

(718) salination _____

(718) water table _____

(719) wastewater treatment _____

### Dichotomous Choice

Circle one of two possible answers given between parentheses in each statement.

1. Conversion of large tracts of grasslands, or rain-fed or irrigated croplands to a more barren state is known as (subsistence agriculture/desertification).
2. Presently, (too many cattle in the wrong places/overgrazing on marginal lands) is the main cause of large-scale desertification.
3. In Africa, (domestic cattle/native wild herbivores) trample grasses and compact the soil surfaces as they wander about looking for water.
4. A 1978 study by biologist David Holpcraft demonstrated that African range conditions improved in land areas where (domestic cattle/native wild herbivores) were ranched.
5. Without irrigation and conservation practices, grasslands that were converted for agriculture often end up as (deserts/forested watersheds).

## Fill-in-the-Blanks

The supply of Earth's seawater is essentially unlimited. The removal of salt from seawater is called (6) _____. For most countries this process is not practical due to the fuel (7) _____ necessary to drive it. Large-scale (8) _____ accounts for nearly two-thirds of the human population's use of freshwater. Irrigation of otherwise useless soil can cause a salt buildup, or (9) _____, due to evaporation in areas of poor soil drainage. Land that drains poorly also becomes waterlogged and raises the (10) _____ _____. When the (10) is too close to the surface, soil becomes saturated with (11) _____ water, which can damage plant roots. A large problem is the fact that water tables are subsiding. For example, overdrafts have depleted half of the Ogallala aquifer that supplies irrigation water for (12) (choose one) ❒ 40  ❒ 20 percent of the croplands in the United States. Human sewage, animal wastes, toxic chemicals, agricultural runoff, sediments, pesticides, and plant nutrients are all sources of water (13) _____, which amplifies the problem of water scarcity. There are three levels of (14) _____ treatment. They are primary, secondary, and (15) _____ treatments. Most of the (14) is not treated adequately. If the current rates of population growth and water depletion hold, the amount of freshwater available for each person on the planet will be (16) (choose one) ❒ 45 - 56 ❒ 55 - 66  ❒ 65 - 76 percent less than it was in 1976. In the future, water, not (17) _____, may become the most important fluid. Unless long overdue planning for the future occurs, wars over water rights are likely to occur.

## Complete the Table

18. Complete the following table, which summarizes three levels of treatment methods for maintaining the water quality of polluted wastewater.

| Treatment | Description |
| --- | --- |
| a. Primary treatment | |
| b. Secondary treatment | |
| c. Tertiary treatment | |

## 39 - III. A QUESTION OF ENERGY INPUTS (pp. 720 - 721)
### ALTERNATIVE ENERGY SOURCES (p. 722)
#### Focus on Bioethics: BIOLOGICAL PRINCIPLES AND THE HUMAN IMPERATIVE (p. 723)

## Boldfaced, Page-Referenced Terms

(720) fossil fuels _____

(720) meltdown _____

(722) solar-hydrogen energy _____

(722) fusion power _____

## Dichotomous Choice

Circle one of two possible answers given between parentheses in each statement.

1. Paralleling the (S-shaped/J-shaped) curve of human population growth is a steep rise in total and per capita energy consumption.
2. The increase in per capita energy consumption is due to (increased numbers of energy users and to extravagant consumption and waste/energy used to locate, extract, transport, store, and deliver energy to consumers).
3. (Total energy/Net energy) is that left over after subtracting the energy used to locate, extract, transport, store, and deliver energy to consumers.
4. Fossil fuels are the carbon-containing remains of (plants/plants and animals) that lived hundreds of millions of years ago.
5. Even with strict conservation efforts, known petroleum and natural gas reserves may be used up during the (current/next) century.
6. The net energy (decreases/increases) as costs of extraction and transportation to and from remote areas increase.
7. World coal reserves can meet human energy needs for several centuries, but burning releases sulfur dioxides into the atmosphere and adds to the global problem of (photochemical smog/acid deposition).
8. By 1990 in the United States, it cost slightly more to generate electricity by nuclear energy than by using coal; today, it costs (more/less).
9. The danger in the use of radioactivity as an energy supply during normal operation is with potential (radioactivity escape/meltdown).
10. After nearly fifty years of research, scientists (have/have not) agreed on the best way to store high-level radioactive wastes.
11. When electrodes in photovoltaic cells exposed to sunlight produce an electric current to split water molecules into oxygen and hydrogen gas (potential fuels), it is known as (fusion power/solar-hydrogen energy).
12. California gets 1 percent of its electricity from (fusion power/wind farms).
13. (Solar-hydrogen energy/Fusion power) involves a mimic of a process occurring in the sun's environment and may provide a good energy source in about fifty years.

## Sequence and Classify

Arrange the consumption of world resources in correct hierarchical order. Enter the letter of the energy source of highest consumption next to 14, the letter of the next highest next to 15, and so on. Write an N in the parentheses following the letter of the resource if it is nonrenewable and an R if the resource is renewable.

14. ____ (  )          A. Hydropower

15. ____ (  )          B. Natural gas

16. ____ (  )          C. Oil

17. ____ (  )          D. Nuclear power

18. ____ (  )          E. Biomass

19. ____ (  )          F. Coal

# Self-Quiz

___ 1. Which of the following processes is not generally considered a component of secondary wastewater treatment?
   a. screens and settling tanks remove sludge
   b. microbial populations are used to break down organic matter
   c. removal of all nitrogen, phosphorus, and toxic substances
   d. chlorine is often used to kill pathogens in the water

___ 2. When fossil-fuel burning gives dust, smoke, soot, ashes, asbestos, oil, bits of lead, other heavy metals, and sulfur oxides, it forms _____.
   a. photochemical smog
   b. industrial smog
   c. a thermal inversion
   d. both a and c

___ 3. _____ result(s) when nitrogen dioxide and hydrocarbons react in the presence of sunlight.
   a. Photochemical smog
   b. Industrial smog
   c. A thermal inversion
   d. Both a and c

___ 4. When weather conditions trap a layer of cool, dense air under a layer of warm air, _____ occurs.
   a. photochemical smog
   b. a thermal inversion
   c. industrial smog
   d. acid deposition

___ 5. The most abundant fossil fuel in the United States is _____.
   a. carbon monoxide
   b. oil
   c. natural gas
   d. coal

___ 6. Sulfur and nitrogen dioxides dissolve in atmospheric water to form a weak solution of sulfuric acid and nitric acid; this describes _____.
   a. photochemical smog
   b. industrial smog
   c. ozone and PANs
   d. acid rain

___ 7. Which of the following statements is false?
   a. Ozone reduction allows more ultraviolet radiation to reach Earth's surface.
   b. CFCs enter the atmosphere and resist breakdown.
   c. Salination of soils aids plant growth and increases yields.
   d. CFCs already in the air will be there for over a century.

___ 8. The most abundant fuel in the United States is _____.
   a. carbon monoxide
   b. oil
   c. natural gas
   d. coal

___ 9. "Photovoltaic cells exposed to sunlight produce an electric current that splits water molecules into oxygen and hydrogen gas" refers to _____.
   a. fusion power
   b. wind energy
   c. water power
   d. solar-hydrogen energy

___ 10. Energy inputs from sunlight and human labor best describes _____.
   a. animal-assisted agriculture
   b. subsistence agriculture
   c. the green revolution
   d. mechanized agriculture

# Chapter Objectives/Review Questions

*Page*          *Objectives/Questions*

(710 - 711)  1.  Identify the principal air pollutants, their sources, their effects, and the possible methods for controlling each pollutant.

(710)        2.  During a _____ _____, weather conditions trap a layer of cool, dense air under a layer of warm air; trapped pollutants may reach dangerous levels.

(710)        3.  Distinguish photochemical smog from industrial smog.

(710 - 712)  4.  Describe the effects of acid rain on an ecosystem. Contrast those effects with the action of CFCs.

(713)        5.  Examine the effects that modern agriculture has wrought on desert and grassland ecosystems.

(713)        6.  Under the banner of the _____ _____ research has been directed towards improving the genetics of crop plants for higher yields and exporting modern agricultural techniques and equipment.

(713)        7.  _____ agriculture runs on energy inputs from sunlight and human labor; _____-_____ agriculture has energy inputs from oxen and other draft animals, and _____ agriculture requires massive inputs of _____, pesticides, and ample irrigation to sustain high-yield crops.

(714)        8.  Explain the repercussions of deforestation that are evident in soils, water quality, and genetic diversity in general.

(714 - 715)  9.  _____ _____ involves cutting and burning trees, tilling ashes with soil, planting crops from one to several seasons, and then abandoning the clear plots.

(717)       10.  The name for the conversion of large tracts of natural grasslands to a more desertlike state is _____.

(718)       11.  Explain why desalination is not a practical solution for the shortage of freshwater.

(719)       12.  Describe the three levels of wastewater treatment, and list some of the methods used in each level.

(720)       13.  State when the U.S. natural gas and petroleum reserves are expected to run out.

(720 - 722) 14.  Describe how our use of fossil fuels, solar energy, and nuclear energy affects ecosystems.

(720 - 721) 15.  Explain what a meltdown is.

(708 - 723) 16.  Be able to discuss the advantages and disadvantages of solar-hydrogen energy, wind energy, and fusion power.

(723)       17.  Demonstrate an understanding of the biological principles that must be applied on a global scale to avoid possible ecological catastrophe.

## Integrating and Applying Key Concepts

1.  If you were Ruler of All People on Earth, how would you encourage people to depopulate the cities and adopt a way of life by which they could supply their own resources from the land and dispose of their waste products safely on their own land?

2.  Explain why some biologists believe that the endangered species list now includes all species.

# 40

# ANIMAL BEHAVIOR

## Interactive Exercises

### 40 - I. THE HERITABLE BASIS OF BEHAVIOR (pp. 726 - 729)
### LEARNED BEHAVIOR (p. 730)

*Selected Words: intermediate* response, *stereotyped* responses

### Boldfaced, Page-Referenced Terms

(727) animal behavior _____

_____

(728) behavior _____

_____

(729) sound system _____

_____

(729) instinctive behavior _____

_____

(729) fixed action patterns _____

_____

(730) learned behavior _____

_____

(730) imprinting _____

_____

(730) classical conditioning _____

_____

(730) operant conditioning _____

_____

(730) habituation _____

_____

(730) spatial or latent learning _____

_____

(730) insight learning _____

_____

## Dichotomous Choice

Circle one of two possible answers given between parentheses in each statement.

1. For snake populations living along the California coast, the food of choice is (the banana slug/tadpoles and small fishes).
2. In Stevan Arnold's experiments, newborn garter snakes that were offspring of coastal parents usually (ate/ignored) a chunk of slug as the first meal.
3. Newborn garter snake offspring of (coastal/inland) parents ignored cotton swabs drenched in essence of slug and only rarely ate the slug meat.
4. The differences in the behavioral eating responses of coastal and inland snakes (were/were not) learned.
5. Hybrid garter snakes with coastal and inland parents exhibited a feeding response that indicated a(n) (environmental/genetic) basis for this behavior.
6. In zebra finches and some other songbirds, singing behavior is an outcome of seasonal differences in the secretion of melatonin, a hormone secreted by the (gonads/pineal gland).
7. In spring, melatonin secretion is suppressed and gonads grow; their secretion of estrogen and testosterone is increased to (indirectly/directly) influence singing behavior.
8. Even before a male bird hatches, a high (estrogen/testosterone) level triggers development of a masculinized brain.
9. Later, at the start of the breeding season, a male's enlarged gonads secrete even more (estrogen/testosterone), which acts on cells in the sound system to prepare the bird to sing when properly stimulated.
10. (Hormones/Genes) influence the organization and activation of mechanisms required for particular forms of behavior.

## Complete the Table

11. Complete the following table to consider examples of instinctive and learned behavior.

| Category | Examples |
|---|---|
| a. Instinctive behavior | |
| b. Learned behavior | |

## Matching

Choose the most appropriate answer for each category of learned behavior.

12. ___ imprinting
13. ___ classical conditioning
14. ___ operant conditioning
15. ___ habituation
16. ___ spatial or latent learning
17. ___ insight learning

A. Birds living in cities learn not to flee from humans or cars, which pose no threat to them.
B. Chimpanzees abruptly stack several boxes and use a stick to reach suspended bananas out of reach.
C. In response to a bell, dogs salivate even in the absence of food.
D. Bluejays store information about dozens or hundreds of places where they have stashed food.
E. Baby geese formed an attachment to Konrad Lorenz if separated from the mother shortly after hatching.
F. A toad learns to avoid stinging or bad-tasting insects after attempts to eat them.

# 40 - II. THE ADAPTIVE VALUE OF BEHAVIOR (p. 731)

## Boldfaced, Page-Referenced Terms

(731) natural selection _____

_____

(731) territory _____

_____

(731) reproductive success _____

_____

(731) adaptive behavior _____

_____

(731) social behavior _____

_____

(731) selfish behavior _____

_____

(731) altruistic behavior _____

_____

## Fill-in-the-Blanks

(1) _____ _____ is the result of differences in survival and reproduction among individuals of a population that differ from one another in heritable traits. Some versions of a trait are better than others at helping the individual survive and reproduce. The frequency of the helpful (2) _____ increase in a population while the others do not. Using the theory of (3) _____ by natural selection as a starting point, one should be able to identify (4) _____ forms of behavior—and to discern how they bestow (5) _____ benefits that offset reproductive costs or disadvantages that might be associated with them. Solitary animals or groups of animals may be studied. If the behavior is adaptive, it must promote the survival and production of offspring of the (6) _____.

## Complete the Table

7. Complete the following table of terms describing forms of individual adaptive behavior.

| Behavior | Description |
|---|---|
| a. Reproductive success | |
| b. Adaptive behavior | |
| c. Social behavior | |
| d. Selfish behavior | |
| e. Altruistic behavior | |

## 40 - III. COMMUNICATION SIGNALS (pp. 732 - 733)

*Selected Words:* *signaling* pheromones, *priming* pheromones, *threat* display

## Boldfaced, Page-Referenced Terms

(732) communication signals _____

_____

(732) signalers _____

_____

(732) signal receivers _____

_____

(732) acoustical signal _____

_____

(732) pheromones _____

_____

(733) visual signals _____

_____

(733) courtship displays _____

_____

(733) tactile signals _____

_____

(733) illegitimate receiver _____

_____

(733) illegitimate signalers _____

_____

## Choice

For questions 1 - 10, choose from the following; some answers may require two letters.

a. chemical signal    b. visual signal    c. acoustical signal

d. tactile signal    e. illegitimate receiver    f. illegitimate signaler

1. ___ Worker termites bang their heads to alert brown soldier termites.

2. ___ A handshake, hug, caress, or shove

3. ___ The bioluminescent signals of male and female fireflies of the same species

4. ___ A soldier termite detects ant scent that serves as a cue; the termite then kills the ant.

5. ___ Signals of some predatory female fireflies seeking a meal of males belonging to other firefly species

6. ___ Information-seeking bees stay in physical contact with a dancing bee who has returned to the bee-hive after foraging.

7. ___ The distinctive night "whining" of a male tungara frog calling to females and rival males

8. ___ Fringe-lipped bat as a receiver of the tungara frog's call

9. ___ A soldier termite shoots thin jets of silvery goo from its nose; the goo emits volatile odors (signaling pheromones) that attract more soldiers to battle danger.

10. ___ Assassin bugs hook dead termite bodies on their dorsal surfaces and acquire termite scent; this deception allows assassin bugs to hunt termite victims more easily.

11. ___ A male zebra finch sings to attract a female and to secure his territory.

12. ___ Exposing formidable canine teeth, male baboons "yawn" at each other when they compete for a receptive female; this is a threat display that may precede an attack.

13. ___ Courtship displays of lekking sage grouse and albatross

14. ___ A chemical odor (priming pheromones) from the urine of male mice will trigger and enhance estrus in female mice.

## 40 - IV. MATING, PARENTING, AND REPRODUCTIVE SUCCESS (pp. 734 - 735)

*Selected Words:* Harpobittacus apicalis, Centrocercus urophasianus

## Boldfaced, Page-Referenced Terms

(734) sexual selection _____

_____

(735) lek _____

_____

## Complete the Table

1.  Complete the following table to supply the common names of the animals that fit the text examples of sexual selection.

| Animals | Descriptions of Sexual Selection |
|---|---|
| a. | Females select the males that offer them superior goods; females permit mating only after they have eaten the "nuptial gift" for about five minutes. |
| b. | Males congregate in a lek or communal display ground; each male stakes out a few square meters as his territory; females are attracted to observe male displays and usually select and mate with only one male. |
| c. | Females of a species cluster in defendable groups at a time they are sexually receptive; males compete for access to the clusters; combative males are favored. |
| d. | Extended parental care improves the likelihood that the current generation of offspring will survive; this behavior comes at a reproductive cost to the adults. |

## 40 - V. COSTS AND BENEFITS OF BELONGING TO SOCIAL GROUPS (pp. 736 - 737)
## SOCIAL LIFE AND SELF-SACRIFICING BEHAVIOR (pp. 738 - 739)
### Focus on Science: ABOUT THOSE SELF-SACRIFICING NAKED MOLE RATS (p. 740)

*Selected Words:* <u>Parus</u> <u>major</u>, <u>Heterocephalus</u> <u>glaber</u>, *DNA fingerprinting*

## Boldfaced, Page-Referenced Terms

(736) cost-benefit approach _____

_____

(737) selfish herd _____

_____

(738) dominance hierarchy _____

_____

(738) theory of indirect selection _____

_____

## Matching

Choose the most appropriate answer for each.

1. ___ cost-benefit approach
2. ___ disadvantages to sociality
3. ___ cooperative predator avoidance
4. ___ the selfish herd

A. Competition for resources, rapid depletion of food resources, cannibalism, and greater vulnerability to disease
B. A simple society brought together by reproductive self-interest; larger, more powerful male bluegills tend to claim the central locations
C. A consideration of social life in terms of reproductive success of the individual
D. Adult musk oxen form a circle around their young while they face outward, and the "ring of horns" successfully deters the wolves; writhing, regurgitating reaction of Australian sawfly caterpillers to a disturbance

# 40 - VI. AN EVOLUTIONARY VIEW OF HUMAN SOCIAL BEHAVIOR (p. 741)

*Selected Words:* adaptive behaviors, *redirected*

## Boldfaced, Page-Referenced Terms

(741) adoption _____

_____

## Dichotomous Choice

Circle one of two possible answers given between parentheses in each statement.

1. Many people seem to believe that attempts to identify the adaptive value of a particular (animal/human) trait is an attempt to define its moral or social advantage.
2. "Adaptive" refers to (a trait with moral value/a trait valuable in gene transmission).
3. In many species, adults that have lost their offspring will, if presented with a substitute, (adopt/reject) it.
4. John Alcock suggests that "husbands and wives who have lost an only child or who fail to produce children themselves might be especially prone to adopt (strangers/relatives)."
5. The human adoption process (can/cannot) be considered adaptive when indirect selection favors adults who direct parenting assistance to relatives.
6. Joan Silk showed that in some traditional societies, children (are not/are) adopted overwhelmingly by relatives.
7. In modern societies in which agencies and other means of adoption exist, adoption of relatives (is/is not) predominant.
8. It (is/is not) possible to test evolutionary hypotheses about the adaptive value of human behaviors.
9. *Adaptive* behavior and *socially desirable* behavior (are not/are) separate issues.
10. Strong parenting mechanisms evolved in the past, and it may be that their redirection toward a (relative/nonrelative) says more about human evolutionary history than it does about the transmission of one's genes.

# Self-Quiz

___ 1.  The observable, coordinated responses that animals make to stimuli are what we call _____.
   a. imprinting
   b. instinct
   c. behavior
   d. learning

___ 2.  In _____, components of the nervous system allow an animal to carry out complex, stereotyped responses to certain environmental cues, which are often simple.
   a. natural selection
   b. altruistic behavior
   c. sexual selection
   d. instinctive behavior

___ 3.  Newly hatched goslings follow any large moving objects to which they are exposed shortly after hatching; this is an example of _____.
   a. homing behavior
   b. imprinting
   c. piloting
   d. migration

___ 4.  A young toad flips its sticky-tipped tongue and captures a bumblebee that stings its tongue; in the future, the toad leaves bumblebees alone. This is _____.
   a. instinctive behavior
   b. a fixed reaction pattern
   c. altruistic
   d. learned behavior

___ 5.  Pavlov's dog experiments represent an example of _____.
   a. classical conditioning
   b. latent learning
   c. selfish behavior
   d. habituation

___ 6.  _____ provides an example of an illegitimate signaler.
   a. A soldier termite killing an ant on cue
   b. An assassin bug with acquired termite odor
   c. A termite pheromone alarm signal
   d. The "yawn" of a dominant male baboon

___ 7.  The claiming of the more protected central locations of the bluegill colony by the largest, most powerful males suggests _____.
   a. cooperative predator avoidance
   b. the selfish herd
   c. a huge parent cost
   d. self-sacrificing behavior

___ 8.  A chemical odor in the urine of male mice triggers and enhances estrus in female mice. The source of stimulus for this response is a _____.
   a. generic mouse pheromone
   b. signaling pheromone
   c. priming pheromone
   d. cue from male mice

___ 9.  Female insects often attract mates by releasing sex pheromones. This is an example of a(n) _____ signal.
   a. chemical
   b. visual
   c. acoustical
   d. tactile

___ 10.  Worker termites bang their heads specifically to attract soldiers; male birds sing to stake out territories, attract females, and discourage males. These are examples of _____ signals.
   a. chemical
   b. visual
   c. acoustical
   d. tactile

___ 11.  An example of dominance hierarchy and self-sacrificing behavior is _____.
   a. cannibalistic behavior of a breeding pair of herring gulls in a huge nesting colony
   b. members of wolf packs helping others by sharing food or fending off predators even though they do not breed
   c. clumps of regurgitating Australian sawfly caterpillars
   d. a huge colony of prairie dogs being ravaged by a parasite

___ 12. When musk oxen form a "ring of horns" against predators, it is _____.
a. a selfish herd
b. cooperative predator avoidance
c. self-sacrificing behavior
d. dominance hierarchy

___ 13. Caring for nondescendant relatives favors the genes associated with helpful behavior and is classified as _____.
a. dominance hierarchy
b. indirect selection
c. altruism
d. both b and c

___ 14. "_____" means only that a given trait has proved beneficial in the transmission of individual's genes.
a. Dominance hierarchy
b. Indirect selection
c. Adaptive
d. Altruism

___ 15. _____ also favors adults who direct parenting behavior toward relatives and so indirectly perpetuate their shared genes.
a. Indirect selection
b. Moral selection
c. Redirected selection
d. Perpetuated selection

---

# Chapter Objectives/Review Questions

| Page | | Objectives/Review Questions |
|---|---|---|
| (727) | 1. | Animal _____ refers to the coordinated, observable responses an animal makes to stimuli. |
| (728 - 729) | 2. | By influencing the development of the nervous system, _____ influence the observable, coordinated responses that animals make to stimuli. |
| (728) | 3. | What explains the fact that coastal and inland garter snakes of the same species have different food preferences? |
| (729) | 4. | Describe the origin and formation of a sound system. |
| (729) | 5. | In _____ behavior, components of the nervous system allow an animal to carry out complex, stereotyped responses to certain environmental cues, which are often simple. |
| (729) | 6. | Describe and cite an example of a fixed action pattern. |
| (730) | 7. | When animals incorporate and process information gained from specific experiences and then use the information to vary or change responses to stimuli, it is _____ behavior. |
| (730) | 8. | A rooster was exposed to a mallard duck during a critical period in the rooster's life and is attracted to the mallard; this time-dependent form of learning is called _____. |
| (730) | 9. | Summarize Pavlov's experiments with dogs and classical conditioning. |
| (730) | 10. | When a toad begins to avoid bad-tasting insects following voluntary attempts to eat them, this represents _____ behavior. |
| (730) | 11. | Give an example of the learned behavior known as habituation. |
| (730) | 12. | An example of _____ or latent learning is the storage of information about hundreds of places in which food is stashed by bluejays. |
| (730) | 13. | An abrupt solution of a problem by an animal (without trial and error attempts) is known as _____ learning. |
| (731) | 14. | What is meant by reproductive success? |
| (731) | 15. | _____ behavior is any behavior by which an individual protects or increases its own chance of producing offspring, regardless of the consequences for the group to which it belongs. |
| (732) | 16. | Examples of _____ signals are chemical, visual, acoustical, and tactile. |
| (732) | 17. | Define the roles of signalers and signal receivers. |
| (732) | 18. | The song of the zebra finch and the "whine" and "chuck" of a male tungara frog are examples of _____ signals. |

(732)  19. Natural selection tends to favor communication signals that promote _____ success.

(733)  20. When soldier termites kill ants on cue, the termites are said to be _____ receivers of a communication signal.

(733)  21. Assassin bugs covered with termite scent are able to use deception to hunt termite victims more easily and as such are acting as _____ signalers.

(733)  22. Distinguish between *signaling* pheromones and *priming* pheromones; cite an example of each.

(733)  23. A baboon threat display, lekking sage grouse, and the bioluminescent flashes of a male firefly are all examples of _____ signals.

(733)  24. An example of a _____ signal is the physical contact of bees in a hive maintaining physical contact during the dance of a returning foraging bee.

(735)  25. List the costs and benefits of parenting in the example of adult Caspian terns.

(736)  26. Explain the cost-benefit approach that evolutionary biologists take to find answers to the questions about social life.

(736)  27. List disadvantages to sociality.

(736 - 737) 28. Studies of Australian sawfly caterpillars indicate _____ predator avoidance.

(737)  29. Define *selfish herd;* cite an example.

(738)  30. List two examples of self-sacrifice in a dominance hierarchy.

(738)  31. With _____ behavior, an individual behaves in a self-sacrificing way that helps others but decreases its own chance of reproductive success.

(738)  32. Hamilton's theory of _____ selection relates to caring for nondescendant relatives and how this favors genes associated with helpful behavior.

(741)  33. It is possible to test evolutionary hypotheses about the _____ value of human behaviors; discuss human adoption practices in this context.

(741)  34. "_____" means only that a given trait has proved beneficial in the transmission of an individual's genes.

# Integrating and Applying Key Concepts

Think about communication signals that humans use and list them. Do you believe a dominance hierarchy exists in human society? Think of examples.

# ANSWERS

---

## Chapter 1  Methods and Concepts in Biology

**1 - I.  ORGANIZATION IN NATURE (pp. 2 - 5)**
1. G; 2. F; 3. I; 4. L; 5. H; 6. K; 7. C; 8. N;
9. B; 10. O; 11. D; 12. E; 13. M; 14. A; 15. J;
16. Metabolism; 17. photosynthesis; 18. ATP;
19. respiration; 20. producers; 21. consumers;
22. Decomposers; 23. interdependencies

**1 - II.  SENSING AND RESPONDING TO THE
ENVIRONMENT (p. 6)
CONTINUITY AND CHANGE (pp. 6 - 7)**
1. receptors; 2. Homeostasis; 3. reproduction; 4. Inheritance; 5. variations; 6. mutation; 7. adaptive

**1 - III.  SO MUCH UNITY, YET SO MANY SPECIES
(pp. 8 - 9)**
1. species; 2. genus; 3. genus; 4. species; 5. a. Plantae; b. Monera; c. Fungi; d. Animalia; e. Protista;
6. D; 7. F; 8. A; 9. E; 10. B; 11. C; 12. G

**1 - IV.  AN EVOLUTIONARY VIEW OF LIFE'S
DIVERSITY (p. 10)**
1. changing; 2. T; 3. T; 4. T; 5. differences

**1 - V.  THE NATURE OF BIOLOGICAL INQUIRY
(p. 11)**
*Focus on Science:* **DARWIN'S THEORY AND DOING
SCIENCE (p. 12)**
**THE LIMITS OF SCIENCE (p. 13)**
1. G; 2. A; 3. D; 4. B; 5. E; 6. C; 7. F; 8. O;
9. O; 10. C; 11. O; 12. C; 13. a. Hypothesis;
b. Prediction; c. Theory; d. Scientific experiment;
e. Control group; f. Variable; 14. subjective;
15. supernatural; 16. convictions

**Self-Quiz**

1. d; 2. a; 3. b; 4. c; 5. d; 6. b; 7. d; 8. a; 9. c;
10. e

---

## Chapter 2  Chemical Foundations for Cells

**2 - I.  REGARDING THE ATOMS (pp. 16 - 18)**
*Focus on Science:* **USING RADIOISOTOPES TO
DATE FOSSILS, TRACK CHEMICALS, AND SAVE
LIVES (p. 19)**
1. J; 2. C; 3. K; 4. B; 5. F; 6. I; 7. E; 8. G; 9. A;
10. D; 11. H

**2 - II.  WHAT IS A CHEMICAL BOND? (pp. 20 - 21)**
1. equation; 2. formula; 3. yields; 4. reactants;
5. products; 6. 12; 7. mass; 8. balanced; 9. C;
10. E; 11. A; 12. B; 13. D; 14. a. Calcium, Ca;
b. Carbon, C; c. Chlorine, Cl; d. Hydrogen, H;
e. Sodium, Na; f. Nitrogen, N; g. Oxygen, O;
15.

**2 - III.  IMPORTANT BONDS IN BIOLOGICAL
MOLECULES (pp. 22 - 23)**
1.

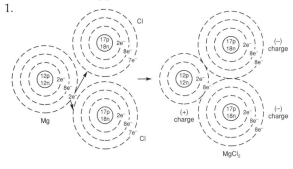

2.

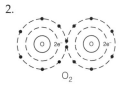
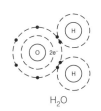

3. In a covalent bond, atoms share electrons to fill their outermost shells. In a nonpolar covalent bond, atoms attract shared electrons equally. An example is the $H_2$ molecule. In a polar covalent bond, atoms do not share electrons equally, and the bond is positive at one end, negative at the other (for example, the water molecule).
4. In the linear DNA molecule, the two nucleotide chains are held together by hydrogen bonds.

## 2 - IV. PROPERTIES OF WATER (pp. 24 - 25)
1. polarity; 2. hydrophilic; 3. hydrophobic;
4. Temperature; 5. hydrogen; 6. evaporation; 7. ice;
8. Cohesion; 9. solvent; 10. solutes; 11. dissolved;
12. hydration

## 2 - V. WATER, DISSOLVED IONS, AND THE WORLD OF CELLS (pp. 26 - 27)
1. protons; 2. hydrogen; 3. hydroxide; 4. pH;
5. neutrality; 6. 6; 7. pH; 8. hydrogen; 9. hydrogen ($H^+$); 10. hydroxide ($OH^-$); 11. hydrogen; 12. hydroxide; 13. hydrogen; 14. hydroxide; 15. 7;
16. salt; 17. Buffers; 18. Bicarbonate; 19. carbonic;
20. acidosis; 21. water; 22. cells; 23. a. 7.3–7.5, slightly basic; b. 6.2–7.4, variable slightly basic or slightly acid; c. 5.0–7.0, slightly acid to neutral; d. 1.0–3.0, strongly acid

## 2 - VI. PROPERTIES OF ORGANIC COMPOUNDS (pp. 28 - 29)
1. methyl; 2. hydroxyl; 3. ketone; 4. amino;
5. phosphate; 6. carboxyl; 7. aldehyde; 8. Enzymes represent a special class of proteins that speed up specific metabolic reactions. Enzymes mediate five categories of reactions by which most of the biological molecules are assembled, rearranged, and broken apart.
9. a. Condensation; b. Electron transfer; c. Cleavage; d. Functional-group transfer; e. Rearrangement
10.

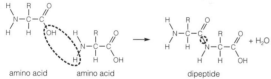

amino acid      amino acid      dipeptide

11. Hydrolysis reactions reverse the chemistry of condensation reactions; in the presence of water, large molecules are split into their component smaller molecules. Both condensation and hydrolysis require the presence of enzymes specific to the particular molecules involved.

## 2 - VII. CARBOHYDRATES (pp. 30 - 31)
1.

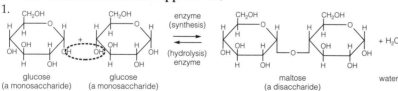

glucose (a monosaccharide)   glucose (a monosaccharide)   maltose (a disaccharide)   water

2. a. Sucrose; b. Ribose; c. Glucose; d. Cellulose; e. Deoxyribose; f. Lactose; g. Chitin; h. Glycogen; i. Starch

## 2 - VIII. LIPIDS (pp. 32 - 33)
1. a. unsaturated; b. saturated
2.

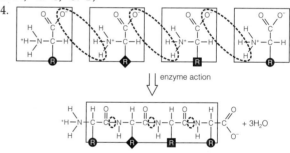

glycerol   three fatty acids   triglyceride (a complete fat molecule)

3. Phospholipids have two fatty acid tails attached to a glycerol backbone; they have hydrophilic heads that dissolve in water. Phospholipids are the main structural materials of cell membranes. 4. B; 5. D; 6. E; 7. A; 8. B; 9. C; 10. D; 11. A; 12. E; 13. A; 14. B; 15. B

## 2 - IX. AMINO ACIDS AND PROTEINS (pp. 34 - 35)
### SOME EXAMPLES OF FINAL PROTEIN STRUCTURE (p. 36)
1. A; 2. C; 3. B;
4.

enzyme action

5. J; 6. F; 7. B; 8. I; 9. A; 10. D; 11. G; 12. E; 13. C; 14. H

## 2 - X. NUCLEOTIDES AND NUCLEIC ACIDS (p. 37)
1. B; 2. A; 3. C; 4. Three as shown:
5. B; 6. A; 7. E; 8. D;
9. C; 10. a. Lipids;
b. Proteins;
c. Proteins; d. Nucleic acids;
e. Carbohydrates; f. Lipids;
g. Lipids;
h. Nucleic acids;
i. Lipids;
j. Carbohydrates

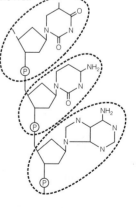

## Self-Quiz
1. a; 2. d; 3. c; 4. e; 5. e;
6. b; 7. c; 8. c; 9. b; 10. b

# Chapter 3    Cell Structure and Function

**3 - I.  BASIC ASPECTS OF CELL STRUCTURE AND FUNCTION (pp. 40 - 43)**

1. plasma membrane (B);  2. cytoplasm (A);  3. DNA-containing region (C);  4. adhesion;  5 - 8. transport;  9. receptor;  10. recognition;  11. J;  12. I;  13. D;  14. C;  15. A;  16. B;  17. H;  18. G;  19. E;  20. F

**3 - II.  CELL SIZE AND CELL SHAPE (p. 44)**
*Focus on Science:* MICROSCOPES—GATEWAYS TO THE CELL (pp. 44 - 45)

1. Cell size is constrained by the surface-to-volume ratio.  Past a certain point of growth, the surface area will not be sufficient to admit enough nutrients and to allow enough wastes to exit the cell.  As a cell grows, the surface area increases with the square and the volume increases with the cube.
2. F;  3. E;  4. G;  5. C;  6. B;  7. A;  8. D

**3 - III.  EUKARYOTIC CELLS (pp. 46 - 49)**

1. Compartmentalization into cell organelles allows a large number of different chemical activities to proceed simultaneously in a very limited space.  Organelles physically separate many incompatible chemical reactions in the cell—in time as well as space.
2. a. Nucleus;  b. Ribosomes;  c. Endoplasmic reticulum;  d. Golgi body;  e. Vesicles;  f. Mitochondria;  g. Cytoskeleton

**3 - IV.  THE NUCLEUS (pp. 50 - 51)**
**THE CYTOMEMBRANE SYSTEM (pp. 52 - 53)**
**VESICLES THAT MOVE OUT OF AND INTO CELLS (p. 54)**

1. a. Nucleoli;  b. Nuclear envelope;  c. Chromatin;  d. Chromosome;  2. K;  3. A;  4. B;  5. J;  6. C;  7. C;  8. G;  9. H;  10. N;  11. E;  12. M;  13. F;  14. D;  15. I;  16. L

**3 - V.  MITOCHONDRIA (p. 55)**
**SPECIALIZED PLANT ORGANELLES (p. 56)**

1. have or possess;  2. T;  3. large;  4. inner;  5. eukaryotic;  6. a;  7. b;  8. d;  9. a;  10. d;  11. a;  12. a;  13. b;  14. a;  15. c

**3 - VI.  CELL SURFACE SPECIALIZATIONS (p. 57)**
**THE CYTOSKELETON (pp. 58 - 59)**

1. E;  2. C;  3. A;  4. G;  5. D;  6. B;  7. H;  8. F;  9. eukaryotic;  10. Protein;  11. tubulin;  12. temporary;  13. contraction;  14. controlled assembly and disassembly of microtubule or microfilament subunits;  15. cytoplasmic streaming;  16. Flagella;  17. Cilia;  18. cilia;  19. cilia;  20. basal bodies;  21. microtubules;  22. microtubules

**3 - VII.  PROKARYOTIC CELLS—THE BACTERIA (pp. 60 - 61)**

1. Bacteria;  2. nucleus;  3. flagella;  4. wall;  5. cytoplasm;  6. ribosomes;  7. circular

**Self-Quiz**

1. Golgi body (K);  2. Vesicle (H);  3. microfilaments (J);  4. Chloroplast (N);  5. Mitochondrion (P);  6. central vacuole (C);  7. Nuclear envelope (A);  8. DNA and nucleoplasm (D);  9. nucleolus (L);  10. rough endoplasmic reticulum (Q);  11. nucleus (M);  12. microtubules (I);  13. plasma membrane (O);  14. cell wall (B);  15. microtubules (I);  16. microfilaments (J);  17. plasma membrane (O);  18. mitochondrion (P);  19. centrioles (G);  20. nuclear envelope (A);  21. nucleoplasm and DNA (D);  22. nucleolus (L);  23. nucleus (M);  24. vesicle (H);  25. lysosome (F);  26. rough endoplasmic reticulum (Q);  27. smooth endoplasmic reticulum (E);  28. Golgi body (K);  29. vesicle (H);  30. c;  31. C;  32. d;  33. b;  34. a;  35. c;  36. a;  37. b;  38. b;  39. d;  40. b,c,d,e;  41. a,b,c,d;  42. b,c,d,e  43. b,c,d,e;  44. c,d;  45. a,b,d;  46. b,c,d,e;  47. b,c,d,e;  48. b,c,d,e;  49. b,c,d,e

---

# Chapter 4    Ground Rules of Metabolism

**4 - I.  ENERGY AND LIFE (pp. 64 - 67)**

1. The world of life maintains a high degree of organization only because it is being resupplied with energy lost from someplace else.  2. second;  3. T;  4. T;  5. increasing;  6. II;  7. II;  8. I;  9. I;  10. II

**4 - II.  DIFFUSION IN THE CELLULAR WORLD (pp. 68 - 69)**
**MOVEMENT THROUGH TRANSPORT PROTEINS (pp. 70 - 71)**

1. Concentration;  2. gradient;  3. diffusion;  4. Osmosis;  5. gated channel protein;  6. channel protein;

7. lipid bilayer; 8. T; 9. hypotonic; 10. isotonic;
11. less; 12. hypertonic; 13. T; 14. osmosis;
15. diffusion; 16. active transport; 17. facilitated
diffusion; 18. diffusion; 19. active transport;
20. facilitated diffusion; 21. active transport; 22. facilitated diffusion; 23. diffusion

## 4 - III. CHARACTERISTICS OF METABOLIC
REACTIONS (pp. 72 - 73)

1. Exergonic reactions show a net loss of energy (energy
out); an example is the breakdown of food molecules in
the human body as the reactants become products.
2. Endergonic reactions show a net gain in energy
(energy in); an example is the construction of starch and
other large molecules from smaller, energy-poor molecules. 3. metabolic pathway; 4. enzyme; 5. endergonic; 6. exergonic; 7. exergonic; 8. B; 9. A;
10. A; 11. B; 12. C; 13. B; 14. D; 15. F; 16. E;
17. A

## 4 - IV. ENZYMES (pp. 74 - 75)
## MEDIATORS OF ENZYME FUNCTION (p. 76)

1. Enzymes are usually protein molecules with enormous catalytic power (a few RNA forms have been
found to act as enzymes). Enzymes do not cause reactions that would not happen on their own; enzymes are
not permanently changed in reactions—they can be used
over and over; each enzyme is highly selective about its
substrates; the same enzyme usually catalyzes both
forward and reverse directions of a reaction. Enzymes
work most effectively at specific temperatures and pH
values. Enzymes lower the energy of activation of the
reactions they catalyze. 2. D; 3. F; 4. C; 5. B; 6. A;
7. E; 8. Enzymes; 9. equilibrium; 10. substrate; 11.
active site; 12. induced-fit; 13. activation energy; 14.
Temperature (pH); 15. pH (temperature); 16. metabolism; 17. Allosteric; 18. feedback inhibition; 19.
cofactors; 20. coenzymes; 21. $NAD^+$ (FAD); 22. FAD
($NAD^+$); 23. $NADP^+$; 24. NADPH; 25. ions

## 4 - V. ATP—THE MAIN ENERGY CARRIER (p. 77)
## ENERGY AND THE FLOW OF ELECTRONS (p. 78)
Focus on Science: YOU LIGHT UP MY LIFE—
   VISIBLE EFFECTS OF METABOLIC ACTIVITY
   (p. 79)

1. adenine; 2. ribose; 3. phosphate; 4. water;
5. enzyme; 6. ADP; 7. phosphate; 8. energy;
9. energy; 10. metabolic; 11. ATP/ADP; 12. phosphorylation; 13. energy; 14. reaction; 15. ATP/ADP;
16. three phosphate groups; 17. ribose; 18. adenine;
19. adenosine triphosphate; 20. intermediate;
21. enzyme; 22. electron transport; 23. cofactors;
24. oxidized; 25. reduced; 26. electron; 27. energy;
28. most; 29. energy; 30. work; 31. phosphate;
32. E; 33. B; 34. C; 35. A; 36. D

## Self-Quiz

1. d; 2. c; 3. e; 4. d; 5. a; 6. c; 7. d; 8. c; 9. d;
10. a

# Chapter 5   Energy-Acquiring Pathways

## 5 - I. PHOTOSYNTHESIS: AN OVERVIEW
   (pp. 82 - 85)
## LIGHT-TRAPPING PIGMENTS (pp. 86 - 87)

1. Autotrophs; 2. carbon dioxide; 3. Photosynthetic;
4. Chemosynthetic; 5. electrons; 6. Heterotrophs;
7. animals; 8. aerobic respiration; 9. $12 H_2O + 6 CO_2$
$\rightarrow 6 O_2 + C_6H_{12}O_6 + 6 H_2O$; 10. (a) Twelve (b) carbon
dioxide (c) oxygen (d) glucose (e) six; 11. light-dependent (light-independent); 12. light-independent (light-
dependent); 13. Carbon dioxide; 14. water;
15. glucose; 16. thylakoid; 17. grana; 18. hydrogen;
19. stroma; 20. thylakoid; 21. photon; 22. pigments;
23. chlorophylls; 24. red (blue); 25. blue (red);
26. Carotenoids; 27. G; 28. F; 29. D; 30. E; 31. B;
32. A; 33. C; 34. H; 35. I

## 5 - II. LIGHT-DEPENDENT REACTIONS
   pp. 88 - 89)
## A CLOSER LOOK AT ATP FORMATION IN
   CHLOROPLASTS (p. 90)

1. photosystem; 2. light (photon); 3. electron;
4. acceptor; 5. Phosphorylation; 6. P700; 7. ATP
synthases (channel proteins); 8. ADP; 9. chemiosmotic; 10. a. Photosystem I: a pigment cluster domi-
nated by P700; b. Electrons: electrons representing
energy are ejected from P700 to an electron acceptor but
move over the electron transport system, where some of
the energy is used to produce ATP;  c. P700: a special
chlorophyll molecule that absorbs wavelengths of 700
nanometers and then ejects electrons;  d. Electron acceptor: a molecule that accepts electrons ejected from chlorophyll P700 and then passes electrons down the electron
transport system;  e. Electron transport system: electrons flow through this system, which is composed of a
series of molecules bound in the thylakoid membrane
that drive the phosphorylation of ADP to produce ATP;
f. ADP: ADP undergoes phosphorylation in cyclic pathway to become ATP;  11. electron acceptor;  12. electron
transport system;  13. photosystem II;  14. photosystem
I; 15. photolysis;  16. NADPH;  17. ATP;  18 - 53. The
following numbers should have a check mark ( √ ): 20,
22, 25, 28, 30, 32, 33, 34, 35, 37, 40, 41, 42, 45, 50, and 51.
All others should be blank.

## 5 - III. LIGHT-INDEPENDENT REACTIONS (p. 91)
## FIXING CARBON—SO NEAR, YET SO FAR (p. 92)
Focus on the Environment: AUTOTROPHS,
   HUMANS, AND THE BIOSPHERE (p. 93)

1. carbon dioxide (D); 2. carbon dioxide fixation (E);
3. phosphoglycerate (F); 4. adenosine triphosphate (H);

5. NADPH (G);  6. phosphoglyceraldehyde (A);  7. sugar phosphates (B);  8. Calvin-Benson cycle, (light-independent reactions) (I);  9. ribulose bisphosphate (C);  10. ATP (NADPH);  11. NADPH (ATP);  12. carbon dioxide;  13. ribulose bisphosphate;  14. PGA;  15. fixation;  16. PGA;  17. PGAL;  18. six;  19. RuBP;  20. carbon dioxide;  21. PGALs;  22. sugar phosphate;  23. fixation;  24. light-dependent;  25. ATP (NADPH);  26. NADPH (ATP);  27. Sugar phosphate;  28. photorespiration;  29. C3;  30. food (glucose);  31. oxygen;  32. oxaloacetate;  33. CAM;  34. carbon dioxide;

35. same cells;  36. chemosynthetic;  37. organic (food);  38. protons (electrons);  39. electrons (protons); 40 - 83. The following numbers should have a check mark ( √ ): 40, 45, 50, 52, 53, 54, 55, 57, 59, 63, 64, 69, 70, 71, 80, and 81. All others should be blank.

## Self-Quiz

1. a;  2. b;  3. c;  4. a;  5. c;  6. d;  7. a;  8. b;  9. c; 10. d

---

# Chapter 6   Energy-Releasing Pathways

**6 - I.   HOW CELLS MAKE ATP (pp. 96 - 99)**
1. Adenosine triphosphate (ATP);  2. Oxygen withdraws electrons from the electron transport system and joins with $H^+$ to form water;  3. Glycolysis followed by some end reactions (called fermentation), and anaerobic electron transport.  Some organisms (including humans) use fermentation pathways when oxygen supplies are low; many microbes rely exclusively on anaerobic pathways.  4. ATP;  5. photosynthesis;  6. aerobic respiration;  7. glycolysis;  8. pyruvate;  9. Krebs;  10. water;  11. ATP;  12. electrons;  13. transport;  14. phosphorylation;  15. ATP;  16. Oxygen;  17. anaerobic;  18. Fermentation;  19. electron transport;  20. $C_6H_{12}O_6 + 6O_2 \rightarrow 6CO_2 + 6H_2O$;  21. One molecule of glucose plus six molecules of **oxygen** (in the presence of appropriate enzymes) yield **six** molecules of carbon dioxide plus **six** molecules of water.

**6 - II.   GLYCOLYSIS:  FIRST STAGE OF THE ENERGY-RELEASING PATHWAYS (pp. 100 - 101)**
1. Autotrophic;  2. Glucose;  3. pyruvate;  4. NADH;  5. ATPs;  6. glucose;  7. ATP;  8. PGAL;  9. phosphate;  10. hydrogen;  11. ATP;  12. water;  13. phosphate;  14. ATP;  15. substrate-level;  16. glucose;  17. pyruvate;  18. three;  19. D;  20. F;  21. B;  22. H;  23. G;  24. A;  25. E;  26. C

**6 - III.   SECOND STAGE OF THE AEROBIC PATHWAY (pp. 102 - 103)**
1. acetyl CoA;  2. Krebs;  3. electron transport;  4. thirty-four;  5. ATP;  6. carbon dioxide;  7. electrons;  8. $NAD^+$ (FAD);  9. FAD ($NAD^+$);  10. inner compartment;  11. inner membrane;  12. outer compartment;  13. outer membrane;  14. cytoplasm;  15. ATP;  16. oxygen ($O_2$);  17. $FADH_2$;  18. NADH;  19. electron transport system

**6 - IV.   THIRD STAGE OF THE AEROBIC PATHWAY (pp. 104 - 105)**
1. three;  2. two;  3. transport;  4. chemiosmotic;  5. synthases;  6. ATP;  7. oxygen;  8. mitochondria;

9. thirty-six (thirty-eight);  10. thirty-eight (thirty-six);  11. c;  12. a, b;  13. c;  14. a;  15. b;  16. a;  17. b;  18. b;  19. a, b, c;  20. b;  21. c;  22. a, b;  23. b;  24. c;  25. a;  26. c;  27. c;  28. a;  29. b;  30. a;  31. c;  32. c

**6 - V.   ANAEROBIC ROUTES (pp. 106 - 107)**
1. oxygen ($O_2$);  2. fermentation;  3. lactate;  4. ethanol;  5. carbon dioxide;  6. Anaerobic;  7. electron;  8. Glycolysis;  9. pyruvate;  10. NADH;  11. pyruvate;  12. lactate;  13. acetaldehyde;  14. carbon dioxide;  15. ethanol;  16. glycolysis;  17. $NAD^+$;  18. bacteria;  19. ATP;  20. sulfate;  21. nitrite;  22. nitrogen;  23 - 74. With a check mark ( √ ): 23, 25, 31, 32, 34, 35, 37, 42, 47, 50, 54, 56, 59, 61, 63, 65, 66, 71; all others lack a check mark.

**6 - VI.   ALTERNATIVE ENERGY SOURCES IN THE HUMAN BODY (pp. 108 - 109)**
*Commentary:* PERSPECTIVE ON LIFE (p. 110)
1. Figure 6.3 shows how any complex carbohydrate or fat can be broken down and at least part of those molecules can be fed into the glycolytic pathway. 2. T;  3. T;  4. Page 110 (last paragraph) tells us that energy flows through time in one direction—from organized to less organized forms; thus energy cannot be completely recycled. 5. Page 110 tells us that Earth's first organisms were anaerobic fermenters. 6. fatty acids;  7. glycerol;  8. glycolysis;  9. amino acids;  10. Krebs cycle;  11. acetyl-CoA;  12. pyruvate;  13 - 102. With a check mark ( √ ): 13, 15, 16, 17, 19, 22, 23, 25, 29, 31, 32, 39, 46, 47, 48, 49, 53, 54, 65, 66, 70, 71, 72, 80, 84, 87, 95, 102; all others lack a check mark. 103. e;  104. b, (c);  105. c, d, e;  106. a;  107. c;  108. h;  109. e;  110. b;  111. b;  112. h;  113. i;  114. e;  115. c;  116. h;  117. g

## Self-Quiz

1. c;  2. c;  3. d;  4. b;  5. d;  6. d;  7. a;  8. d;  9. d;  10. d;  11. C;  12. A, B, D;  13. B, (C), D; 14.   C, E;  15. B, C;  16. A, D;  17. A;  18. B, D;  19. E;  20. A, E

# Chapter 7    Cell Division and Mitosis

**7 - I. DIVIDING CELLS: THE BRIDGE BETWEEN GENERATIONS (pp. 114 - 116)**
1. G; 2. H; 3. C; 4. E; 5. A; 6. B; 7. J; 8. I; 9. F; 10. D

**7 - II. MITOSIS AND THE CELL CYCLE (p. 117)**
1. interphase; 2. mitosis; 3. $G_1$; 4. S; 5. $G_2$; 6. prophase; 7. metaphase; 8. anaphase; 9. telophase; 10. cytokinesis (cytoplasmic division); 11. 5; 12. 2; 13. 4; 14. 3; 15. 1; 16. 10; 17. 1

**7 - III. STAGES OF MITOSIS (pp. 118 - 119)**
1. interphase-daughter cells (F); 2. anaphase (A); 3. late prophase (G); 4. metaphase (D); 5. interphase-parent cell (E); 6. early prophase (C); 7. transition to metaphase (B); 8. telophase (H)

**7 - IV. DIVISION OF THE CYTOPLASM (pp. 120 - 121)**
*Focus on Science:* HENRIETTA'S IMMORTAL SOUL (p. 122)
1. a; 2. b; 3. a; 4. a; 5. b

**Self-Quiz**

1. a; 2. d; 3. d; 4. a; 5. d; 6. c; 7. e; 8. c; 9. c; 10. d

---

# Chapter 8    Meiosis

**8 - I. COMPARISON OF ASEXUAL AND SEXUAL REPRODUCTION (pp. 124 - 126)**
**MEIOSIS AND THE CHROMOSOME NUMBER (pp. 126 - 127)**
1. b; 2. a; 3. b; 4. b; 5. a; 6. b; 7. a; 8. b; 9. a; 10. b; 11. T; 12. gamete; 13. diploid; 14. T; 15. T; 16. T; 17. T

**8 - II. A VISUAL TOUR OF THE STAGES OF MEIOSIS (pp. 128 - 129)**
1. E ($2n = 2$); 2. D ($2n = 2$); 3. B ($2n = 2$); 4. A ($n = 1$); 5. C ($n = 1$); 6. anaphase II (H); 7. metaphase II (F); 8. metaphase I (A); 9. prophase II (B); 10. telophase II (C); 11. telophase I (G); 12. prophase I (E); 13. anaphase I (D)

**8 - III. KEY EVENTS OF MEIOSIS I (p. 130 - 131)**
1. D; 2. B; 3. E; 4. C; 5. A

**8 - IV. FROM GAMETES TO OFFSPRING (pp. 132 - 133)**
1. b; 2. c; 3. a; 4. c; 5. b; 6. b; 7. b; 8. b; 9. c; 10. b; 11. 2 ($2n$); 12. 5 ($n$); 13. 4 ($n$); 14. 1 ($2n$); 15. 3 ($n$); 16. A; 17. B; 18. E; 19. D; 20. C; 21. During prophase I of meiosis, crossing over and genetic recombination occur. During metaphase I of meiosis, the two members of each homologous chromosome assort independently of the other pairs. Fertilization is a chance mix of different combinations of alleles from two different gametes.

**8 - V. MEIOSIS AND MITOSIS COMPARED (pp. 134 - 135)**
1. a. Mitosis; b. Mitosis; c. Meiosis; d. Meiosis; e. Meiosis; f. Meiosis; g. Meiosis; h. Mitosis; i. Meiosis; 2. C; 3. F; 4. D; 5. A; 6. B; 7. E; 8. 4; 9. 8; 10. 4; 11. 8; 12. 2

**Self-Quiz**

1. a; 2. a; 3. d; 4. b; 5. c; 6. b; 7. d; 8. a; 9. b; 10. c

---

# Chapter 9    Observable Patterns of Inheritance

**9 - I. MENDEL'S INSIGHTS INTO THE PATTERNS OF INHERITANCE (pp. 138 - 141)**
**MENDEL'S THEORY OF SEGREGATION (pp. 142 - 143)**
**INDEPENDENT ASSORTMENT (pp. 144 - 145)**
1. F; 2. A; 3. C; 4. G; 5. E; 6. I; 7. D; 8. H; 9. J; 10. B; 11. monohybrid; 12. probability; 13. segregation; 14. testcross; 15. 1:1; 16. dihybrid; 17. independent assortment; 18. genotype: ½ *Tt*; ½ *tt*, phenotype: ½ tall; ½ short; 19. a. 1 tall : 1 short; 1 heterozygous tall : 1 homozygous short. b. All tall; 1 homozygous tall : 1 heterozygous tall. c. All short; all homozygous short. d. 3 tall : 1 short; 1 homozygous tall : 2 heterozygous tall : 1 homozygous short. e. 1 tall : 1 short; 1 homozygous short : 1 heterozygous tall. f. All tall; all heterozygous tall. g. All tall; all homozygous tall. h. All tall; 1 het-

erozygous tall : 1 homozygous tall.  20. a. $\frac{9}{16}$; pigmented eyes, right-handed.  b. $\frac{3}{16}$ pigmented eyes, left-handed. c. $\frac{3}{16}$ blue-eyed, right-handed.  d $\frac{1}{16}$ blue-eyed, left-handed (note Punnett square below).

21.  Albino = *aa*, normal pigmentation = *AA* or *Aa*.  The woman of normal pigmentation with an albino mother is genotype *Aa*; the woman received her recessive gene (*a*) from her mother and her dominant gene (*A*) from her father.  It is likely that half of the couple's children will be albinos (*aa*) and half will have normal pigmentation but be heterozygous (*Aa*).  22. a. $F_1$: black trotter; $F_2$: nine black trotters, three black pacers, three chestnut trotters, one chestnut pacer.  b. Black pacer.  c. *BbTt*; d. *bbtt*, chestnut pacers and *BBTT*, black trotters.

## 9 - II.  DOMINANCE RELATIONS (p. 146)
## MULTIPLE EFFECTS OF SINGLE GENES (p. 147)
## INTERACTIONS BETWEEN GENE PAIRS
##    (pp. 148 - 149)

1. a. Incomplete dominance;  b. Codominance; c. Multiple alleles;  d. Epistasis;  e. Pleiotropy
2. a. phenotype: all pink, genotype: all *RR'*.  b. Phenotype: all white, genotype: all *R'R'*.  c. Phenotype: ½ red; ½ pink, genotype: ½ *RR*; ½ *RR'*.  d. Phenotype: all red, genotype: all *RR*.
3. a. The man must have sickle-cell trait with the genotype $Hb^A Hb^S$, and the woman he married would have a normal genotype, $Hb^A Hb^A$.  The couple could be told that the probability is ½ that any child would have sickle-cell trait and ½ that any child would have the normal genotype.
b. Both the man and the woman have the genotype $Hb^S Hb^S$.  The probability of children from this marriage is:  ¼ normal, $Hb^A Hb^A$; ½ sickle-cell trait, $Hb^A Hb^S$; ¼ sickle-cell anemia, $Hb^S Hb^S$.

4. a. Genotypes: ¼ $I^A I^A$; ¼ $I^A I^B$; ¼ $I^A i$; ¼ $I^B i$, phenotypes: ½ A; ¼ AB; ¼ B.
b. Genotypes: ¼ $I^A I^B$; ¼ $I^B i$; ¼ $I^A i$; ¼ *ii*, phenotypes: ¼ AB; ¼ B; ¼ A; ¼ O.
c. Genotypes:  all $I^A i$, phenotypes: all A.
d. Genotypes:  all *ii*, phenotypes: all O.
e. Genotypes:  ¼ $I^A I^A$; ½ $I^A I^B$; ¼ $I^B I^B$, phenotypes: ¼ A; ½ AB; ¼ B.
5. a. ¼ color; ¾ white.  b. ¾ color; ¼ white.  c. ¼ color; ¾ white.
d. The genotype of the male parent is *RrPp* and the genotype of the female parent is *rrpp*.  The offspring are ¼ walnut comb, *RrPp*; ¼ rose comb, *Rrpp*; ¼ pea comb, *rrPp*; ¼ single comb, *rrpp*.
e. The genotype of the walnut-combed male is *RRpp* and the genotype of the single-combed female is *rrpp*.  All offspring are rose comb with the genotype *Rrpp*.

## 9 - III.  LESS PREDICTABLE VARIATIONS IN
##     TRAITS (pp. 150 - 151)
## EXAMPLES OF ENVIRONMENTAL EFFECTS ON
##     PHENOTYPE (p. 152)

1. c;  2. b;  3. a;  4. c;  5. a

## Self-Quiz

1. d;  2. b;  3. a;  4. c;  5. d;  6. e;  7. a;  8. a;  9. c;  10. d

## Integrating and Applying Key Concepts

*DdPp* × *Ddpp*

# Chapter 10    Chromosomes and Human Genetics

## 10 - I.  THE CHROMOSOMAL BASIS OF
##     INHERITANCE—AN OVERVIEW (pp. 156 - 158)
*Focus on Science:* PREPARING A KARYOTYPE
    DIAGRAM (pp. 159 - 160)

1. genes;  2. homologous;  3. Alleles;  4. crossing over; 5. sex;  6. autosomes;  7. karyotype;  8. metaphase

## 10 - II.  SEX DETERMINATION IN HUMANS
##     (pp. 160 - 161)
## EARLY QUESTIONS ABOUT GENE LOCATIONS
##     (pp. 162 - 163)

1. I;  2. J;  3. F;  4. A;  5. E;  6. B;  7. H;  8. C;  9. G; 10. D;  11. Two blocks of the Punnett square should be XX and two blocks should be XY. 12. sons;  13. mothers;  14. daughters;  15. twice;  16. X;  17. d;  18. a; 19. a. $F_1$ flies all have red eyes: 1/2 heterozygous red females : ½ red-eyed males.  b. $F_2$ Phenotypes: females all have red eyes; males ½ red eyes, ½ white eyes; Genotypes: females are $X^W X^W$, $X^W X^w$; males are $X^W Y$ and $X^w Y$.

## 10 - III.  HUMAN GENETIC ANALYSIS
##     (pp. 164 - 165)
## PATTERNS OF INHERITANCE (pp. 166 - 167)

1. d;  2. a and c;  3. b;  4. c;  5. a;  6. b;  7. b;  8. b and d;  9. a;  10. c;  11. c;  12. b;  13. b;  14. a;  15. b; 16. c;  17. a;  18. d;  19. The woman's mother is heterozygous normal, *Aa*, the woman is also heterozygous normal, *Aa*.  The albino man, *aa*, has two heterozygous normal parents, *Aa*.  The two normal children are heterozygous heterozygous normal, *Aa*; the albino child is *aa*.  20. Assuming the father is heterozygous with Huntington's disorder and the mother normal, the chances are ½ that the son will develop the disease.
21.  If only male offspring are considered, the probability is ½ that the couple will have a color-blind son.
22.  The probability is that ½ of the sons will have hemophilia; the probability is 0 that a daughter will express hemophilia; the probability is that ½ of the daughters will be carriers.  23. If the woman marries a normal male, the

chance that her son would be color-blind is ½. If she marries a color-blind male, the chance that her son would be color-blind is also ½. 24. a. Autosomal recessive; b. Autosomal dominant; c. X-linked recessive; d. Autosomal dominant; e. X-linked recessive; f. X-linked dominant; g. X-linked recessive; h. Autosomal dominant; i. Autosomal recessive

## 10 - IV.  CHANGES IN CHROMOSOME NUMBER (pp. 168 - 169)
## CHANGES IN CHROMOSOME STRUCTURE (p. 170)

1. All gametes will be abnormal. 2. One-half of the gametes will be abnormal. 3. Polyploid individuals have three or more chromosomes of each type. Polyploidy in humans is lethal. Polyploidy may cause disruptive interactions between the genes of autosomes and sex chromosomes in key steps of reproduction and development pathways. All but 1 percent of human polyploids die before birth. The rare newborns die within a month. Polyploid plants exhibit larger organs than normal diploid plants. 4. a. Due to nondisjunction, an abnormal $n + 1$ gamete (one extra chromosome) unites with a normal gamete at fertilization; the new individual will be trisomic ($2n + 1$) with three copies of one type of chromosome. b. Due to nondisjunction, an abnormal $n - 1$ gamete (short one chromosome) unites with a normal gamete at fertilization; the new individual will be monosomic ($2n - 1$) with only one copy of one type of chromosome. 5. c; 6. b; 7. c; 8. d; 9. a; 10. b; 11. d; 12. c; 13. a; 14. d; 15. duplication (B); 16. inversion (C); 17. deletion (A); 18. translocation (D)

## 10 - V.  *Focus on Bioethics:* PROSPECTS IN HUMAN GENETICS (p. 171)

1. a. Prenatal diagnosis; b. Genetic counseling; c. Phenotypic treatments; d. Genetic screening

## Self-Quiz

1. d; 2. d; 3. b; 4. c; 5. b; 6. a; 7. c; 8. b; 9. b; 10. c

---

# Chapter 11  DNA Structure and Function

## 11 - I.  DISCOVERY OF DNA FUNCTION (pp. 174 - 177)

1. a. Miescher: identified "nuclein" from nuclei of pus cells and fish sperm; discovered DNA. b. Griffith: discovered the transforming principle in *Streptococcus pneumoniae*; live, harmless R cells were mixed with dead S cells; R cells became S cells. c. Avery: reported that the transforming substance in Griffith's bacteria experiments was probably DNA, the substance of heredity. d. Hershey and Chase: worked with radioactive sulfur (protein) and phosphorus (DNA) labels; T4 bacteriophage and *E. coli* demonstrated that labeled phosphorus was in bacteriophage DNA and contained hereditary instructions for new bacteriophages.

## 11 - II.  DNA STRUCTURE (pp. 178 - 179)

1. A five-carbon sugar called deoxyribose, a phosphate group, and one of the four nitrogen-containing bases. 2. guanine (pu); 3. cytosine (py); 4. adenine (pu); 5. thymine (py) 6. deoxyribose (B); 7. phosphate group (G); 8. purine (C); 9. pyrimidine (A); 10. purine (E); 11. pyrimidine (D); 12. nucleotide (F); 13. T; 14. T; 15. F, sugar; 16. T; 17. T; 18. pairing; 19. constant; 20. sequence; 21. different; 22. Living organisms have so many diverse body structures and behave in different ways because the many different habitats of Earth have selected those genotypes most able to survive in those habitats. The remaining genotypes have perished. The directions that code for the building of those body structures and that enable the specific successful behaviors reside in DNA or in a few cases, RNA. All living organisms follow the same rules for base pairing between the two nucleotide strands in DNA; adenine always pairs with thymine in undamaged DNA and cytosine always pairs with guanine. All living organisms must extract energy from food molecules and the reactions of glycolysis occur in virtually all of Earth's species. That means that similar enzyme sequences enable similar metabolic pathways to occur. While virtually all living organisms on Earth use the same code and the same enzymes during replication, transcription and translation, the particular array of proteins being formed differs from individual to individual even of the same species according to the sequences of nitrogenous bases that make up an individual's chromosome(s), and therein lies the key to the enormous diversity of life on Earth: no two individuals have the exact same array of proteins in their phenotypes.

## 11 - III.  A CLOSER LOOK AT DNA (pp. 180 - 181)
*Focus on Health:* WHEN DNA CAN'T BE FIXED (p. 182)

1.

| | | | |
|---|---|---|---|
| T - | A | T | - A |
| G - | C | G | - C |
| A - | T | A | - T |
| C - | G | C | - G |
| C - | G | C | - G |
| C - | G | C | - G |
| old | new | new | old |

2. F, adenine bonds to thymine (during replication) or uracil (during transcription). 3. F, it is a conserving process because each "new" DNA molecule contains one "old" strand from the parent cell attached to a strand of "new" complementary nucleotides that were assembled from stockpiles in the cell. 4. T; 5. T; 6. DNA; 7. histones; 8. histone; 9. DNA; 10. nucleosome;

11. genes; 12. metaphase; 13. protein; 14. nucleosome; 15. DNA; 16. histones

# Chapter 12    From DNA to Proteins

### 12 - I.  TRANSCRIPTION OF DNA INTO RNA (pp. 184 - 187)

1. sequence; 2. gene; 3. transcription (translation); 4. translation (transcription); 5. transcription; 6. translation; 7. protein; 8. folded; 9. structural (functional); 10. functional (structural); 11. a. ribosomal RNA; rRNA; RNA molecule that associates with certain proteins to form the ribosome, the "workbench" on which polypeptide chains are assembled; b. messenger RNA; mRNA; RNA molecule that moves to the cytoplasm, complexes with the ribosome where translation will result in polypeptide chains; c. transfer RNA; tRNA; RNA molecule that moves into the cytoplasm, picks up a specific amino acid, and moves it to the ribosome where tRNA pairs with a specific mRNA code word for that amino acid; 12. RNA molecules are single-stranded, while DNA has two strands; uracil substitutes in RNA molecules for thymine in DNA molecules; ribose sugar is found in RNA, while DNA has deoxyribose sugar. 13. Both DNA replication and transcription follow base-pairing rules; nucleotides are added to a growing RNA strand one at a time as in DNA replication. 14. Only one region of a DNA strand serves as a template for transcription; transcription requires different enzymes (three types of RNA polymerase); the results of transcription are single-stranded RNA molecules, but replication results in DNA, a double-stranded molecule. 15. C; 16. B; 17. E; 18. A; 19. D; 20. A-U-G-U-U-C-U-A-U-U-G-U-A-A-U-A-A-A-G-G-A-U-G-G-C-A-G-U-A-G; 21. DNA (E); 22. introns (B); 23. cap (F); 24. exons (A); 25. tail (D); 26. mature mRNA transcript (C)

### 12 - II.  DECIPHERING mRNA TRANSCRIPTS (pp. 188 - 189)
### STAGES OF TRANSLATION (pp. 190 - 191)

1. F; 2. B; 3. G; 4. H; 5. C; 6. A; 7. E; 8. D; 9. a. initiation; b. chain elongation; c. chain termination; 10. mRNA transcript: AUG UUC UAU UGU AAU AAA GGA UGG CAG UAG; 11. tRNA anticodons: UAC AAG AUA ACA UUA UUU CCU ACC GUC—; 12. amino acids: (start) met phe tyr cys asn lys gly try gln (stop); 13. amino acids; 14. three; 15. one; 16. mRNA; 17. codon; 18. mRNA; 19. assembly (synthesis); 20. Transfer; 21. amino acid; 22. protein (polypeptide); 23. codon; 24. anticodon

### 12 - III.  HOW MUTATIONS AFFECT PROTEIN SYNTHESIS (p. 192)
### SUMMARY OF PROTEIN SYNTHESIS (p. 193)

1. Viruses, ultraviolet radiation, and certain chemicals are examples. 2. mutagens; 3. substitution; 4. amino acid; 5. hemoglobin; 6. transposable; 7. DNA (H);

8. new mRNA transcript (J); 9. intron (E); 10. exon (A); 11. mature mRNA transcript (L); 12. tRNAs (C); 13. rRNA subunits (G); 14. mRNA (B); 15. anticodon (K); 16. amino acids (D); 17. tRNA (F); 18. ribosome-mRNA complex (I); 19. polypeptide (M)

### 12 - IV.  THE NATURE OF CONTROLS OVER GENE EXPRESSION (p. 194)
### EXAMPLES OF GENE CONTROL IN PROKARYOTIC CELLS (pp. 194 - 195)

1. a. Repressor protein; prevents transcription enzymes (RNA polymerases) from binding to DNA; this is negative transcription control. b. Hormones; major agents of vertebrate gene control; signaling molecules that move through the bloodstream to affect gene expression in target cells. c. Promoter; specific base sequences on DNA that serve as a binding site for a control agent; before RNA assembly can occur on DNA, the enzymes must bind with the promoter site. d. Operator; short DNA base sequences between promoter and the start of a gene; a binding site for control agents. 2. regulator gene (K); 3. genes that code for synthesizing proteins (G); 4. repressor protein (E); 5. promoter (J); 6. operator (B); 7. lactose operon (A); 8. RNA polymerase (D); 9. repressor-lactose complex (F); 10. lactose (C); 11. mRNA transcript (I); 12. lactose-metabolizing enzymes (H); 13. *Escherichia coli*; 14. operon; 15. transcription controls; 16. regulator; 17. promoter; 18. negative control; 19. low; 20. RNA polymerase (mRNA transcription); 21. blocks; 22. repressor protein; 23. operator; 24. needed (required)

### 12 - V.  GENE CONTROL IN EUKARYOTIC CELLS (pp. 196 - 197)
### *Focus on Science:* GENES, PROTEINS, AND CANCER (pp. 198 - 199)

1. All cells in the body descend from the same zygote; as cells divide to form the body, they become specialized in composition, structure, and function—they differentiate through selective gene expression. 2. DNA (genes); 3. differentiation; 4. selective; 5. controls; 6. regulatory; 7. activators; 8. Barr; 9. mosaic; 10. anhidrotic ectodermal dysplasia; 11. selective; 12. F, increase, high, start; 13. T; 14. T; 15. F, bring about; 16. T; 17. F, abnormal

### Self-Quiz

1. c; 2. b; 3. c; 4. a; 5. c; 6. a; 7. a; 8. d; 9. b; 10. d; 11. b; 12. b; 13. c; 14. d

# Chapter 13    Recombinant DNA and Genetic Engineering

**13 - I.   RECOMBINATION IN NATURE—AND IN THE LABORATORY (pp. 202 - 205)**
**WORKING WITH DNA FRAGMENTS (pp. 206 - 207)**
*Focus on Science:* RIFF-LIPS AND DNA FINGERPRINTS (p. 208)

1. The bacterial chromosome, a circular DNA molecule, contains all the genes necessary for normal growth and development. Plasmids, small, circular molecules of "extra" DNA, carry only a few genes and are self-replicating.  2. F, small circles of DNA in bacteria;  3. T;  4. mutations;  5. recombinant DNA;  6. species;  7. amplify;  8. protein;  9. research;  10. Genetic engineering;  11. plasmid;  12. viruses;  13. a. Bacterial enzymes that cut apart DNA molecules injected into the cell by viruses; several hundred have been identified.  b. A replication enzyme that joins the short fragments of DNA.  c. A collection of DNA fragments produced by restriction enzymes and incorporated into plasmids.  d. After a DNA library is inserted into a host cell's cloning vector (often a plasmid), repeated replications and divisions of the host cells produce multiple, identical copies of DNA fragments, or cloned DNA.  e. A method for making many identical copies of any DNA sequences.  f. Several procedures allow researchers to determine the nucleotide sequence of a DNA fragment (see Fig. 13.6).  g. RFLPs are variations in the banding patterns of DNA fragments from different individuals; the variations occur because no two individuals (except identical twins) have identical base sequences in their entire DNA set. Restriction enzymes cut different DNA molecules at different sites and into different numbers of DNA fragments.  h. A DNA fingerprint is a unique array of RFLPs.  14. B;  15. F;  16. E;  17. C;  18. A;  19. D;  20. T;  21. F;  22. F;  23. T;  24. F;  25. B;  26. D;  27. G;  28. H;  29. C;  30. I;  31. E;  32. F;  33. A;  34. fingerprint;  35. splicing enzymes;  36. introns;  37. sequencing;  38. "Good" alleles that code for producing normal LDL receptors on cell surfaces exist in DNA libraries. Through PCR, they can be copied over and over again. Through recombinant DNA techniques, the good alleles can be spliced into harmless viruses that are allowed to infect liver cells surgically removed from the patient and cultured in the laboratory. Later about a billion of the modified cells can be infused into the patient's hepatic portal vein, which leads directly to the liver. Some modified cells with the substituted "good" alleles may take up residence, produce the receptors and begin removing cholesterol from the bloodstream.  39. TACAAGATAACATTAGTCATC.

**13 - II.   MODIFIED HOST CELLS (p. 209)**
**BACTERIA, PLANTS, AND THE NEW TECHNOLOGY (pp. 210 - 211)**
**GENETIC ENGINEERING OF ANIMALS (p. 212)**
*Focus on Bioethics:* REGARDING HUMAN GENE THERAPY (p. 213)

1. a. A nucleic acid hybridization technique used to identify bacterial colonies harboring the DNA (gene) of interest; a short nucleotide sequence is assembled from radioactively labeled subunits (part of the sequence must be complementary to that of the desired gene).  b. Base pairing between nucleotide sequences from different sources can indicate that the two different sources carry the same gene(s).  c. Any DNA molecule "copied" from mRNA.  d. A process that uses a viral enzyme to transcribe mRNA into DNA, which can then be inserted into a plasmid for amplification; the same cDNA can be inserted into bacteria, which they will command to make a specific protein.  2. probes (hybridization;  3. expression (translation);  4. engineered;  5. introns;  6. transcriptase;  7. mRNA;  8. cDNA;  9. Enzyme;  10. DNA;  11. cDNA;  12. template (transcript);  13. Researchers are working to sequence the estimated 3 billion nucleotides present in human chromosomes.  14. If human collagen can be produced, it may be used to correct various skin, cartilage, and bone disorders.  15. The plasmid of *Agrobacterium* can be used as a vector to introduce desired genes into cultured plant cells; *Agrobacterium* was used to deliver a firefly gene into cultured tobacco plant cells.  16. Certain cotton plants have been genetically engineered for resistance to worm attacks.  17. In separate experiments the rat and human somatotropin genes became integrated into the mouse DNA. The mice grew much larger than their normal littermates.  18. Bacteria that have specific proteins on their cell surfaces facilitate ice crystal formation on whatever substrate the bacteria are located; bacteria without the ability to synthesize those proteins ("ice-minus") have been genetically engineered and were sprayed upon strawberry plants. Nothing bad happened.  19. Genetically engineered bacteria, harmless to begin with, have had specific genes spliced into them that allow them to degrade oil. If "fail-safe" genes are spliced in also, specific cues in the environment will cause the fail-safe gene to trip into action, membrane function will be destroyed and the genetically engineered bacteria will die.  20. ice-minus;  21. body cells (genetic material);  22. gene therapy;  23. eugenic engineering

## Self-Quiz

1. a;  2. a;  3. b;  4. c;  5. d;  6. a;  7. c;  8. b;  9. a;  10. d

# Chapter 14 Microevolution

**14 - I. EARLY BELIEFS, CONFOUNDING DISCOVERIES (pp. 216 - 219)**
**A FLURRY OF NEW THEORIES (pp. 220 - 221)**
**DARWIN'S THEORY TAKES FORM (pp. 222 - 223)**
1. G; 2. H; 3. J; 4. I; 5. D; 6. E; 7. F; 8. C; 9. B; 10. A; 11. b; 12. a; 13. b; 14. a; 15. a; 16. b; 17. a; 18. b; 19. a; 20. b; 21. b; 22. b; 23. a. John Henslow; b. Cambridge University; c. H. M. S. *Beagle*; d. Charles Lyell; e. Thomas Malthus; f. Galápagos Islands; g. Natural selection; h. Alfred Wallace; i. *Archaeopteryx*

**14 - II. INDIVIDUALS DON'T EVOLVE; POPULATIONS DO (pp. 224 - 225)**
*Focus on Science:* **WHEN IS A POPULATION *NOT* EVOLVING? (pp. 226 - 227)**
1. M; 2. P; 3. M; 4. B; 5. P; 6. M; 7. B; 8. P; 9. B; 10. M; 11. D; 12. C; 13. E; 14. B; 15. A; 16. Phenotypic variation is due to the effects of genes and the environment. The environment can mediate how the genes governing traits are expressed in an individual. For example, ivy plants grown from cuttings of the same parent all have the same genes, but the amount of sunlight affects the genes governing leaf growth. Ivy plants growing in full sun have smaller leaves than those growing in full shade. Offspring inherit genes, not phenotype.
17. There are five conditions: no genes are undergoing mutation; a very, very large population; the population is isolated from other populations of the species; all members of the population survive and reproduce; and mating is random.
18. a. 0.64 *BB*, 0.16 *Bb*, 0.16 *Bb*, and 0.04 *bb*; b. genotypes: 0.64 *BB*, 0.32 *Bb*, and 0.04 *bb*; phenotypes: 96% black, 4% gray.
c.

| Parents | B sperm | b sperm |
|---|---|---|
| 0.64 *BB* | 0.64 | 0 |
| 0.32 *Bb* | 0.16 | 0.16 |
| 0.04 *bb* | 0 | 0.04 |
| Totals = | 0.80 | 0.20 |

19. Find (b) first, then (c), and finally (a). a. $2pq = 2 \times (0.9) \times (0.1) = 2 \times (0.09) = 0.18 = 18\%$, which is the percentage of heterozygotes. b. $p^2 = \sqrt{0.81}$, $p = 0.81 = 0.9$ = the frequency of the dominant allele. c. $p + q = 1$, $q = 1.0 - 0.9 = 0.1$ = the frequency of the recessive allele.
20. a. homozygous dominant = $p^2 \times 200 = (0.8)^2 \times 200 = 0.64 \times 200 = 128$ individuals. b. q = (1.00 - p) = 0.20; homozygous recessive = $q^2 \times 200 = (0.2)^2 \times 200 = (0.04) (200) = 8$ individuals. c. heterozygotes = $2pq \times 200 = 2 \times 0.8 \times 0.2 \times 200 = 0.32 \times 200 = 64$ individuals. Check: 128 + 8 + 64 = 200.

21. If $p = 0.70$, since $p + q = 1$, $0.70 + q = 1$; then $q = 0.30$, or 30 percent.
22. If $p = 0.60$, since $p + q = 1$, $0.60 + q = 1$; then $q = 0.40$; thus, $2pq = 0.48$, or 48 percent.
23. D; 24. H; 25. I; 26. J; 27. G; 28. C; 29. A; 30. F; 31. B; 32. E; 33. population; 34. Hardy-Weinberg; 35. gene frequency; 36. genetic equilibrium; 37. mutations; 38. genetic drift; 39. Gene; 40. Natural selection; 41. Natural selection

**14 - III. A CLOSER LOOK AT NATURAL SELECTION (p. 227)**
**DIRECTIONAL CHANGE IN THE RANGE OF VARIATION (pp. 228 - 229)**
**SELECTION AGAINST OR IN FAVOR OF EXTREME PHENOTYPES (pp. 230 - 231)**
**SPECIAL OUTCOMES OF SELECTION (pp. 232 - 233)**
1. a. The most common phenotypes are favored; the example of phenotypic variation within the population of butterflies, and flies that induce the formation of intermediate-sized galls for survival success. b. Allele frequencies shift in a steady, consistent direction in response to a new environment or a directional change in an old one; light to dark forms of peppered moths. c. Forms at both ends of the phenotypic range are favored, and intermediate forms are selected against; finches in Cameroon, West Africa. 2. a. directional; b. disruptive; c. stabilizing; 3. natural; 4. Stabilizing; 5. Directional; 6. Disruptive; 7. Gallflies; 8. Sickle-cell anemia; 9. stabilizing; 10. allele; 11. balanced polymorphism; 12. sexual dimorphism; 13. selection; 14. Sexual; 15. sexual; 16. c, d; 17. e; 18. c; 19. b; 20. d; 21. b; 22. a; 23. c, d; 24. b; 25. b; 26. b; 27. e; 28. a; 29. e.

**14 - IV. GENE FLOW (p. 233)**
**GENETIC DRIFT (pp. 234 - 235)**
1. a; 2. b; 3. a; 4. b; 5. a; 6. a; 7. a; 8. b; 9. a; 10. b; 11. In the founder effect, a few individuals leave a population and establish a new one; by chance, allele frequencies will differ from the original population. In bottlenecks, disease, starvation, or some other stressful situation nearly eliminates a population; relative allele frequencies are randomly changed.

**Self-Quiz**

1. c; 2. d; 3. d; 4. b; 5. d; 6. c; 7. b; 8. a; 9. c; 10. e

# Chapter 15    Speciation

## 15 - I.  ON THE ROAD TO SPECIATION
   (pp. 238 - 241)
1. c;  2. a;  3. b;  4. a;  5. b;  6. a;  7. c;  8. c;  9. a;
10. c;  11. C (pre);  12. A (pre);  13. B (post);  14. G
(pre);  15. H (post);  16. E (pre);  17. D (pre);  18. F
(post);  19. f;  20. c;  21. a;  22. e;  23. d;  24. b;
25. f;  26. a. one;  b. two;  c. Barely between A and B,
but considerable divergence begins between B and C.
d. D;  e. B and C

## 15 - II.  MODELS OF SPECIATION (pp. 242 - 243)
PATTERNS OF SPECIATION (pp. 244 - 245)
1. a. Sympatric;  b. Allopatric;  c. Parapatric;  2. sympatric;  3. parapatric;  4. allopatric;  5. allopatric;
6. allopatric;  7. a. Seven pairs;  b. AA;  c. Seven
pairs;  d. BB;  e. The chromosomes fail to pair in meiosis.  f. Fertilization of nonreduced gametes produced a

fertile tetraploid.  g. Sterile hybrid, ABD.  h. Fertilization of nonreduced gametes resulted in a fertile hexa-
ploid.  i. *Triticum monococcum*; the unknown wild wheat,
*T. tauschii*;  8. a. Horizontal branching;  b. Many
branchings of the same lineage at or near the same point
in geologic time.  c. A branch that ends before the present.  d. Softly angled branching.  e. Vertical continuation
of a branch.  f. A dashed line.  9. I;  10. H;  11. E;  12.
C;  13. J;  14. D;  15. A;  16. F;  17. B;  18. G;
19. Speciation;  20. species;  21. Divergence;
22. reproductive isolating;  23. geographic;  24. polyploidy (hybridization);  25. hybridization (polyploidy)

### Self-Quiz

1. c;  2. d;  3. d;  4. a;  5. c;  6. c;  7. b;  8. a;  9. c;
10. d;  11. e;  12. c;  13. c;  14. b

# Chapter 16    The Macroevolutionary Puzzle

## 16 - I.  FOSSILS—EVIDENCE OF ANCIENT LIFE
   (pp. 248 - 251)
EVIDENCE FROM COMPARATIVE EMBRYOLOGY
   (pp. 252 - 253)
1. F;  2. J (D);  3. I;  4. A;  5. E;  6. H;  7. C;  8. D (J);
9. B;  10. G

## 16 - II.  EVIDENCE OF MORPHOLOGICAL
   DIVERGENCE (pp. 254 - 255)
EVIDENCE FROM COMPARATIVE
   BIOCHEMISTRY (pp. 256 - 257)
1. T;  2. F, low;  3. F, convergence;  4. T;  5. F, differences;  6. d;  7. c;  8. a;  9. d;  10. b;  11. M;  12. M;
13. M;  14. B;  15. B;  16. B;  17. M;  18. B;  19. M;
20. M;  21. Systematics relies on taxonomy, the naming
and identification of organisms.  It also depends heavily
on phylogenetic reconstruction, the identification of
evolutionary patterns that unite different organisms.
Systematics also uses classification, the retrieval systems

that consist of many hierarchical levels, or ranks.  The
work of classification feeds back into taxonomy, since
naming species or higher taxa should reflect their place
in the hierarchy.

## 16 - III.  ORGANIZING THE EVIDENCE—
   CLASSIFICATION SCHEMES (pp. 258 - 259)
1. B;  2. E;  3. A;  4. D;  5. C.;  6. kingdom; phylum
(or division); class; order; family; genus; species;
7. kingdom: Plantae; division: Anthophyta; class:
Dicotyledonae; order: Asterales; family: Asteraceae;
genus: *Archibaccharis*; species: *linearilobis*;  8. G;  9. C;
10. J;  11. H;  12. B;  13. A;  14. E;  15. D;  16. I;
17. F

### Self-Quiz

1. c;  2. b;  3. b;  4. a;  5. d;  6. b;  7. a;  8. b;  9. a;
10. e;  11. d;  12. c

# Chapter 17    The Origin and Evolution of Life

## 17 - I.  CONDITIONS ON THE EARLY EARTH
   (pp. 262 - 265)
EMERGENCE OF THE FIRST LIVING CELLS
   (pp. 266 - 267)
LIFE ON A CHANGING GEOLOGIC STAGE
   (pp. 268 - 269)
1. f;  2. e;  3. a;  4. c;  5. d;  6. b;  7. f;  8. c;  9. b;
10. a;  11. d;  12. f;  13. e;  14. c;  15. a;  16. b;  17. e;
18. c;  19. a;  20. b;  21. D;  22. C;  23. E;  24. A;

25. B;  26. E;  27. F;  28. I;  29. B;  30. G;  31. H;
32. C;  33. A;  34. J;  35. D

## 17 - II.  ORIGIN OF PROKARYOTIC AND
   EUKARYOTIC CELLS (pp. 270 - 271)
*Focus on Science:* WHERE DID ORGANELLES COME
   FROM? (pp. 272 - 273)
LIFE IN THE PALEOZOIC ERA (pp. 274 - 275)
1. b;  2. b;  3. a;  4. b;  5. a;  6. a;  7. b;  8. a;  9. a;
10. b;  11. a;  12. b;  13. a;  14. a;  15. b;  16. organ-

elles; 17. gene; 18. plasma; 19. channel; 20. ER;
21. genes; 22. plasmids; 23. prokaryotic; 24. endosymbiosis; 25. eukaryotes; 26. Electron; 27. oxygen;
28. bacteria; 29. ATP; 30. host; 31. incapable;
32. mitochondria; 33. ATP; 34. DNA; 35. chloroplasts; 36. host; 37. eubacteria; 38. protistans;
39. F; 40. A; 41. G; 42. C; 43. E; 44. D; 45. B;
46. F; 47. A, B, C, D, E, G

## 17 - III.  LIFE IN THE MESOZOIC ERA
(pp. 276 - 277)
*Focus on the Environment:* HORRENDOUS END TO
DOMINANCE (p. 278)
LIFE IN THE CENOZOIC ERA (p. 279)
SUMMARY OF EARTH AND LIFE HISTORY
(pp. 280 - 281)
1. D; 2. E; 3. B; 4. F; 5. A; 6. G; 7. C; 8. D, E;
9. B; 10. A, C, F, G; 11. a. Mesozoic, 240 - 205;

b. Archean, 4,600 - 3,800;  c. Paleozoic, 435 - 360;
d. Paleozoic, 550 - 500;  e. Archean, 4,600 - 3,800;
f. Mesozoic, 135 - 65;  g. Paleozoic, 360 - 280;
h. Proterozoic, 2,500 - 570;  i. Archean, 3,800 - 2,500;
j. Paleozoic, 290 - 240;  k. Proterozoic, 700 - 550 mya;
l. Mesozoic, 181 - 65;  m. Proterozoic, 2,500 - 570;
n. Cenozoic, 65 - 1.65;  o. Cenozoic, 1.65 - present;
p. Paleozoic, 440 - 435;  q. Paleozoic, 435 - 360

## Self-Quiz
1. b; 2. c; 3. c; 4. b; 5. e; 6. d; 7. a; 8. d; 9. c;
10. b

---

# Chapter 18    Bacteria, Viruses, and Protistans

## 18 - I.  CHARACTERISTICS OF BACTERIA
(pp. 284 - 287)
BACTERIAL REPRODUCTION (p. 288)
1. b; 2. d; 3. c; 4. e; 5. a; 6. prokaryotic; 7. plasmids; 8. wall; 9. capsule; 10. micrometers; 11. binary
fission; 12. cocci; 13. bacilli; 14. spiral; 15. positive

## 18 - II.  BACTERIAL CLASSIFICATION (p. 289)
MAJOR GROUPS OF BACTERIA (pp. 290 - 291)
1. Archaebacteria; 2. peptidoglycan; 3. eubacteria;
4. cyanobacteria; 5. nitrogen-fixation; 6. phosphorus;
7. nitrogen; 8. *Rhizobium*; 9. endospores; 10 pressure;
11. *Clostridium botulinum* (*C. tetani*); 12. *Clostridium
tetani* (*C. botulinum*); 13. Lyme disease; 14. spirochete;
15. membrane receptors; 16. sunlight; 17. decomposers; 18. streptomycins; 19. penicillins; 20. K;
21. d, L; 22. c, C; 23. c, J; 24. a, G; 25. c, E; 26. a,
A; 27. b, I; 28. c, H; 29. e, B; 30. c, K; 31. c, N;
32. c, D; 33. c, M; 34. a, F

## 18 - III.  THE VIRUSES (pp. 292 - 293)
VIRAL MULTIPLICATION CYCLES (pp. 294 - 295)
*Focus on Health:* EBOLA AND OTHER EMERGING
PATHOGENS (p. 295)
1. a. Nonliving, infectious agents, smaller than the
smallest cells; require living cells to act as hosts for their
replication; not acted upon by antibiotics.
b. The core can be DNA or RNA; the capsid can be protein and/or lipid.
c. Bacteriophage viruses may use the lytic pathway, in
which the virus quickly subdues the host cells and replicates itself and descendants are released as the cell
undergoes lysis; or they may use a temperate pathway,
in which viral genes remain inactive inside the host cell

during a period of latency, which may be a long time,
before activation and lysis.
2. There can be multiple answers (see Table 18.3 in text).
a. Possible answers include Herpes simplex (a Herpesvirus), Varicella-zoster (a Herpesvirus), Rhinovirus (a
Picornavirus), Poliovirus (an enterovirus of the Picorna
group), and HIV (a Retrovirus).
b. Herpes simplex: DNA virus. Initial infection is a lytic
cycle that causes herpes (sores) on mucous membranes
on mouth or genitals. Recurrent infections are temperate. Most cells are in nerves and skin. No immunity. No
cure. Varicella-zoster: DNA virus. Initial infection is a
lytic cycle that causes sores on skin. Generally, immunity is conferred by one infection, but in some people
subsequent infections follow temperate cycles and cause
"shingles." Rhinovirus: RNA virus. Causes the *common
cold*. Host cells are generally mucus-producing cells of
respiratory tract. Poliovirus: RNA virus causes *polio*.
Host cells are in motor nerves that lead to the diaphragm
and other important muscles. Destruction of these nerve
cells causes paralysis that may be temporary or permanent. Recurrences can occur. Immunize your children!
HIV: RNA virus. Host cells are specific white blood
cells. Temperate cycle has a latency period that may last
longer than a year before host tests positive for HIV. As
white blood cells are destroyed, the host's immune system is progressively destroyed (AIDS). No cure exists.
3. virus; 4. nucleic acid; 5. protein coat (viral capsid);
6. Viruses; 7. *Herpesvirus* (or *Varicella*); 8. viroids;
9. Retroviruses (HIV); 10. lysogenic; 11. Nanometers;
12. micrometers; 13. 86,000; 14. latency; 15. prions;
16. Rhinoviruses, RNA; 17. Retroviruses, RNA;
18. Herpesviruses, DNA; 19. C; 20. F; 21. H; 22. E;
23. D; 24. K; 25. I; 26. G; 27. L; 28. B; 29. A;
30. J

## 18 - IV. PROTISTAN CLASSIFICATION (p. 296) PREDATORY AND PARASITIC MOLDS (pp. 296 - 297)

1. a. E; b. E; c. P; d. E; e. P; f. E; g. E; 2. (+): a, c, d, e, g; (–) b, f, h, i, j; 3. Chytrids; 4. water molds (Oomycetes); 5. decomposers; 6. slime molds; 7. reproductive structures; 8. spores; 9. plasmodial; 10. meters; 11. late blight; 12. fish; 13. downy mildew; 14. chytrids; 15. enzymes

## 18 - V. ANIMAL-LIKE PROTISTANS (pp. 298 - 299)
*Focus on Health:* MALARIA AND THE NIGHT-FEEDING MOSQUITOES (p. 300)

1. pseudopods; 2. Foraminiferans; 3. spines; 4. radiolarians; 5. trypanosomes; 6. freshwater; 7. contractile vacuoles; 8. gullet; 9. enzyme-filled vesicles; 10. *Plasmodium*; 11. mosquito; 12. gametes; 13. C, E, I, J; 14. C, E, K; 15. B, F, J; 16. A; 17. D, H; 18. B, F; 19. B, F, G; 20. B; 21. A; 22. D; 23. C; 24. E

## 18 - VI. THE (MOSTLY) PHOTOSYNTHETIC SINGLE-CELLED PROTISTANS (p. 301) THE (MOSTLY) MULTICELLED PHOTOSYNTHETIC PROTISTANS (pp. 302 - 303)

1. chloroplasts; 2. eyespot; 3. heterotrophic; 4. algae; 5. Chrysophytes; 6. diatoms; 7. fucoxanthin; 8. silica (glass); 9. diatomaceous earth; 10. filters; 11. phytoplankton; 12. Dinoflagellates; 13. agar; 14. marine; 15. walls; 16. brown; 17. algin; 18. photosynthetic; 19. cellulose; 20. starch;

21. a. Chlorophyll $a$, phycobilins. b. Cyanobacteria. c. Agar, used as a moisture-preserving agent and culture medium; carrageenan is a stabilizer of emulsions. d. *Bonnemaisonia, Euchema.* e. Chlorophylls $a$ and $c_1$ and $c_2$ + various carotenoids such as fucoxanthin. f. Chrysophytes. g. Algin, used as a thickener, emulsifier and stabilizer of foods, cosmetics, medicines, paper and floor polish; also are sources of mineral salts and fertilizer. h. *Macrocystis* (sea palm), *Sargassum, Laminaria.* i. Chlorophylls $a$ and $b$. j. A heterotrophic prokaryote fusing with endosymbiont autotrophic prokaryote. k. Chlorophytes form much of the phytoplankton base of many food webs that support humans. l. *Volvox, Ulva, Spirogyra, Chlamydomonas.* 22. zygote, F; 23. resistant zygote, B; 24. meiosis and germination, H; 25. asexual reproduction, C; 26. gamete production, E; 27. gametes meet, G; 28. cytoplasmic fusion, A; 29. fertilization, D

### Self-Quiz

1. d; 2. a; 3. b; 4. c; 5. b; 6. a; 7. b; 8. c; 9. a; 10. b; 11. d; 12. A, E; 13. A, I; 14. A, B; 15. A, E, J; 16. A, C, D; 17. A, H; 18. A, F, G; 19. A, C; 20. B; 21. C; 22. A; 23. D; 24. E; 25. F; 26. D; 27. B; 28. C; 29. A; 30. A, C; 31. A, D; 32. A; 33. B, E; 34. B, F; 35. A, D; 36. A, D

# Chapter 19   Plants and Fungi

## 19 - I. EVOLUTIONARY TRENDS AMONG PLANTS (pp. 306 - 309)

1. C; 2. D; 3. E; 4. A; 5. B; 6. a. Well-developed root systems; b. Well-developed shoot systems; c. Xylem; d. Phloem; e. Lignin production; f. Cuticle; g. Stomata; h. Interaction of young roots and mycorrhizal fungi; i. Sporophyte (diploid) dominance of the life cycle; j. Heterospory; k. Haploid microspores mature to become pollen grains; l. Seeds; 7. C; 8. A; 9. B

## 19 - II. BRYOPHYTES (pp. 310 - 311)

1. air; 2. T; 3. rhizoids; 4. T; 5. Mosses; 6. sporophytes; 7. T; 8. Peat; 9. water; 10. T; 11. gametophyte; 12. gametophyte; 13. sperm; 14. Fertilization; 15. zygote; 16. sporophyte; 17. Meiosis; 18. spores; 19. gametophytes

## 19 - III. *Focus on the Environment:* ANCIENT PLANTS, CARBON TREASURES (p. 311) EXISTING SEEDLESS VASCULAR PLANTS (pp. 312 - 313)

1. c; 2. b; 3. e; 4. b; 5. d; 6. d; 7. e; 8. c; 9. a;

10. d; 11. e; 12. d; 13. c; 14. c; 15. e; 16. b; 17. d; 18. d; 19. b; 20. e; 21. rhizome; 22. Meiosis; 23. spores; 24. spores; 25. gametophyte; 26. gametophyte; 27. sperms; 28. egg; 29. fertilization; 30. zygote; 31. embryo

## 19 - IV. GYMNOSPERMS—PLANTS WITH "NAKED" SEEDS (pp. 314 - 315)

1. b; 2. d; 3. a; 4. b; 5. c; 6. a; 7. e; 8. b; 9. e; 10. c; 11. d; 12. e; 13. sporophyte; 14. cones; 15. cones; 16. ovule; 17. meiosis; 18. gametophyte; 19. eggs; 20. meiosis; 21. pollen; 22. Pollination; 23. tube; 24. Sperm; 25. Fertilization; 26. embryo

## 19 - V. ANGIOSPERMS—FLOWERING, SEED-BEARING PLANTS (p. 316)

1. B; 2. C; 3. E; 4. D; 5. A; 6. G; 7. F. 8. a. Gametophyte, none or simple vascular tissue, no. b. Sporophyte, yes, no. c. Sporophyte, yes, no. d. Sporophyte, yes, no. e. Sporophyte, yes, yes. f. Sporophyte, yes, yes.

**19 - VI. CHARACTERISTICS OF FUNGI (p. 318)**
1. C; 2. D; 3. B; 4. E; 5. A

**19 - VII. CONSIDER THE CLUB FUNGI**
   **(pp. 318 - 319)**
1. spore; 2. Club; 3. Nuclear; 4. diploid;
5. Meiosis; 6. spores; 7. nuclear; 8. dikaryotic;
9. C; 10. D; 11. E; 12. B; 13. A

**19 - VIII. SPORES AND MORE SPORES**
   **(pp. 320 - 321)**
1. hyphae; 2. gametangia; 3. zygosporangium;
4. spores; 5. mycelium; 6. multicelled; 7. T;
8. *Neurospora crassa*; 9. truffles; 10. *Aspergillis*;
11. *Penicillium*; 12. T; 13. Ascocarps; 14. T; 15. T;
16. If the sexual phase is unknown for a particular fungus, it is referred to the informal group known as the "imperfect fungi." If mycologists later detect a sexual phase in the fungus, it is then assigned to (usually) the sac fungi or the club fungi. An example of an imperfect fungus is the fungus that is predatory on roundworms, *Arthropbotrys dactyloides.*

**19 - IX. *Commentary:* A LOOK AT THE UNLOVED**
   **FEW (p. 322)**
**BENEFICIAL ASSOCIATIONS BETWEEN FUNGI**
   **AND PLANTS (pp. 322 - 323)**
1. a. *Histoplasma capsulatum,* Parasite of human respiratory system, infection can damage all organs.

b. *Claviceps purpurea,* Parasite of rye and other grains, affects humans, vomiting, diarrhea, hallucinations, hysteria, and convulsions. c. *Epidermophyton fluccosum,* Parasite of human skin, feeds on warm, damp tissues between toes, turns skin scaly, red, and cracked. d. *Venturia inequalis,* Apple, causes scabs to form on apples. 2. Symbiosis; 3. mutualism; 4. lichen; 5. parasitism; 6. hostile; 7. air; 8. soil; 9. environmental; 10. mycorrhizae; 11. carbohydrates; 12. minerals; 13. exomycorrhiza; 14. temperate; 15. endomycorrhizae; 16. soil; 17. air pollution

**Self-Quiz**

1. *Polyporus* (C). 2. Fly agaric mushroom (C). 3. Cup fungus (A). 4. Algae and fungi—lichen (E). 5. Algae and fungi—lichen (E). 6. Mushroom (C). 7. *Eupenicillium* (A). 8. c; 9. a; 10. c; 11. a; 12. d; 13. b; 14. a; 15. d; 16. c; 17. c; 18. b; 19. d; 20. d; 21. c; 22. a; 23. a; 24. c; 25. c; 26. d; 27. a; 28. a and c; 29. b.

---

# Chapter 20   Animals: The Invertebrates

**20 - I. OVERVIEW OF THE ANIMAL KINGDOM**
   **(pp. 326 - 329)**
**PUZZLES ABOUT ORIGINS (p. 330)**
**SPONGES—SUCCESS IN SIMPLICITY**
   **(pp. 330 - 331)**
1. K; 2. F; 3. L; 4. A; 5. D; 6. H; 7. G; 8. C;
9. E; 10. J; 11. I; 12. B; 13. a. Placozoa;
b. Sponges; c. Cnidaria; d. Tubellarians, flukes, tapeworms; e. Nematoda; f. Rotifera; g. Snails, slugs, clams, squids, octopuses; h. Annelida; i. Crustaceans, spiders, insects; j. Echinodermata; k. Chordata; l. Vertebrates; 14. c; 15. a; 16. c; 17. d; 18. c; 19. b; 20. c; 21. a; 22. c; 23. b; 24. mollusks; 25. chordates; 26. coelomate; 27. bilateral; 28. radial; 29. J; 30. G; 31. K; 32. I; 33. B; 34. C; 35. H; 36. D; 37. A; 38. F; 39. E

**20 - II. CNIDARIANS—TISSUES EMERGE**
   **(pp. 332 - 333)**
1. cnidarians; 2. nematocysts; 3. medusa; 4. polyp;
5. gastrodermis; 6. epidermis; 7. epithelium;
8. Nerve; 9. sensory; 10. contractile; 11. mesoglea;
12. hydrostatic; 13. corals; 14. gonads; 15. planulas;
16. feeding polyp; 17. reproductive polyp; 18. female medusa; 19. planula

**20 - III. FLATWORMS, ROUNDWORMS,**
   **ROTIFERS—AND SIMPLE ORGAN SYSTEMS**
   **(pp. 334 - 335)**
*Focus on Health:* **A ROGUE'S GALLERY OF**
   **PARASITIC WORMS (p. 336)**
1. b; 2. a; 3. c; 4. a; 5. c; 6. b; 7. c; 8. b; 9. b;
10. a; 11. c; 12. a; 13. b; 14. a; 15. a; 16. c; 17. c;
18. a; 19. a; 20. a; 21. b; 22. a; 23. b; 24. b; 25. b;
26. branching gut; 27. pharynx; 28. brain; 29. nerve cord; 30. ovary; 31. testis; 32. planarian (genus name = *Dugesia*); 33. no; 34. yes; 35. no coelom or acoelomate; 36. roundworm; 37. no; 38. a "false" coelom (pseudocoelomate); 39. sexually; 40. larvae; 41. snail; 42. larvae; 43. human; 44. Larvae; 45. intermediate; 46. human; 47. small intestine; 48. proglottids; 49. organs; 50. proglottids; 51. feces; 52. larval; 53. intermediate

**20 - IV. A MAJOR DIVERGENCE (p. 337)**
**MOLLUSKS—A WINNING BODY PLAN**
   **(pp. 338 - 339)**
1. b; 2. b; 3. a; 4. b; 5. a; 6. b; 7. a; 8. a; 9. b;
10. a; 11. III; 12. I; 13. II; 14. mouth; 15. anus;
16. gill; 17. heart; 18. radula; 19. foot; 20. shell;
21. stomach; 22. mouth; 23. gill; 24. mantle;

25. muscle; 26. foot; 27. stomach; 28. internal shell; 29. mantle; 30. reproductive organ; 31. gill; 32. ink sac; 33. tentacle; 34. soft; 35. bilateral; 36. mantle; 37. shell; 38. gills; 39. torsion; 40. foot; 41. head; 42. radula; 43. cephalopods; 44. gastropods; 45. slugs; 46. bivalves; 47. Humans; 48. foot; 49. respiration; 50. mucus; 51. palps; 52. siphons; 53. wastes; 54. Cephalopods; 55. smartest; 56. tentacles; 57. radula; 58. jet; 59. oxygen; 60. closed; 61. hearts; 62. nervous; 63. brains; 64. humans; 65. invertebrates

## 20 - V. ANNELIDS—SEGMENTS GALORE (pp. 340 - 341)

1. F; 2. J; 3. B; 4. H; 5. D; 6. L; 7. C; 8. I; 9. A; 10. G; 11. E; 12. K; 13. brain; 14. pharynx; 15. nerve cord; 16. hearts; 17. blood vessel; 18. crop; 19. gizzard; 20. earthworm; 21. Annelida; 22. segmented and closed circulatory system; 23. Protostome; 24. yes; 25. bilateral; 26. yes

## 20 - VI. ARTHROPODS—THE MOST SUCCESSFUL ORGANISMS ON EARTH (p. 342)

1. e; 2. f; 3. a; 4. b; 5. f; 6. a; 7. f; 8. d; 9. b; 10. c; 11. e; 12. d; 13. b; 14. a; 15. a;
16. Arthropods, as a group, have the highest number of different species, occupy the most habitats, and have very efficient defenses against predators and competitors, and the capacity to exploit the greatest amounts and kinds of foods.

## 20 - VII. A LOOK AT SPIDERS AND THEIR KIN (p. 343)
## A LOOK AT THE CRUSTACEANS (pp. 344 - 345)
## VARIABLE NUMBERS OF MANY LEGS (p. 345)

1. E; 2. G; 3. F; 4. D; 5. C; 6. B; 7. A; 8. an exoskeleton; 9. lobsters and crabs; 10. similar; 11. Lobsters and crabs; 12. Barnacles; 13. Copepods; 14. barnacles; 15. barnacles; 16. molts; 17. millipedes; 18. centipedes; 19. Millipedes; 20. Centipedes; 21. cephalothorax; 22. abdomen; 23. swimmerets; 24. legs; 25. cheliped; 26. antennae; 27. lobster; 28. crustaceans; 29. exoskeleton and jointed legs; 30. Protostome; 31. yes; 32. bilateral; 33. yes; 34. poison gland; 35. brain; 36. heart; 37. spinnerets; 38. book lung; 39. chelicerates

## 20 - VIII. A LOOK AT INSECT DIVERSITY (pp. 346 - 347)

1. C; 2. E; 3. A; 4. D; 5. B; 6

## 20 - IX. THE PUZZLING ECHINODERMS (pp. 348 - 349)

1. deuterostomes; 2. echinoderms; 3. calcium; 4. radial; 5. bilateral; 6. brain; 7. tube; 8. water; 9. ampulla; 10. muscle; 11. whole; 12. digesting; 13. lower stomach; 14. upper stomach; 15. anus; 16. gonad; 17. coelom; 18. digestive gland; 19. eyespot; 20. tube feet; 21. starfish; 22. yes; 23. deuterostome; 24. water vascular system; 25. a. brittle stars; b. sea urchin; c. sea cucumber; d. feather star (crinoid); 26. Echinodermata; 27. Water vascular system and body wall with spines, spicules, or plates. 28. radial

## Self-Quiz

1. a; 2. c; 3. d; 4. d; 5. b; 6. c; 7. d; 8. c; 9. d; 10. a; 11. a; 12. b; 13. h, I; 14. b, FJL; 15. c, ACD; 16. g, BG; 17. i, H; 18. e, M; 19. j, O; 20. f, EN; 21. a, K; 22. d, P

# Chapter 21   Animals: The Vertebrates

## 21 - I. THE CHORDATE HERITAGE (pp. 352 - 354)
## INVERTEBRATE CHORDATES (pp. 354 - 355)
## EVOLUTIONARY TRENDS AMONG THE VERTEBRATES (pp. 356 - 357)
## EXISTING JAWLESS FISHES (p. 357)
## EXISTING JAWED FISHES (pp. 358 - 359)

1. nerve cord; 2. pharynx; 3. notochord; 4. invertebrate; 5. vertebrates; 6. lancelets; 7. filter-feeding; 8. gill slits; 9. Tunicates (sea squirts); 10. tadpoles; 11. notochord; 12. torsion bar; 13. lungs; 14. circulatory; 15. larva; 16. mutation; 17. sex organs; 18. predators; 19. cephalochordates (lancelets); 20. vertebrae; 21. predators; 22. jaws; 23. brain; 24. paired (fleshy); 25. fleshy; 26. adult tunicate (sea squirt); 27. lancelet; 28. notochord; 29. pharynx with gill slits; 30. mouth; 31. sievelike pharynx; 32. stomach; 33. intestine; 34. anus; 35. atrial opening (exit); 36. oral opening; 37. pharyngeal gill slits; 38. anus; 39. notochord; 40. dorsal, tubular nerve cord; 41. early jawless fish (agnathan); 42. supporting structure; 43. gill slit; 44. placoderm; 45. jaws; 46. acorn worms; 47. tunicates; 48. lancelets; 49. lampreys, hagfishes; 50. jawed, armored fishes; 51. cartilaginous fishes; 52. bony fishes; 53. Amphibia; 54. Reptilia; 55. Aves; 56. Mammalia; 57. sharks; 58. gill slits; 59. bony; 60. ray-finned; 61. C; 62. G; 63. J; 64. F; 65. M; 66. H; 67. N; 68. I; 69. B; 70. O; 71. D; 72. E; 73. E; 74. K; 75. L; 76. A; 77. all jawless; 78. filter-feeders; 79. jaws; 80. cartilage; 81. bone; 82. ray-finned fishes; 83. lobe-finned fishes; 84. branch ⑤; 85. branch ④; 86. Silurian; 87. About 375 million years ago.

**21 - II. AMPHIBIANS (pp. 360 - 361)**
**REPTILES (pp. 362 - 363)**
**BIRDS (pp. 364 - 365)**
**MAMMALS (pp. 366 - 367)**
1. lungs;  2. fins;  3. skeletal elements;  4. brains;
5. balance;  6. circulatory;  7. blood;  8. insects;
9. salamanders;  10. water;  11. reproduce;  12. toxins;
13. insects;  14. reptilian;  15. limb;  16. amniote;
17. turtles (lizards);  18. lizards (turtles);  19. internal;
20. Carboniferous;  21. turtles;  22. Carboniferous;
23. Triassic;  24. Jurassic;  25. Birds;  26. four;
27. synapsid;  28. Carboniferous;  29. G;  30. D;
31. E;  32. B;  33. C;  34. A;  35. F;  36. A;  37. A;
38. A;  39. C;  40. C;  41. D;  42. D;  43. a. dry, scaly
skin;  b. four- chambered heart;  c. feather development;  d. hair development;  e. loss of limbs;
44. embryo (notochord);  45. albumin;  46. yolk sac;
47. reptiles;  48. feathers;  49. sternum (breastbone);
50. air cavities;  51. shelled amniote;  52. temperature;
53. teeth;  54. platypus;  55. pouched (marsupials);
56. placental;  57. four;  58. hair

**21 - III. EVOLUTIONARY TRENDS AMONG THE**
**PRIMATES (pp. 368 - 369)**
**FROM EARLY PRIMATES TO HOMINIDS**
**(pp. 370 - 371)**
**EMERGENCE OF HUMANS (pp. 372 - 373)**
1. mammals;  2. hair (mammary glands);  3. orders;
4. Primates;  5. depth;  6. prehensile;  7. grasping;

8. smell;  9. brains;  10. culture;  11. hand bones;
12. teeth;  13. 60;  14. rodents;  15. insects;
16. Miocene;  17. drier;  18. grasslands;  19. dryopiths
(hominoids);  20. Asia;  21. 10;  22. 5;  23. b;  24. c;
25. c;  26. c;  27. c;  28. b;  29. d;  30. a;  31. a;  32. c;
33. c;  34. c;  35. d;  36. a;  37. e;  38. e;  39. b;  40. b;
41. apes;  42. 10;  43. 5;  44. 4.5;  45. australopiths;
46. bipedal;  47. 2.5;  48. *Homo erectus*;  49. fire;
50. tool;  51. Neanderthals;  52. *Homo erectus*;
53. biochemical;  54. immunological;  55. 40,000;
56. *Homo sapiens*;  57. B, G, H, K;  58. A, G, K;  59. C,
G, H, L;  60. D, J, N;  61. B, G, H, K;  62. E, I, M

**Self-Quiz**

1. d;  2. a;  3. d;  4. c;  5. early amphibian, Amphibia;
6. Arctic fox, Mammalia;  7. soldier fish, Osteichthyes;
8. Ostracoderm, Agnatha;  9. owl, Aves;  10. sea turtle,
Reptilia;  11. shark, Chondrichthyes;  12. coelacanth,
Osteichthyes (lobe-finned fish);  13. reef ray, Chondrichthyes;  14. tunicate, Urochordata;  15. lancelet,
Cephalochordata;  16. d, H;  17. b, B;  18. i, F;  19. g,
D;  20. h, A;  21. c, E;  22. a, I;  23. e, G;  24. f, C;
25. c;  26. a;  27. a;  28. a;  29. a;  30. b;  31. d;
32. B;  33. H;  34. C;  35. D;  36. A;  37. E;  38. G;
39. F

---

# Chapter 22   Plant Tissues

**22 - I. OVERVIEW OF THE PLANT BODY**
**(pp. 376 - 379)**
1. ground tissues (C);  2. vascular tissues (E);  3. dermal tissues (D);  4. shoot system (A);  5. root system
(B);  6. H;  7. J;  8. E;  9. G;  10. F;  11. B;  12. A;
13. D;  14. C;  15. I;  16. apical meristem;  17. primary
tissues derived from the apical meristem;  18. vascular
cambium;  19. cork cambium

**22 - II. TYPES OF PLANT TISSUES (pp. 380 - 381)**
1. collenchyma;  2. sclerenchyma;  3. parenchyma;
4. b;  5. c;  6. a;  7. a;  8. b;  9. b;  10. a;  11. c;
12. a;  13. a;  14. c;  15. pits (xylem);  16. cytoplasm
(xylem);  17. tracheids (xylem);  18. vessel (xylem);
19. vessel (xylem);  20. sieve (phloem);  21. companion
(phloem);  22. sieve (phloem);  23. sieve (phloem);
24. a. Vessel members and tracheids; no; conduct water
and dissolved minerals absorbed from soil, mechanical
support.  b. Sieve tube members and companion cells;
yes; transports sugar and other solutes.  25. a. Primary
plant body; cutin in the cuticle layer over epidermal cells
restricts water loss and resists microbial attack; openings
(stomates) permit water vapor and gases to enter and
leave the plant.  b. Secondary plant body; replaces epidermis to cover roots and stems.  26. a. One; in threes or

multiples thereof, usually parallel; one pore or furrow;
distributed throughout ground stem tissue.  b. Two, in
fours or fives or multiples thereof; usually netlike; three
pores or pores with furrows;  positioned in a ring in the
stem.

**22 - III. SHOOT PRIMARY STRUCTURE**
**(pp. 382 - 383)**
1. primordium;  2. apical meristem;  3. bud;  4. blade;
5. petiole;  6. node;  7. stem;  8. bud;  9. blade;
10. sheath;  11. node;  12. dicot;  13. monocot;
14. epidermis;  15. ground tissue;  16. vascular bundle;
17. sclerenchyma cells;  18. air space;  19. vessel;
20. sieve tube;  21. companion cell;  22. epidermis;
23. cortex;  24. vascular bundle;  25. pith;  26. vessels;
27. vascular cambium;  28. sieve tube;  29. fibers

**22 - IV. A CLOSER LOOK AT LEAVES**
**(pp. 384 - 385)**
*Commentary:* **USES AND ABUSES OF LEAVES**
**(p. 385)**
1. palisade mesophyll (D);  2. spongy mesophyll (B);
3. lower epidermis (A);  4. stomata (C);  5. vein(s) (E);
6. G;  7. K;  8. C;  9. L;  10. E;  11. I;  12. D;  13. J;
14. B;  15. A;  16. F;  17. M;  18. H

## 22 - V.  ROOT PRIMARY STRUCTURE (pp. 386 - 387)

1. root hair (H); 2. endodermis (G); 3. pericycle (B); 4. epidermis (E); 5. cortex (D); 6. root apical meristem (F); 7. root cap (A); 8. endodermis (G); 9. pericycle (B); 10. primary phloem (C); 11. taproot; 12. adventitious; 13. fibrous; 14. hairs; 15. vascular; 16. cortex; 17. oxygen; 18. endodermis; 19. endodermal; 20. control; 21. pericycle; 22. lateral

## 22 - VI.  WOODY PLANTS (pp. 388 - 389)

1. M; 2. C; 3. K; 4. N; 5. O; 6. P; 7. L; 8. E; 9. I; 10. D; 11. A; 12. H; 13. F; 14. B; 15. J;

16. G; 17. vascular cambium (C); 18. bark (E); 19. periderm (A); 20. phloem (D); 21. xylem (B)

### Self-Quiz

1. b; 2. a; 3. b; 4. c; 5. a; 6. d; 7. d; 8. d; 9. c; 10. c; 11. b; 12. b

---

# Chapter 23    Plant Nutrition and Transport

## 23 - I.  UPTAKE OF WATER AND NUTRIENTS AT THE ROOTS (pp. 392 - 395)

1. hydrogen; 2. thirteen; 3. mineral ions; 4. macronutrients; 5. micronutrients; 6. a. iron, micronutrient; b. potassium, macronutrient; c. magnesium, macronutrient; d. chlorine, micronutrient; e. manganese, micronutrient; f. molybdenum, micronutrient; g. nitrogen, macronutrient; h. sulfur, macronutrient; i. boron, micronutrient; j. zinc, micronutrient; k. calcium, macronutrient; l. copper, micronutrient; m. phosphorus, macronutrient; 7. C; 8. E; 9. A; 10. B; 11. F; 12. D; 13. vascular cylinder (G); 14. exodermis (B); 15. endodermis (F); 16. cytoplasm (A); 17. water movement (E); 18. Casparian strip (C); 19. endodermal cell wall (D); 20. mutualism; 21. Nitrogen "fixed" by bacteria; 22. root nodules; 23. scarce minerals; 24. Root hairs

## 23 - II.  Focus on the Environment: PUTTING DOWN ROOTS (p. 396)
## CONSERVATION OF WATER IN STEMS AND LEAVES (pp. 396 - 397)

1. 90 percent; 2. T; 3. T; 4. stomata; 5. opens; 6. T; 7. T; 8. decrease; 9. opens; 10. T; 11. night; 12. day

## 23 - III.  A THEORY OF WATER TRANSPORT (pp. 398 - 399)

1. water; 2. xylem; 3. transpiration; 4. Cohesion; 5. tension

## 23 - IV.  DISTRIBUTION OF ORGANIC COMPOUNDS THROUGH THE PLANT (pp. 400 - 401)

1. photosynthesis; 2. starch; 3. fats; 4. seeds; 5. Starch; 6. Fats; 7. Storage; 8. hydrolysis; 9. sucrose; 10. sucrose; 11. D; 12. F; 13. B; 14. G; 15. A; 16. C; 17. E; 18. B; 19. D; 20. A; 21. F; 22. C; 23. A

### Self-Quiz

1. c; 2. e; 3. c; 4. d; 5. b; 6. d; 7. b; 8. a; 9. b; 10. c

---

# Chapter 24    Plant Reproduction and Development

## 24 - I.  REPRODUCTIVE STRUCTURES OF FLOWERING PLANTS (pp. 404 - 407)
## Focus on the Environment: POLLEN SETS ME SNEEZING (p. 407)

1. sporophyte (C); 2. flower (B); 3. meiosis (D); 4. gametophyte (F); 5. gametophyte (E); 6. fertilization (A); 7. petal; 8. sepal; 9. anther; 10. stigma; 11. ovary; 12. stamen; 13. carpel; 14. ovule; 15. I; 16. C; 17. G; 18. F; 19. B; 20. E; 21. H; 22. A; 23. D

## 24 - II.  A NEW GENERATION BEGINS (pp. 408 - 409)

1. anther; 2. pollen sac; 3. microspore mother; 4. meiosis; 5. microspores; 6. pollen tube; 7. sperm-producing; 8. pollen; 9. stigma; 10. male gametophyte; 11. ovule; 12. integuments; 13. meiosis; 14. megaspores; 15. megaspore; 16. megaspore; 17. mitosis; 18. eight; 19. embryo sac; 20. female gametophyte; 21. endosperm; 22. egg; 23. Chemical and molecular cues guide a pollen tube's growth

through tissues of the ovary, toward the egg chamber and sexual destiny. 24. The embryo sac is the site of double fertilization. 25. a. Fusion of one egg nucleus (*n*) with one sperm nucleus (*n*); the plant embryo (2*n*); eventually develops into a new sporophyte plant. b. Fusion of one sperm nucleus (*n*) with the endosperm mother cell (2*n*); endosperm tissues (3*n*); nourishes the embryo within the seed.

## 24 - III. FROM ZYGOTE TO SEED (pp. 410 - 411)
1. a. "Seed leaves" that develop as part of the embryo; plants with large cotyledons absorb endosperm and store food, and thin cotyledons may produce enzymes for transferring food from the embryo to the germinating seedling. b. Ovules within the ovary; contains a 2*n* embryo sporophyte and 3*n* endosperm, these wrapped within a seed coat. c. Integuments of ovule; protection for embryo with its endosperm (stored food). d. Usually a mature, ripened ovary; seed protection and dispersal in specific environments. 2. nucleus; 3. vacuole; 4. zygote; 5. embryo; 6. embryo; 7. seed coat; 8. shoot tip; 9. cotyledons; 10. embryo; 11. endosperm; 12. root tip; 13. ovule; 14. fruit; 15. E; 16. A; 17. C; 18. F; 19. G; 20. B; 21. D; 22. f; 23. a; 24. g; 25. b; 26. e; 27. c; 28. d; 29. fruits; 30. animals; 31. parent; 32. fruits; 33. animals; 34. germination

## 24 - IV. ASEXUAL REPRODUCTION OF FLOWERING PLANTS (pp. 412 - 413)
1. E; 2. G; 3. I; 4. F; 5. H; 6. D; 7. B; 8. C; 9. A; 10. D; 11. E; 12. A; 13. B; 14. C

## 24 - V. PATTERNS OF GROWTH AND DEVELOPMENT (pp. 414 - 415)
1. embryo; 2. germination; 3. genes; 4. soil; 5. rains; 6. imbibition; 7. coat; 8. aerobic; 9. root;

10. primary; 11. divisions (enlargements); 12. enlargements (divisions); 13. dicot (monocot); 14. monocot (dicot); 15. heritable; 16. differentiation; 17. environmental; 18. growth

## 24 - VI. PLANT HORMONES (pp. 416 - 417)
*Focus on Science:* FOOLISH SEEDLINGS, GORGEOUS GRAPES (p. 417)
1. e; 2. c; 3. a; 4. d; 5. a; 6. e; 7. a; 8. b; 9. a; 10. c; 11. e; 12. c; 13. e; 14. c; 15. E; 16. F; 17. D; 18. H; 19. B; 20. G; 21. A; 22. C

## 24 - VII. ADJUSTMENTS IN THE RATE AND DIRECTION OF GROWTH (pp. 418 - 419)
1. a; 2. c; 3. d; 4. b; 5. d; 6. b; 7. a

## 24 - VIII. BIOLOGICAL CLOCKS AND THEIR EFFECTS (pp. 420 - 421)
1. D; 2. I; 3. H; 4. J; 5. B; 6. A; 7. G; 8. C; 9. F; 10. E; 11. Pr; 12. Pfr; 13. Pfr; 14. Pr; 15. response; 16. flowering; 17. longer; 18. shorter; 19. mature; 20. long-day; 21. short-day

## 24 - IX. LIFE CYCLES END, AND TURN AGAIN (p. 422)
1. b; 2. d; 3. c; 4. b; 5. a; 6. a; 7. c; 8. b; 9. a; 10. b; 11. a; 12. a

## Self-Quiz
1. c; 2. c; 3. b; 4. c; 5. c; 6. d; 7. b; 8. c; 9. d; 10. a; 11. d; 12. d

---

# Chapter 25   Tissues, Organ Systems, and Homeostasis

## 25 - I. EPITHELIAL TISSUE (pp. 426 - 429)
1. F, Anatomy; 2. F, unlike tissues; 3. T; 4. T; 5. F, impermeable; 6. F, Exocrine; 7. F, Exocrine; 8. tissue; 9. organ; 10. metabolic; 11. internal environment; 12. organ; 13. organ system (urinary system, excretory system); 14. organism; 15. homeostasis; 16. tissue; 17. epithelial; 18. connective (epithelial)

## 25 - II. CONNECTIVE TISSUE (pp. 430 - 431)
## MUSCLE TISSUE (p. 432)
## NERVOUS TISSUE (p. 433)
*Focus on Science:* FRONTIERS IN TISSUE RESEARCH (p. 433)
1. F, groups of; 2. T; 3. F, synchronously with; 4. T; 5. T; 6. T; 7. ground substance; 8. Loose; 9. Tendons (Ligaments); 10. ligaments (tendons); 11. connective, D, 9, 11; 12. epithelial, G, 1, 6, 12; 13. muscle, I, 8, (11); 14. muscle, J, 5, (11); 15. connective, E, 7, 11, (14); 16. gametes, 2; 17. connective, B, 10; 18. epithelial, H, 1, 6, 12; 19. connective, 1, 6, 15; 20. nervous, 3; 21. muscle, C, 13; 22. epithelial, F, 1, 6, 12; 23. connective, A, 4, 14

## 25 - III. ORGAN SYSTEMS (pp. 434 - 435)
1. germ cells; 2. gametes; 3. meiosis; 4. somatic; 5. mitosis; 6. tissues; 7. ectoderm; 8. endoderm; 9. mesoderm; 10. mesoderm; 11. endoderm; 12. ectoderm; 13. circulatory; 14. respiratory; 15. urinary (= excretory); 16. skeletal; 17. endocrine; 18. immune; 19. reproductive; 20. digestive; 21. muscular; 22. nervous; 23. integumentary; 24. G; 25. I; 26. H; 27. B; 28. F; 29. E; 30. C; 31. J; 32. A; 33. D; 34. K; 35. thoracic; 36. cranial; 37. pelvic; 38. midsagittal; 39. transverse; 40. anterior; 41. frontal; 42. ventral; 43. fluid; 44. chemical/hormonal; 45. nervous

1. F, positive; 2. T; 3. F, an effector; 4. F, *Plasma* is the fluid portion of blood located *in* blood vessels. *Interstitial fluid* is located outside of blood vessels, lymphatic vessels and body cells, but within the body proper. 5. positive; 6. negative; 7. integrator; 8. glands; 9. receptors; 10. hypothalamus; 11. effectors; 12. sweat

## Self-Quiz

1. d; 2. b; 3. c; 4. c; 5. d, 6. c; 7. c; 8. d; 9. a; 10. d; 11. d; 12. a; 13. A; 14. H; 15. E; 16. F; 17. K; 18. I; 19. J; 20. G; 21. B; 22. D; 23. C

# Chapter 26    Protection, Support, and Movement

**26 - I. INTEGUMENTARY SYSTEMS (pp. 440 - 443)**
*Focus on Health:* **SUNLIGHT AND SKIN (p. 443)**

1. epidermis; 2. dermis; 3. hypodermis; 4. adipose; 5. hair; 6. sensory neuron; 7. oil gland; 8. smooth muscle; 9. hair follicle; 10. sweat gland; 11. blood vessels; 12. epidermis; 13. dermis; 14. hypodermis; 15. a; 16. c; 17. d; 18. b; 19. c; 20. c; 21. e; 22. Anabolic steroids; 23. muscle; 24. acne; 25. testes; 26. aggression; 27. facial; 28. menstrual periods; 29. clitoris

**26 - II. SKELETAL SYSTEMS (pp. 444 - 445)**
**A CLOSER LOOK AT BONES AND THE JOINTS BETWEEN THEM (pp. 446 - 447)**

1. contract (relax); 2. relax (contract); 3. contractile; 4. antagonistic; 5. hydrostatic; 6. exoskeleton; 7. Bones; 8. muscles; 9. Haversian; 10. Red marrow; 11. calcium (phosphate); 12. phosphate (calcium); 13. osteoblasts; 14. sex; 15. osteoporosis; 16. axial; 17. appendicular; 18. synovial; 19. cartilage; 20. Synovial; 21. osteoarthritis; 22. rheumatoid arthritis; 23. nutrient canal; 24. yellow marrow; 25. compact bone; 26. spongy bone; 27. connective tissue covering (periosteum); 28. Haversian system; 29. Haversian canal (blood vessel); 30. mineral deposits (calcium phosphate); 31. osteocyte (bone cell); 32. cranial; 33. clavicle; 34. sternum; 35. scapula; 36. radius; 37. carpal bones; 38. femur; 39. tibia; 40. tarsal bones; 41. metatarsals

**26 - III. SKELETAL-MUSCULAR SYSTEMS
        (pp. 448 - 449)**
**MUSCLE STRUCTURE AND FUNCTION
        (pp. 450 - 451)**

1. Skeletal; 2. bones; 3. tendons; 4. Skeletal; 5. bones; 6. joints; 7. biceps brachii; 8. triceps brachii; 9. bones; 10. Tendons; 11. D; 12. C; 13. A; 14. E; 15. B; 16. F; 17. triceps brachii; 18. pectoralis major; 19. external oblique; 20. rectus abdominis; 21. adductor longus; 22. quadriceps femoris; 23. tibialis anterior; 24. gastrocnemius; 25. deltoid; 26. biceps brachii; 27. biceps contracts; 28. triceps relaxes; 29. biceps relaxes; 30. triceps contracts; 31. muscle; 32. muscle cell (muscle fiber); 33. myofibril; 34. sarcomere;

35. myosin filament; 36. actin filament; 37. myofibrils; 38. sarcomeres; 39. myosin; 40. actin; 41. sliding-filament; 42. myosin; 43. sarcomere

**26 - IV. CONTROL OF MUSCLE CONTRACTION
        (p. 452)**
**PROPERTIES OF WHOLE MUSCLES (p. 453)**
**ATP FORMATION AND LEVELS OF EXERCISE
        (p. 454)**

1. excitability; 2. neuron; 3. action potential; 4. motor; 5. ATP; 6. creatine phosphate; 7. Contraction; 8. bone; 9. calcium; 10. sarcoplasmic reticulum; 11. actin; 12. active transport; 13. action potentials; 14. sarcoplasmic reticulum; 15. diameter; 16. motor; 17. motor unit; 18. muscle twitch; 19. twitch; 20. Tetanus; 21. weak; 22. frequency (rate); 23. axon of motor neuron serving one motor unit; 24. stimulation points; 25. axon of another motor neuron serving another motor unit; 26. individual muscle cells; 27. time that stimulus is applied; 28. contraction phase; 29. relaxation phase; 30. time (seconds); 31. force; 32. Its oxygen-requiring reactions provide most of the ATP needed for muscle contraction during prolonged, moderate exercise. 33. It provides the energy to make myosin filaments slide along actin filaments. 34. Used to clear the actin binding sites of any obstacles to cross-bridge formation with myosin heads. 35. Supplies phosphate to ADP → ATP, which powers muscle contraction for a short time because creatine phosphate supplies are limited. 36. Stored in muscles and in the liver, glucose is stored by the animal body in this form of starch. 37. An anaerobic pathway in which glucose is broken down to lactate with a small yield of ATP; this pathway operates during intense exercise. 38. Supplies commands (signals) to muscle cells to contract and relax. 39. Projections from the thick filaments of myosin that bind to actin sites and form temporary cross-bridges. Making and breaking these cross-bridges cause myosin filaments to be pulled to the center of a sarcomere. 40. The repetitive unit of muscle contraction. Many sarcomeres constitute a myofibril. Many myofibrils constitute a muscle cell.
41. The endoplasmic reticulum of a muscle cell. Stores calcium ions and releases them in response to incoming signals from motor neurons; uses active transport to

bring the calcium ions back inside. 42. dephosphorylation of creatine phosphate; 43. glycolysis alone; 44. aerobic respiration; 45. ATP

1. e; 2. c; 3. c; 4. d; 5. b; 6. d; 7. a; 8. c; 9. d; 10. b; 11. e; 12. d; 13. a; 14. e; 15. b; 16. c; 17. b; 18. d; 19. a; 20. e

---

# Chapter 27   Circulation

### 27 - I. CIRCULATORY SYSTEM—AN OVERVIEW (pp. 456 - 459)

1. nutrients (food); 2. wastes; 3. closed; 4. Blood; 5. heart; 6. interstitial fluid; 7. heart; 8. rapidly; 9. capillary; 10. solutes; 11. lymphatic; 12. digestive (respiratory); 13. respiratory (digestive); 14. respiratory; 15. urinary; 16. pulmonary; 17. systemic; 18. oxygen; 19. oxygen; 20. heart; 21. ventricles; 22. atrial; 23. one; 24. heart(s); 25. blood vessels; 26. hearts; 27. Open circulatory system; blood is pumped into short tubes that open into spaces in the body's tissues, mingles with tissue fluids, then is reclaimed by open-ended tubes that lead back to the heart. 28. Closed circulatory system; blood flow is confined within blood vessels that have continuously connected walls and is pumped by five pairs of "hearts."

### 27 - II. CHARACTERISTICS OF BLOOD (pp. 460 - 461)
### Focus on Health: BLOOD DISORDERS (p. 462)
### BLOOD TYPING AND BLOOD TRANSFUSIONS (pp. 462 - 463)

1. connective; 2. pH; 3. closed; 4. 4 - 5; 5. iron; 6. cell count; 7. 50 - 60; 8. red bone marrow; 9. Stem cells; 10. Neutrophils; 11. Platelets; 12. nucleus; 13. four; 14. nucleus; 15. nine; 16. AB; 17. O; 18. women; 19. Rh⁻; 20. a. Plasma proteins; b. Red blood cells; c. Neutrophils; d. Lymphocytes; e. Platelets; 21. stem; 22. red blood (E); 23. platelets (C); 24. neutrophils (A); 25. B (D); 26. T (D); 27. monocytes; 28. macrophages (A)

### 27 - III. HUMAN CARDIOVASCULAR SYSTEM (pp. 464 - 465)
### THE HEART IS A LONELY PUMPER (pp. 466 - 467)

1. pulmonary; 2. systemic; 3. jugular; 4. superior vena cava; 5. pulmonary; 6. hepatic; 7. renal; 8. inferior vena cava; 9. iliac; 10. femoral; 11. femoral; 12. iliac; 13. abdominal; 14. renal; 15. brachial; 16. coronary; 17. pulmonary; 18. ascending aorta (aortic arch); 19. carotid; 20. aorta; 21. left pulmonary veins; 22. semilunar valve; 23. left ventricle; 24. inferior vena cava; 25. atrioventricular valve; 26. right pulmonary artery; 27. superior vena cava

### 27 - IV. BLOOD PRESSURE IN THE CARDIOVASCULAR SYSTEM (pp. 468 - 469)
### Focus on Health: CARDIOVASCULAR DISORDERS (pp. 470 - 471)
### HEMOSTASIS (p. 472)

1. vein; 2. artery; 3. arteriole; 4. capillary; 5. smooth muscle, elastic fibers; 6. valve; 7. artery; 8. capillary; 9. endothelial; 10. capillary bed (diffusion zone); 11. interstitial; 12. Valves; 13. veins (venules); 14. venules (veins); 15. atrium; 16. ventricle; 17. systole; 18. diastole; 19. Arteries; 20. ventricles; 21. Arterioles; 22. pressure; 23. stroke; 24. coronary occlusion; 25. Plaque; 26. low; 27. thrombus; 28. hemostasis; 29. platelet plug formation; 30. coagulation; 31. collagen; 32. insoluble; 33. F, pressure; 34. F, blood pressure cannot remain constant because it passes through various kinds of vessels that have varied structures. 35. aorta; 36. pressure; 37. resistance; 38. elastic; 39. little; 40. does not drop much; 41. arterioles; 42. nervous; 43. medulla oblongata; 44. beat more slowly; 45. relax; 46. vasodilation

### 27 - V. LYMPHATIC SYSTEM (p. 472 - 473)

1. tonsils; 2. right lymphatic duct; 3. thymus gland; 4. thoracic duct; 5. spleen; 6. bone marrow; 7. Lymph; 8. fats; 9. small intestine

### Self-Quiz

1. d; 2. e; 3. c; 4. e; 5. e; 6. a; 7. b; 8. a; 9. a; 10. a

---

# Chapter 28   Immunity

### 28 - I. THREE LINES OF DEFENSE (pp. 476 - 478)
### COMPLEMENT PROTEINS (p. 478)
### INFLAMMATION (pp. 480 - 481)

1. mucous; 2. Lysozyme; 3. Gastric; 4. bacterial; 5. clotting; 6. Phagocytic (Macrophage); 7. complement system; 8. Antibodies; 9. histamine; 10. capillaries; 11. Basophils secrete histamine and prostaglandins, which change permeability of blood vessels in damaged or irritated tissues. 12. Eosinophils attack parasitic worms by secreting corrosive enzymes.

13. Neutrophils are the most abundant white blood cells; they quickly phagocytize bacteria and reduce them to molecules that can be used for other purposes.
14. Monocytes mature into macrophages, which slowly go about engulfing foreign agents and cleaning out dead and damaged cells.

## 28 - II. THE IMMUNE SYSTEM (pp. 482 - 483)
## LYMPHOCYTE BATTLEGROUNDS (p. 484)
## CELL-MEDIATED RESPONSES (pp. 484 - 485)

1. nonspecific; 2. immune system; 3. MHC marker; 4. nonself; 5. lymphocytes; 6. B cell; 7. T cell; 8. thymus; 9. viruses; 10. cancer; 11. antigen-presenting cells (macrophages); 12. virgin helper T cells; 13. memory T cells; 14. intracellular; 15. virgin B cells; 16. effector B cells; 17. antibodies; 18. extracellular; 19. a; 20. d; 21. e; 22. b; 23. c; 24. cytotoxic T; 25. helper T; 26. antigen-MHC; 27. macrophage; 28. cell-; 29. primary immune response; 30. memory cells; 31. antigens; 32. natural killer; 33. perforins

## 28 - III. ANTIBODY-MEDIATED RESPONSES (pp. 486 - 487)
*Focus on Health:* CANCER AND IMMUNOTHERAPY (p. 487)
## IMMUNE SPECIFICITY AND MEMORY (pp. 488 - 489)

1. immunoglobulin; 2. binding sites; 3. antigen; 4. destruction; 5. extracellular pathogens; 6. Antibodies; 7. helper T; 8. antigen-MHC; 9. macrophage; 10. antibody; 11. effector B; 12. B cell; 13. c; 14. e; 15. a; 16. b; 17. d; 18. antibodies; 19. antigen; 20. virgin B; 21. memory cells; 22. effector cells; 23. clonal selection hypothesis; 24. nonself; 25. Recombination; 26. billion; 27. Cancer; 28. monoclonal (pure) antibodies

## 28 - IV. IMMUNITY ENHANCED, MISDIRECTED, OR COMPROMISED (pp. 490 - 491)
*Focus on Health:* AIDS—THE IMMUNE SYSTEM COMPROMISED (pp. 492 - 493)

1. immunization; 2. active; 3. primary immune response; 4. memory cells; 5. Allergy; 6. autoimmune disease; 7. Myasthenia gravis; 8. Rheumatoid arthritis; 9. human immunodeficiency virus; 10. male homosexuals; 11. enveloped retroviruses; 12. reverse transcriptase; 13. one; 14. 22,200,000; 15. six; 16. a, b, c, d; 17. a, b; 18. a, b; 19. b, c, (d); 20. b, c, (d); 21. a, b, c; 22. b, c, (d); 23. a, b, c, d; 24. a, b, c; 25. Estimates will vary.

## Self-Quiz

1. d; 2. b; 3. b; 4. a; 5. a; 6. e; 7. e; 8. e; 9. a; 10. a; 11. H; 12. D; 13. E; 14. J; 15. B; 16. G; 17. C; 18. I; 19. F; 20. A

---

# Chapter 29 Respiration

## 29 - I. THE NATURE OF RESPIRATION (pp. 496 - 498)
## INVERTEBRATE RESPIRATION (p. 499)
## VERTEBRATE RESPIRATION (pp. 500 - 501)

1. gill; 2. countercurrent flow; 3. tracheas; 4. 21; 5. aerobic metabolism; 6. $O_2$ (oxygen); 7. carbon dioxide ($CO_2$); 8. respiration; 9. pressure gradient; 10. high; 11. lowest; 12. high; 13. lower; 14. surface area; 15. partial pressure; 16. Hemoglobin; 17. lung; 18. airways; 19. blood; 20. Hypoxia; 21. blood vessel in gill filament; 22. oxygen-poor blood; 23. oxygen-rich blood; 24. water; 25. blood

## 29 - II. HUMAN RESPIRATORY SYSTEM (pp. 502 - 503)

1. diaphragm; 2. rib cage; 3. increases; 4. drops; 5. ventilating; 6. pleural sac; 7. larynx; 8. glottis; 9. bronchi; 10. bronchioles; 11. alveoli; 12. intercostal muscles; 13. diaphragm; 14. pharynx; 15. epiglottis; 16. vocal cords; 17. trachea; 18. bronchus; 19. bronchioles; 20. chest cavity; 21. abdominal cavity; 22. alveolar duct; 23. bronchiole; 24. alveolus, alveoli; 25. Capillary; 26. a. Alveoli; b. Bronchial tree; c. Diaphragm; d. Larynx; e. Pharynx

## 29 - III. GAS EXCHANGE AND TRANSPORT (pp. 504 - 505)
*Focus on Health:* WHEN THE LUNGS BREAK DOWN (pp. 506 - 507)

1. partial pressure; 2. Diffusion; 3. carbon dioxide; 4. Hemoglobin; 5. bicarbonate; 6. oxyhemoglobin (hemoglobin); 7. systemic (low-pressure); 8. partial pressure; 9. brain; 10. Emphysema; 11. lung cancer; 12. N2 (nitrogen gas); 13. joints; 14. decompression

## Self-Quiz

1. c; 2. e; 3. a; 4. c; 5. a; 6. a; 7. a; 8. a; 9. a; 10. d; 11. H; 12. E; 13. J; 14. K; 15. G; 16. N; 17. F; 18. L; 19. B; 20. D; 21. A; 22. C; 23. M; 24. I

# Chapter 30 Digestion and Human Nutrition

## 30 - I. THE NATURE OF DIGESTIVE SYSTEMS (pp. 510 - 513)
## OVERVIEW OF THE HUMAN DIGESTIVE SYSTEM (p. 514)

1. Nutrition; 2. carbohydrates; 3. particles; 4. molecules; 5. absorbed; 6. incomplete; 7. circulatory; 8. complete; 9. opening; 10. Motility; 11. secretion; 12. stomach; 13. small intestine; 14. anus; 15. accessory; 16. pancreas; 17. circulatory; 18. respiratory; 19. carbon dioxide; 20. urinary; 21. salivary glands; 22. liver; 23. gallbladder; 24. pancreas; 25. anus; 26. large intestine; 27. small intestine; 28. stomach; 29. esophagus; 30. pharynx; 31. mouth (oral cavity)

## 30 - II. INTO THE MOUTH, DOWN THE TUBE (p. 515)
## DIGESTION IN THE STOMACH AND SMALL INTESTINE (pp. 516 - 517)

1. a. Mouth; b. Salivary glands; c. Stomach; d. Small intestine; e. Pancreas; f. Liver; g. Gallbladder; h. Large intestine; i. Rectum; 2. salivary amylase; 3. epiglottis; 4. esophagus; 5. peristalsis; 6. Pepsin; 7. starches; 8. salivary amylase; 9. disaccharide; 10. stomach; 11. small intestine; 12. amylase; 13. small intestine; 14. disaccharidases; 15. stomach; 16. pepsins; 17. small intestine; 18. pancreas; 19. amino acids; 20. Lipase; 21. small intestine; 22. fatty acid; 23. Bile; 24. gallbladder; 25. lipase; 26. small intestine; 27. small intestine; 28. Carboxypeptidase; 29. Pancreatic nucleases; 30. T; 31. F, cellular respiration

## 30 - III. ABSORPTION IN THE SMALL INTESTINE (pp. 518 - 519)
## DISPOSITION OF ABSORBED ORGANIC COMPOUNDS (p. 520)
## THE LARGE INTESTINE (p. 521)
### Focus on Health: CANCER IN THE SYSTEM (p. 521)

1. Constructing hormones, nucleotides, proteins, and enzymes; 2. Monosaccharides, free fatty acids, and glycerol; 3. The three uses are (a) to construct components of cells and storage forms (such as glycogen) and specialized derivatives such as steroids and acetylcholine; (b) to convert to amino acids as needed; and (c) to serve as a source of energy; 4. glucose; 5. fats (lipids); 6. glycogen; 7. liver; 8. proteins; 9. hormones; 10. nucleotides; 11. fat; 12. ammonia; 13. urea; 14. T; 15. T; 16. F, small intestine

## 30 - IV. HUMAN NUTRITIONAL REQUIREMENTS (pp. 522 - 523)
## VITAMINS AND MINERALS (pp. 524 - 525)
### Focus on Science: TANTALIZING ANSWERS TO WEIGHTY QUESTIONS (pp. 526 - 527)

1. food pyramid; 2. carbohydrates; 3. bread; 4. 6 - 11; 5. vegetable; 6. 3 - 5; 7. fruit; 8. 2 - 4; 9. apples; 10. berries; 11. meat; 12. proteins; 13. 2 - 3; 14. amino acids; 15. milk; 16. 2 - 3; 17. 0; 18. a. 2,070; b. 2,900; c. 1,230; 19. women; 20. Complex carbohydrates; 21. 58 to 60; 22. Phospholipids; 23. energy reserves; 24. 30; 25. essential fatty acids; 26. Proteins; 27. essential; 28. milk (eggs); 29. eggs (milk); 30. Vitamins; 31. Minerals; 32. a. Consult Figure 30.15 (men's column, 6'1"). 178 + 6 = 184 lbs. If he is 195 lbs., he is 11 lbs. overweight; 33. To maintain that weight, he can ingest 195 × 10 = 1,950 kilocalories each day. To reach the desired weight (184 lbs.) he must increase his exercise level and eat less of the required food groups until he reaches 184 lbs. Thereafter, to maintain that weight, he should ingest 1,840 kilocalories each day. The excess 11 pounds should be lost gradually by adopting an everyday exercise program that over many months would gradually eliminate the excess kilocalories that are stored mostly in the form of fat. The smallest range of serving sizes shown in Fig. 30.13 will help keep the total caloric intake to about 1600 kcal; 34. a. 6 servings; b. bread, cereal, rice , pasta; 35. a. 2 servings; b. fruits; 36. a. 3 servings; b. vegetables; 37. a. 2 servings; b. milk, yogurt or cheese; 38. a. 2 servings; b. legume, nut, poultry, fish or meats; 39. a. Scarcely any; b. added fats and simple sugars

## Self-Quiz

1. b; 2. b; 3. a; 4. e; 5. d; 6. c; 7. e; 8. b; 9. a; 10. b; 11. c; 12. D; 13. C; 14. H; 15. J; 16. B; 17. A; 18. I; 19. F; 20. E; 21. G

# Chapter 31   The Internal Environment

## 31 - I.   URINARY SYSTEM OF MAMMALS
### (pp. 530 - 533)
1. metabolism; 2. urine; 3. lungs; 4. sweating; 5. Thirst; 6. metabolism; 7. ammonia; 8. urea; 9. uric acid; 10. kidney; 11. ureter; 12. urinary bladder; 13. urethra; 14. cortex; 15. medulla; 16. ureter; 17. glomerular capillaries; 18. proximal tubule; 19. Bowman's capsule; 20. distal tubule; 21. collecting duct; 22. loop of Henle

## 31 - II.   URINE FORMATION (pp. 534 - 535)
*Focus on Health:* WHEN THE KIDNEYS BREAK
  DOWN (p. 536)
1. T; 2. T; 3. kidneys; 4. nephrons; 5. bloodstream; 6. arteriole; 7. Bowman's capsule; 8. glomerulus; 9. blood pressure; 10. solutes; 11. blood capillaries (peritubular capillaries); 12. secretion; 13. Ureters; 14. urinary bladder; 15. urethra; 16. aldosterone; 17. sodium; 18. less; 19. blood pressure; 20. hypertension; 21. table salt (sodium chloride); 22. ADH; 23. inhibited; 24. volume

## 31 - III.   THE BODY'S ACID-BASE BALANCE
### (p. 536)
### ON FISH, FROGS, AND KANGAROO RATS (p. 537)
### MAINTAINING THE BODY'S CORE
  TEMPERATURE (pp. 538 - 539)
1. kidneys; 2. $H^+$; 3. 7.43; 4. Acids; 5. bases; 6. lowered; 7. $H^+$; 8. bicarbonate ($HCO_3^-$); 9. urinary; 10. loops of Henle; 11. water; 12. water; 13. solutes; 14. very dilute; 15. hypothalamus; 16. core; 17. pilomotor response; 18. Peripheral vasoconstriction; 19. hypothermia; 20. T; 21. T; 22. T; 23. D; 24. B; 25. A; 26. C; 27. G; 28. F; 29. E

## Self-Quiz

1. d; 2. c; 3. e; 4. a; 5. d; 6. d; 7. d; 8. b; 9. e; 10. b; 11. b

---

# Chapter 32   Neural Control and the Senses

## 32 - I.   INVERTEBRATE BEGINNINGS
### (pp. 542 - 545)
### NEURONS — THE COMMUNICATION
  SPECIALISTS (pp. 546 - 547)
### ACTION POTENTIALS (pp. 548 - 549)
1. neurons; 2. Neuroglial; 3. Sensory; 4. interneurons; 5. motor; 6. cell body; 7. Dendrites; 8. signals (stimuli); 9. axon; 10. cell body; 11. axon; 12. endings; 13. voltage differential; 14. resting membrane potential; 15. action potential (nerve impulse); 16. disturbance; 17. neurotransmitters; 18. contract; 19. potassium; 20. Na; 21. Channel; 22. Transport; 23. potassium; 24. sodium; 25. 30; 26. 10; 27. gates; 28. sodium–potassium pumps; 29. localized; 30. Graded; 31. duration; 32. trigger zone; 33. action potential; 34. nerve impulse; 35. all-or-nothing; 36. threshold; 37. trigger zone (axonal membrane); 38. channel proteins; 39. gates; 40. sodium-potassium pump; 41. lipid bilayer; 42. action potential; 43. threshold; 44. resting membrane potential; 45. milliseconds; 46. millivolts

## 32 - II.   CHEMICAL SYNAPSES (pp. 550 - 551)
### PATHS OF INFORMATION FLOW (pp. 552 - 553)
*Focus on Health:* CASES OF SKEWED
  INFORMATION FLOW (p. 554)
1. chemical synapse; 2. neurotransmitters; 3. Acetylcholine (ACh); 4. excitatory; 5. inhibitory; 6. postsynaptic; 7. Endorphins; 8. Synaptic integration; 9. summed; 10. myelin; 11. node unsheathed; 12. reflex; 13. stretch reflex; 14. acetylcholine; 15. axon; 16. myelin sheath; 17. blood vessels; 18. axons; 19. unsheathed node; 20. Schwann cell (myelin sheath); 21. sensory neuron; 22. inter; 23. motor neuron; 24. receptor; 25. cell body; 26. axon; 27. axon endings; 28. dendrites; 29. H, F; 30. E; 31. K; 32. I; 33. J; 34. B; 35. C; 36. G; 37. A, D

## 32 - III.   FUNCTIONAL DIVISIONS OF
  VERTEBRATE NERVOUS SYSTEMS (pp. 554 - 555)
### THE MAJOR EXPRESSWAYS (pp. 556 - 557)
1. somatic; 2. autonomic; 3. sympathetic; 4. parasympathetic; 5. central nervous system; 6. vertebral column; 7. cranial; 8. B; 9. D; 10. E; 11. C; 12. A; 13. spinal cord; 14. cervical nerves; 15. thoracic nerves; 16. lumbar nerves; 17. sacral nerves; 18. coccygeal nerves; 19. autonomic; 20. sympathetic; 21. parasympathetic; 22. increases; 23. decreases; 24. spinal cord; 25. white matter; 26. gray matter; 27. reflexes; 28. spinal cord; 29. ganglion; 30. vertebra; 31. spinal nerve; 32. meninges; 33. gray matter; 34. white matter

**32 - IV.  THE HUMAN BRAIN (pp. 558 - 559)**
**MEMORY (p. 560)**
*Focus on Health:* **DRUGS, THE BRAIN, AND**
  **BEHAVIOR (pp. 560 - 561)**
1. medulla oblongata;  2. cerebellum;  3. pons;
4. midbrain;  5. reticular formation;  6. forebrain;
7. hypothalamus;  8. sexual;  9. thalamus;  10. cere-
brospinal fluid;  11. memory;  12. short-term memory;
13. long-term memory;  14. Endorphins (Enkephalins);
15. pain;  16. neurotransmitters;  17. I;  18. H;  19. A;
20. L;  21. K;  22. D;  23. F;  24. B;  25. C;  26. J;
27. G;  28. E;  29. cerebellum;  30. medulla oblongata;
31. pons;  32. hypothalamus;  33. thalamus;  34. corpus callosum;  35. D;  36. A;  37. D;  38. D;  39. A;
40. B;  41. C;  42. C;  43. D

**32 - V.  SENSORY SYSTEMS (p. 562)**
**SOMATIC SENSATIONS (p. 563)**
**HEARING AND BALANCE (pp. 564 - 565)**
1. receptors;  2. stimulus;  3. Chemoreceptors;
4. mechanoreceptors;  5. photoreceptors;  6. thermoreceptors;  7. sensation;  8. perception;  9. nerve pathways;  10. brain regions;  11. D;  12. C;  13. A;  14. B;
15. A;  16. E;  17. E;  18. B;  19. D;  20. B (F);  21. B;
22. A;  23. C;  24. B;  25. touch;  26. cold;  27. skin;
28. skeletal;  29. mechanoreceptors;  30. Free;
31. Pain;  32. referred pain;  33. Mechanoreceptors;
34. skin;  35. free nerve endings (C);  36. Ruffini endings (A);  37. Meissner corpuscle (D);  38. epidermis;
39. dermis;  40. Pacinian corpuscle (B);  41. brain;
42. frequency;  43. number;  44. hearing;  45. amplitude;  46. frequency;  47. higher;  48. mechanoreceptors;  49. middle;  50. cochlea;  51. organ of Corti;
52. motion sickness;  53. middle earbones (malleus,
incus, stapes);  54. cochlea;  55. auditory nerve;
56. tympanic membrane/eardrum;  57. oval window;
58. basilar membrane;  59. tectorial membrane

**32 - VI.  VISION (pp. 566 - 567)**
**CASE STUDY—FROM SIGNALING TO VISUAL**
  **PERCEPTION (pp. 568 - 569)**
1. Light;  2. photoreceptor;  3. Vision;  4. Eyespots;
5. Eyes;  6. cornea;  7. retina;  8. focal point;
9. Visual accommodation;  10. Farsighted;  11. Cone;
12. fovea;  13. vitreous body;  14. cornea;  15. iris;
16. lens;  17. aqueous humor;  18. ciliary muscle;
19. retina;  20. fovea;  21. optic nerve;  22. blind
spot/optic disk;  23. sclera

**Self-Quiz**

1. a;  2. c;  3. d;  4. a;  5. d;  6. b;  7. c;  8. e;  9. b;
10. a;  11. M;  12. G;  13. D;  14. F;  15. K;  16. H;
17. I;  18. N;  19. J;  20. B;  21. A;  22. L;  23. C;
24. E;  25. O

---

# Chapter 33   Endocrine Control

**33 - I.  THE ENDOCRINE SYSTEM (pp. 574 - 575)**
1. D;  2. A;  3. E;  4. B;  5. C;  6. a. 1, six releasing
and inhibiting hormones; synthesizes ADH, oxytocin.
b. 2, ACTH, TSH, FSH, LH, GSH.  c. 2, stores and
secretes two hypothalamic hormones, ADH and oxytocin.  d. 3, sex hormones of opposite sex, cortisol, aldosterone.  e. 3, epinephrine, norepinephrine.
f. 4, estrogen, progesterone.  g. 5, testosterone.  h. 6,
melatonin.  i. 7, thyroxine and triiodothyronine.  j. 8,
parathyroid hormone (PTH).  k. 9, thymosins.  l. 10,
insulin, glucagon, somatostatin.

**33 - II.  SIGNALING MECHANISMS (pp. 576 - 577)**
1. a;  2. b;  3. a;  4. b;  5. b;  6. a;  7. b;  8. b

**33 - III.  THE HYPOTHALAMUS AND PITUITARY**
  **GLAND (pp. 578 - 579)**
1. A (H);  2. P (G);  3. A (A);  4. A (I);  5. A (D);
6. I (B);  7. P (E);  8. A (C);  9. A (F);  10. hypothalamus;  11. posterior;  12. anterior;  13. intermediate;
14. releasers;  15. inhibitors

**33 - IV.  EXAMPLES OF ABNORMAL PITUITARY**
  **OUTPUT (p. 580)**
1. a. Gigantism;  b. Pituitary dwarfism;  c. Diabetes
insipidus;  d. Acromegaly

**33 - V.  SOURCES AND EFFECTS OF OTHER**
  **HORMONES (p. 581)**
1. D(c);  2. I(e);  3. F(h);  4. A(g);  5. H(k);  6. E(d);
7. C(a);  8. K(j);  9. B(i);  10. J(f);  11. G(b);  12. F (l);
13. A (m);  14. B (o);  15. C (n)

**33 - VI.  FEEDBACK CONTROL OF HORMONE**
  **SECRETIONS (pp. 582 - 583)**
1. E;  2. F;  3. D;  4. H.  5. I;  6. J;  7. A;  8. G;  9. B;
10. C;  11. thyroid;  12. metabolic (metabolism);
13. iodine;  14. TSH;  15. goiter;  16. Hypothyroidism;
17. hypothyroid;  18. Hyperthyroidism;  19. gonads;
20. hormones;  21. testes;  22. ovaries;  23. gametes;
24. hormones;  25. secondary

**33 - VII.  RESPONSES TO LOCAL CHEMICAL**
  **CHANGES (pp. 584 - 585)**
1. parathyroid;  2. PTH;  3. calcium;  4. calcium;
5. enzymes;  6. calcium;  7. $D_3$;  8. intestinal;  9. rickets;  10. a. Glucagon, causes glycogen (a storage polysaccharide) and amino acids to be converted to glucose
in the liver (glucagon raises the glucose level).
b. Insulin, stimulates glucose uptake by liver muscle,
and adipose cells; promotes synthesis of proteins and
fats, and inhibits protein conversion to glucose (lowers
the glucose level).  c. Somatostatin, helps control diges-

tion; can block secretion of insulin and glucagon.
11. increases; 12. excessive; 13. energy; 14. ketones;
15. insulin; 16. glucagon; 17. type 1 diabetes; 18. type
1 diabetes; 19. type 2 diabetes; 20. Type 2 diabetes

**33 - VIII. HORMONAL RESPONSES TO
ENVIRONMENTAL CUES (pp. 586 - 587)**
1. b; 2. a; 3. a; 4. a; 5. b; 6. b; 7. a; 8. a; 9. a;
10. b

Self-Quiz

**Self-Quiz**
1. a; 2. e; 3. d; 4. a; 5. b; 6. e; 7. d; 8. b; 9. a;
10. c; 11. H; 12. F; 13. K; 14. L; 15. N; 16. D;
17. E; 18. A; 19. G; 20. B; 21. I; 22. Q; 23. O;
24 C; 25. J; 26. M; 27. P

# Chapter 34    Reproduction and Development

**34 - I.  THE BEGINNING: REPRODUCTIVE MODES
(pp. 590 - 592)**
**STAGES OF DEVELOPMENT—AN OVERVIEW
(p. 593)**
1. asexual; 2. environmental; 3. variation; 4. Gamete
formation; 5. Fertilization; 6. zygote; 7. Cleavage;
8. embryo; 9. blastula; 10. gastrulation; 11. germ;
12. Ectoderm; 13. nervous; 14. Endoderm; 15. gut's;
16. Mesoderm; 17. organ; 18. specialization;
19. adulthood; 20. stage; 21. F; 22. B; 23. C; 24. A;
25. E; 26. D; 27. a. mesoderm; b. ectoderm;
c. endoderm; d. mesoderm; e. ectoderm; f. mesoderm; g. endoderm; h. mesoderm; i. mesoderm

**34 - II.  A VISUAL TOUR OF FROG AND CHICK
DEVELOPMENT (pp. 594 - 595)**
**EARLY MARCHING ORDERS (pp. 596 - 597)**
1. D; 2. H; 3. B; 4. G; 5. J; 6. I; 7. E; 8. F; 9. A;
10. C; 11. gray crescent; 12. animal; 13. vegetal;
14. Cleavage; 15. blastocoel; 16. gastrula;
17. archenteron; 18. yolk; 19. Neural; 20. blastula;
21. cytoplasm; 22. T; 23. two; 24. inner cell mass;
25. T

**34 - III.  HOW SPECIALIZED TISSUES AND
ORGANS FORM (pp. 598 - 599)**
1. c; 2. b; 3. b; 4. a; 5. c; 6. c; 7. c; 8. b; 9. b;
10. a; 11. b; 12. c; 13. c; 14. c; 15. a

**34 - IV.  REPRODUCTIVE SYSTEM OF HUMAN
MALES (pp. 600 - 601)**
**MALE REPRODUCTIVE FUNCTION (pp. 602 - 603)**
1. testes; 2. secondary; 3. scrotum; 4. twelve;
5. sixteen; 6. cooler; 7. mitotic; 8. seminiferous;
9. meiosis; 10. sperms; 11. Sertoli; 12. epididymis;
13. vas deferens; 14. urethra; 15. seminal vesicles;
16. Prostate; 17. bulbourethral; 18. Leydig;
19. Testosterone; 20. testosterone; 21. anterior;
22. hypothalamus; 23. decrease; 24. LH; 25. Sertoli;
26. increase; 27. DNA; 28. mitochondria; 29. microtubules; 30. testosterone; 31. hypothalamus;
32. GnRH; 33. Sertoli; 34. Sertoli; 35. testosterone;
36. testosterone; 37. Leydig; 38. testosterone;
39. GnRH

**34 - V.  REPRODUCTIVE SYSTEM OF HUMAN
FEMALES (pp. 604 - 605)**
**FEMALE REPRODUCTIVE FUNCTION
(pp. 606 - 607)**
**VISUAL SUMMARY OF THE MENSTRUAL CYCLE
(pp. 608 - 609)**
1. ovary; 2. oviduct; 3. uterus; 4. cervix; 5. myo-
metrium; 6. endometrium; 7. vagina; 8. labia major;
9. labia minor; 10. clitoris; 11. urethra; 12. Meiosis;
13. I; 14. 300,000; 15. follicle; 16. hypothalamus;
17. anterior pituitary; 18. zona pellucida; 19. estrogens; 20. meiosis I; 21. ovulation; 22. Ovulation;
23. LH; 24. pituitary; 25. endometrium; 26. zona
pellucida; 27. corpus luteum; 28. Progesterone;
29. corpus luteum; 30. menstruation; 31. endometrial;
32. estrogens; 33. endometrial; 34. menopause

**34 - VI.  PREGNANCY HAPPENS (p. 609)**
**FORMATION OF THE EARLY EMBRYO
(pp. 610 - 611)**
**EMERGENCE OF THE VERTEBRATE BODY PLAN
(p. 612)**
**ON THE IMPORTANCE OF THE PLACENTA (p. 613)**
**EMERGENCE OF DISTINCTLY HUMAN FEATURES
(pp. 614 - 615)**
*Focus on Health:* **MOTHER AS PROTECTOR,
PROVIDER, POTENTIAL THREAT (pp. 616 - 617)**
1. sperm; 2. oviduct; 3. enzymes; 4. ovum;
5. implantation; 6. placenta; 7. second; 8. fetus;
9. F; 10. D; 11. H; 12. C; 13. I; 14. J; 15. B;
16. G; 17. A; 18. E; 19. embryonic disk; 20. amniotic cavity; 21. embryonic disk; 22. amniotic cavity;
23. yolk sac; 24. embryo; 25. amnion; 26. allantois;
27. yolk sac; 28. umbilical cord; 29. four; 30. gill
arches; 31. somites; 32. five; 33. forelimb; 34. C;
35. B; 36. F; 37. H; 38. A; 39. I; 40. G; 41. E;
42. D; 43. C; 44. A; 45. B; 46. C; 47. E; 48. A;
49. D; 50. B; 51. maternal; 52. blood vessels;
53. maternal; 54. uterus; 55. fetal; 56. embryonic;
57. umbilical cord; 58. intervillus; 59. chorionic villus;
60. amniotic; 61. d; 62. b; 63. f; 64. a; 65. e;
66. g; 67. c; 68. a; 69. d; 70. a; 71. b

**34 - VII. FROM BIRTH ONWARD (pp. 618 - 619)**
**CONTROL OF HUMAN FERTILITY (pp. 620 - 621)**
1. thirty-eight;  2. uterus;  3. two to eighteen;  4. cervix;  5. amnion;  6. one hour;  7. fifteen;  8. umbilical cord;  9. estrogen and progesterone;  10. prolactin; 11. oxytoxin;  12. 100,000;  13. postnatal;  14. prenatal; 15. aging;  16. J;  17. F;  18. K;  19. D;  20. H;  21. C; 22. L;  23. A;  24. E;  25. I;  26. B;  27. G

**34 - VIII.** *Focus on Health:* **SEXUALLY**
    **TRANSMITTED DISEASES (pp. 622 - 623)**
*Focus on Bioethics:* **MEDICAL INTERVENTIONS**
    **THAT PROMOTE OR END PREGNANCY (p. 624)**
1. A, F;  2. A, C;  3. D, E;  4. F;  5. E;  6. F;  7. D; 8. C;  9. F;  10. C;  11. B;  12. C;  13. C;  14. B, F; 15. D, F;  16. B;  17. A, B, C, D, E, F

1. b;  2. b;  3. a;  4. c;  5. e;  6. d;  7. b;  8. e;  9. a; 10. d;  11. e;  12. a;  13. b;  14. c;  15. e;  16. d; 17. a;  18. c;  19. a;  20. b;  21. d

---

# Chapter 35   Population Ecology

**35 - I. CHARACTERISTICS OF POPULATIONS**
    **(pp. 628 - 630)**
**POPULATION SIZE AND EXPONENTIAL GROWTH**
    **(pp. 630 - 631)**
1. L;  2. H;  3. E;  4. N;  5. B;  6. J;  7. A;  8. K; 9. C;  10. G;  11. D;  12. M;  13. F;  14. O;  15. I; 16. a. It increases.  b. It decreases.  c. It must increase. 17. population growth rate;  18. It must decrease. 19. 100,000;  20. a. 100,000;  b. 300,000

**35 - II. LIMITS ON THE GROWTH OF**
    **POPULATIONS (pp. 632 - 633)**
**LIFE HISTORY PATTERNS (pp. 634 - 635)**
*Focus on Science:* **THE GUPPIES OF TRINIDAD**
    **(p. 635)**
1. limiting;  2. Carrying capacity;  3. logistic;  4. carrying capacity;  5. increases;  6. decreases;  7. density-dependent;  8. density-independent;  9. Insurance companies;  10. III;  11. dependent;  12. carrying capacity;  13. independent;  14. C;  15. D;  16. E; 17. B;  18. A;  19. F;  20. A, E, H;  21. B, D, C;  22. G

**35 - III. HUMAN POPULATION GROWTH**
    **(pp. 636 - 637)**
**CONTROL THROUGH FAMILY PLANNING**
    **(pp. 638 - 639)**
**POPULATION GROWTH AND ECONOMIC**
    **DEVELOPMENT (pp. 640 - 641)**
**SOCIAL IMPACT OF NO GROWTH (p. 641)**
1. a. 1962 - 1963;  b. 2025 or sooner;  c. Depends on the age and optimism of the reader.  2. 5.7 billion;  3. T;

4. short;  5. T;  6. limiting factors;  7. B;  8. D;  9. A; 10. C;  11. 88.4;  12. 88.4;  13. dependent;  14. cholera; 15. immigration;  16. emigration;  17. birth;  18. demographic transition;  19. industrial;  20. transition; 21. 2.5;  22. 2.1;  23. reproductive;  24. delayed; 25. limiting factors;  26. family planning;  27. fertility; 28. 3.1;  29. D;  30. C;  31. B;  32. A

1. d;  2. a;  3. b;  4. a;  5. d;  6. d;  7. a;  8. b;  9. a; 10. a;  11. c;  12. c;  13. b;  14. d;  15. b;  16. a; 17. a;  18. density;  19. size;  20. distribution; 21. age;  22. reproductive;  23. zero population growth; 24. exponential;  25. reproductive;  26. J-shaped; 27. decreases

# Chapter 36    Community Interactions

**36 - I. FACTORS THAT SHAPE COMMUNITY STRUCTURE (pp. 644 - 646)**
**MUTUALISM (p. 647)**
1. community; 2. habitat; 3. niche; 4. commensalistic; 5. mutualistic; 6. interspecific; 7. predation; 8. parasitism; 9. symbiosis; 10. obligatory; 11. mycorrhiza; 12. D; 13. G; 14. A; 15. F; 16. B; 17. E; 18. C; 19. a. It cannot complete its life cycle in any other plant, and its larvae eat only yucca seeds. b. The yucca moth is the plant's only pollinator.

**36 - II. COMPETITIVE INTERACTIONS (pp. 648 - 649)**
**PREDATION AND PARASITISM (pp. 650 - 651)**
*Focus on the Environment:* **THE COEVOLUTIONARY ARMS RACE (pp. 652 - 653)**
1. interspecific competition; 2. competitive exclusion; 3. niche; 4. resource partitioning; 5. predatory-prey; 6. carrying capacity; 7. Coevolution; 8. camouflage; 9. Camouflage; 10. mimicry; 11. D; 12. G; 13. I; 14. A; 15. B; 16. C; 17. D; 18. E; 19. I; 20. B; 21. H; 22. H; 23. G; 24. F; 25. D; 26. more; 27. reduce; 28. less; 29. the same; 30. more

**36 - III. FORCES CONTRIBUTING TO COMMUNITY STABILITY (pp. 654 - 655)**
**COMMUNITY INSTABILITY (pp. 656 - 657)**
1. succession; 2. Pioneer; 3. pioneers; 4. climax; 5. primary; 6. replacement; 7. secondary; 8. facili-tate; 9. succession; 10. climax-pattern; 11. community; 12. pioneers; 13. fires; 14. b; 15. a; 16. a; 17. b; 18. b; 19. a; 20. b; 21. b; 22. a; 23. a

**36 - IV. PATTERNS OF BIODIVERSITY (pp. 658 - 659)**
1. a. Resource availability tends to be higher and more reliable. Tropical latitudes have more sunlight of greater intensity, rainfall amount is higher, and the growing season is longer. Vegetation grows all year long to support diverse herbivores, etc. b. Species diversity might be self-reinforcing. When a greater number of plant species compete and coexist, a greater number of herbivore species evolve because no herbivore can overcome the chemical defenses of all kinds of plants. Then more predators and parasites evolve in response to the diversity of prey and hosts. c. The rates of speciation in the tropics have exceeded those of background extinction. At higher latitudes, biodiversity has been suppressed during times of mass extinction. 2. tropics; 3. Iceland; 4. biodiversity; 5. Iceland; 6. dispersal; 7. distance; 8. area; 9. Island C

**Self-Quiz**

1. b; 2. b; 3. c; 4. a; 5. b; 6. d; 7. e; 8. b; 9. e; 10. a; 11. d; 12. d; 13. d; 14. a; 15. d

---

# Chapter 37    Ecosystems

**37 - I. THE NATURE OF ECOSYSTEMS (pp. 662 - 665)**
**ENERGY FLOW THROUGH ECOSYSTEMS (pp. 666 - 667)**
*Focus on Science:* **ENERGY FLOW AT SILVER SPRINGS, FLORIDA (p. 668)**
1. producers; 2. heterotrophs; 3. herbivores; 4. carnivores; 5. omnivores; 6. Parasites; 7. decomposers; 8. detritivores; 9. open; 10. energy; 11. nutrient; 12. energy; 13. nutrients; 14. trophic; 15. food chain; 16. food web; 17. productivity; 18. grazing; 19. detrital; 20. pyramid; 21. producers; 22. biomass; 23. biomass; 24. smallest; 25. energy; 26. trophic; 27. large; 28. 1; 29. 6; 30. 16; 31. low; 32. 4; 33. a. The rate at which the ecosystem's producers capture and store a given amount of energy in a given length of time. b. The total rate of photosynthesis for an ecosystem during a specified period. c. The rate of energy storage in plant tissues in excess of the rate of aerobic respiration by the plants themselves. 34. b; 35. c; 36. b; 37. a; 38. b; 39. b; 40. b; 41. b; 42. d; 43. b; 44. a; 45. b; 46. b, c; 47. a; 48. b; 49. open; 50. input; 51. cannot; 52. can; 53. have; 54. E; 5. B; 56. D; 57. C; 58. E; 59. D; 60. E

**37 - II. BIOGEOCHEMICAL CYCLES—AN OVERVIEW (p. 669)**
**HYDROLOGIC CYCLE (pp. 670 - 671)**
**CARBON CYCLE (pp. 672 - 673)**
*Focus on the Environment:* **FROM GREENHOUSE GASES TO A WARMER PLANET? (pp. 674 - 675)**
**NITROGEN CYCLE (pp. 676 - 677)**
**PHOSPHORUS CYCLE (p. 677)**
**PREDICTING THE IMPACT OF CHANGE IN ECOSYSTEMS (p. 678)**
1. Usually as mineral ions such as ammonium ($NH_4^+$). 2. Inputs from the physical environment and the cycling activities of decomposers and detritivores. 3. The amount of a nutrient being cycled through the ecosystem is greater. 4. Common sources are rainfall or snowfall, metabolism (such as nitrogen fixation), and weathering of rocks. 5. Losses of mineral ions occurs by runoff. 6. a. Oxygen and hydrogen move in the form of water molecules. b. A large portion of the nutrient is in the

form of atmospheric gas such as carbon and nitrogen (mainly $CO_2$). c. Nutrients are not in gaseous forms; nutrients move from land to the seafloor and only "return" to land through geological uplifting of long duration; phosphorus is an example. 7. F; 8. H; 9. D; 10. B; 11. A (C); 12. E (B); 13. G (H); 14. C; 15. Hubbard Brook; 16. tree roots; 17. Rainfall; 18. sedimentary; 19. plants; 20. six; 21. plants; 22. C; 23. D; 24. F; 25. B; 26. G; 27. A; 28. E; 29. greenhouse effect; 30. heat; 31. CFCs; 32. carbon dioxide; 33. warming; 34. Bacteria (*Rhizobium*); 35. nitrogen fixation; 36. nitrification; 37. nitrite; 38. nitrous oxide; 39. d; 40. e; 41. c; 42. a; 43. b; 44. Soil nitrogen compounds are vulnerable to being leached and lost from the soil; some fixed nitrogen is lost to air by denitrification; nitrogen fixation comes at high metabolic cost to plants that are symbionts of nitrogen-fixing bacteria; losses of nitrogen are enormous in agricultural regions through the tissues of harvested plants, soil erosion, and leaching processes. 45. reservoir; 46. sedimentary; 47. phosphates; 48. ocean; 49. continental; 50. organisms; 51. ionized; 52. plants; 53. herbivores; 54. Decomposition; 55. ecosystem; 56. mosquitoes; 57. malaria; 58. biological magnification; 59. Ecosystem modeling

## Self-Quiz

1. c; 2. c; 3. d; 4. b; 5. a; 6. a; 7. b; 8. a; 9. a; 10. b; 11. b; 12. d

---

# Chapter 38   The Biosphere

**38 - I.   AIR CIRCULATION PATTERNS AND REGIONAL CLIMATES (pp. 680 - 683)**
**THE OCEAN, LAND FORMS, AND REGIONAL CLIMATES (pp. 684 - 685)**
1. D; 2. J; 3. G; 4. C; 5. E; 6. A; 7. F; 8. B; 9. K; 10. M; 11. L; 12. H; 13. I; 14. equatorial; 15. warm; 16. rises or ascends; 17. moisture; 18. descends; 19. moisture; 20. ascends; 21. moisture; 2. descends; 23. east (easterlies); 24. west (westerlies); 25. tropical; 26. warm; 27. cool; 28. cold; 29. solar (sun's); 30. rotation; 31. F; 32. B; 33. E; 34. D; 35. G; 36. C; 37. A

**38 - II.   THE WORLD'S BIOMES (pp. 686 - 687)**
**SOILS OF MAJOR BIOMES (p. 688)**
**DESERTS (p. 689)**
**DRY SHRUBLANDS, DRY WOODLANDS, AND GRASSLANDS (pp. 690 - 691)**
1. F; 2. I; 3. G; 4. A; 5. H; 6. C; 7. B; 8. E; 9. D; 10. J; 11. H; 12. D; 13. E; 14. F; 15. B; 16. A; 17. C; 18. G; 19. d; 20. a; 21. d; 22. b; 23. c; 24. a; 25. e; 26. d; 27. e; 28. b

**38 - III.   TROPICAL RAIN FORESTS AND OTHER BROADLEAF FORESTS (pp. 692 - 693)**
**CONIFEROUS FORESTS (p. 694)**
**TUNDRA (p. 695)**
1. d; 2. c; 3. c; 4. a; 5. b; 6. b; 7. a; 8. d; 9. c, d; 10. b; 11. tropical rain; 12. 25°C; 13. humidity; 14. decomposition; 15. tropical deciduous; 16. temperate deciduous; 17. coniferous; 18. Boreal; 19. taiga; 20. montane; 21. Spruce; 22. pines; 23. pine barrens; 24. tundra; 25. permafrost; 26. alpine; 27. permafrost

**38 - IV.   FRESHWATER PROVINCES (pp. 696 - 697)**
**THE OCEAN PROVINCES (pp. 698 - 699)**
1. lake; 2. littoral; 3. limnetic; 4. plankton; 5. profundal; 6. overturns; 7. 4°C; 8. spring; 9. thermocline; 10. cools; 11. fall; 12. down; 13. up; 14. higher; 15. short; 16. Oligotrophic; 17. eutrophic; 18. eutrophication; 19. Streams; 20. runs; 21. benthic (C); 2. pelagic (A); 23. neritic (B); 24. oceanic (D); 25. a; 26. c; 27. c; 28. b; 29. c; 30. a; 31. b; 32. a; 33. b; 34. c; 35. a; 36. b; 37. a; 38. b; 39. c; 40. a; 41. b; 42. b; 43. c

**38 - V.   CORAL REEFS AND BANKS (pp. 700 - 701)**
**LIFE ALONG THE COASTS (pp. 702 - 703)**
*Focus on the Environment:* **EL NIÑO AND SEESAWS IN THE WORLD'S CLIMATES (pp. 704 - 705)**
1. a. Atolls are ring-shaped coral reefs that enclose or almost enclose a shallow lagoon. b. Fringing reefs form next to the land's edge in regions of limited rainfall, as on the leeward side of tropical islands. c. Barrier reefs form around islands or parallel with the shore of a continent. A calm lagoon forms behind them. 2. d; 3. a; 4. c; 5. e; 6. c; 7. a; 8. b; 9. d; 10. b; 11. a; 12. b; 13. e; 14. d; 15. a; 16. d; 17. e; 18. a; 19. c; 20. e; 21. c

## Self-Quiz

1. d; 2. d; 3. b; 4. d; 5. a; 6. b; 7. c; 8. c; 9. d; 10. c; 11. c; 12. b

# Chapter 39   Human Impact on the Biosphere

**39 - I.  AIR POLLUTION—PRIME EXAMPLES (pp. 708 - 711)**
**OZONE THINNING—GLOBAL LEGACY OF AIR POLLUTION (p. 712)**
1. Pollutants;  2. thermal inversion;  3. winters;
4. industrial smog;  5. photochemical smog;  6. oxygen;  7. nitrogen dioxide;  8. photochemical;  9. gasoline;  10. PANs;  11. sulfur;  12. nitrogen;  13. sulfur;
14. nitrogen;  15. nitrogen;  16. acid deposition;
17. sulfuric (nitric);  18. nitric (sulfuric);  19. acid rain;
20. 5;  21. Chlorofluorocarbons;  22. c;  23. d;  24. e;
25. b;  26. a;  27. b;  28. f;  29. e;  30. a;  31. b;
32. b;  33. d;  34. a (b);  35. d;  36. f;  37. c;  38. d;
39. e;  40. f;  41. e;  42. c;  43. d

**39 - II.  WHERE TO PUT SOLID WASTES, WHERE TO PRODUCE FOOD (p. 713)**
**DEFORESTATION—CONCERTED ASSAULT ON FINITE RESOURCES (pp. 714 - 715)**
*Focus on the Environment:* YOU AND THE TROPICAL RAIN FOREST (p. 716)
1. F;  2. H (G);  3. I;  4. E;  5. J;  6. B;  7. C;  8. A;
9. D;  10. G

**39 - III.  TRADING GRASSLANDS FOR DESERTS (p. 717)**
**A GLOBAL WATER CRISIS (pp. 718 - 719)**
1. desertification;  2. overgrazing on marginal lands;
3. domestic cattle;  4. native wild herbivores;
5. deserts;  6. desalination;  7. energy;  8. agriculture;
9. salination;  10. water table;  11. saline;  12. 20;
13. pollution;  14. wastewater;  15. tertiary;  16. 55 - 66;  17. oil;  18. a. Screens and settling tanks remove sludge, which is dried, burned, dumped in landfills, or treated further; chlorine is often used to kill pathogens in water, but does not kill them all. b. Microbial populations are used to break down organic matter after primary treatment but before chlorination. c. It removes nitrogen, phosphorus, and toxic substances, including heavy metals, pesticides, and industrial chemicals; it is largely experimental and expensive.

**39 - III.  A QUESTION OF ENERGY INPUTS (pp. 720 - 721)**
**ALTERNATIVE ENERGY SOURCES (p. 722)**
*Focus on Bioethics:* BIOLOGICAL PRINCIPLES AND THE HUMAN IMPERATIVE (p. 723)
1. J-shaped;  2. Increased numbers of energy users and to extravagant consumption and waste. 3. Net energy;
4. plants;  5. next;  6. decreases;  7. acid deposition;
8. less;  9. meltdown;  10. have not;  11. solar-hydrogen energy;  12. wind farms;  13. Fusion power;
14. C (N);  15. F (N);  16. B (N);  17. E (R);  18. A (R);
19. D (N)

**Self-Test**

1. c;  2. b;  3. a;  4. b;  5. d;  6. d;  7. c;  8. d;  9. d;
10. b

# Chapter 40   Animal Behavior

**40 - I.  THE HERITABLE BASIS OF BEHAVIOR (pp. 726 - 729)**
**LEARNED BEHAVIOR (p. 730)**
1. the banana slug;  2. ate;  3. inland;  4. were not;
5. genetic;  6. pineal gland;  7. directly;  8. estrogen;
9. testosterone;  10. Hormones;  11. a. Cuckoo birds are social parasites in that adult females lay eggs in the nests of other bird species; young cuckoos instinctively eliminate the natural-born offspring (eggs are maneuvered onto their backs and pushed out of the nest) and then receive the undivided attention of their unsuspecting foster parents. b. Young toads instinctively capture edible insects with sticky tongues; if a bumblebee is captured and then stings the tongue, the toad learns to leave bumblebees alone. 12. E;  13. C;  14; F;  15. A;
16. D;  17. B

**40 - II.  THE ADAPTIVE VALUE OF BEHAVIOR (p. 731)**
1. Natural selection;  2. alleles;  3. evolution;  4. adaptive;  5. reproductive;  6. individual;  7. a. Consideration of individual survival and production of offspring. b. Any behavior that promotes propagation of an individual's genes and tends to occur at increased frequency in future generations. c. Cooperative, interdependent relationships among individuals of the same species. d. Within a population, any behavior that increases an individual's chances to produce or protect offspring of its own, regardless of the consequences for the population. e. Within a population, a self-sacrificing behavior. The individual behaves in a way that helps others but that decreases its own chances to produce offspring.

**40 - III.  COMMUNICATION SIGNALS (pp. 732 - 733)**
1. c;  2. d;  3. b;  4. a, e;  5. b, f;  6. d;  7. c;  8. c, e;
9. a;  10. a, f;  11. c;  12. b;  13. b;  14. a

**40 - IV.  MATING, PARENTING, AND REPRODUCTIVE SUCCESS (pp. 734 - 735)**
1. a. Hangingflies;  b. Sage grouse;  c. Lions, sheep, elk, and bison;  d. Caspian terns